Molecular and Cellular Biology of the Eye Lens

Molecular and Cellular Biology of the Eye Lens

Edited by

Hans Bloemendal

University of Nijmegen
Nijmegen, The Netherlands

A WILEY-INTERSCIENCE PUBLICATION
JOHN WILEY & SONS, New York • Chichester • Brisbane • Toronto

Library of Congress Cataloging in Publication Data

Main entry under title:

Molecular and cellular biology of the eye lens.

"A Wiley-Interscience publication."
Includes indexes.
1. Crystalline lens. 2. Molecular biology.
3. Cataract. 4. Cytology. I. Bloemendal, Hans.
[DNLM: 1. Cataract—Pathology. 2. Lens, Crystalline
—Cytology. 3. Molecular biology. WW 260 M718]
QP478.M64 612'.844 80-26815
ISBN 0-471-05171-3

Printed in the United States of America
10 9 8 7 6 5 4 3 2 1

Preface

The significance of the vertebrate eye lens as attractive tissue for fundamental cell-biochemical and cell-biological studies has been recognized for almost a century. During that time, advances in molecular biology, immunology, electron microscopy, and the other disciplines of the basic sciences have gradually and sometimes drastically altered the concepts underlying the fundamental processes occurring in lens, such as induction, differentiation, growth, and aging. These advances have frequently led to a clearer understanding of the ever existing interrelationship between structure and function of biomacromolecules, which governs all living systems. One must, however, admit that this understanding, even after a century of increasingly intensified research, is far from being complete, both in normal and disordered states of the eye lens.

Lens investigations apparently follow the fashionable trends in molecular biology. After a long period of empirical work, isolation and characterization of the composite proteins and lipids, and comparative studies between species, new topics such as protein biosynthesis and nucleic acid and membrane research penetrated the field. From experiments carried out in the relatively recent past, it is evident that the following tenets are essential to a better understanding of lens molecular and cellular biology:

Unraveling the problem of the *in vivo* organization of the crystallins.

Rigorous analysis of lens plasma membrane composition, assembly, and function.

Elucidation of the nature of interaction between lens membranes and cytoskeleton.

Careful determination of the functions of the lenticular cytoskeleton (possibly also its involvement in protein biosynthesis and/or in post-translational processes).

Studies on the regulation of transcription and translation of lenticular tissue.

Analysis of the gene organization both of the soluble crystallins and the cytoskeletal and membranous protein subunits.

Improved isolation and analytical techniques, high-resolution morphology, and molecular cloning are the necessary tools that one hopes will enable lens researchers to gain these objects.

This book is intended to be helpful in this respect. Since some topics have been dealt with from more than one viewpoint, inevitable overlapping and duplication have resulted. Although these have been kept to a minimum, they have been retained purposely to ensure a better understanding of certain chapters as entities and to avoid the necessity of repeatedly consulting sections from other chapters.

In this book relatively much attention has been paid to hitherto rather unexplored topics in lens research, in particular the plasma membrane and the cytoskeleton, since it is felt that insight into the biology of these cell structures will be of major importance in the characterization of abnormalities that occur in both artificially induced and "spontaneous" cataracts. A number of chapters included in this volume aim to provide the required background information for fruitful studies in this direction.

HANS BLOEMENDAL

Nijmegen, The Netherlands
February 1981

Contents

CHAPTER
1

The Lens Proteins

Hans Bloemendal

Department of Biochemistry, University of Nijmegen, Nijmegen, The Netherlands

1.1 INTRODUCTION

The lens is one of the most ideal tissues for the study of various fundamental biological processes because of its ability to preserve a complete record of one of the highest forms of morphological (see Chapter 2) and biochemical specialization. Not only can terminal differentiation (see Chapter 3) be studied fruitfully, but also some aspects of the process of aging, still poorly understood, can be examined in this system. The reason is that the lens never sheds its cells. The differentiated fibers are displaced toward the center of the organ and the oldest protein is therefore found in the lens core. This permits a simple comparison of the protein derived from the outer lens fibers with the proteins from the inner part of the lens to determine whether changes in structure or packing have occurred. In Chapter 7 evidence is provided that progressive but slow degradation and deamidation take place while the organ ages. Obviously one can arrive at such conclusions only if sufficient structural data exist and the characteristics of lens proteins are well determined. Thus sound knowledge of the major lenticular proteins is important for a proper understanding of the molecular biology of the eye lens. In this respect major does not necessarily mean abundant in quantity, for proteins that occur in relatively low concentration may play major roles in various fundamental processes.

The lens contains about 35% by weight of structural proteins. These have been subdivided into the water-soluble crystallins and a water-insoluble fraction, formerly designated albuminoid (1). The crystallins form the bulk of lens proteins and most studies until recently have been devoted to these constituents of the lens. Moreover, because civilized people are carnivores, cattle lenses have always been easily obtainable as by-products of the slaughterhouses. Hence calf or bovine lenses have been investigated more frequently and in greater detail than those isolated from other animals. For this reason most data given in this chapter are related to calf (or cow) crystallins and noncrystallin protein. Only incidentally are results of studies on lenticular proteins from rat, rabbit, chick, and other animals also included.

1.2 CRYSTALLINS

1.2.1 Preparation of Crystallins

Various isolation procedures and descriptions of the properties of lens proteins have been summarized in other published books and reviews (2–11). In this chapter only the most commonly applied purification methods of crystallins are reviewed. Moreover, only the properties reported recently are emphasized here. The preparation methodology described, although originally applied for the isolation of bovine lens proteins, can in most cases be adopted for the purification of a great variety of other vertebrate lenticular proteins.

Immediately after the animals are killed the eyes are collected, put on ice, and transported to the laboratory. After an incision in the cornea, the lens can be squeezed out of the eye by light pressure with the fingertip on the edge of the organ. Lenses can be freed easily from adhering iris and remainders of the epithelium by rolling them carefully over a sheet of filter paper.

To obtain the so-called water-soluble fraction lenses are stirred either in distilled water, adjusted with ammonia to pH 7.5, or in one of the numerous alkaline or neutral buffers mentioned in the literature (compare Table 1). A volume of 2 to 3 ml of buffer per gram wet weight of lens tissue is convenient, but other ratios are permissible. Depending on the time of stirring, varying concentrations of protein are obtained. About half of a cow lens dissolves in about 3 hours. Remaining nuclei must be homogenized and stirred for a longer period to get soluble crystallins from this fraction.

Alternatively, if only the proteins from the peripheral part of the lens are

Table 1 Buffers Used for Gel Filtration of Crystallins by Various Investigators

Buffer composition	pH	Reference
0.05 *M* Tris-Cl, 0.05 *M* NaCl, 0.001 *M* EDTA	7.5	12
0.1 *M* Tris-Cl, 1 *M* NaCl, 0.005 *M* EDTA	7.5	13
0.01 *M* phosphate, 0.01 *M* 2-mercaptoethanol	7.2	14
0.1 *M* Tris-Cl	7.3	15
0.1 *M* Tris-Cl, 0.01 *M* EDTA	9.0	16
0.05 *M* Tris-Cl, 0.0514 *M* NaCl, 0.001 *M* EDTA	8.5	17
0.01 *M* Tris-Cl, 0.1 *M* KCl	7.6	18
0.04 *M* phosphate, 0.003 *M* 2-mercaptoethanol	6.8	19
0.02 *M* Tris-Cl, 0.08 *M* NaCl, 0.01 *M* EDTA, 0.2 m*M* phenylmethylsulfonyl fluoride	7.3	20
0.02 *M* phosphate, 0.1 *M* NaCl, 0.01 *M* EDTA	7.4	21
1% ammonium bicarbonate, 0.010 *M* 2-mercaptoethanol	8.3	22
0.1 *M* ammonium formate	7.3	23

desired, the lens cortex is isolated with the aid of a glass tube or a cork borer (approx. 8 mm in diameter) and the nucleus is discarded. The solutions are centrifuged for 30 min at 10,000 to 15,000 $\times g$. The opalescent supernatant fraction containing the crystallins can be kept frozen at −20°C for many months. The pellet, consisting of insoluble lens proteins, the plasma membranes, and the cytoskeleton, can be further fractionated by extraction with 5.5 to 8 *M* urea, eventually followed by extraction with 1% sodium dodecylsulfate.

1.2.2 Separation of Crystallins by Gel Filtration

The crystallins are most conveniently separated by gel filtration on materials such as Sephadex G200, Ultrogel AcA34, Bio-Gel P300, Sepharose 6B-Cl, or Agarose Bio-Gel A-5m. Equilibration of the gel filtration columns may be carried out in various buffers (12–23) (see Table 1). In order to avoid time-consuming dialysis before gel electrophoretic analyses the use of volatile buffers is recommended. In our studies 100 m*M* ammonium formate adjusted to pH 7.5 with ammonia appeared to be very satisfactory despite its rather low buffer capacity. Whereas as a rule four size classes of crystallins are obtained upon gel filtration (Figure 1; see following section for nomenclature), subdivision of the β_L fraction can be achieved (23) with the aid of the volatile buffer (compare Figure 14).

1.2.3 Nomenclature of the Crystallins

1.2.3.1 Crystallin Aggregates

In order to avoid confusion in discussions of lens protein components between students of identical problems, it is desirable to have a generally accepted nomenclature for both the polymeric proteins and their subunits. As far as the classification of the polymeric proteins is concerned Mörner's original designation—α-, β-, and γ-crystallin—is still used, but for γ-crystallin not completely tenable anymore. α-Crystallin has the highest molecular weight and migrates with the highest mobility both in free electrophoresis and various stabilizing media used for zone electrophoresis, with the exception of polyacrylamide and starch gel, which the large aggregates cannot penetrate. β-Crystallin is intermediate in size, whereas the γ-crystallin fraction, consisting of a mixture of monomeric proteins, has the lowest molecular weight and the lowest electrophoretic mobility at neutral or alkaline pH. We stressed previously (24) that the general classification fails in the case of γ-crystallin. Van Dam (25) observed in 1966 that a 28,000 M_r component with features of the β-crystallins coincides with the γ fraction on Sephadex G200 columns. Hence we suggested, in accordance with a proposal of Rabaey et al. (26) for bird γ-crystallins, the

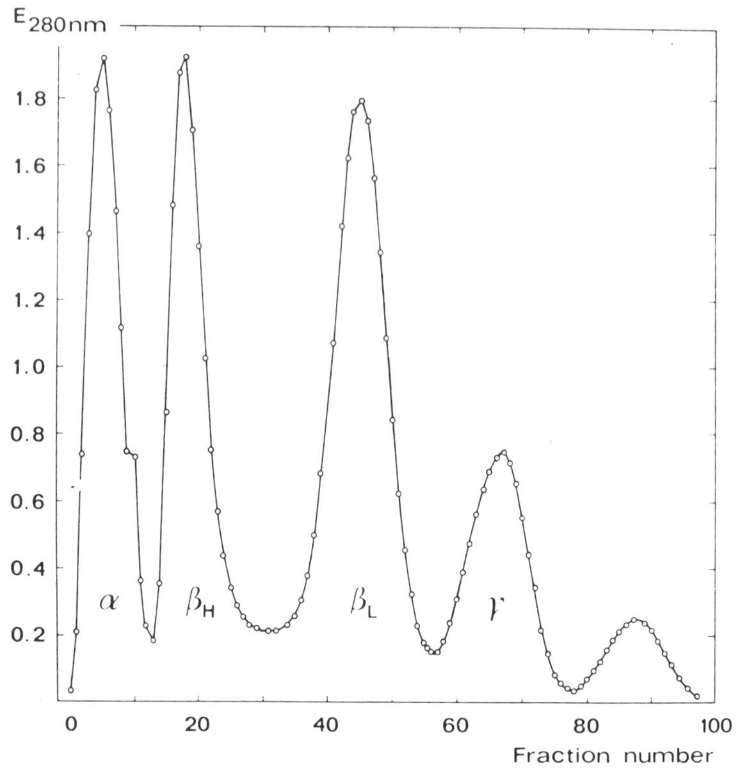

Figure 1 Separation of the water-soluble proteins from calf eye lenses. A column (100 × 2.5 cm) loaded with Sephadex G200 was used. The packing material was equilibrated in a buffer containing 0.05 *M* Tris-Cl, pH 7.6, 0.05 *M* sodium chloride, and 0.001 *M* EDTA and eluted with the same buffer. 200 mg of protein was applied. Elution was for 40 hours.

general designation LM (low molecular weight) crystallin, rather than γ-crystallins for all water-soluble lens proteins with molecular weights lower than 30,000 irrespective of the source. The name γ-crystallin is restricted to that class of lens proteins that can be separated from α- and β-crystallin by gel filtration on a Sephadex G75 column followed by fractionation on sulfethyl Sephadex (24) (see Figure 2).

1.2.3.2 Crystallin Subunits

With regard to the subunits of α-crystallin there is now general agreement about their nomenclature. This is based on the electropboretic behavior of the subunits in alkaline polyacrylamide gels containing 5.5 to 8 *M* urea (27, 28) (Figure 3). Here the symbols A and B stand for acidic and basic, respectively. αA_2, the major α-crystallin polypeptide chain, and αB_2 are the only primary gene products (see Chapter 5). αA_1 and αB_1 arise from these polypeptides by the postsynthetic process of deamidation (29, 30). In addition to these "intact" subunits, degradation products may arise on aging or maturation by cleavage of defined numbers of amino acid residues from the *C*-terminal end (31–33). For instance, $\alpha A_2{}^{1-169}$ represents a chain identical to αA_2 but lacking the four amino acid residues at the *C*-terminus.

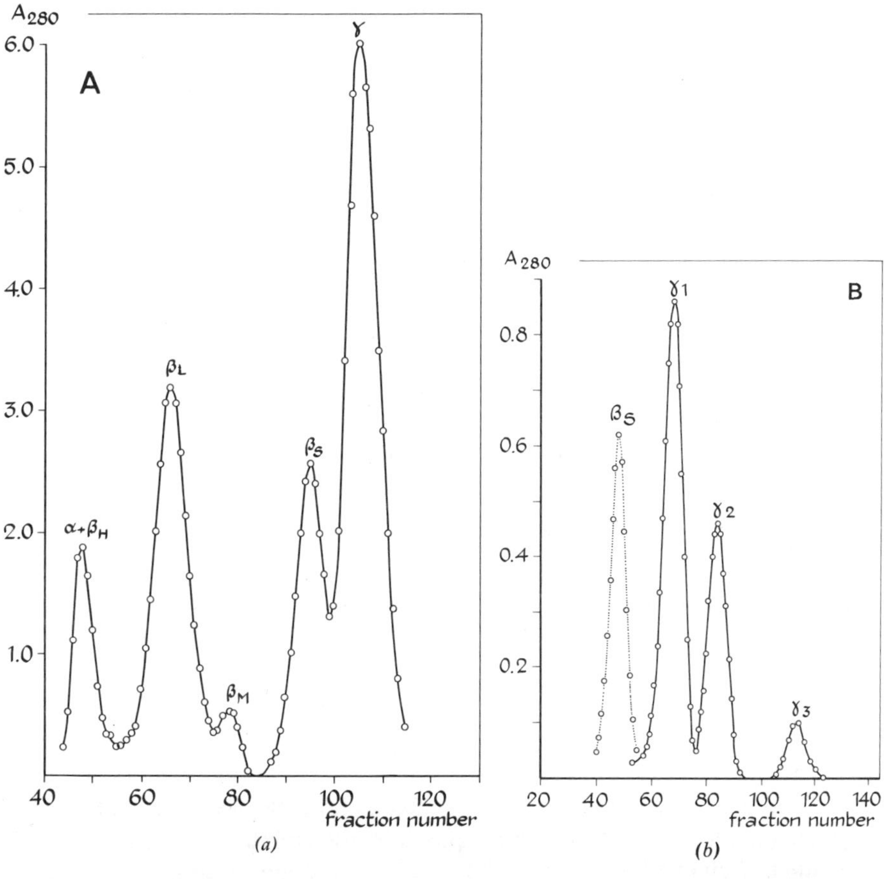

Figure 2 (*a*) Separation of the water-soluble lens proteins on Sephadex G75. A column of 100 × 3.0 cm was used. The buffer was the same as described in Figure 1. Elution was carried out at a flow rate of 30 ml/hour. Gel filtration was preceded by centrifugation at 220,000 × *g* for 14 hours. The upper two-thirds of the supernatant fraction were used. 260 OD units of protein were applied. (*b*) SE Sephadex chromatography of the LM crystallin fraction. 100 OD units of protein were applied to a column packed as described by Björk (40). Column dimensions were 40 × 1.5 cm. Elution was carried out at a flow rate of 30 ml/hour. Reprinted with permission from ref. 24.

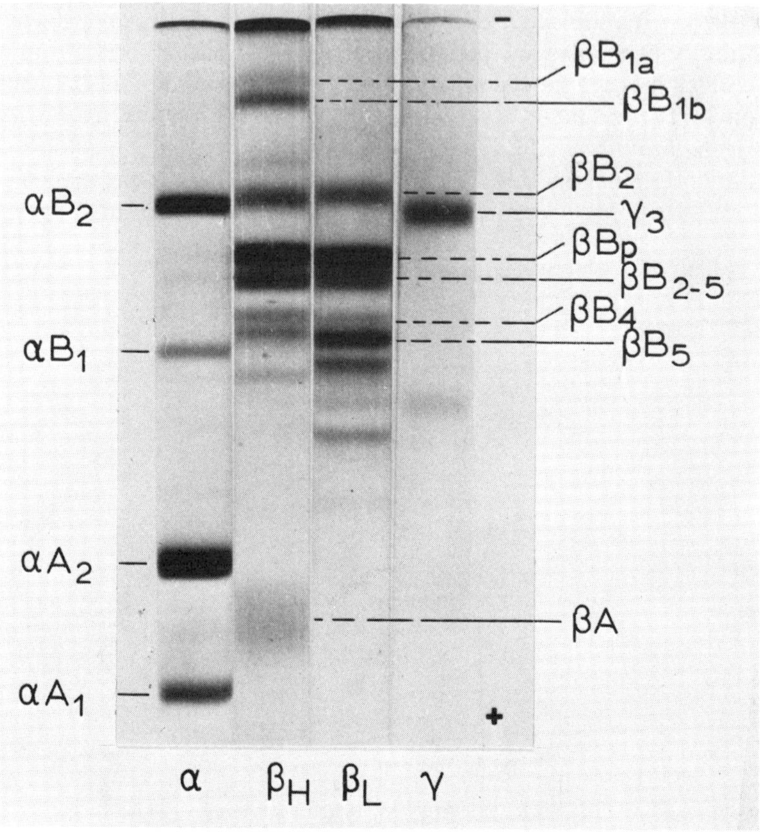

Figure 3 Polyacrylamide gel electrophoresis of isolated α, β_H, β_L, and γ-crystallin. Separation on basis of charge. Electrophoresis was performed in 10% acrylamide gels, containing 6 *M* urea and 100 mg/l DTT at pH 8.6 as described by Bloemendal (212). After a preelectrophoretic run for 1 hour, samples containing 40 to 50 μg of protein were layered on top of the gel and subjected to electrophoresis for 5 hours at 5 mA/gel. Gels were stained with amido black.

We have also proposed a nomenclature for the subunits of the β-crystallins, although at this time possible interrelationships between all individual β-chains have not yet been resolved (compare Figure 3).

The major polypeptide has been called βB_P (principal basic β-crystallin chain (34, 35). Its complete amino acid sequence has recently been elucidated (36). Herbrink et al. demonstrated by means of peptide mapping that great similarities exist between the βB_P and βB_4 chains (37). As in the case of α-crystallin the classification is related to electrophoretic behavior on alkaline–urea gels. However, contrary to α-crystallin chains the suffix increases with increasing acidity (compare Figure 3). Thus βB_1 is more basic than βB_2. In β_N, N stands for neutral. β_H-crystallin contains an acidic polypeptide β_A and is further characterized by two highly basic chains called βB_{1a} and βB_{1b}, respectively. It cannot be excluded that β_A which always appears as a broad band or even as a smear represents more than just one component. From biosynthetic experiments we provided evidence that βB_{1b} arises from βB_{1a} by some post-translational process (38).

Also the monomeric Υ-crystallin can be subdivided into different chains of increasing acidity: γ_1, γ_2, γ_3, γ_4, and γ_5 (39). This is different from the nomenclature originally proposed by Björk (40). Since the concentrations of the individual γ-crystallin polypeptides differ markedly, only γ_3 (Björk's γ_{II}) is easily visualized on electropherograms of the Sephadex G200 γ-crystallin fraction.

In sodium dodecylsulfate gel electrophoresis, which resolves polypeptides according to size, a number of bands, which migrate with different mobilities in urea polyacrylamide, coincide (Figure 4). In fact, according to their molecular weight αA and αB chains should also coincide on SDS gels (41).

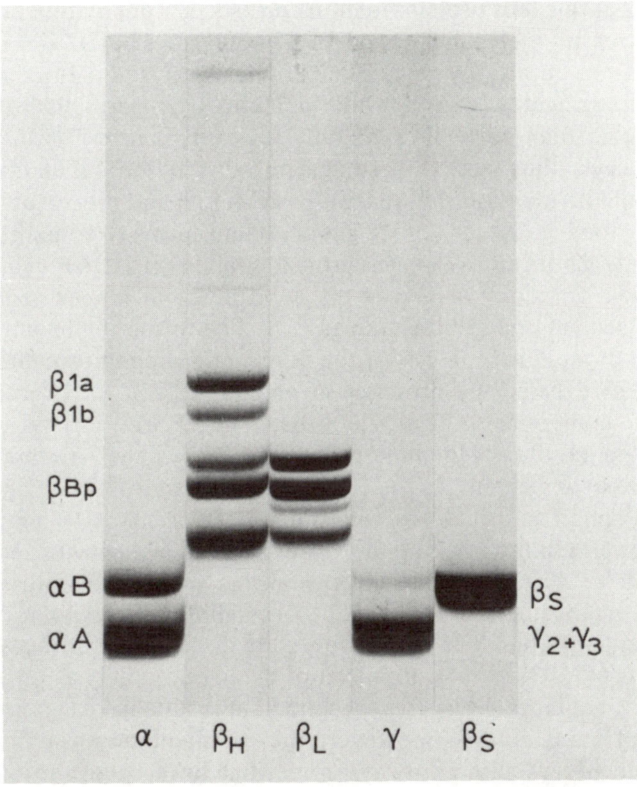

Figure 4 SDS gel electrophoresis of lens proteins (separation on basis of size). Ten percent acrylamide gels were prepared according to Laemmli (213) Electrophoresis was performed at 5 mA/gel. Gels were stained with Coomassie brilliant blue. Note the presence of β_s in the γ-crystallin fraction, visualized after isolation of the corresponding peak in Figure 2*b*.

However, these chains appear as two bands with different apparent molecular weight despite their almost identical molecular weights of 20,000 as revealed by elucidation of the primary structure (42, 43). The electrophoretic behavior of the αB polypeptide chain shows that differences in mobility observed between proteins on SDS gels should be interpreted with great caution (41) and do not necessarily reflect differences in molecular size. De Jong et al. (44) have studied this phenomenon in more detail. It was demonstrated that the substitution Thr → Ala, Gln → Leu, Pro → Thr or Pro → Ala in mammalian α-crystallin A chains increased the mobility in SDS gel electrophoresis.

1.2.4 α-Crystallin

The α-crystallin fraction has been the most extensively studied of the water-soluble proteins (3, 4, 6, 8, 9, 45, 46). Ultracentrifugal studies revealed that this lens protein fraction consists of a population of aggregates with an average molecular weight of 800,000 (6, 45–47). Gel filtration on Bio-Gel A-5m results in separation of the α-crystallin fraction into two groups of aggregates, one of which contains aggregates with a molecular weight higher than 50×10^6 (48, 49). However, it is well-established now that all α-crystallins basically are made up of the four subunits A_1, A_2, B_1, B_2. It should be mentioned here that in rabbit (50) and presumably in human lenses (51) the high molecular weight crystallin seems to contain β-crystallin constituents also. Ion exchange chromatography on DEAE cellulose of the rabbit HM-α fraction resulted in the appearance of a peak sedimenting at 4 S and a second one sedimenting at 20 S. The former had some characteristics of β-crystallin. Analysis of the polypeptides composing human HMw α-crystallin revealed the presence of 45,000, 27,000, 22,000, 20,000, and 11,000 M_r components. The 27,000 band might well be a β component, whereas the 11,000 polypeptide presumably arises by *C*-terminal degradation of preexisting chains. In the electron microscope α-crystallin aggregates appear as spherical structures with a diameter of about 17 nm (47, 49, 52, 53) (Figure 5). On the other hand, α_{HM}-crystallin species with sedimentation coefficients of 40 to 240 S do not reveal this spherelike feature but rather a random arrangement of individual α-crystallin particles (49).

Treatment with 6 to 8 *M* urea induces the dissociation of α-crystallin into its subunits (47, 52, 54, 55). Removal of the urea by dialysis or gel filtration on columns equilibrated in urea-free medium results in reaggregation of the subunits (47, 54, 56–58). However, the sedimentation coefficient of the recombined α-crystallin is always somewhat lower than the original S_{20W} value of about 18 S (59, 60).

It has been suggested (61–63) that components other than the four polypeptides are also involved in the aggregate formation. These constituents might be removed by the urea treatment followed by extensive dialysis.

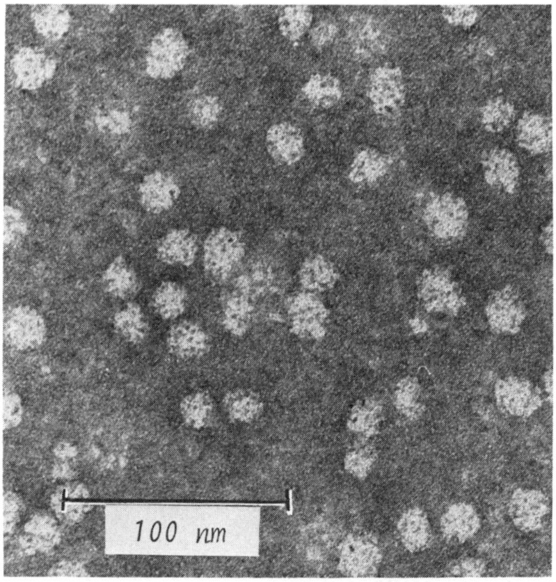

Figure 5 Electron micrograph of α-crystallin particles.

At high concentration hybrid formation between α- and β-crystallin polypeptides may take place upon removal of urea (17) (Figure 6). We were able to show that selective reassociation of all soluble lens proteins including α-, β- and γ-crystallins may be achieved provided the concentration of the composing chains is rather low (17, 64) (Figure 7).

Manski and co-workers (65–67) concluded from selective filtration and immunochemical studies that complex formation occurs between the different crystallin classes. With progressing age an initial α-β crystallin complex binds γ-crystallin. This complex formation is dependent on ionic conditions.

Microinjection studies of α-crystallin mRNA into living oocytes of *Xeropus laevis* (68) suggest that aggregation of α-crystallin subunits is triggered by the αA chains, which are subsequently joined by αB polypeptides. This model is in accord with a similar proposal by Bindels et al. (69).

The isolated αA and αB chains also can form aggregates (70), which can again be dissociated in urea or SDS. αA and αB aggregates are significantly smaller than native α-crystallin. According to Li and Spector (70), α-crystallin has virtually no α-helical conformation. The conclusion was drawn from CD and ORD studies. The same result was obtained after laser Raman spectroscopy (71, 72).

Horowitz (73) concluded on the basis of physicochemical parameters that bovine and human α-crystallin, in spite of some differences in amino acid composition, have similar secondary and tertiary structure.

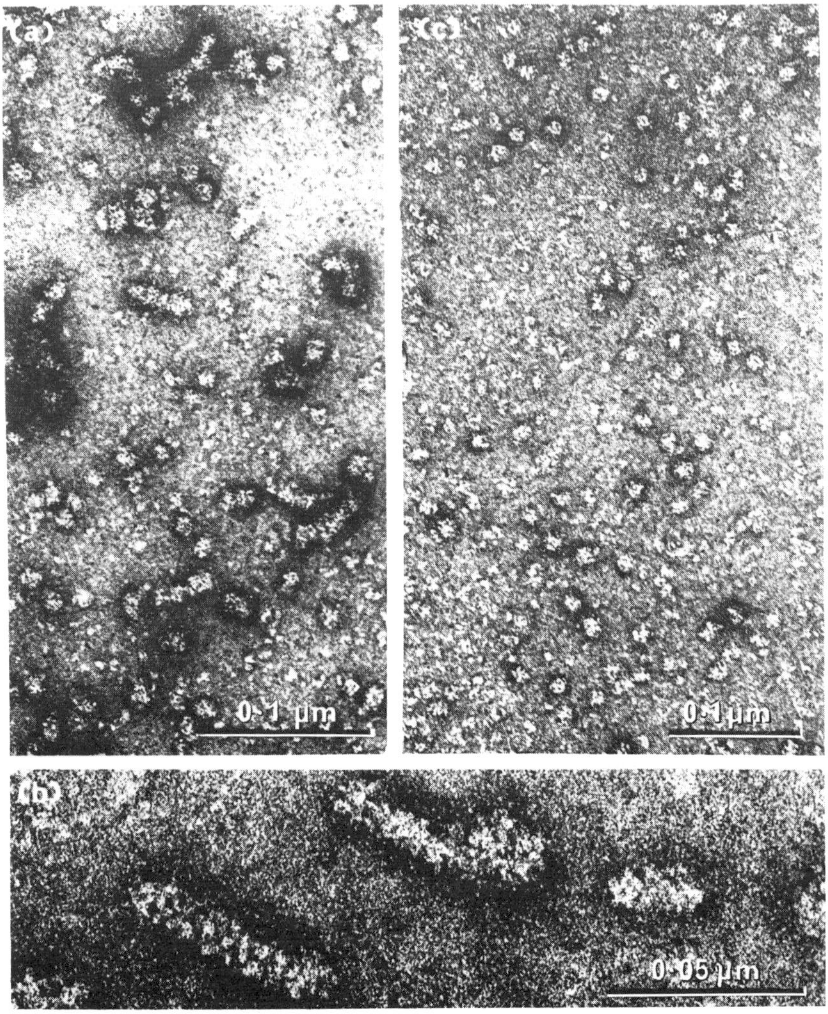

Figure 6 *(a, b)* Electron micrographs of $\alpha\beta$ hybrids obtained after reassociation of the crystallin subunits at high protein concentration (200 mg/ml). *(c)* Correctly recombined α particles (13 S) (reassociation at 2 mg/ml protein). Reprinted with permission from ref. 17.

1.2.4.1 Preparation of the α-Crystallin Polypeptide Chains

S-Carboxymethylated α-crystallin can be resolved into two subunit fractions by chromatography on sulfethyl Sephadex C50 columns equilibrated in 0.2 *M* sodium formate containing 7 *M* urea, pH 3.2 (74, 75). The first peak emerging from the column comprises the acidic chains αA_2 and αA_1, respectively. Further fractionation is possible on the same columns, which should now be equilibrated, however, in 0.05 *M* sodium acetate containing 7 *M* urea, pH 5.5 (76). Alternatively the separation into αA_1 and αA_2 is also possible by preparative isoelectric focusing in a pH range from 4 to 6 (77). The latter method is also convenient for resolving the αB chains. This can be achieved with ampholytes in the pH range 6 to 8 (Figure 8).

Fractionation of α-crystallin on DEAE cellulose equilibrated in 0.01 *M*

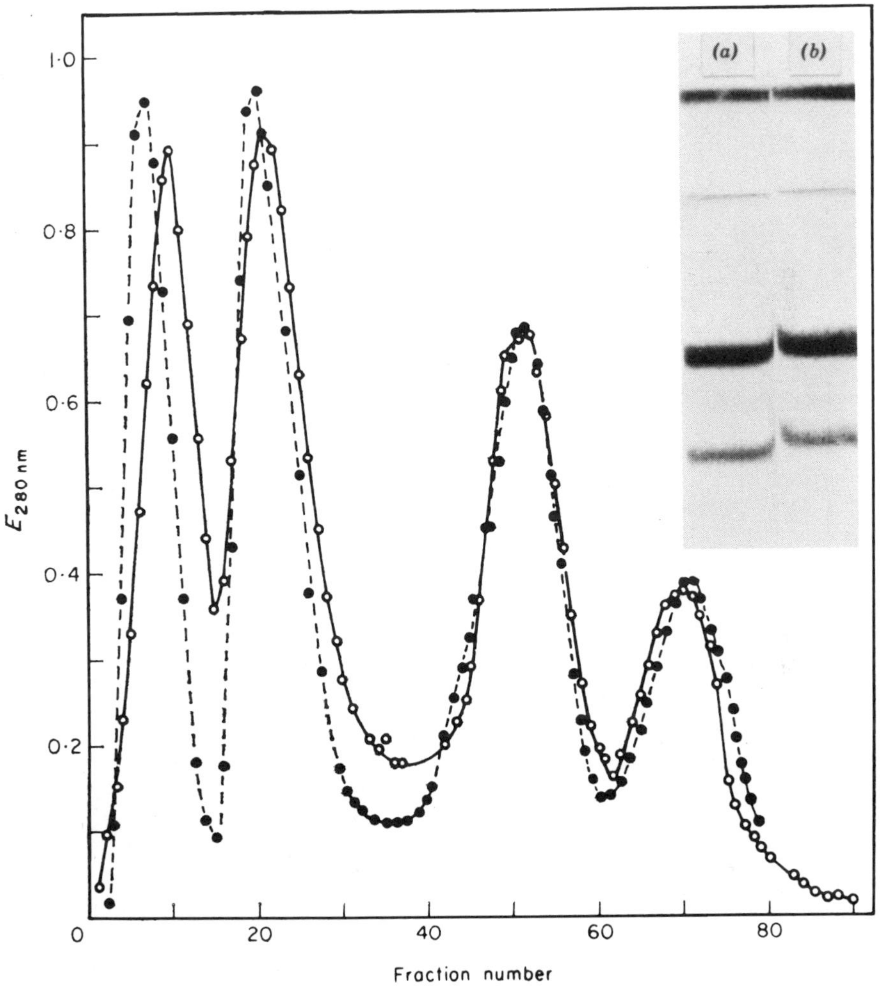

Figure 7 Selective reassociation of water-soluble crystallins as revealed by gel filtration on Sephadex G200. Control, •----•; recombined crystallins, ○——○. Inset: polyacrylamide–urea gel electrophoresis of the control α-crystallin (A) and the reaggregated α-crystallin (B). Note that this reassociated α-crystallin does not contain any β-crystallin subunit. Reprinted with permission from ref. 17.

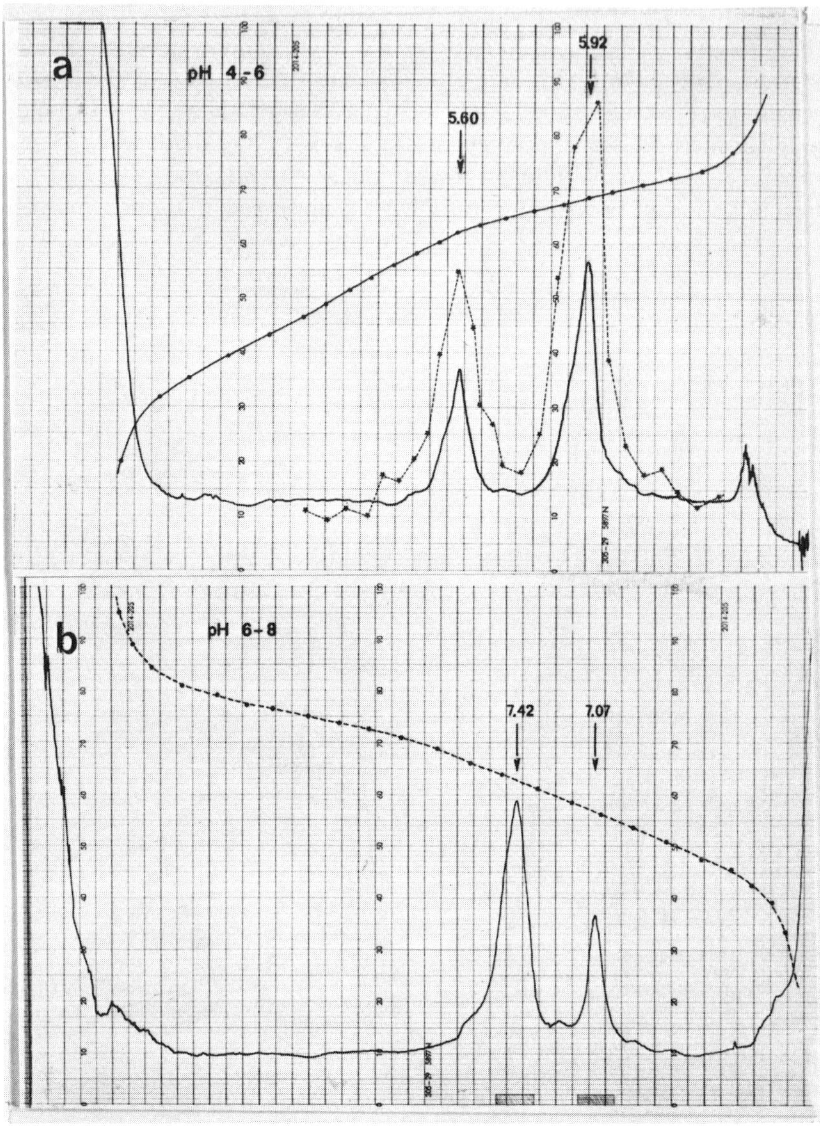

Figure 8 Separation of the units of ^{14}C-carboxymethylated α-crystallin by isoelectric focusing. The sucrose gradient ranging from 0 to 50% contained 1% ampholytes. (*a*) Elution pattern of the acidic subunits of α-crystallin in a pH gradient 4 to 6. —— absorbance at 280 nm (★) radioactivity; (•) pH gradient. (*b*) Elution pattern of the basic subunits of α-crystallin in a pH gradient 6 to 8. —— absorbance at 280 nm (•) pH gradient. Electrophoresis was carried out for 60 to 72 hours at 20°C with a load never exceeding 1 to 2 W until a final potential of 800 V was reached. A Gilford 2000 spectrophotometer gave an automatic recording of the effluent. Reprinted with permission from ref. 77.

α-CRYSTALLIN SEQUENCE : HOMOLOGY BETWEEN αA_2 AND αB_2 POLYPEPTIDE CHAIN

```
αA2  ac-Met-Asp-Ile-Ala-Ile-Gln-His-Pro-Trp-Phe-Lys-Arg-Thr-Leu-Gly-Pro-Phe-    -Tyr-Pro-Ser-Arg-Leu-Phe-Asp-Gln-Phe-Phe-Gly-Glu-
αB2  ac-Met-Asp-Ile-Ala-Ile-His-His-Pro-Trp-Ile-Arg-Arg-Pro-Phe-Phe-Pro-Phe-His-Ser-Pro-Ser-Arg-Leu-Phe-Asp-Gln-Phe-Phe-Gly-Glu-

-Gly-Leu-Phe-Glu-Tyr-Asp-Leu-Leu-Pro-Phe-Leu-Ser-Ser-Thr-Ile-Ser-Pro-Tyr-Tyr-    -Arg-Gln-    -Ser-Leu-Phe-Arg-            -Thr-Val-
-His-Leu-Leu-Glu-Ser-Asp-Leu-Phe-Pro-    -Ala-Ser-Thr-Ser-Leu-Ser-Pro-Phe-Tyr-Leu-Arg-Pro-Pro-Ser-Phe-Leu-Arg-Ala-Pro-Ser-Trp-

-Leu-Asp-Ser-Gly-Ile-Ser-Glu-Val-Arg-Ser-Asp-Arg-Asp-Lys-Phe-Val-Ile-Phe-Leu-Asp-Val-Lys-His-Phe-Ser-Pro-Glu-Asp-Leu-Thr-Val-
-Ile-Asp-Thr-Gly-Leu-Ser-Glu-Met-Arg-Leu-Glu-Lys-Asp-Arg-Phe-Ser-Val-Asn-Leu-Asn-Val-Lys-His-Phe-Ser-Pro-Glu-Glu-Leu-Lys-Val-

-Lys-Val-Gln-Glu-Asp-Phe-Val-Glu-Ile-His-Gly-Lys-His-Asn-Glu-Arg-Gln-Asp-Asp-His-Gly-Tyr-Ile-Ser-Arg-Glu-Phe-His-Arg-Arg-Tyr-
-Lys-Val-Leu-Gly-Asp-Val-Ile-Glu-Val-His-Gly-Lys-His-Glu-Glu-Arg-Gln-Asp-Glu-His-Gly-Phe-Ile-Ser-Arg-Glu-Phe-His-Arg-Lys-Tyr-

-Arg-Leu-Pro-Ser-Asn-Val-Asp-Gln-Ser-Ala-Leu-Ser-Cys-Ser-Leu-Ser-Ala-Asp-Gly-Met-Leu-Thr-Phe-Ser-Gly-Pro-Lys-Ile-Pro-Ser-Gly-
-Arg-Ile-Pro-Ala-Asp-Val-Asp-Pro-Leu-Ala-Ile-Thr-Ser-Ser-Leu-Ser-Ser-Asp-Gly-Val-Leu-Thr-Val-Asp-Gly-Pro-Arg-Lys-Gln-

-Val-Asp-Ala-Gly-His-Ser-Glu-Arg-Ala-Ile-Pro-Val-Ser-Arg-Glu-Glu-Lys-Pro-            -Ser-Ser-Ala-Pro-Ser-Ser-COOH  173 RESIDUES
-        -Ala-Ser-Gly-Pro-Glu-Arg-Thr-Ile-Pro-Ile-Thr-Arg-Glu-Glu-Lys-Pro-Ala-Val-Thr-Ala-Ala-Pro-Lys-Lys-COOH  175 RESIDUES
```

HOMOLOGY : 55 %

Figure 9 The primary structure of the α-crystallin subunits αA_2 and αB_2.

Tris-Cl containing 6 *M* urea, pH 7.6, yields in a single purification step the four separated polypeptides when the column is eluted with a linear gradient ranging from 0.01 *M* to 0.15 *M* Tris-Cl in 6 *M* urea and 0.01% dithiothreitol at pH 7.6 (78). The subunit αA_1 is not present in embryonic calf lens (77, 79). The αA_2 chain revealed about 55% homology with the αB_2 chain. The complete amino acid sequence of both types of polypeptide is shown in Figure 9.

1.2.4.2 Some Rodents Have an Additional αA Polypeptide

Rat lens and a number of other rodent α-crystallins are made up of, in addition to the αA and αB subunits, a minor chain that can be isolated upon ion exchange chromatography on DEAE cellulose (Figure 10). On SDS gels this polypeptide migrates in the region of M_r 24,000, whereas in alkaline urea gel electrophoresis it is detected between αB_1 and αA_2 (80). We propose to call this subunit αA^{Ins} since sequence studies revealed that its primary structure is very similar to the αA_2 chain. Actually, the complete sequence of αA^{Ins} is identical to that of αA_2, with the notable exception that between amino acid residues 63 and 64 an insertion of 22 amino acids has been found. Apparently the antigenic determinants are not located in this region because the isolated polypeptides can be precipitated with antibodies raised against purified calf lens α-crystallin, which does not contain the rat insertion. At the moment one can only speculate whether this insertion arose after duplication of the αA_2 gene followed by crossing-over and some displacement of the coding region. Alternatively, we may have a gene product of incomplete splicing* of a messenger RNA precursor.

Extended polypeptides have been detected owing to incomplete processing of precursor chains or by read-through of the messenger in the case of point mutations whereby a stop codon is altered, as in the case of hemoglobin Constant Spring (83). However, so far no other proteins with comparable structural relationship as observed in αA_2 and αA^{Ins} have been reported.

The sequence of the insertion is given in Figure 11. The presence of three methionyl residues (a rather rare amino acid in normal α-crystallin chains) is striking. The absence of any acidic residue and the presence of five basic ones cause the migration behavior on alkaline urea gels and isofocusing electrophoresis (80, 84). The concentration of rat αA^{Ins} is about 20 times lower than that of normal αA chains. αA^{Ins} occurs also in a number of other rodents (compare Figure 12).

*From the large primary RNA transcript, the precursor of the messenger, certain internal nucleotide sequences are "spliced out," whereas the residual regions to be used as coding segments are covalently linked to form the functional mRNA (see refs. 81 and 82).

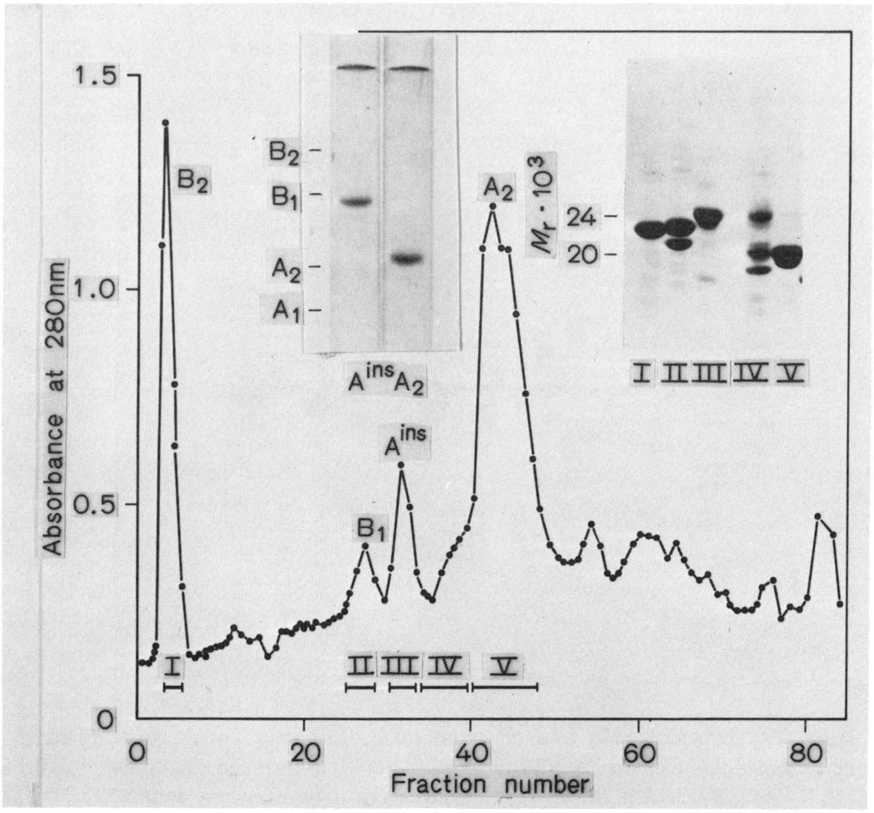

Figure 10 Purification of αA^{Ins} and αA_2 by ion exchange chromatography on DEAE cellulose. Subunits of 250 mg of rat α-crystallin were separated by chromatography on a column (1.1 × 20 cm) of DEAE cellulose. The fractions were pooled as indicated and analyzed by polyacrylamide gel electrophoresis in Tris/glycine buffer, pH 8.5, containing 6 *M* urea (left inset), and in the presence of sodium dodecylsulfate (right inset). Reprinted with permission from ref. 80.

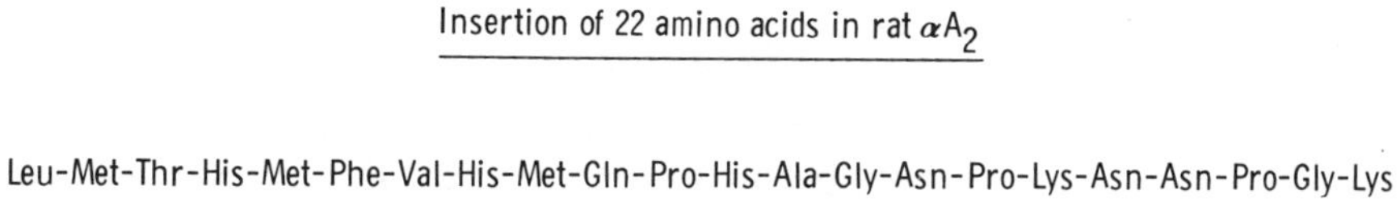

Figure 11 Sequence of the insertion found in the specific rat chain αA^{Ins}.

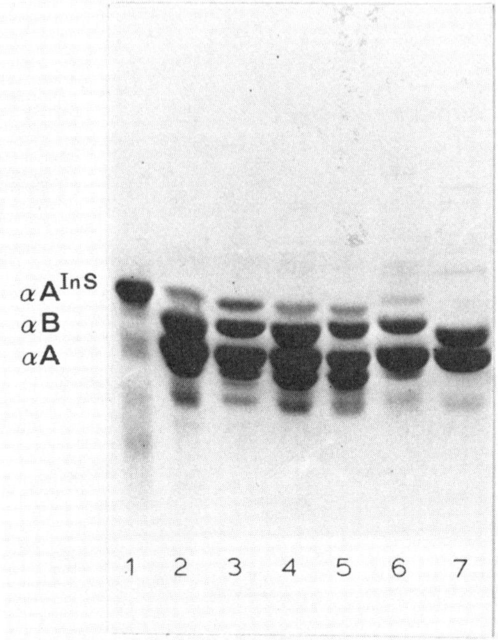

Figure 12 The αA^{Ins} chain is found in a number of rodents. Owing to its apparently higher molecular weight (24,000) instead of the 20,000 of the αA chain) it can be detected on polyacrylamide gel electrophoresis in sodium dodecylsulfate. 1, Purified rat αA^{Ins}; 2–, α-crystallin patterns from (2) rat, (3) mouse, (4) gerbil, (5) hamster, and (6) guinea pig. 7, For comparison calf α-crystallin is also shown.

1.2.4.3 β-Crystallin

β-Crystallin is the most heterogeneous population of the soluble lens proteins. As can be seen in Figure 1 the β-crystallin aggregates appear as two major size classes upon gel filtration (13, 34, 35, 37). Figure 13 shows that β-crystallins from various species give similar elution patterns. However, under well-chosen conditions the two β-fractions tend to split further (23). The result depicted in Figure 14 was obtained by using a gel filtration column (125 $\times$ 2 cm) loaded with Sephadex G200 and equilibrated in 100 m*M* ammonium formate, pH 7.3. On such a column the β_L-fraction could be separated into a β_{L1} and a β_{L2} peak, respectively. Gel electrophoresis reveals that β_H and β_L share the major polypeptide βB_P (35). β_{L1} and β_{L2} also share this chain (compare Figure 15). A conversion of β_H- to β_L-crystallin and a reduction in the amount of β-crystallin has been demonstrated by Li (85) upon the introduction of sucrose into dilute solutions of crystallin (see Figure 16).

As early as 1964 Björk (86) obtained four different β-fractions on gel filtration. Whether these were identical to the four peaks observed by Spector and Katz (55) after chromatography on DEAE cellulose has not been established. Armand et al. (87) were able to fractionate β_H further into three fractions on an organomercurial Sephadex column. Liem-The and Hoenders (88) found three β-fractions in rabbit lens. Isolated β-crystallins tend to aggregate particularly on concentration or lyophilization (19, 47, 89). Separation of rabbit β-crystallin on Sephacryl S-200 yielded four fractions eluting between α- and γ-crystallin. The estimated molecular weights were 250,000, 130,000, 60,000, and 37,000, respectively (90).

The β-crystallin subunits can be separated by DEAE-52 Sephadex chromatography (35). By this method the major β chain βB_p has been obtained pure. Recently we elucidated the complete primary structure of this polypeptide. Surprisingly a duplication was found in that the *N*-terminal half of βB_p appeared to be homologous with the *C*-terminal half. Moreover, comparison with the known amino acid sequence of αA, αB, and γ_{II} revealed clear-cut sequence homology with the latter protein (cf. Figure 11

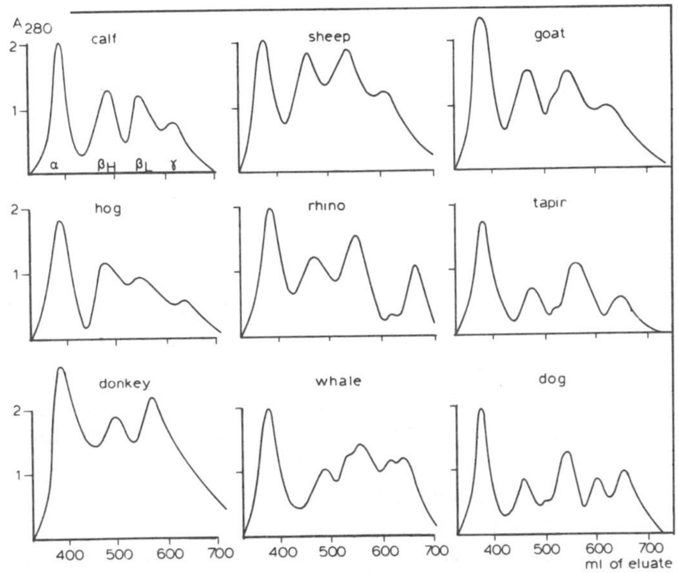

Figure 13 Gel filtration patterns of β-crystallin chains from various species. Separation was carried out on Sephadex G200 columns. Reprinted with permission from ref. 92.

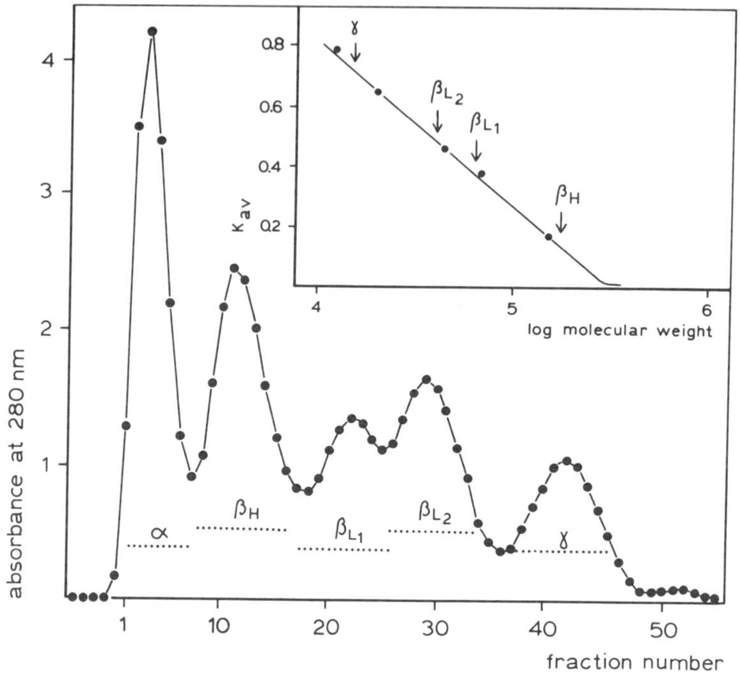

Figure 14 Resolution of the β_L peak into β_{L1} and β_{L2} on gel filtration on a Sephadex G200 column equilibrated in 100 m*M* ammonium formate, pH 7.3. Absorbance of 5 ml fractions, starting with the void volume fraction (after 150 ml), was measured at 280 nm. Inset: $K_{av} = (V_e - V_0)\ (V_1 - V_0)$ in which V_e = elution volume of the peak, V_1 = column volume, and V_0 = void volume. Marker proteins used were cytochrome C (M_r 13,000), soybean trypsin inhibitor (M_r 20,000), ovalbumin (M_r 45,000), bovine serum albumin, (M_r 67,000), and rabbit immunoglobulin G (M_r 150,000). Molecular weights (peak values) observed were α-crystallin larger than 6 $\times 10^6$, $\beta_H = 1.8 \times 10^5$, $\beta_{L1} = 6.2 \times 10^4$, $\beta_{L2} = 4.2 \times 10^4$, and $\gamma = 1.6 \times 10^4$ daltons. Reprinted with permission from ref. 23.

in Chapter 6). The genetic implication of the observation that a β-crystallin constituent (36) reveals partial homology with γ-crystallin is not yet understood.

From the work of others (91) and from investigations of our group (92) it is established that the βB_p chain is a characteristic β-crystallin polypeptide in a great variety of mammalian species. Zigler and Sidbury (93) studying β-crystallins from six different mammalian species, found that in large part the structural features are similar to calf β-crystallin. These authors provide some evidence that primate β_H crystallin appeared to possess a much larger polypeptide (M_r 66,000 on SDS gel) than other known β-crystallin subunits. We have recently extended these studies to other mammals and found that βB_p is again the major constituent of both β_H and β_L in a variety of species (92). Zigler (94) found some age-related changes in the polypeptide composition of bovine lens. In particular, shifting patterns of polypeptide abundance were observed which were ascribed to postsynthetic modifications.

β_s-Crystallin, which elutes together with the γ-crystallins upon gel filtration on Sephadex G200, has, as do the other β-crystallin polypeptides, a blocked *N*-terminus, reported to be *N*-acetyltryptophan (95). This amino acid, however, has hitherto never been found as *N*-terminus in other proteins. β_s is separated from the γ-crystallins by chromatography on SE Sephadex columns (compare Figure 2*B*).

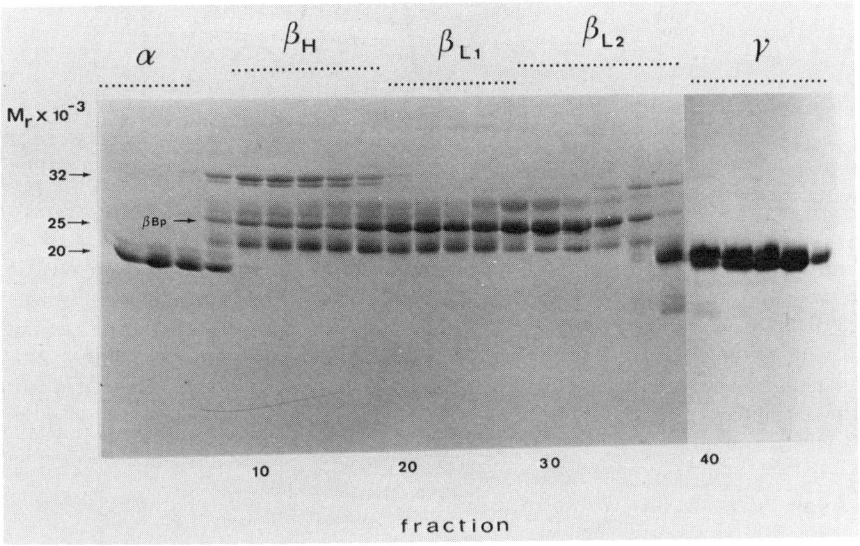

Figure 15 Subunit composition of β_{L1} and β_{L2} from Figure 14. Aliquots of alternate (even) fractions of the gel filtration run (depicted in Figure 14) were lyophilized. The protein concentration in the fractions was calculated from the absorbance at 280 nm: 0.4 units for 1 mg/ml α-crystallin, 2.2 for β-crystallin, and 2.3 for γ-crystallin. Samples of 0.025 mg protein were applied onto a 0.75-mm-thick SDS–polyacrylamide gel. The position of the main fractions of the lens crystallins is indicated at the top of the figure in a manner analogous to Figure 14. On the left the molecular weights of characteristic crystallin polypeptides are indicated: αA (M_r 20,000 daltons), βB_p (M_r 25,000), and βB_{1a} (M_r 32,000). Reprinted with permission from ref. 23.

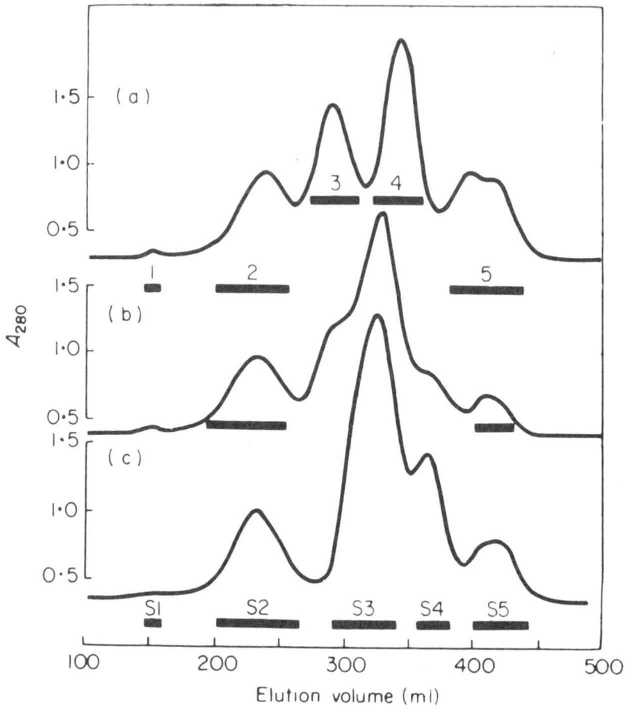

Figure 16 Fractionation of calf crystallins by gel filtration on a Sepharose GB-CL column. Extraction and elution were performed in a buffer containing 0.1 *M* KCl, 0.001 *M* DTE, and 0.05 *M* Tris. *(a)* Without sucrose, *(b)* with 0.5 *M* sucrose, and *(c)* with 1.0 *M* sucrose. Fractions were pooled and are indicated as follows: 2, α-crystallin; 3 and 4, β-crystallin; 5, γ-crystallin. Courtesy of Dr. Li. Reprinted with permission from ref. 85.

1.2.4.4 Y-Crystallin

The fraction that elutes after β_L from the gel filtration column has been called γ-crystallin. Since several workers do not distinguish between, for instance, the low molecular weight β-crystallin species β_s (25) and the true γ-crystallin, we previously proposed to designate these proteins LM (low molecular weight) crystallins. In fact, this fraction comprises all water-soluble monomeric lens proteins ranging in molecular weight from about 20,000 to 27,000. In calf lens γ-crystallin represents about one-fifth of the whole water-soluble protein. By free isoelectric focusing calf lens γ-crystallin can be resolved into five components with isoelectric points greater than 7.0. Bours therefore defined all lens proteins with this electrophoretic migration behavior as γ-crystallin (96). Björk (40, 97, 98), the first to systematically investigate these proteins, obtained four homogeneous fractions, namely, γ_{II}, γ_{IIIa}, γ_{IIIb}, and γ_{IVb}. Furthermore, he succeeded in crystallizing fraction Y_{II}. The X-ray diffraction pattern of the latter component was published recently (99, 100). The molecule appeared to be an ellipsoid with main axes of approximately 55 × 30 × 25 Å (101). The amino acid sequence of γ_{II} has been reported by Croft (102, 103). As can be seen from Figure 17 there is a relatively high cysteine content. Although these residues are rather randomly distributed there is a clustering of four cysteines at the *N*-terminal site (between residues 15 and 32) (104). Y-Crystallin has also been isolated from rat lens (105, 106, 107), but the *N*-terminal amino acid is Ala instead of Gly. Human γ-crystallin strongly resembles calf lens γ-crystallin as judged on the basis of amino acid composition, electrophoretic mobility, *N*-terminus, and fingerprinting (108).

It is γ that is responsible for the so-called cold cataract. If the temperature at neutral pH is lowered below 10°C, γ-crystallin precipitates. When the temperature is raised the γ-crystallin goes back into solution. Proteins that reveal such a behavior are called cryoprotein. Like α-crystallin, γ-crystallin has almost no α-helical structure. ORD and CD studies showed the major conformation is in β sheets and some random coil (102, 109).

In order to quantitate lens crystallins present in minute amounts a radioimmunotest has been developed by Russell et al. (10). This assay was developed for mouse α and γ-crystallins which were iodinated by the lactoperoxidase method (111). Almost concomitantly a specific radioimmunoassay for human α- and γ-crystallin has been reported by Sandberg and Closs (112). Borkman and Lerman (113) categorized human α- and γ- and bovine α- and β-crystallin solutions by fluorescence data. According to the tabulation of Burnstein (114) the results were for the four species 80% class I, 10% class II, and 10% class III. This means that 80% of their

GLY-LYS-ILE-THR-PHE-TYR-GLU-ASP-ARG-GLY-PHE-GLN-GLY-HIS-CYS-TYR-GLN-
CYS-SER-SER-ASN-ASN-CYS-LEU-GLN-PRO-PRO-TYR-PHE-SER-ARG-CYS-ASN-SER-
ILE-ARG-VAL-ASP-VAL-HIS-SER-LEU-LEU-TYR-MET-LEU-GLN-ARG-PRO-ASP-TYR-
GLY-HIS-TYR-GLN-ARG-GLY-ASN-TYR-PRO-GLN-TYR-GLN-TRP-MET-GLY-PHE-ASP-
ASP-SER-ILE-ARG-SER-CYS-ARG-LEU-ILE-PRO-GLN-HIS-THR-GLY-THR-PHE-ARG-
MET-ARG-ILE-TYR-GLU-ARG-ASP-ASP-PHE-ARG-GLY-GLN-MET-SER-GLU-ILE-THR-
ASP-PRO-CYS-ASP-SER-LEU-GLN-ASP-ARG-PHE-HIS-LEU-THR-GLU-VAL-ASN-TYR-
VAL-LEU-GLU-GLY-SER-TRP-GLU-MET-PRO-SER-TYR-ARG-GLY-ARG-GLN-TYR-LEU-
LEU-ARG-PRO-GLY-GLU-TYR-ARG-ARG-TYR-LEU-ASP-TRP-GLY-ALA-MET-ASN-ALA-
LYS-VAL-GLY-SER-LEU-ARG-ARG-VAL-MET-ASP-PHE-TYR

Figure 17 The primary structure of γ_{II}-crystallin as determined by Croft. Courtesy of Dr. Croft.

tryptophan residues are buried in the interior of the molecule, whereas 20% are in contact with water. On the other hand, human β- and bovine γ-crystallin seemed to have all Trp residues buried and none emitting exposed to water.

Comparison of β_S, cattle lens cortex γ, and calf lens γ-crystallin revealed that these proteins differ not only in molecular size but also in immunochemical properties. The presence of these LM crystallins appeared to be age-related (115). The low molecular weight fraction of some avian lens crystallin seems to lack γ-crystallin. For instance, pigeon LM crystallin (26) has the characteristics of β_s- rather than those of γ components. Amino acid composition, *N*-terminus, and peptide mapping suggested that it satisfies some of the criteria for a β-crystallin species (116).

1.2.4.5 δ-Crystallin

Lenses of reptiles and birds possess a crystallin species that is different from α-, β-, and γ-crystallin (7, 117, 118). Although these reptilian and avian crystallins were thought to belong to the β-crystallin population Rabaey (119) recognized its specific nature and proposed the designation FISC (first important soluble crystallin) owing to its prevalence in embryonic lens tissue. Later on the name δ-crystallin was introduced by Zwaan and Ikeda (120).

Most studies on δ-crystallin have been undertaken with chicken in which it represents about 50% of the total protein in the adult lens and approximately 75% in embryonic lenses (119, 121). Like α- and β-crystallin, δ-crystallin is a polymeric protein consisting of four closely related (if not similar) subunits (122–127).

Laser Raman spectroscopy of chick and pigeon lenses showed that δ-crystallins exist mainly in the α-helical configuration in contrast with bovine lens proteins (128, 129). Delta crystallin is the first soluble lens protein whose corresponding cDNA has been subjected to molecular cloning (130, 131).

1.2.4.6 FM Crystallin

In addition to α-, β-, and γ-crystallin, mammalian lenses contain a fast-moving fraction that was called pre-α-crystallin by van Dam and ten Cate (13) because this protein migrates ahead of α-crystallin in agar gel electrophoresis. We proposed to change this name to FM (fast-migrating) crystallin since the protein did not reveal any biochemical relation with α-crystallin subunits. For instance, its amino acid composition is completely different from α-crystallin (compare Table 2), it is a monomeric protein, and it does not reveal any immunochemical cross reaction with αA- or αB-chains.

Table 2 Amino Acid Composition of FM Crystallin and α-Crystallin Subunits[a]

Amino acid	FM Crystallin	αA_2-Crystallin[b]	αB_2-Crystallin[c]
Lysine	2.8	4.0	5.7
Histidine	7.6	4.0	5.1
Arginine	5.7	7.5	8.0
Aspartic acid	6.1	9.2	7.4
Theronine	2.1	2.9	4.0
Serine	6.9	13.3	9.7
Glutamic acid	13.2	9.8	9.7
Proline	4.2	6.9	9.7
Glycine	9.2	5.8	4.6
Alanine	11.6	3.5	5.1
Cysteine	[d]	0.6	—
Valine	7.2	5.8	5.7
Methionine	1.4	1.2	1.1
Isoleucine	2.9	5.2	5.7
Leucine	9.5	8.1	5.7
Tyrosine	0.4	3.5	1.1
Phenylalanine	8.1	8.1	7.4
Tryptophan	1.2	0.6	1.1

[a]Values are given as residues per 100 residues.
[b]Calculated from the data of Van der Ouderaa et al. (42).
[c]Calculated from the data of Van der Ouderaa et al. (43).
[d]Not determined.

A fast-migrating acidic crystallin has also been observed in some salientian and caudate amphibians. This protein, which is confined to fiber cells, has also been called pre-α-crystallin (133–135). Brahma and van Doorenmaalen (136) and Brahma (137) were able to show that calf FM crystallin and Rana esculenta pre-α-crystallin are antigenically different proteins. The function of these fast-moving "crystallins" has not been elucidated yet. FM crystallin can be obtained in a pure form by the following procedure.

Lens cortices are homogenized in a medium containing 50 m*M* Tris-Cl, pH 7.4, 25 m*M* KCl, 5 m*M* $MgAc_2$, and 0.35 *M* sucrose. The homogenate is centrifuged at 12,000 × *g* for 20 min and the supernatant is treated with deoxycholate to a final concentration of 0.5%. Thereafter polyribosomes are removed by ultracentrifugation as described by Bloemendal et al. (138) and the supernatant is fractionated by $(NH_4)_2SO_4$ precipitation. FM crystallin precipitates in the fraction between 30 and 40% saturation. $(NH_4)_2SO_4$ is removed and the desalted preparation is further purified on a DEAE Sephadex A-50 column, equilibrated in a medium consisting of 0.05 *M* Tris-Cl, pH 7.6, 0.1 *M* NaCl, 0.001 *M* EDTA, and 0.001 *M* β-mercaptoethanol.

The column is developed with a gradient of 0.15 to 1 *M* NaCl containing all other constituents of the buffer mentioned above.

Final purification is achieved by preparative polyacrylamide gel electrophoresis (Leget and Bloemendal, unpublished). On SDS gels FM crystallin has an apparent molecular weight of 14,500. The nature and function of this lens protein are still obscure.

1.3 NONCRYSTALLIN PROTEIN IN THE LENS

1.3.1 Membrane and Cytoskeleton Proteins

As mentioned earlier, about 90% of the total mammalian lens protein is contributed by the crystallins. The noncrystallin proteins found mainly in the so-called albuminoid fraction (1) have received until recently much less attention than the water-soluble crystallins.

The water-insoluble lens fraction can be further fractionated by 6 to 8 *M* urea extraction into urea-soluble and urea-insoluble fractions. The first fraction contains mainly constituents of the intercellular matrix, whereas the latter comprises the intrinsic plasma membrane proteins.

The urea-soluble fraction can be further separated into four new fractions on a Sephadex G200 column equilibrated in 6 *M* urea (139) (Figure 18). Polyacrylamide gel electrophoretic analysis of the separated fractions (Fig-

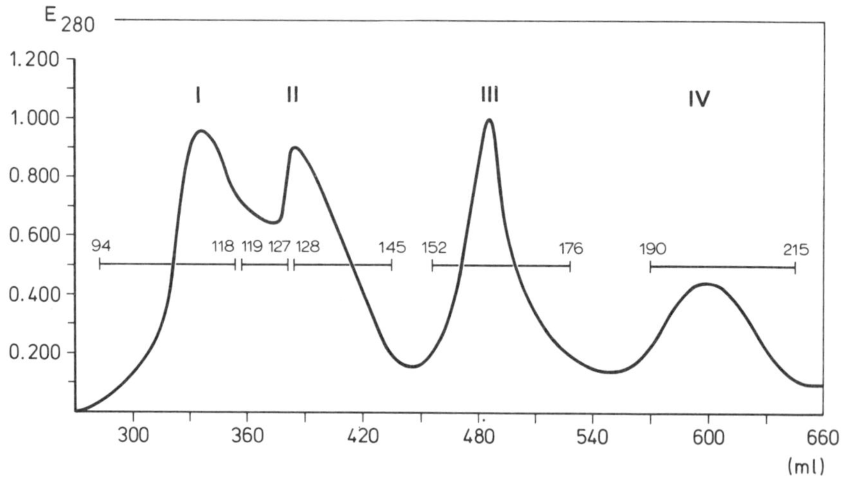

Figure 18 Gel filtration of the urea-soluble part of the water-insoluble lens fraction on a Sephadex G200 column equilibrated in 6 *M* urea. Reprinted with permission from ref. 139.

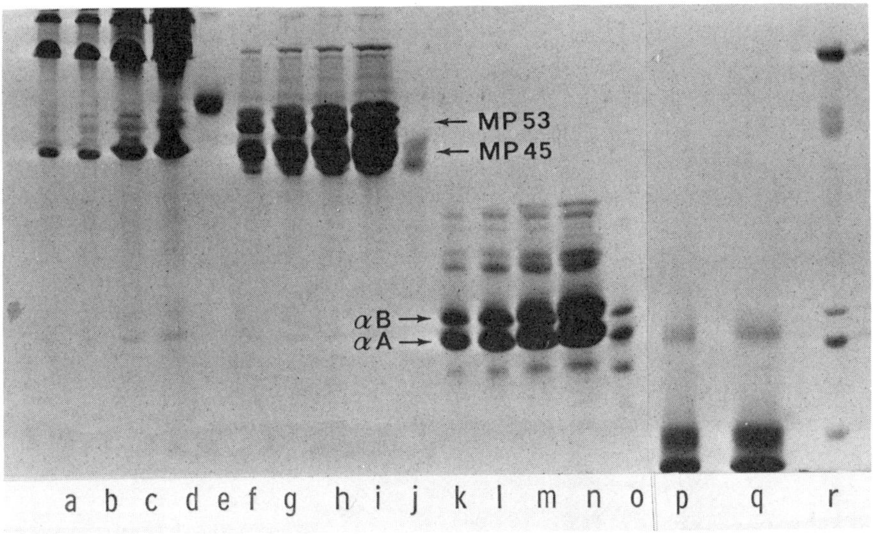

Figure 19 Electrophoretic analysis of the four fractions from Figure 18 on a polyacrylamide gel containing 0.1% SDS. *a–d*, *f–i*, *k–n*, and *p–q* represent fractions of peaks I, II, III, and IV, respectively. Markers at mol. wt. *(e)* 68,000, *(j)* 45,000, and *(o)* 20,000, and *(r)* mixture of markers containing also cytochrome and C (13,500). Reprinted with permission from ref. 139.

ure 19) reveals that the second fraction contains cytoskeletal proteins, in particular actin, whereas in the third fraction the "water-insoluble" crystallins are found. Two hitherto unidentified polypeptides of M_r 13,000 and M_r 10,000, respectively, are found under peak IV. The latter polypeptide may well correspond to an M_r 9600 chain isolated from human lens (140, 141).

After initial studies on calf lens plasma membranes (142, 143), it seemed necessary to investigate systematically those proteins thhat are extrinsic or intrrinsic parts of the lipid bilayer of lens cell membranes. We proposed a rational nomenclature for both extrinsic and intrinsic plasma membrane proteins (144). The abbreviation MP stands for membrane protein, followed by the apparent molecular weight in kilodaltons. For example, the principal fiber plasma membrane protein with a molecular weight of about 26,000 is indicated by MP26. Another important lens membrane component, typical for both epithelial and fiber cell membranes, is MP34 (compare Figure 20).

As described in Chapter 3 lenticular plasma membranes can be prepared in two ways. One method is based on density criteria (142, 145, 146) and the other on solubility (143, 147, 152). From Figure 21 it may be concluded that membranes isolated by either buoyant density or their specific solubility properties yield almost identical protein patterns.

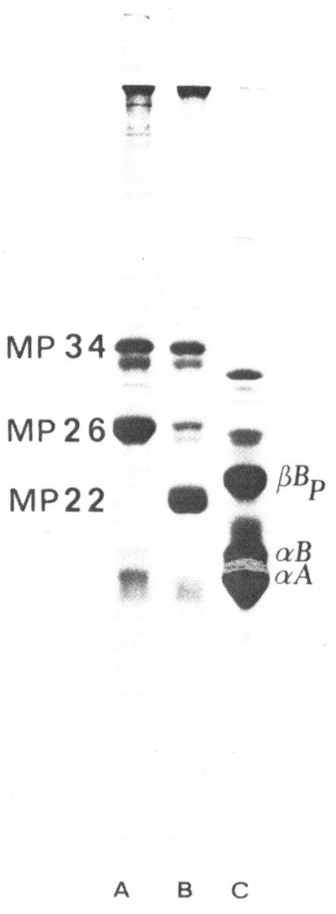

Figure 20 Nomenclature of the major lens plasma membrane proteins related to their electrophoretic behavior on polyacrylamide gels containing 0.1% SDS. A, Lens fiber plasma membranes ("fresh"). B, Lens fiber plasma membranes ("aged"). Note the transition from MP26 to MP22. C, Water-soluble lens proteins (for comparison).

It can also be seen that in addition to some crystallin polypeptides major bands are localized at M_r 26,000, M_r 34,000, and in the molecular weight range of 42,000 to 55,000 (Figure 21).

The former polypeptide has been identified by Kibbelaar et al. (153) as actin. The latter protein was tentatively indicated as desminlike. Mousa and Trevithick (154) demonstrated that the ratio of β- and γ-actin changes during differentiation. The localization of actin in the lens was determined with the aid of immunofluorescence by Kibbelaar et al. (155). In cells of epithelial origin Franke et al. (156) find as a rule intermediate filaments of the so-called prekeratin type. We were able to show that lens intermediate filaments apparently form an exception to this rule in that the 55,000 mol. wt. protein is of the vimentin type. Therefore, the designation desminlike must be replaced by vimentin. Proteins of identical molecular weight occur in the cytoskeletons of pickerel, chick frog, and human lens (157).

A highly purified M_r 42,000 protein was obtained by extraction of the water-insoluble fraction of the calf lens with *N*-tris(hydroxymethyl)methyl-2-aminoethane sulfonic acid (TES buffer) followed by gel filtration. Polyacrylamide gel electrophoresis, amino acid analyses, peptide mapping, and electron microscopy show that this protein is identical to nonmuscle actin (compare Figure 22). On the other hand, evidence was provided that a similar protein could be isolated from the water-soluble fraction. It binds specifically to deoxyribonuclease I (DNAse I). On the basis of this property this protein could be purified applying DNAse I-affinity chromatography.

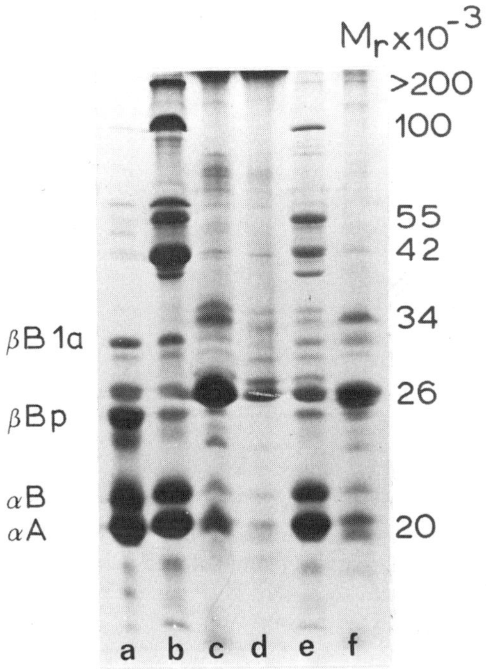

Figure 21 Comparison of the gel electrophoretic patterns of lens plasma membrane proteins obtained after solubilization in SDS of the urea-insoluble fraction *(c)* with those obtained after density gradient centrifugation. (a) Water-soluble crystallins (for comparison); (b) urea-insoluble lens proteins with the cytoskeletal components; (c)urea-insoluble fraction; (d) fraction c (after deoxycholate treatment in the electron microscope mainly junctions are seen!); (e) total water-insoluble fraction (represents fractions *b* plus *c*); (f) purified membranes after sucrose density gradient centrifugation (note the similarity between *c* and *f*).

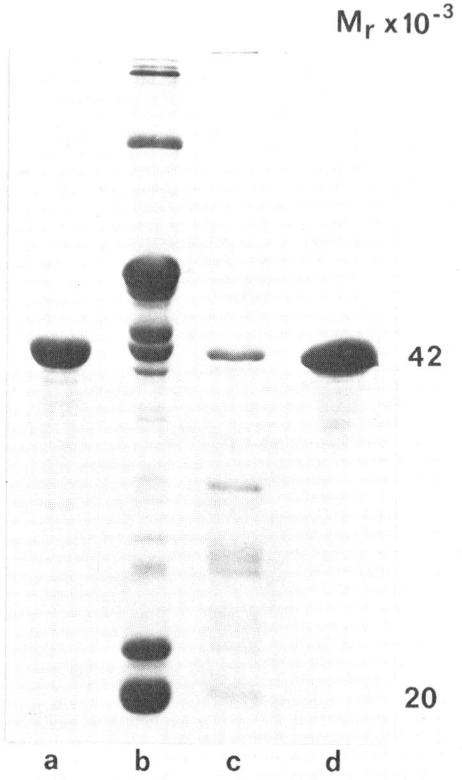

Figure 22 Polyacrylamide gel electrophoresis of F-actin from bovine eye lens. *(a)* Rabbit muscle actin (for comparison); *(b)* water-insoluble fraction; *(c)* tris(hydroxymethyl)methyl-2-aminoethane sulfuric acid (TES) extract from the so-called crude membrane fraction; *(d)* lens cortex actin purified by gel filtration of the TES extract. Reprinted with permission from 153.

Concomitantly the protein of M_r 55,000 and a rather large amount of α-crystallin copurify with the M_r 42,000 protein. A schematic representation of the isolation of lens actin from the water-soluble fraction is given in Figure 23. The results show that the lens fiber cell, as do other nonmuscle cells, comprises F (filamentous) actin in a water-insoluble form and a pool of G (globular) actin in solution. From Table 3 it can be concluded that there is a great similarity in amino acid composition of lens-derived actin as compared to actin from a number of other tissues. The same holds true for the 55,000 mol. wt. protein (compare Table 4). Meanwhile several other

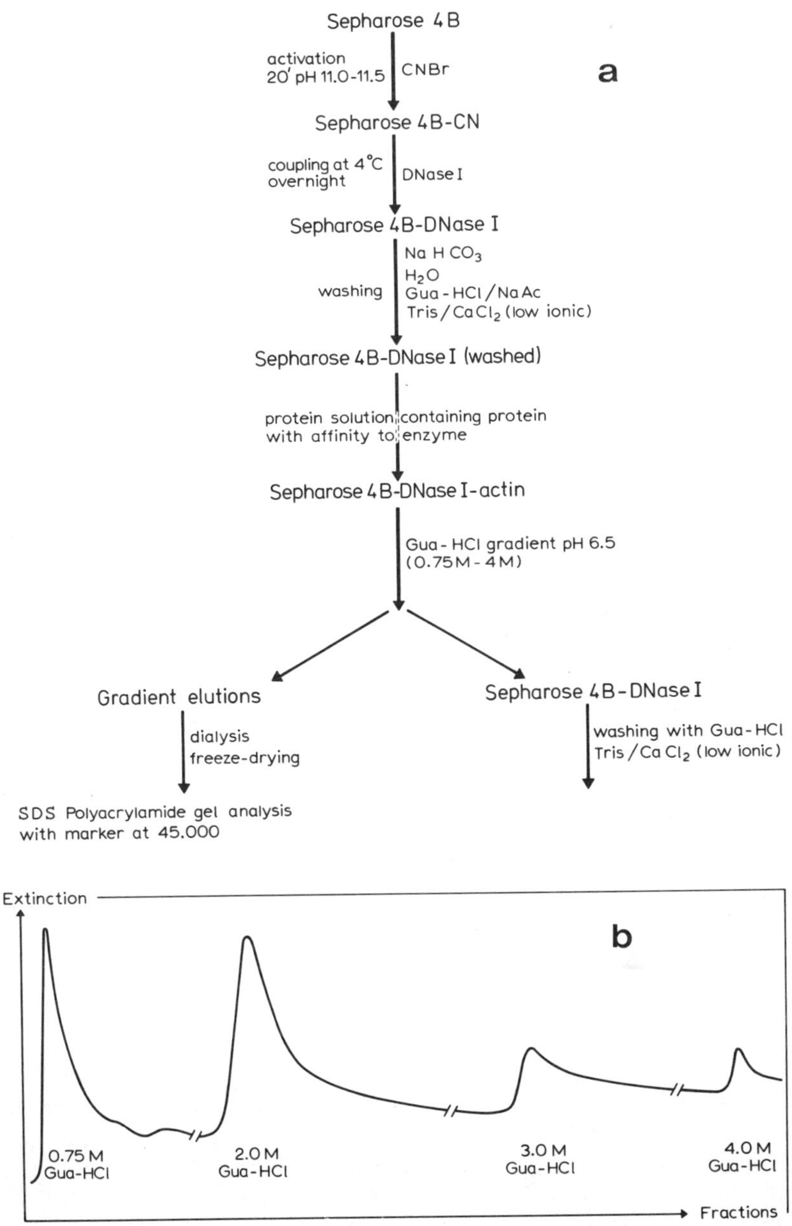

Figure 23 Schematic representation of the isolation of lens G-actin by affinity chromatography on a DNAse I column. *(a)* preparation and regeneration of the affinity column; *(b)* elution profile after affinity chromatography.

Table 3 Comparison of Amino Acid Composition of Actin from Different Tissues and Species

Amino acid residue	Calf lens[a]	Calf lens[b]	Calf thymus[c]	Rabbit skeletal muscle[d]	Human platelets[e]
Lysine	19	20	21	22	19
Histidine	11	10	11	11	10
Arginine	18	26	18	16	18
Asparagine	35	31	35	33	33
Threonine	27	24	29	29	26
Serine	27	37	24	25	25
Glutamic acid	44	53	47	42	46
Proline	23	20	20	26	20
Glycine	28	37	29	31	32
Alanine	30	31	29	34	32
Cysteīne	4	—	4	—	4
Valine	20	14	15	17	16
Methionine	7	10	10	15	10
Isoleucine	17	15	20	20	23
Leucine	33	25	30	27	30
Tyrosine	10	12	12	15	12
Phenylalanine	14	10	13	13	13
Tryptophan	—	—	—	—	4

Reprinted with permission from ref. 153.
[a]Isolated from the water-soluble fraction on affinity chromatography followed by preparative dodecylsulfate gel electrophoresis.
[b]Isolated from the water-insoluble fraction by TES buffer extraction.
[c]Isolated by Dr. M. Kibbelaar.
[d]Gift of Dr. A. Fagraeus, Karolinska Institute.
[e]Booyse et al. (166).

authors have studied the proteins of calf lens plasma membranes (148–151) and those derived from chick (158–161). Similar investigations have recently been initiated with human lenses (162–164). In all studies the MP26 component emerges as the most important polypeptide common to lens fiber plasma membranes irrespective of their source. It cannot be ruled out that this component may undergo some degradation or transition to an M_r 22,000 component upon aging (see also Figure 20) (144, 162). That MP26 is definitely different from βB_p, which has almost the same molecular size, can be concluded from amino acid analysis and peptide mapping data (Table 5). Immunoprecipitation studies by Alcalá and Maisel resulted in the isolation of an antiserum against chick MP26 (158). Species specificity was

Table 4 Comparison of Amino Acid Compositions of intermediate-sized Filament Protein from Different Tissues and Species

Amino acid	Calf lens component M_r55,000[a]	Bovine epidermis[b]	Calf brain[c]	Human astrocytes[d]	Hog stomach[e]	Chicken gizzard[e]	Baby hamster kidney cells[f]
Asparagine	9.5	9.7	9.5	9.1	9.2	8.8	10.3
Threonine	5.5	3.9	4.8	4.7	5.2	5.7	4.8
Serine	7.2	10.5	3.4	5.9	7.6	7.4	6.5
Glutamic acid	17.5	14.8	16.6	18.4	19.5	19.7	18.7
Proline	3.1	1.3	0.0	3.0	2.9	2.3	2.9
Glycine	5.1	12.9	6.0	4.9	5.1	5.0	5.7
Alanine	10.4	6.6	9.8	10.2	9.0	8.6	8.4
Valine	4.7	5.7	4.5		4.9	5.0	5.4
Cysteīne	—	0.9	0.0	0.0	0.7	0.5	0.8
Methionine	1.8	2.0	2.1	2.0	2.0	2.5	2.0
Isoleucine	3.5	3.8	3.9	3.2	3.9	4.2	3.8
Leucine	10.8	10.2	11.6	11.8	10.0	9.9	11.2
Tyrosine	2.6	2.4	4.1	2.8	2.8	3.0	2.3
Phenylalanine	2.3	3.7	3.8	2.7	2.7	2.7	2.3
Lysine	5.7	4.6	5.7	6.0	4.7	4.7	5.5
Histidine	2.1	1.2	2.0	1.9	1.3	1.6	1.5
Arginine	9.2	6.5	9.7	8.5	8.6	8.7	8.1

Reprinted with permission from ref. 153.

[a]Data expressed in residues per 100 residues being averages of four independent determinations.

[b]Steinert and Idler (166a)

[c]Davison and Winslow (166b).

[d]Eng et al. (166c).

[e]Small and Sobieszek (167).

[f]Strager et al. (168).

Table 5 Comparison of Amino Acid Compositions of MP26, MIP[a] and βB_p[b]

Amino acid	MP26	MIP	βB_p
Asparagine	4.9	4.7	8.3
Threonine	4.5	4.9	3.4
Serine	7.8	8.4	8.4
Glutamic acid	16.3	15.8	15.8
Proline	6.7	5.7	6.8
Glycine	10.7	10.6	9.3
Alanine	14.4	12.2	4.2
Valine	8.2	8.4	6.9
Cysteine	0.0	0.6	1.0
Methionine	1.8	2.3	1.0
Isoleucine	2.6	2.7	3.1
Leucine	13.2	13.7	5.0
Tyrosine	2.9	2.8	4.4
Phenylalanine	5.2	6.2	4.0
Tryptophan	—	—	2.0
Lysine	2.8	1.5	6.0
Histidine	3.1	2.4	4.1
Arginine	5.3	5.7	5.0

Data for the calf lens 26,000 M_r component are expressed in residues per 100 residues.
[a]Values were obtained by Broekhuyse et al. (149). These authors designated MP26 as main intrinsic polypeptide.
[b]Values were obtained by Herbrink et al. (37).

demonstrated by showing that this antiserum was negative against human lens plasma membranes (165).

The differences between protein patterns of epithelial and fiber plasma membranes (169) apparently reflect changes in molecular organization of these structures during differentiation (170). The most remarkable difference is the accumulation of MP26 in lens fiber membranes (Figure 24).

1.3.2 RNAse Inhibitor

A protein that inhibits the activity of alkaline RNAse *in vitro* has been found in many different tissues (171–174). This inhibitor appears to be a potent protective factor for the preparation of polyribosomes (175–179) and of messenger (180, 181).

The presence of this inbibitor could also be demonstrated in calf lens (173, 182, 183). Actually lens tissue provides one of the most suitable sources for the isolation of this RNAse inhibitor. Even though the inhibitor represents only a minute fraction (less than 0.1 part per 1000) of the total water-soluble lens protein, it can be freed from the bulk of contaminants (mainly crystallins) in only two simple purification steps, namely, DEAE-cellulose cbromatography followed by gel filtration on Ultrogel AcA 34. The inhibitor fraction obtained can be analyzed by polyacrylamide gel electrophoresis. In Figure 25*d* it can be seen that the inhibitor is freed from the bulk of α-crystallin. Remaining contaminations can be removed by electro-elution (184) (Compare Figure 26). The quantitative evaluation of the results after the two purification steps is shown in Table 6. It can be concluded that the inhibitor loses activity after lyophilization.

There are indications that the lens inhibitor consists of two subunits, although this has not been observed in alkaline RNAse inhibitors derived from other tissues.

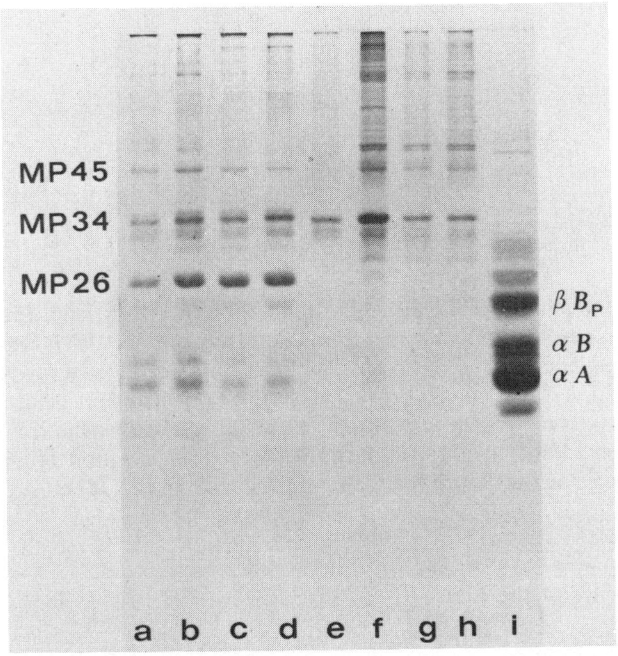

Figure 24 Sodium dodecylsulfate–gel electrophoresis of lens plasma membrane proteins. *(a–d)* Isolated fiber plasma membrames; *(e–h)* isolated epithelial plasma membranes; *(i)* pattern of water-soluble crystallins for comparison.

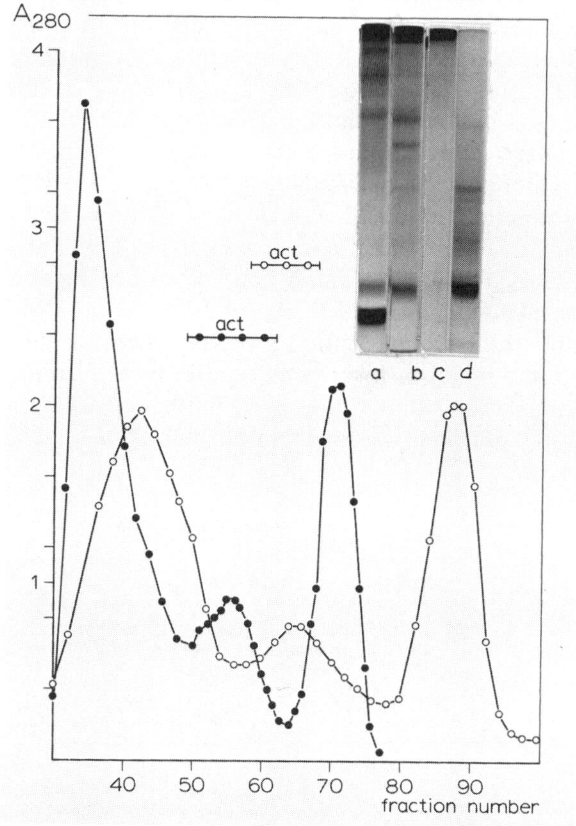

Figure 25 Isolation of alkaline RNAse inhibitor from calf lens. Fractionation of calf lens homogenate (75,000 × *g* supernatant fraction) on DEAE Sephades A50 at pH 7 (●——●) and at pH 6 (○——○). The localization of the activity is indicated by the abbreviation "act." Inset: polyacrylamide gel electrophoresis at pH 8.9 of the isolated active fractions *(a)* at pH 7, *(b)* at pH 6, *(c)* α-crystallin (for comparison), and *(d)* after additional purification of Ultrogel. The position of the inhibitor is indicated by an arrow. Reprinted with permission from ref. 184).

1.3.3 Enzymes of the Lens

Various lenticular enzymes bave been described in tbe literature. For instance, all the enzymes of the glycolytic pathway have been found in the lens. Furthermore, since the protein-synthesizing machinery is operative both in lens epithelium and in the outer cortex all the enzymes known to be involved in protein biosynthesis have to be present. We were also able to show that an enzyme that is responsible for α-N-terminal acetylation of the crystallins can be isolated from lens tissue (185, 186). ecause these enzymes are not specific in the sense of being exclusive for lenticular tissue, they are not dealt with in this chapter (see Chapter 9).

1.3.3.1 Neutral Protease

A neutral protease, which may be involved in the degradation of the crystallins, has been described to occur in ox lens cortex (187–191). However, neither fractionation to purity nor unraveling of the exact function of this protease have been achieved yet. The neutral protease activity has also been demonstrated in a variety of other species, including monkey, pig, sheep, rabbit, rat, and mouse (164). Human lens seems to contain only 25% of the amount of protease found in cattle (192). Those comparisons, however, are conclusive only if they are made between tissues of similar age and if the preparations are processed under standard conditions.

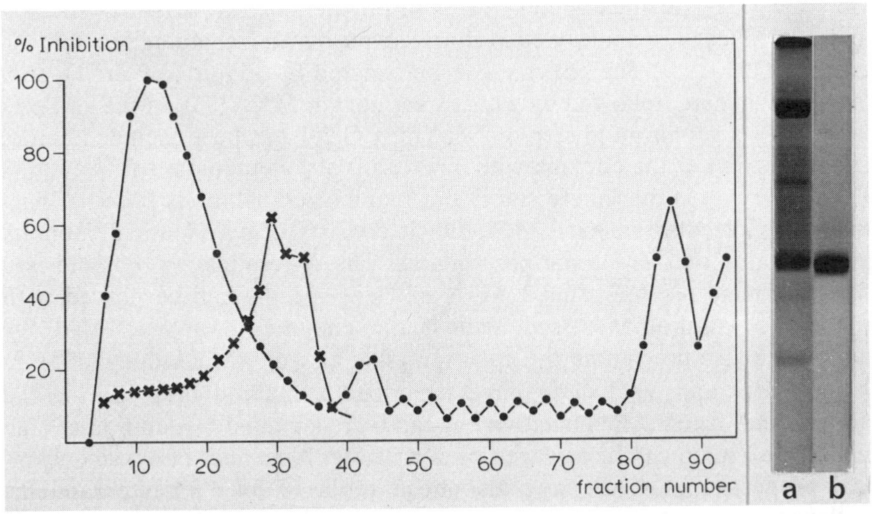

Figure 26 RNAse inhibitor from bovine eye lens. Electro-elution of the purified inhibitor and its complex. Inhibitor activity determined in the absence of (•——•) and in the presence (×——×) of parachloromercuribenzoate. Protein patterns *(a)* after polyacrylamide gel electrophoresis of the starting material purified on DEAE Sephadex and *(b)* after electro-elution. Reprinted with permission from ref. 184.

Table 6 Recovery of the RNAse Inhibitor Activity after Various Purification Steps

Fraction	Absorbance (280 nm)	Total activity	Specific activity	Percent recovery
30,000 × *g* supernatant	20,800	171,000	8.4	100
DEAE				
Wet	51.6	64,500	1,290	40
Lyophilized	22.8	37,000	1,740	20
Ultrogel				
Wet	51.1	44,700	10,000	32
Lyophilized	3.9	15,300	3,290	10.4

1.3.3.2 Leucine Aminopeptidase

The enzyme leucine aminopeptidase (LAP), although not strictly lens-specific, is described in somewhat more detail. This enzyme, localized in the cytosol, presumably exists in numerous, if not all, tissues, but as far as we know never in such abundance as in bovine lens. In particular Hanson and his colleagues made great efforts to purify and characterize this Zn enzyme (193–198). The enzyme can be isolated by addition of $ZnSO_4$ to a lens homogenate, followed by a heat treatment at 54°C (196). In the author's laboratory a gel filtration step on Sephadex G 200 has been included before crystallization of the enzyme with 57% (wt./vol.) ammonium sulfate.

The enzyme is extremely stable but lyophilyzed solutions become inactive. Its molecular weight is 326,000; it consists of six identical subunits and contains two Zn atoms per subunit. The *N*-terminal amino acid is a free threonine residue. That LAP is not lens-specific can be proved with the aid of antibodies raised against the enzyme. It appears that the antibodies also precipitate the corresponding enzyme from kidney (195).

Carpenter and Vahl (199) have described a method of preparing the apoenzyme. Apo-LAP is inactive but can be reactivated by addition of zinc ions. One Zn ion can be replaced by Mn or Mg. A number of physicochemical properties of LAP have also been reported by Carpenter and his colleagues (200, 201). The Zn enzyme crystallizes in the hexagonal space group $P6_322$ with unit cell dimensions $a = 132$ Å and $c = 122$ Å (202). The asymmetric unit consists of one protomer of molecular weight 54,000. Limited digestion of LAP by trypsin treatment results in splitting of only one bond, namely, between Arg-137 and Lys-138. Despite this cleavage the aggregate molecule remains intact and retains its full catalytic activity (203). The complete amino acid sequence (204, 105) is depicted in Figure 27.

Studies upon the active center of the enzyme have been carried out by Müller-Frohne (206) and Fittkau et al. (207). The findings provide evidence for the presence of 6 catalytically active sites which means one active center per subunit. Electron microscopy revealed that six subunits are arranged with 32 symmetry at the vertexes of a distorted prism. The subunits are elongated along the sides of triangles (208).

```
                                            10                                         20
Thr-Lys-Gly-Leu-Val-Leu-Gly-Ile-Tyr-Ser-Lys-Glu-Lys-Glu-Glu-Asp-Glu-Pro-Gln-Phe-
                                            30                                         40
Thr-Ser-Ala-Gly-Glu-Asn-Phe-Asn-Lys-Leu-Val-Ser-Gly-Lys-Leu-Arg-Glu-Ile-Leu-Asn-
                                            50                                         60
Ile-Ser-Gly-Pro-Pro-Leu-Lys-Ala-Gly-Lys-Thr-Arg-Thr-Phe-Tyr-Gly-Leu-His-Glu-Asp-
                                            70                                         80
Phe-Pro-Ser-Val-Val-Val-Val-Gly-Leu-Gly-Lys-Lys-Thr-Ala-Gly-Ile-Asp-Glu-Gln-Glu-
                                            90                                        100
Asn-Trp-His-Glu-Gly-Lys-Glu-Asn-Ile-Arg-Ala-Ala-Val-Ala-Ala-Gly-Cys-Arg-Gln-Ile-
                                           110                                        120
Gln-Asp-Leu-Glu-Ile-Pro-Ser-Val-Glu-Val-Asp-Pro-Cys-Gly-Asp-Ala-Gln-Ala-Ala-Ala-
                                           130                                        140
Glu-Gly-Ala-Val-Leu-Gly-Leu-Tyr-Glu-Tyr-Asp-Asp-Leu-Lys-Gln-Lys-Arg-Lys-Val-Val-
                                           150                                        160
Val-Ser-Ala-Lys-Leu-His-Gly-Ser-Glu-Asp-Gln-Glu-Ala-Trp-Gln-Arg-Gly-Val-Leu-Phe-
                                           170                                        180
Ala-Ser-Gly-Gln-Asn-Leu-Ala-Arg-Arg-Leu-Met-Glu-Thr-Pro-Ala-Asn-Glu-Met-Thr-Pro-
                                           190                                        200
Thr-Lys-Phe-Ala-Glu-Ile-Val-Glu-Glu-Asn-Leu-Lys-Ser-Ala-Ser-Ile-Lys-Thr-Asp-Val-
                                           210                                        220
Phe-Ile-Arg-Pro-Lys-Ser-Trp-Ile-Glu-Glu-Gln-Glu-Met-Gly-Ser-Phe-Leu-Ser-Val-Ala-
                                           230                                        240
Lys-Gly-Ser-Glu-Glu-Pro-Pro-Val-Phe-Leu-Glu-Ile-His-Tyr-Lys-Gly-Ser-Pro-Asn-Ala-
                                           250                                        260
Ser-Glu-Pro-Pro-Leu-Val-Phe-Val-Gly-Lys-Gly-Ile-Thr-Phe-Asp-Ser-Gly-Gly-Ile-Ser-
                                           270                                        280
Ile-Lys-Ala-Ala-Ala-Asn-Met-Asp-Leu-Met-Arg-Ala-Asp-Met-Gly-Gly-Ala-Ala-Thr-Ile-
                                           290                                        300
Cys-Ser-Ala-Ile-Val-Ser-Ala-Ala-Lys-Leu-Asp-Leu-Pro-Ile-Asn-Ile-Val-Gly-Leu-Ala-
                                           310                                        320
Pro-Leu-Cys-Glu-Asn-Met-Pro-Ser-Gly-Lys-Ala-Asn-Lys-Pro-Gly-Asp-Val-Val-Arg-Ala-
                                           330                                        340
Arg-Asn-Gly-Lys-Thr-Ile-Gln-Val-Asp-Asn-Thr-Asp-Ala-Glu-Gly-Arg-Leu-Ile-Leu-Ala-
                                           350                                        360
Asp-Ala-Leu-Cys-Tyr-Ala-His-Thr-Phe-Asn-Pro-Lys-Val-Ile-Ile-Asn-Ala-Ala-Thr-Leu-
                                           370                                         380
Thr-Gly-Ala-Met-Asp-Ile-Ala-Leu-Gly-Ser-(Thr,Gly,Gly,Ala,Val,Phe)-Thr-Asn-Ser-Ser
                                           390                                        400
Trp-Met-Asn-Lys-Leu-Phe-Glu-Ala-Ser-Ile-Glu-Thr-Gly-Asp-Arg-Val-Trp-Arg-Met-Pro-
                                           410                                        420
Leu-Phe-Glu-His-Tyr-Thr-Arg-Gln-Val-Ile-Asp-Cys-Gln-Leu-Ala-Asp-Val-Asn-Asn-Ile-
                                           430                                        440
Gly-Lys-Tyr-Arg-Ser-Ala-Gly-Ala-Cys-Thr-Ala-Ala-Ala-Phe-Leu-Lys-Glu-Phe-Val-Thr-
                                           450                                        460
His-Pro-Lys-Trp-Ala-His-Leu-Asp-Ile-Ala-Gly-Val-Met-Thr-Asn-Lys-Asp-Glu-Val-Pro-
                                           470
Tyr-Leu-Arg-Lys-Gly-Met-Ala-Gly-Arg-Pro-Thr-Arg-Phe-Ser-Gln-Asp-Ser-Ala.
```

Figure 27 The primary structure of bovine leucine aminopeptidase.

The function of LAP *in vivo* is still only speculative. It might well be that peptides formed by the action of an endopeptidase are further degraded by exopeptic action of LAP. For instance, in fibrogenesis the proteins of cell organelles such as mitrochondria and nuclei have to be removed. LAP may well play a role in the degradation of these proteins. Moreover, upon aging of the eye lens peptides of defined length are split from the native crystallins. Hydrolysis of the amino acids starting from the *N*-terminus might be catalyzed by LAP. While this manuscript was in preparation the complete amino acid sequence of the LAP subunits was established in the author's laboratory (204, 205).

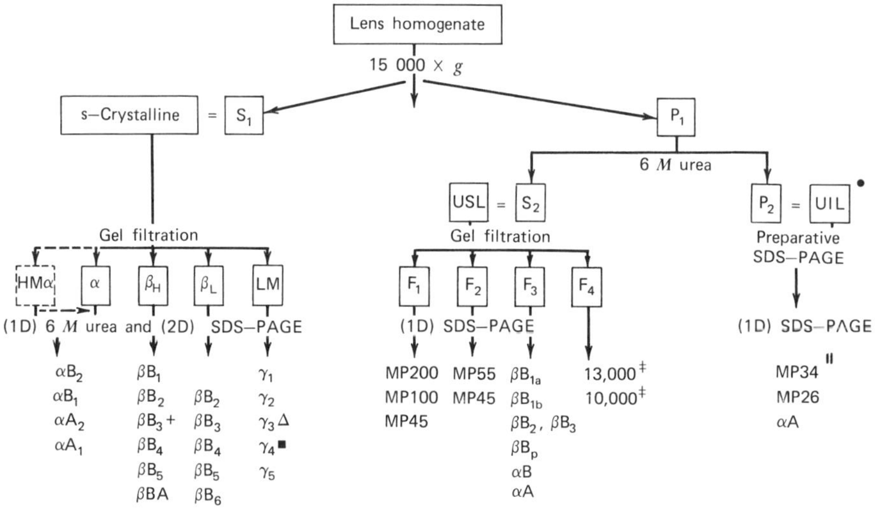

Figure 28 Topography of major calf lens proteins and their subunits. Key:

S_1,S_2	Supernatant fractions	1D	First dimension
s-crystallins	Soluble crystallins	2D	Second dimension
P_1,P_2	Pellets	βB_1	Splits into βB_{1a} and βB_{1b}
USL	Urea-soluble proteins	βB_3+	βB_p (according to nomenclature proposed by Bloemendal and Herbrink, ref. 34)
UIL	Urea-insoluble proteins		
•	Yield depends strongly on centrifugation rate	$\gamma_3\Delta$	B_S (can also be separated from LM by SE Sephades chromatography)
PAGE	Polyacrylamide gel electrophoresis		
		γ_4■	Björk's Y_{II} fraction
F	Fraction	MP	Membrane and cytoskeleton proteins
=	Unidentified	‖	Doubles (compare ref.11)

1.4 CONCLUDING REMARKS

Until 1972 the major efforts of students of lens proteins were devoted to the soluble crystallins. Although by no means have all problems of the soluble lens proteins been solved (e.g., the complexity of β-crystallin), one can conclude from the foregoing section on noncrystallin proteins that there is now an obvious shift in interest toward these proteins, in particular the plasma membrane and cytoskeleton proteins.

We have seen the great similarities that exist between crystallins of various mammalian species. It seems that similarities among species characterize the insoluble lens proteins as well. For instance, we observed that the major intrinsic lens fiber plasma membrane protein MP26 undergoes degradation to an MP22 polypeptide in calf lens fiber cells (144). The analogous situation exists in human lens (162, 163, 165, 209).

A question that needs elucidation is whether the "soluble" crystallins found in membrane preparations are contaminations, as suggested frequently, or represent, at least in some cases, components of a defined structure. In this context recent studies in my laboratory may shed some light on this problem. It appears that newly synthesized α-crystallin A chains can be incorporated into both heterologous (210) and homologous (211) membrane structures. These observations support the idea that soluble crystallin may serve as (extrinsic) constituent of membranes, thus becoming an insoluble crystallin.

In conclusion, a scheme is given in Figure 28 showing the topography of the major calf lens proteins at the present stage of investigation. This scheme may be helpful for comparative studies with species other than calf. No doubt the scheme will require extensions and corrections in the near future. However, such changes can only be welcome, since they will reflect progress in lens protein research.

REFERENCES

1 C. T. Mörner, *Z. Physiol. Chem.*, **18**, 61 (1894).

2 A. Spector, *Invest. Ophthalmol.*, **3**, 182 (1964).

3 A. Spector, *Invest. Ophthalmol.*, **4**, 579 (1965).

4 S. G. Waley, in H. Davson, Ed., *The Eye*, Vol. 1, Academic Press, London, 1969, p. 299.

5 J. F. R. Kuck, in C. N. Graymore, Ed., *Biochemistry of the Eye*, Academic Press, New York, 1970, p. 183.

6 H. Bloemendal, *Acta Morphol. Neerl-Scand.*, **10**, 197 (1972).

7 R. M. Clayton, in H. Davson and L. T. Graham, Eds., *The Eye*, Vol. 5, Academic Press, New York, 1974, p. 399.

8 J. J. Harding and K. J. Dilley, *Exp. Eye Res.*, **20,** 1 (1976).

9 H. Bloemendal, in J. Paul, Ed., *Biochemistry of Cell Differentiation II,* Vol. 15, University Park Press, Baltimore, 1977, p. 195.

10 H. Bloemendal and W. W. de Jong, *Trends in Biochem. Sci.*, **4,** 137 (1979).

11 H. Bloemendal, *Crit. Rev. Chem.*, in press (1981).

12 M. Testa, G. Armand, and E. Balazs, *Exp. Eye Res.*, **4,** 327 (1965).

13 A. F. van Dam and G. ten Cate, *Biochim. Biophys. Acta,* **121,** 183 (1966).

14 D. E. S. Truman, A. G. Brown, and K. V. Rao, *Exp. Eye Res.*, **12,** 304 (1971).

15 H. J. Hoenders and G. J. van Kamp, *Acta. Morphol. Neerl-Scand.*, **10,** 215 (1972).

16 H. Watanabe and I. Kawakami, *Exp. Eye Res.*, **17,** 205 (1973).

17 H. Bloemendal, A. Zweers, E. L. Benedetti, and H. Walters, *Exp. Eye Res.*, **20,** 463 (1975).

18 D. Roy and A. Spector, *Biochemistry,* **15,** 1188 (1976).

19 P. Ocken, S. C. Fu, R. Hart, J. White, B. Wagner, and K. Lewis, *Exp. Eye Res.*, **24,** 355 (1977).

20 R. Siezen and H. Hoenders, *FEBS Lett.*, **80**(1), 75 (1977).

21 F. Giblin, B. Chakrapani, and V. N. Reddy, *Exp. Eye Res.*, **26,** 507 (1978).

22 I. Thomson, C. Wilkinson, A. Burns, D. Truman, and R. Clayton, *Exp. Eye Res.*, **26,** 351 (1978).

23 F. Asselbergs, M. Koopmans, W. van Venrooij, and H. Bloemendal, *Exp. Eye Res.*, **28,** 223 (1979).

24 H. Bloemendal and A. Zweers, in O. Hockwin, Ed., *Progress of Lens Biochemistry Research, Docum. Ophthalmol. Proc. Series 8,* W. Junk, The Hague, 1976, p. 91.

25 A. F. van Dam, *Exp. Eye Res.*, **5,** 255 (1966).

26 M. Rabaey, I. Rikkers, and M. de Mets, *Exp. Eye Res.*, **14,** 208 (1972).

27 H. Bloemendal, *Exp. Eye Res.*, **8,** 227 (1969).

28 S. G. Waley, *Exp. Eye Res.*, **8,** 477 (1969).

29 H. Bloemendal, A. Berns, F. van der Ouderaa, and W. de Jong, *Exp. Eye Res.*, **14,** 80 (1972).

30 J. Delcour and H. Bouchet, *Exp. Eye Res.*, **26,** 191 (1978).

31 W. W. de Jong, F. van Kleef, and H. Bloemendal, *Eur. J. Biochem.*, **48,** 271 (1974).

32 F. S. M. van Kleef, M. J. C. M. Nijzink-Maas, and H. J. Hoenders, *Eur. J. Biochem.*, **48,** 563 (1974).

33 J. A. Kramps, W. W. de Jong, J. Wollensak, and H. J. Hoenders, *Biochim. Biophys. Acta,* **533,** 487 (1978).

34 H. Bloemendal and P. Herbrink, *Ophthal. Res.*, **6,** 81 (1974).

35 P. Herbrink and H. Bloemendal, *Biochim. Biophys. Acta,* **336,** 370 (1974).

36 H. P. C. Driessen, P. Herbrink, H. Bloemendal, and W. W. de Jong, *Exp. Eye Res.*, 31, 243 (1980).

37 P. Herbrink, H. van Westreenen, and H. Bloemendal, *Exp. Eye Res.,* **20,** 541 (1975).

38 F. Vermorken, P. Herbrink, and H. Bloemendal, *Eur. J. Biochem.,* **78,** 617 (1977).

39 M. Kibbelaar and H. Bloemendal, *Exp. Eye Res.,* **21,** 25 (1975).

40 I. Björk, *Exp. Eye Res.,* **3,** 254 (1964).

41 F. J. van der Ouderaa, W. W. de Jong, and H. Bloemendal, *Mol. Biol. Rep.,* **1,** 365 (1974).

42 F. J. van der Ouderaa, W. W. de Jong, and H. Bloemendal, *Eur. J. Biochem.,* **39,** 207 (1973).

43 F. J. van der Ouderaa, W. W. de Jong, A. Hilderink, and H. Bloemendal, *Eur. J. Biochem.,* **49,** 157 (1974).

44 W. W. de Jong, A. Zweers, and L. H. Cohen, *Biochem. Biophys. Res. Commun.,* **82,** 532 (1978).

45 A. Spector, *Arch. Ophthalmol.,* **81,** 127 (1969).

46 H. Bloemendal, *Science,* **197,** 127 (1977).

47 H. Bloemendal, W. S. Bont, J. F. Jongkind, and J. H. Wisse, *Exp. Eye Res.,* **1,** 300 (1962).

48 A. Spector, L. K. Li, R. C. Augusteyn, A. Schneider, and T. Freund, *Biochem. J.,* **124,** 337 (1971).

49 H. Kramps, A. Stols, H. Hoenders, and K. de Groot, *Eur. J. Biochem.,* **50,** 503 (1975).

50 K. N. Liem-The, A. L. H. Stols, and H. J. Hoenders, *Exp. Eye Res.,* **20,** 307 (1975).

51 D. Roy and A. Spector, *Exp. Eye Res.,* **22,** 273 (1976).

52 H. Bloemendal, W. Bont, E. Benedetti, and J. Wisse, *Exp. Eye Res.,* **4,** 319 (1965).

53 R. J. Siezen, J. G. Bindels, and H. J. Hoenders, *Eur. J. Biochem.,* **91,** 387 (1978).

54 I. Björk, *Exp. Eye Res.,* **3,** 239 (1964).

55 A. Spector and E. Katz, *J. Biol. Chem.,* **240,** 1979 (1965).

56 W. S. Bont, J. F. Jongkind, J. H. Wisse, and H. Bloemendal, *Biochim. Biophys. Acta,* **59,** 512 (1962).

57 L. K. Li and A. Spector, *Exp. Eye Res.,* **13,** 110 (1972).

58 S. Lerman, P. Mehta, and H. A. Jones, *Ophthal. Res.,* **3,** 30 (1972).

59 J. H. Wisse, A. Zweers, J. F. Jongkind, W. S. Bont, and H. Bloemendal, *Biochem. J.,* **99,** 179 (1966).

60 G. J. van Kamp, F. S. M. van Kleef, and H. J. Hoenders, *Biochim. Biophys. Acta,* **342,** 89 (1974).

61 A. Spector, T. Freund, L. K. Li, and R. C. Augusteyn, *Invest. Ophthalmol.,* **10,** 677 (1971).

62 J. A. Jedziniak, J. H. Kinoshita, E. M. Yates, L. O. Hoeker, and G. B. Benedek, *Invest. Ophthalmol.,* **11,** 905 (1972).

63 L. K. Li, *Exp. Eye Res.*, **18,** 383 (1974).

64 H. Bloemendal, A. Zweers, and H. Walters, *Nature*, **255,** 426 (1975).

65 W. Manski, K. Malinowski, and G. Bonitsis, *Exp. Eye Res.*, **29,** 635 (1980).

66 K. Malinowski and W. Manski, *Exp. Eye Res.*, **30,** 527 (1980).

67 K. Malinowski and W. Manski, *Exp. Eye Res.*, **30,** 537 (1980).

68 F. A. M. Asselbergs, M. Koopmans, W. J. van Venrooij, and H. Bloemendal, *Eur. J. Biochem.*, **91,** 65 (1978).

69 J. G. Bindels, R. J. Siezen, and J. J. Hoenders, *Ophthal. Res.*, **11,** 441 (1979).

70 L. K. Li and A. Spector, *Exp. Eye Res.*, **19,** 49 (1974).

71 N. T. Yu and E. J. East, *J. Biol. Chem.*, **250,** 2196 (1975).

72 L. H. Cohen, R. A. Schachar, and S. A. Solin, *Invest. Ophthalmol.*, **14,** 380 (1975).

73 J. Horowitz, *Exp. Eye Res.*, **23,** 471 (1976).

74 J. G. G. Schoenmakers and H. Bloemendal, *Biochem. Biophys. Acta*, **31,** 257 (1968).

75 J. G. G. Schoenmakers, R. Matze, M. van Poppel, and H. Bloemendal, *Protein Res.*, **1**(1), 19 (1969).

76 J. G. G. Schoenmakers, J. J. T. Gerding, and H. Bloemendal, *Eur. J. Biochem.*, **11,** 472 (1969).

77 J. G. G. Schoenmakers and H. Bloemendal, *Nature*, **220,** 790 (1968).

78 G. J. van Kamp, H. J. Hoenders, and H. Bloemendal, *Biochim. Biophys. Acta*, **243,** 149 (1971).

79 W. G. Palmer and J. Papaconstantinou, *Proc. Natl. Acad. Sci. U.S.*, **64,** 404 (1969).

80 L. H. Cohen, L. W. Westerhuis, W. W. de Jong, and H. Bloemendal, *Eur. J. Biochem.*, **89,** 259 (1978).

81 A. J. Jeffreys and R. A. Flavell, *Cell*, **12,** 1097 (1977).

82 R. Breathnach, J. L. Mandel, and P. Chambon, *Nature*, **270,** 314 (1977).

83 J. B. Chegg, D. J. Weatherhall, and P. F. Milner, *Nature*, **234,** 337 (1971).

84 H. Bloemendal, L. H. Cohen, and W. W. de Jong, *Interdiscipl. Topics Gerontol.*, **12,** 261 (1978).

85 L. K. Li, *Exp. Eye Res.*, **27,** 553 (1978).

86 I. Björk, *Exp. Eye Res.*, **3,** 248 (1964).

87 G. Armand, E. A. Balazs, and M. Testa, *Exp. Eye Res.*, **10,** 143 (1970).

88 K. N. Liem-The and H. J. Hoenders, *Exp. Eye Res.*, **18,** 143 (1974).

89 A. L. Shapiro, *Invest. Ophthalmol.*, **7,** 544 (1968).

90 M. K. Mostafapour and V. N. Reddy, *Invest. Ophthalmol.*, **7,** 660 (1978).

91 J. S. Zigler, Jr. and J. B. Sidbury, Jr., *Invest. Ophthalmol.*, **8,** 673 (1976).

92 F. C. S. Ramaekers, P. L. E. van Kan, and H. Bloemendal, *Opthal. Res.*, **11,** 143 (1979).

93 J. S. Zigler, Jr. and J. B. Sidbury, Jr., *Comp. Biochem. Physiol.*, **53B,** 349 (1976).

94 J. S. Zigler, Jr., *Exp. Eye Res.*, **26,** 537 (1978).

95 L. R. Croft, *Ciba Foundation Symposium,* Vol. 19, Elsevier, Amsterdam, 1973, p. 207.

96 J. Bours, *Exp. Eye Res.*, **16,** 501 (1973).

97 I. Björk, *Exp. Eye Res.*, **1,** 145 (1961).

98 I. Björk, *Exp. Eye Res.*, **9,** 152 (1970).

99 C. H. Carlisle, P. F. Lindley, D. S. Moss, and C. Slingsby, *J. Mol. Biol.*, **110,** 417 (1977).

100 Y. N. Chirgadze, S. V. Nikonov, M. B. Garber, and L. S. Reshetnikova, *J. Mol. Biol.*, **110,** 619 (1977).

101 T. L. Blundell, P. F. Lindley, D. S. Moss, C. Slingsby, I. J. Tickle, and W. G. Turnell, *Acta Cryst.*, **B34,** 3653 (1978).

102 L. R. Croft, *Chem. Commun.*, 437 (1972).

103 L. R. Croft, *Biochem. J.*, **128,** 961 (1972).

104 C. Slingsby and L. R. Croft, *Exp. Eye Res.*, **26,** 291 (1978).

105 S. Zigman, J. Schultz, and T. Yulo, *Biochim. Biophys. Acta,* **221,** 576 (1970).

106 S. Zigman, J. Schultz, and T. Yulo, *Ophthal. Res.*, **2,** 110 (1971).

107 B. J. Wagner and S. C. J. Fu, *Exp. Eye Res.*, **26,** 255 (1978).

108 K. J. Dilley and J. J. Harding, *Biochim. Biophys. Acta,* **386,** 391 (1975).

109 J. Horwitz, I. Kabasawa, and J. H. Konshita, *Exp. Eye Res.*, **25,** 199 (1977).

110 P. Russell, P. Larper, and J. H. Kinoshita, *Exp. Eye Res.*, **27,** 673 (1978).

111 J. I. Thorell and B. G. Johansson, *Biochim. Biophys. Acta,* **251,** 363 (1971)

112 H. O. Sandberg and O. Closs, *Exp. Eye Res.*, **27,** 701 (1978).

113 R. F. Borkman and S. Lerman, *Exp. Eye Res.*, **26,** 705 (1978).

114 E. A. Burstein, N. S. Vedenkina, and M. N. Ivkova, *Photochem. Photobiol.*, **18,** 263 (1973).

115 I. Kabasawa, Y. Tsunematsu, G. W. Barber, and J. H. Kinoshita, *Exp. Eye Res.*, **24,** 437 (1977).

116 D. S. Mc Devitt and L. R. Croft, *Exp. Eye Res.*, **25,** 473 (1977).

117 W. Manski and S. P. Halbert, in H. Peeters, Ed., *Protides of the Biological Fluids, Proceedings of the Twelfth Colloquium, Bruges,* Elsevier, Amsterdam, 1964, p. 117.

118 M. Rabaey, *Invest. Ophthalmol.*, **4,** 560 (1965).

119 M. Rabaey, *Am. J. Ophthalmol.*, **59,** 697 (1965).

120 J. Zwaan and A. Ikeda, in J. W. Rohen, Ed., *Eye Structure, II, Symp.*, Schatlauer-Verlag, Stuttgart, 1965, p. 419.

121 J. Piatigorsky, H. Webster, and S. P. Craig, *Develop. Biol.*, **27,** 176 (1972).

122 R. M. Clayton and D. E. S. Truman, *Nature,* **214,** 1201 (1967).

123 D. E. S. Truman, A. G. Brown, and K. V. Rao, *Exp. Eye Res.*, **12,** 304 (1971).

124 J. Piatigorsky, P. Zelenka, and R. I. Simpson, *Exp. Eye Res.*, **18,** 435 (1974).

125 J. Piatigorsky, *J. Biol. Chem.*, **251,** 4416 (1976).

126 P. Zelenka and J. Piatigorsky, *Exp. Eye Res.*, **22,** 115 (1976).
127 R. Reszelbach, T. Shinohara, and J. Piatigorsky, *Exp. Eye Res.*, **25,** 583 (1977).
128 J. F. R. Kuck, E. J. East, and N. T. Yu, *Exp. Eye Res.*, **23,** 9 (1976).
129 N. T. Yu, E. J. East, R. C. C. Chang, and J. F. R. Kuck, *Exp. Eye Res.*, **24,** 321 (1977).
130 S. P. Bhat and J. Piatigorsky, *Proc. Natl. Acad. Sci. USA*, **76,** 3299 (1979).
131 S. P. Bhat, R. E. Jones, M. A. Sullivan, and J. Piatigorsky, *Nature*, **284,** 234 (1980).
132 W. G. M. van den Broek, J. N. Leget, and H. Bloemendal, *Biochim. Biophys. Acta*, **310,** 278 (1973).
133 J. C. Campbell, R. M. Clayton, and D. E. S. Truman, *Exp. Eye Res.*, **7,** 4 (1968).
134 S. K. Brahma and W. J. van Doorenmaalen, *Exp. Eye Res.*, **8,** 168 (1969).
135 D. S. Mc Devitt and C. R. Collier, *Exp. Eye Res.*, **21,** 1 (1975).
136 S. K. Brahma and W. J. van Doorenmaalen, *Experienta*, **32,** 930 (1976).
137 S. K. Brahma, *Exp. Eye Res.*, **25,** 311 (1977).
138 H. Bloemendal, J. G. G. Schoenmakers, A. Zweers, R. Matze, and E. L. Benedetti, *Biochim. Biophys. Acta*, **123,** 217 (1966).
139 M. A. Kibbelaar and H. Bloemendal, *Exp. Eye Res.*, **29,** 679 (1979).
140 D. Roy and A. Spector, *Exp. Eye Res.*, **26,** 429 (1978).
141 D. Roy and A. Spector, *Exp. Eye Res.*, **26,** 445 (1978).
142 H. Bloemendal, A. Zweers, F. Vermorken, I. Dunia, and E. L. Benedetti, *Cell Diff.*, **1,** 91 (1972).
143 A. Lasser and E. A. Balazs, *Exp. Eye Res.*, **13,** 292 (1972).
144 H. Bloemendal, A. J. M. Vermorken, and M. Kibbelaar, *Exp. Eye Res.*, **24,** 413 (1977).
145 I. Dunia, C. Sen Ghosh, E. L. Benedetti, A. Zweers, and H. Bloemendal, *FEBS Lett.*, **45,** 139 (1974).
146 E. L. Benedetti, I. Dunia, C. J. Bentzel, A. J. M. Vermorken, M. Kibbelaar, and H. Bloemendal, *Biochim. Biophys. Acta*, **457,** 353 (1976).
147 Z. Dische, *Exp. Eye Res.*, **11,** 338 (1971).
148 R. M. Broekhuyse and E. D. Kuhlmann, *Exp. Eye Res.*, **19,** 297 (1974).
149 R. M. Broekhuyse, E. D. Kuhlmann, and A. L. H. Stols, *Exp. Eye Res.*, **23,** 365 (1976).
150 R. M. Broekhuyse and E. D. Kuhlmann, *Exp. Eye Res.*, **26,** 305 (1978).
151 J. Alcalá, N. Lieska, and H. Maisel, *Exp. Eye Res.*, **21,** 581 (1975).
152 H. Maisel, M. Perry, J. Alcalá, and P. R. Waggoner, *Ophthalmol. Res.*, **8,** 55 (1976).
153 M. Kibbelaar, A. M. E. Selten-Verteegen, I. Dunia, E. L. Benedetti, and H. Bloemendal, *Eur. J. Biochem.*, **95,** 543 (1979).
154 G. Y. Mousa and J. R. Trevithick, *Exp. Eye Res.*, **29,** 71 (1979).
155 M. A. Kibbelaar, F. C. S. Ramaekers, P. J. Ringens, A. M. E. Selten-

Versteegen, L. G. Poels, P. H. K. Yap, A. L. van Rossum, T. E. W. Feltkamp, and H. Bloemendal, *Nature,* **285,** 506 (1980).

156 W. W. Franke, E. Schmid, K. Weber, and M. Osborn, *Exp. Cell Res.,* **118,** 95 (1979).

157 M. Katar, W.-K. Lo, M. Nagpal, and H. Maisel, *Experientia,* **36,** 416 (1980).

158 J. Alcalá and H. Maisel, *Exp. Eye Res.,* **26,** 219 (1978).

159 P. G. C. Odeigah, R. M. Clayton, and D. E. S. Truman, *Exp. Eye Res.,* **28,** 311 (1979).

160 H. Maisel, J. Alcalá, and N. Lieska, *Docum. Ophthalmol. Proc. Res.,* **8,** 121 (1976).

161 J. Alcalá, H. Maisel, and N. Lieska, *Exp. Cell Res.,* **109,** 63 (1977).

162 D. Roy, A. Spector, and P. Farnsworth, *Exp. Eye Res.,* **28,** 353 (1979).

163 J. Horwitz, N. P. Robertson, M. M. Wong, J. S. Zigler, and J. H. Kinoshita, *Exp. Eye Res.,* **28,** 359 (1979).

164 J. Alcalá, J. Valentine, and H. Maisel, *Abstr. Ann. Meet. Assoc. Res. Vision Ophthalmol.,* 1978, p. 165.

165 J. Alcalá, J. Valentine, and H. Maisel, *Exp. Eye Res.,* **30,** 659 (1980).

166 F. M. Booyse, T. P. Hoveke, and M. E. Rafelson, Jr., *J. Biol. Chem.,* **248,** 4083 (1973).

166a P. Steinert and W. Idler, *Biochem. J.,* **151,** 603 (1975).

166b P. Davison and B. Winslow, *J. Neurobiol.,* **53,** 119 (1974).

166c L. Eng, J. Vanderhaegen, A. Bignami, and B. Gersil, *Brain Res.,* **28,** 351 (1971).

167 J. V. Small and A. Sobieszek, *J. Cell Sci.,* **23,** 234 (1977).

168 J. M. Starger, W. E. Brown, A. E. Goldman, and R. D. Goldman, *J. Cell Biol.,* **78,** 93 (1978).

169 A. J. M. Vermorken, J. M. H. C. Hilderink, I. Dunia, E. L. Benedetti, and H. Bloemendal, *FEBS Lett.,* **83,** 301 (1977).

170 H. Bloemendal, *Ophthal. Res.,* **11,** 243 (1979).

171 J. Roth, *Biochim. Biophys. Acta,* **21,** 34 (1956).

172 N. Kraft and K. Shortman, *Aust. J. Biol. Sci.,* **23,** 175 (1970).

173 W. G. M. van den Broek, M. A. G. Koopmans, and H. Bloemendal, *Mol. Biol. Rep.,* **1,** 295 (1974).

174 J. Blackburn, *J. Biol. Chem.,* **254,** 12484 (1979).

175 W. S. Bont, G. Rezelman, and H. Bloemendal, *Biochem. J.,* **95,** 15c (1965).

176 W. S. Bont, G. Rezelman, I. Meisner, and H. Bloemendal, *Arch. Biochem. Biophys.,* **119,** 36 (1967).

177 G. Blobel and V. R. Potter, *Proc. Natl. Acad. Sci. U.S.,* **55,** 1283 (1966).

178 A. A. M. Gribnau, J. G. G. Schoenmakers, and H. Bloemendal, *Arch. Biochem. Biophys.,* **130,** 48 (1969).

179 J. Th. M. Burghouts, A. L. H. Stols, and H. Bloemendal, *Biochem. J.,* **119,** 749 (1970).

180 G. Spohr, B. Kayibanda, and K. Scherrer, *Eur. J. Biochem.,* **31,** 194 (1972).

181 A. L. J. Gielkens, M. H. L. Salden, and H. Bloemendal, *Proc. Natl. Acad. Sci. U.S.*, **71,** 1093 (1974).
182 B. J. Ortwerth and R. J. Byrnes, *Exp. Eye Res.*, **12,** 120 (1971).
183 B. J. Ortwerth and R. J. Byrnes, *Exp. Eye Res.*, **14,** 114 (1972).
184 H. Bloemendal, A. Zweers, M. Koopmans, and W. van den Broek, *Biochem. Biophys. Res. Commun.*, **77**(1), 416 (1977).
185 M. Granger, G. I. Tesser, W. W. de Jong, and H. Bloemendal, *Proc. Natl. Acad. Sci. U.S.*, **9,** 3010 (1976).
186 P. Smeets, M. Granger, J. W. van Nispen, H. Bloemendal, and G. I. Tesser, *Int. J. Peptide Protein Res.*, **9,** 52 (1977).
187 R. van Heyningen and S. G. Waley, *Exp. Eye Res.*, **1,** 336 1962).
188 S. G. Waley and R. van Heyningen, *Biochem. J.*, **83,** 274 (1962).
189 S. G. Waley and R. van Heyningen, *Exp. Eye Res.*, **1,** 343 (1962).
190 R. van Heyningen and S. G. Waley, *Biochem. J.*, **86,** 92 (1963).
191 A. M. J. Blow, R. van Heyningen, and A. J. Barrett, *Biochem. J.*, **145,** 591 (1975).
192 P. Trayburn and R. van Heyningen, *Exp. Eye Res.*, **22,** 251 (1976).
193 D. Glässer and H. Hanson, *Naturwissenschaften*, **50,** 595 (1963).
194 D. Glässer, M. John, and H. Hanson, *Z. Physiol. Chem.*, **351,** 1337 (1970).
195 H. Hanson, D. Glässer, M. Ludewig, H. G. Mannsfeldt, M. John, and H. Nesvadba, *Z. Physiol. Chem.*, **348,** 689 (1967).
196 H. Hanson, D. Glässer, and H. Kirschke, *Z. Physiol. Chem.*, **340,** 107 (1965).
197 D. Glässer, U. Kettmann, and H. Hanson, *Z. Physiol. Chem.*, **351,** 1329 (1970).
198 H. Hanson and M. Frohne, *Methods Enzymol.*, **42,** 504 (1976).
199 F. H. Carpenter and J. M. Vahl, *J. Biol. Chem.*, **248,** 294 (1973).
200 S. W. Melby and F. H. Carpenter, *J. Biol. Chem.*, **246,** 2459 (1971).
201 F. H. Carpenter and K. T. Harrington, *J. Biol. Chem.*, **247,** 5580 (1972).
202 F. Jurnak, A. Rich, L. van Loon-Klaassen, H. Bloemendal, A. Taylor, and F. H. Carpenter, *J. Mol. Biol.*, **112,** 149 (1977).
203 L. A. H. van Loon-Klaassen, H. Th. Cuypers, and H. Bloemendal, *FEBS Lett.*, **107,** 366 (1979).
204 L. A. H. van Loon-Klaassen, H. Th. Cuypers, H. van Westreenin, W. W. de Jong, and H. Bloemendal, *Biochem. Biophys. Res. Commun.*, **95,** 334 (1980).
205 H. Th. Cuypers, L. A. H. van Loon-Klaassen, W. W. de Jong, and H. Bloemendal, manuscript submitted for publication.
206 M. Müller-Frohne, *Ophthal. Res.*, **11,** 377 (1979).
207 S. Fittkau, G. Kämmerer, and W. Damerau, *Ophthal. Res.*, **77,** 381 (1979).
208 N. A. Kiselev, V. Ya. Stel'mashchuk, V. L. Tsuprun, M. Ludewig, and H. Hanson, *J. Mol. Biol.*, **115,** 33 (1977).
209 R. M. Broekhuyse and E. D. Kuhlmann, *Exp. Eye Res.*, **30,** 305 (1980).

210 A. J. M. Vermorken, M. A. Kibbelaar, J. M. H. C. Hilderink, and H. Bloemendal, *Biochem. Biophys. Res. Commun.*, **88**(2), 597 (1979).

211 F. C. S. Ramaekers, A. M. E. Selten-Versteegen, and H. Bloemendal, *Biochim. Biophys. Acta*, **596,** 57 (1980).

212 H. Bloemendal, *Zone Electrophoresis in Blocks and Columns,* Elsevier Monographs, Chemistry Section No. 28, 1963.

213 U. K. Laemmli, *Nature,* **227,** 680 (1970).

CHAPTER
2

The Morphology of the Lens

Harry Maisel, C. V. Harding, J. R. Alcalá, J. Kuszak, and R. Bradley

Department of Anatomy and the Kresge Eye Institute, Wayne State University, Detroit, Michigan

2.1 INTRODUCTION

The lens is a transparent, highly refractive structure located between the pupillary portion of the iris and the vitreous. It is held in a fairly fixed position by the zonular fibers that extend from the lens to the ciliary body, and by the vitreous body posteriorly. The lens is enclosed by a capsule that is actually a thick basement membrane, and to which the zonular fibers attach. Blood vessels, lymphatics, and nerves are not found in the lens, which continues to increase in volume, weight, and size throughout life (1).

The development of the lens is basic to an understanding of lens structure, and is briefly reviewed here. The posterior cells of the lens vesicle elongate to form the *primary lens fibers,* which obliterate the cavity of the vesicle and abut on the anterior epithelium. This process is completed early in development, and thereafter *secondary fibers* are added throughout life by elongation of epithelial cells circumferentially at the lens equator. As a result, previously formed fibers become progressively more internalized. Because all cells are retained in the lens, the older fibers are always found toward the central region (nucleus) and the younger ones toward the periphery (cortex). Thus at any age the internalized concentric layers of lens reflect the processes of fiber cell differentiation, maturation, and aging of the lens cells.

2.2 LENS CELL DIFFERENTIATION AND MATURATION

As a result of primary fiber formation, the epithelium of the posterior part of the lens disappears and cell division in the anterior epithelium becomes restricted to a circumferential germinative zone located slightly anterior to the equator (Figure 1). Upon completion of a terminal cell division one or both of the daughter cells enter the equatorial zone, forming

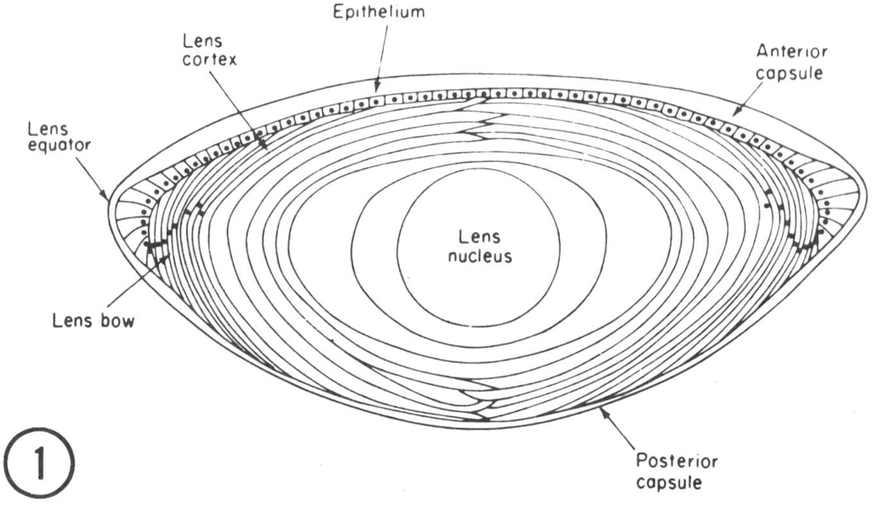

Figure 1 Diagrammatic section of dog lens (106). Reprinted by permission from Williams and Wilkins.

meridional rows, and begin to elongate (Figures 2 and 3). As the cell elongates, its basal (capsular) aspect extends posteriorly along the inner surface of the capsule, while its apical end extends toward the apical surface of the anterior epithelial cells. The equatorial cell thus rotates through 90°, relative to the epithelial cell, as it elongates and takes its place at the periphery of the contiguous fiber mass (2). This movement results in the juxtaposition of the apical surfaces of the anterior epithelial and fiber cells.

The processes of cell elongation, rotational change, and the anterior shift of nuclei in the elongating fibers result in the bow configuration of nuclei at the equatorial region of the sectioned lens (Figures 1 and 4). Thus the *bow zone* of the adult lens consists of the elongating fiber cells which contain nuclei, and which extend from the anterior epithelium to the postequatorial capsule. The nucleated bow fiber cells of the adult rhesus monkey form a layer 300 to 500 μm thick anteriorly beneath the epithelium (3).

With further elongation the basal aspect of the cell reaches the central area of the posterior capsule and its apical end reaches the central region beneath the anterior epithelium (3). As new cells are added the basal end of the fiber detaches from its loose attachment to the capsule and interdigitates with fibers from the other side, thus forming the *posterior suture* (Figure 5). Anteriorly the fibers are displaced from contact with the anterior epithelium and interdigitate with fibers from the other side to form the *anterior suture*. According to Kuwabara (3) the fibers detach from the posterior capsule concurrently with the elimination of their nuclei. Thus a

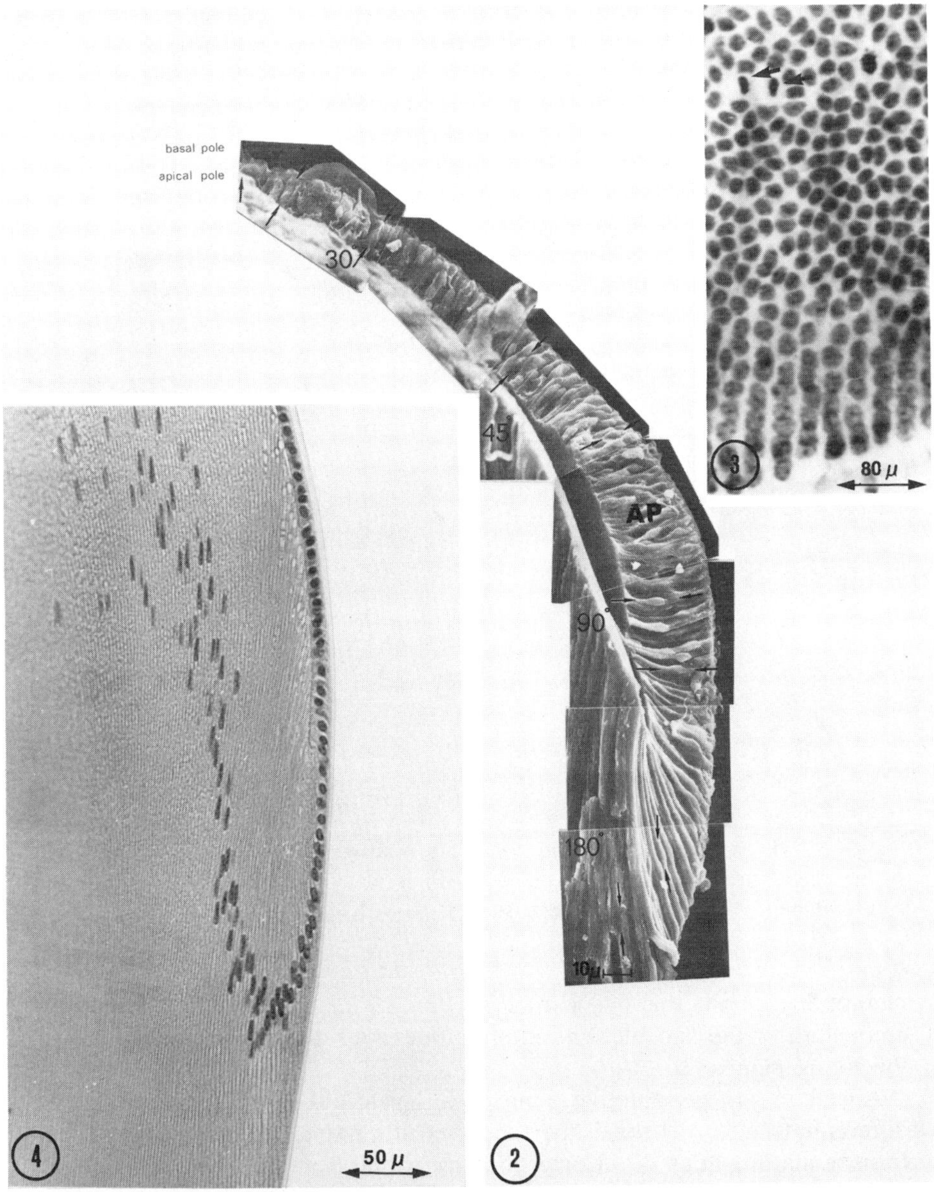

Figure 2 SEM of 7-day chick embryo showing the rotation of cells at the lens equator (×370). AP, annular pad.

Figure 3 Portion of whole mount preparation of mouse lens epithelium showing meridional rows (bottom) and mitoses (arrows) in the proliferative zone (×153.75) (107). Reprinted by permission from Wistar Institute Press.

Figure 4 Section of rat lens showing the bow zone (×250). (Courtesy of B. Worgul, Columbia University, New York.)

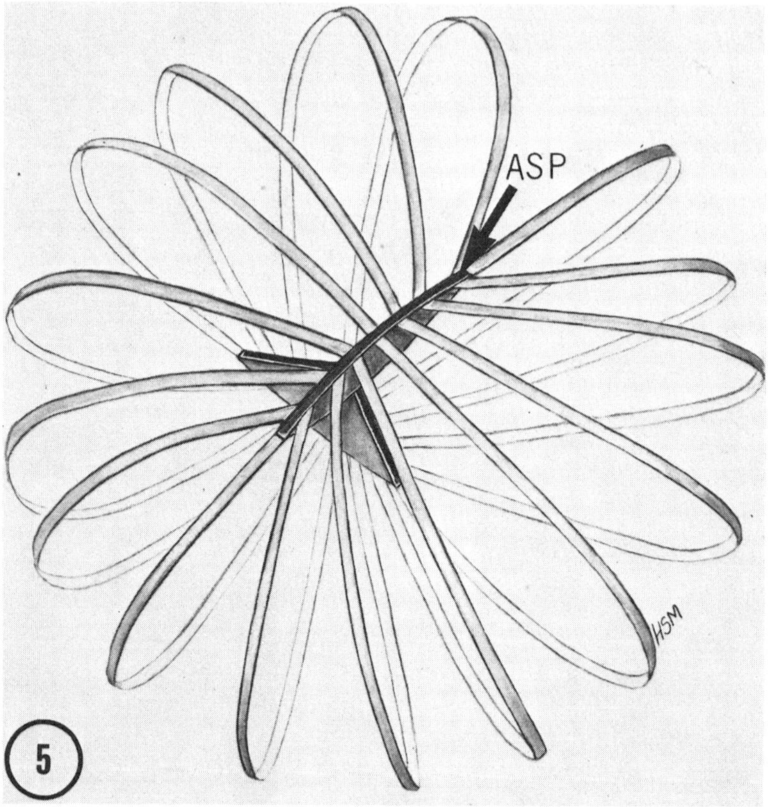

Figure 5 The arrangement of *mature* fibers relative to the sutures in the adult rabbit lens. The anterior suture plane (ASP) is at a right angle to the posterior suture plane (59). Reprinted by permission from S. Karger AG, Basel.

mature fiber is one that has lost its nucleus, no longer abuts on the anterior epithelium or the capsule, and interdigitates with the contralateral fiber at the suture planes (3).

A cell in the germinative zone of a 1-year-old rat takes 6 weeks to differentiate into a cortical fiber cell, and the latter cell takes 6 weeks to form a mature fiber (4). Corresponding values for the 1-year-old albino rabbit are 6 and 10 weeks, respectively. The rate of fibrogenesis decreases with age (5), and in the neonatal rat one fiber lamella is formed daily (6).

The terms *cortex* and *nucleus* are used to delineate zones of the fiber mass, with the nucleus representing the more central and oldest cells. The human *embryonic* nucleus represents the primary fibers (1). The *fetal nucleus* corresponds to the secondary fibers added during fetal life: these meet anteriorly to form the Y sutures. Fibers added during childhood and adolescence, at first cortical in position, are incorporated into the *adult nucleus,* the anatomical size of which increases throughout life. The fibers that arise after birth form complicated branched sutures.

Although the nucleated fibers of the bow zone can be designated as the *superficial cortex,* the extent of the *deep cortex* and the junction between the cortex and nucleus is controversial. The optical properties of the lens (7) as well as morphological and biochemical criteria have been used as markers for the boundary zone. In some species, the transition in consistency from a soft superficial zone to a harder central zone serves to delineate cortex from nucleus (8). The extent of the different lens zones, as depicted by Kuwabara (3) and Hogan et al. (1), is shown in Figure 6.

Using a slit-image photographic technique Brown (7) observed that the width of the human nucleus (antero-posterior axis) remains constant with age whereas that of the cortex increases. In an 11-year-old subject the width of the nucleus represents 80% of the lens, and in a 45-year-old subject it is 66%. However, the height (superior–inferior axis) of the nucleus decreases with age. Thus the anatomical nucleus increases with age, but the optical nucleus remains constant.

2.3 THE LENS CAPSULE AND THE ZONULAR ATTACHMENTS

The lens capsule is a noncellular optically clear basement membrane that encloses the lens and anchors the zonular insertions (Figures 7–10). It consists of collagen and glycoprotein (9, 10) secreted by the lens epithelial cells. The major part of the anterior capsule consists of numerous parallel

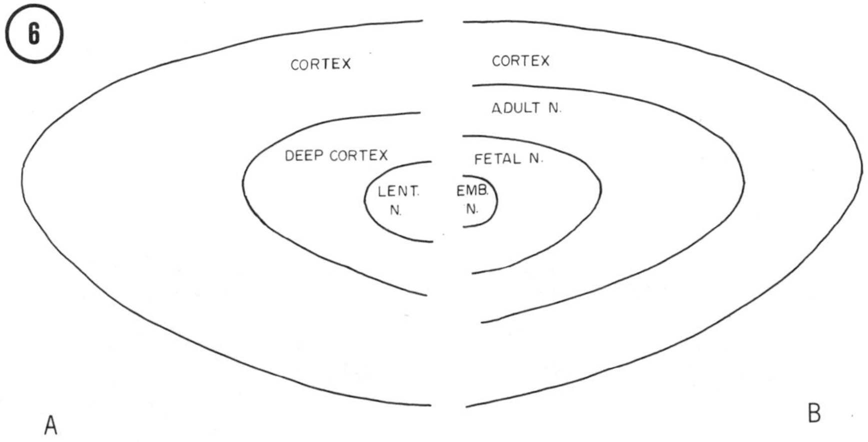

Figure 6 Lens zones (A) according to Kuwabara (3) (reprinted by permission from Academic Press Inc. (London) Ltd.) and (B) according to Hogan et al. (1) (reprinted by permission from W. B. Saunders).

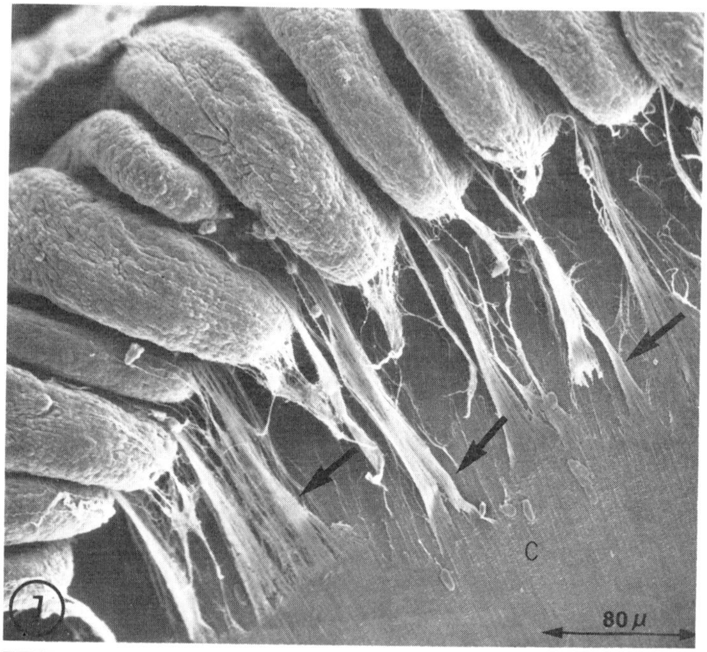

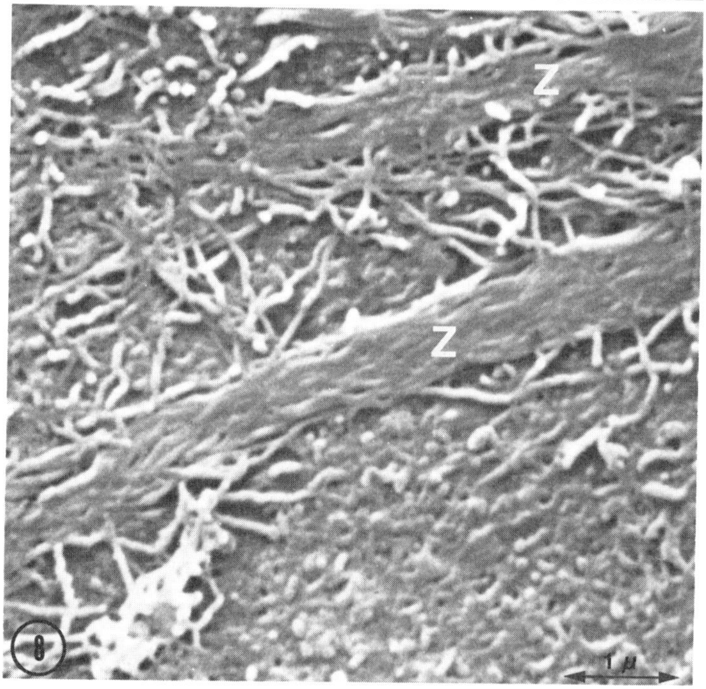

Figure 7 Zonular fibers of the 18-day chick embryo extending from ciliary processes to the lens capsule (×265).

Figure 8 Final termination of smallest anterior zonular bundles (Z) of the human lens (× 14,750) (15). Reprinted by permission from The C. V. Mosby Company.

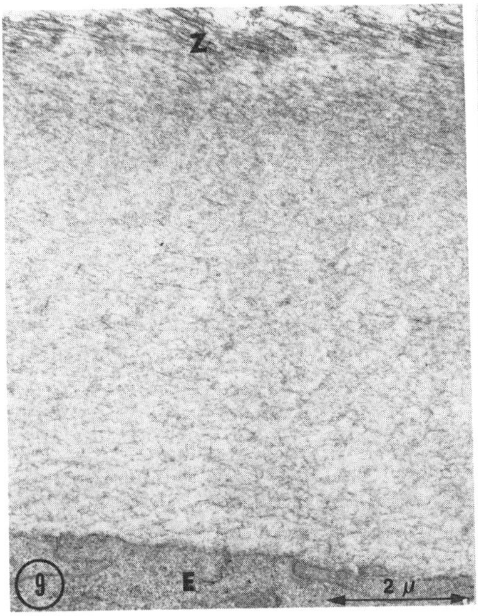

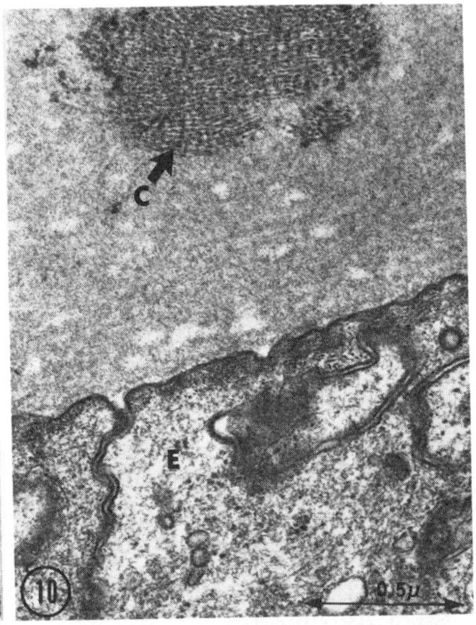

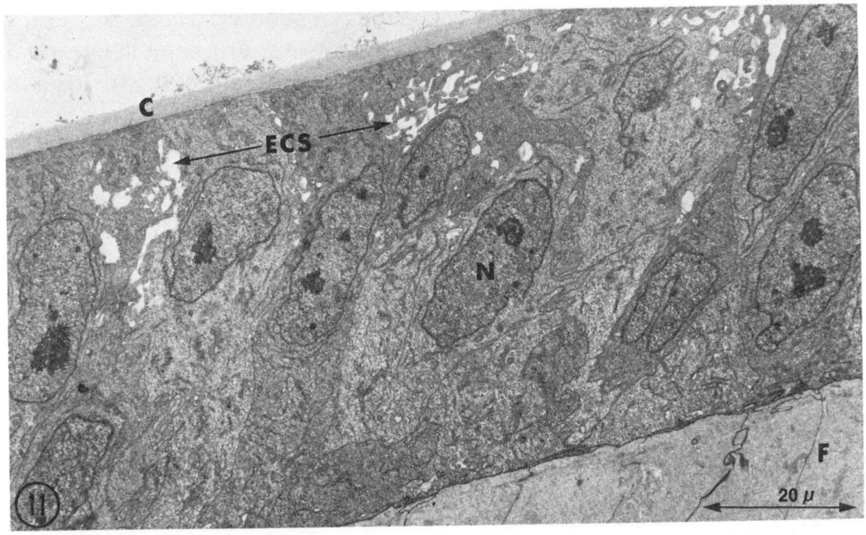

Figure 9 Anterior lens capsule of gray squirrel. Z, zonular fibers; E, epithelium (×9125). (Courtesy of N. Rafferty, Northwestern University, Evanston, Ill.)

Figure 10 Collagen inclusions (C) in 60-year-old human anterior capsule (×42,500).

Figure 11 Anterior lens epithelium of 8-day chick embryo. The extracellular space (ECS) between cell interdigitations is prominent toward the base (×2950). C, capsule; F, fiber cells; N, nucleus. (64). Reprinted by permission from Academic Press Inc. (London) Ltd.

laminae (40 nm thick) regularly repeated at about 70 nm intervals (11). Negative stain of the disrupted capsule of the adult cat shows many filaments in long parallel arrays. The filaments have a helical configuration with the angle of the helix being about 50° to the long axis of the filament. The filament strands are 0.9 nm in diameter with an elasticity modulus similar to that of collagen filaments. The elastic properties of the capsule depend on the extension of the superhelices of the filaments (12).

Superficially zonular fibers 10 nm in diameter penetrate about 0.6 μm into the capsule. With increase in age in man there is an increase in capsular thickness, loss of zonular fibers at the anterior central area, a progressive loss of lamination, and the appearance of electron-dense fibrillar inclusions (11). The inclusions first appear at age 17, and are collagenlike fibers with a periodicity of cross-striation of 48 to 52 nm (Figure 10); refs. 11, 13).

The anatomy and insertion of zonular fibers of the human and rhesus monkey lens have been described by Farnsworth et al. (14), Streeten (15), and Farnsworth and Burke (16), using scanning electron microscopy (SEM). They extend from the ciliary body and attach to the lens capsule as anterior, equatorial, and posterior zonules. The zonules initially insert as bundles 10 to 60 μ in diameter which then separate out as smaller components. The component fibrils of the zonules are 35 to 55 nm in diameter and they blend into a fibrillar ridged meshwork on the capsular surface. Beyond the insertion of the zonular fibrils the capsule has a pebbly granular appearance. A fibrillar meshwork of 10-nm fibrils composes the superficial zonular lamella of the lens capsule, and fibrils embed to a depth of 0.5 to 1 μm into the capsular matrix proper (Figure 8).

2.4 ULTRASTRUCTURE OF LENS CELLS

The fine structure of the cells of the adult lens has been studied by many authors, including Brini and Porte (17), Cohen (18, 19), Dickson and Crock, (20), Futagami (21), Gorthy (22–24), Hogan et al. (1), Harding et al. (25), Kuwabara (3, 26, 27), Leeson (28), Porte et al. (29), Rafferty and Esson (30), Rafferty and Goossens (31–33), Wanko and Gavin (34–36), and Willis et al. (37). Reference to these studies provides extensive observations on the lens morphology of different species.

2.4.1 Lens Epithelium

The epithelium, which consists of a single layer of cells beneath the anterior capsule, is divided into regional subpopulations on the basis of their proliferative kinetics (25, 38). The cells of the central zone overlying the anterior pole of the lens do not normally undergo mitosis but can divide in response to various stimuli including injury (39) and hormones (40). The germinative zone located in a preequatorial position consists of actively dividing cells (41). After a terminal division the cells enter the meridional rows at the equator of the lens and elongate as the young fiber cells of the bow region. A transitional zone between the germinative zone and the meridional rows has been described (41a).

In birds, the epithelium and the fiber mass are topographically separated by the *annular pad* (Figure 2), which consists of pseudostratified cells adjoining the epithelium and progressively elongating cells toward the fiber area (42). The peripheral epithelium and the outer annular pad region contain proliferating cells and constitute the germinative zone (43).

2.4.2 Cells of Central Zone

The cells are roughly cuboidal in shape, with basal surfaces attached to the capsule, and apical surfaces attached to the underlying fiber cells. They contain organelles typical of metabolically active cells. The nuclei are large, elliptical, and frequently lobed with one or two small nucleoli (Figure 11). The cytoplasm is granular and contains mitochondria, free ribosomes, and polysomes (Figures 12 and 15). The Golgi apparatus (compare Figure 24) is usually located in the apical region between the nucleus and underlying fibers (30). The endoplasmic reticulum consists of numerous long and short profiles of rough-surfaced cisternae most often in the vicinity of mitochondria. The mitochondria are often twisted with irregular cristae, and are concentrated apically. Centrioles are seen frequently in cells of the preequatorial region where mitosis occurs, but less often in the central cells. The cytoplasm contains a variable amount of glycogen particles (29) and small dense bodies containing lysosomal enzymes (23, 44). Acid phosphatase is also present in the Golgi apparatus and endoplasmic reticulum (24). γ-Glutamyl transpeptidase has been demonstrated histochemically (45) as well as gluthathione (46). Cytoskeletal elements (discussed below) consist of microtubules, intermediate (about 12 nm) filaments, and actin (Figures 15 and 24). See also Chapters 3 and 4.

The lateral plasma membranes are markedly infolded, with few gap junctions and desmosomes (Figures 11–15). The basal membranes of the cells are infolded, but from a flat basal line, whereas the apical membrane shows many folds and interdigitations with processes of underlying fiber cells (Figure 11, ref. 3). Intercellular spaces and pinocytotic vesicles are primarily confined to the basal region of the cell (Figure 14).

As seen by scanning electron microscopy (SEM) the lateral interdigitations of the mature human lens epithelial cells are relatively uniform in diameter (0.2 μm) but vary in length (0.8–2.5 μm) (47). Many small pores and larger craterlike formations have been seen on the basal surface of

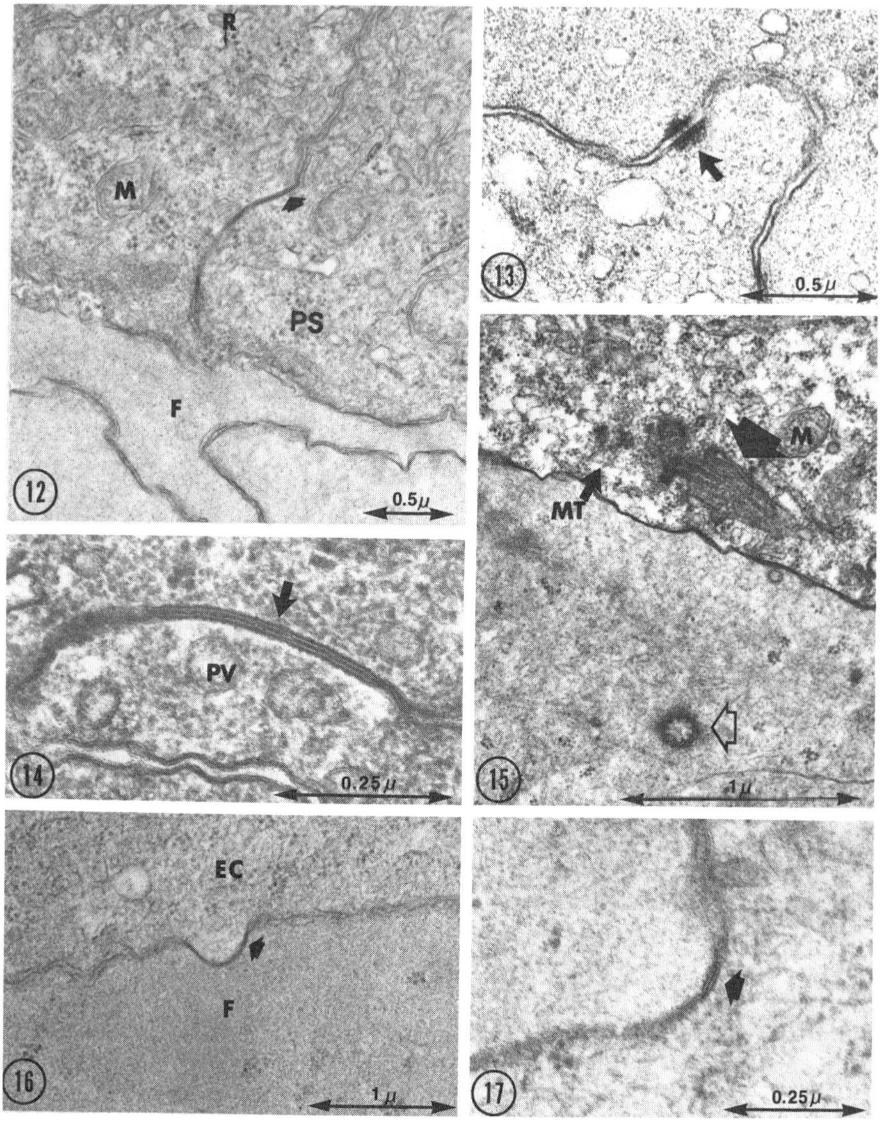

Figure 12 Gap junction (arrow) between chick lens epithelial cells. (×35,000). M, mitochondrion; R, ribosomes; F, fiber cell.

Figure 13 Desmosome (arrow) between human lens epithelial cells (×36,500).

Figure 14 Tight junction (arrow) between chick lens epithelial cells toward basal end of cells (×92,500).

Figure 15 A cilium is seen in the chick lens epithelial cell (solid arrow) and the fiber cell (open arrow) (×27,250). MT, microtubule; M, mitochondrion.

Figure 16 Gap junction (arrow) between apical surface of a chick lens epithelial cell (EC) and fiber cell (F) (×20,250).

Figure 17 Gap junction of Figure 16 (×80,000).

cells in the germinative-central zone. Gap junctions between the apical membrane of epithelial cells with the membrane of the underlying fiber cells have been reported (3, 30), although their frequency is unclear (Figures 16, 17). The presence of desmosomes (masculae adherentes) varies with the species. They are absent in the mouse lens (32) but are present in the human, monkey, rat, and frog lens (3, 22, 29, 31, 32). Only one example of a tight junction (zonulae occludentes) was found between chick lens epithelial cells toward the basal end of the cells near the capsule (Figure 14). Indeed Gorthy (22) has shown that horseradish peroxidase (mol. wt. 40,000) can pass through the capsule and intercellular space penetrating even between fiber cells. Cytochemically Na-K-ATPase is present in the lateral cell membranes, but is not found in the apical and basal membranes or in the capsule or fiber cells (48).

SEM studies indicate that the central epithelial cells are maximally adherent to the capsule along their margins (49). In the human lens (47) the epithelial cells are in general flattened and polygonal with numerous interdigitations (0.2 μm diameter) along their lateral borders. The size of the cells increases from the equatorial region (8×9 μm), to a more elongate form in the intermediate zone (12×20 μm). The cells of the central zone are larger, with sizes varying from 30×45 μm to 30×75 μm. Since the epithelial cells are firmly bound to the basal lamina of the capsule along their cell borders and at various points throughout the cell surface, the basal laminae of the capsule cannot slide freely over them (47).

2.4.3 Cells of Germinative Zone

The cells of the germinative zone show no important structural differences from the central zone cells other than those associated with mitosis.

2.5 FIBER CELLS OF THE EQUATORIAL AND BOW REGION OF THE LENS

The maturation of the epithelial cell is reflected morphologically in changes involving the shape of the cell, its plasma membrane, nucleus, cytoplasm, and other cell organelles (Figures 2, 18, and 19).

The pyramidal-shaped cells (wider base) of the meridional rows or superficial bow zone are enriched in ribosomes, endoplasmic reticulum, and multivesicular bodies. Mitochondria are present and microtubules and microfilaments are abundant (Figure 19). The latter cytoskeletal structures aligned along the long axis of the cell are important in cell elongation and differentiation (50, 51). Cells enter the meridional rows in staggered fashion, producing potential furrows between alternating cells (52). These authors suggest that the tension of the capsule, in conjunction with the arrangement

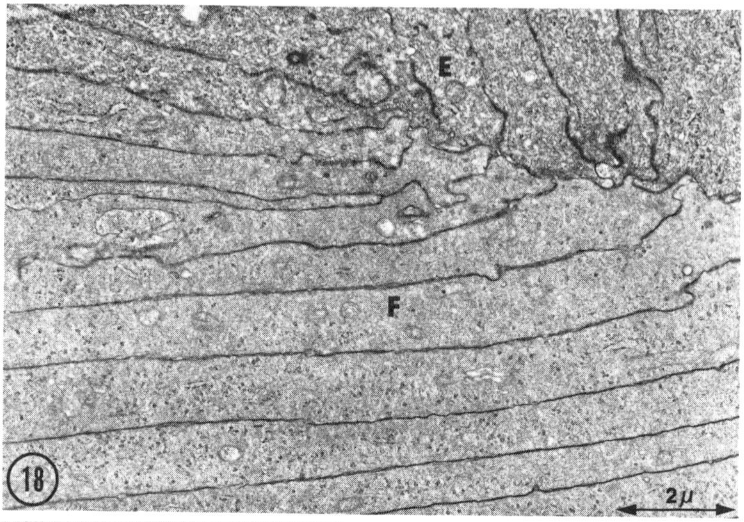

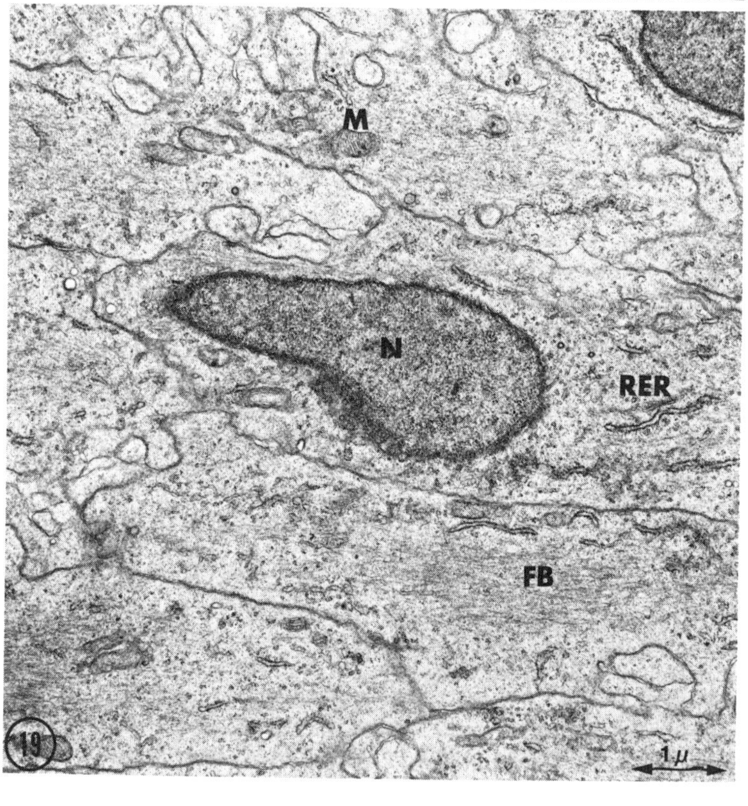

Figure 18 Fiber cells (F) of the chick lens bow zone interdigitating with the overlying epithelium (E) (×8875).

Figure 19 Fiber cells of the chick lens bow zone showing intermediate filament (FB), rough endoplasmic reticulum (RER), and mitochondrion (M) (×12,500). Reprinted by permission from Academic Press Inc. (London) Ltd.

of adjacent fibers, dictates the manner in which the elongating cells align in the meridional rows. The apical portion of the elongating cell is frequently ciliated (compare Figure 15). In the deeper bow zone there is a progressive dispersal and loss of organelles and the cytoplasm becomes more granular and electron-dense.

The nucleus of the cell undergoes a progressive series of changes culminating in its disappearance (53). The rounded nucleus of the anterior epithelial cell elongates in the superficial bow zone, and many large nucleoli are evident. The nucleoli of elongating fiber cells are much larger than those of the epithelial cells and consist of a granular component (about 15 nm in diameter and a fibrillar component (54). In the deeper bow zone the nuclear chromatin substance becomes more homogeneous and the RNA granules of the nucleoli become denser. The dense nucleolar substance often shows a spoke-like configuration. The nuclei become rounder and the nuclear substance becomes almost indistinguishable from the surrounding cytoplasm. Subsequently the nuclear membrane fragments into vesicles that disappear from the cytoplasm. At this stage the nuclear and cytoplasmic substance are indistinguishable. Kuwabara and Imaizumi (53) emphasize that the first step in the denucleation process is the transformation in which the appearance of the nucleus becomes more like that of the cytoplasm. There is no evidence of nuclear extrusion into the extracellular space, or pycnosis. Scattered remnants of dense nucleolar substance may persist for some time in mature cells.

Worgul et al. (54a) described bispheroid structures containing RNA and protein in the cytoplasm at the ends of rat cortical fiber cells. These "terminal bodies," which consist of two spherical subunits, a large granular sphere about 1.0 μm in diameter and a smaller fibrillar sphere about 0.5 μm in diameter, may represent sequestered ribonucleoprotein particles.

The cell elongation is accompanied by the elaboration of new plasma membrane, particularly in the form of gap junction and the associated 26,000 dalton polypeptide (55, 56; Figure 20). Desmosomes are still present (3), and interdigitations of the ball-and-socket type become apparent.

2.5.1 Mature Fiber Cells

2.5.1.1 Cortical Fiber Cells

The mature cortical fiber cells lack nuclei, extend from one suture plane to the other, and are not attached to the capsule. They are elongated cells, hexagonal in cross-section, that taper slightly as they extend anteriorly and posteriorly. The hexagonal fiber has two long sides and four short sides (Figure 21). The angles at the junction of the short sides are referred to here as apexes. The wide sides are parallel to the lens surface (Figure 20).

The SEM measurements of mature rat fibers are 4.9 μm from apex to

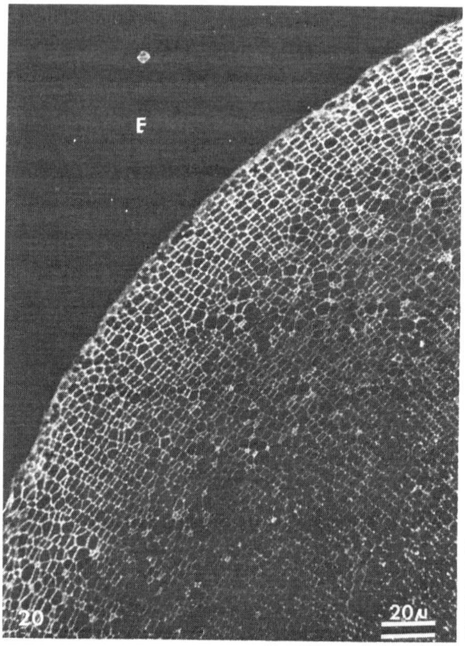

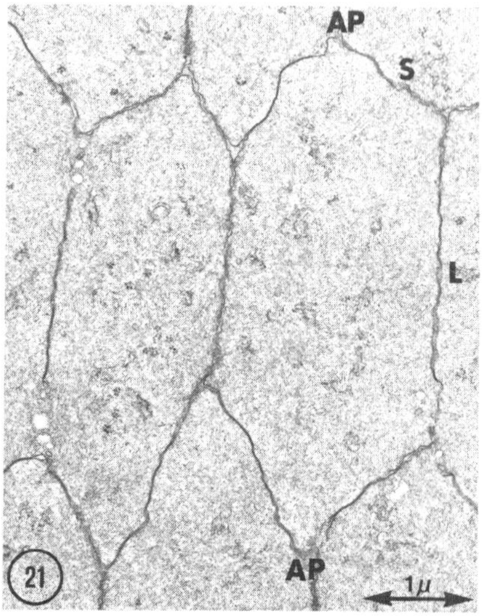

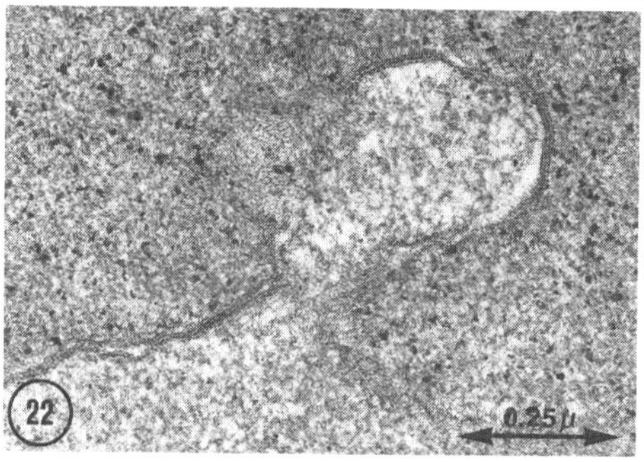

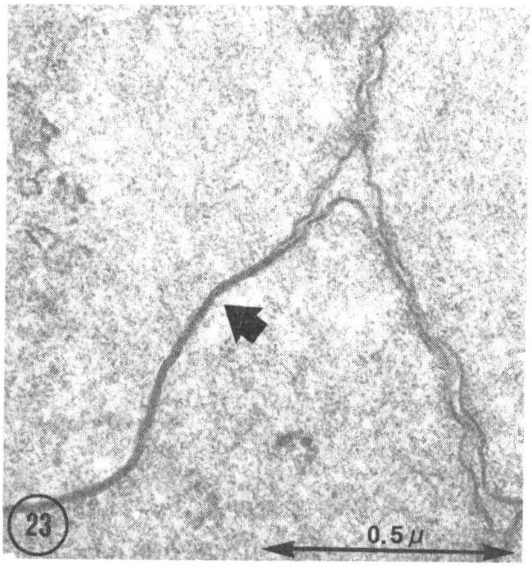

Figure 20 Immunofluorescence of a frontal section of a 20-day chick embryo lens using an antiserum to membrane protein 26K. There is no reaction in the epithelium (E) or cytoplasm of the fiber cells. The long sides of the fibers are parallel to the surface of the lens (×360) (56). Reprinted by permission from Academic Press Inc. (London) Ltd.

Figure 21 Hexagonal shape of chick lens cortical fiber cells. Cells are closely packed (×12,500). AP, apex; S, short side; L, long side.

Figure 22 Ball-and-socket junction of mature chick lens fiber cell. The membrane consists almost entirely of gap junction (×54,000).

Figure 23 The short side of one chick lens mature fiber cell consists almost entirely of gap junction (arrow) (×60,000).

apex and 1.3 μm between long sides (57). Comparable figures for the mouse lens are 2 to 4 μm and 1.0 to 1.5 μm (58). Kuwabara (3), however, found that the majority of the cortical fibers are flat and beltlike in shape, and are hexagonal only in certain areas of the superficial cortex. The fiber cells make a sharp bend before terminating at the uneven suture planes, where the ends of the cells are often lobulated and irregular. The anterior suture plane of the rabbit is vertical and the posterior is horizontal (59).

The surface of the fiber has many ball-and-socket (Figure 22) and other forms of interdigitations. The plasma membrane exists mainly in the form of gap junctions, which account for at least 65% of the plasma membrane of the wide side of a mature chick cortical fiber cell (60) (compare Figure 23). The packing pattern of the 9 nm gap junction connexons is pleomorphic even after routine ultrastructural fixation procedures. This implies that the fiber cells always remain in a metabolically coupled state (92).

The cytoplasm consists of granular substance uniformly distributed within the cell. In favorable sections cytoplasmic filaments and many irregularly beaded structures are seen. Cytoplasmic organelles such as mitochondria, nuclei, and microtubules are absent. The cells are closely packed and the intercellular space is limited to 2 to 12% of the lens water (61, 62).

2.5.1.2 Nuclear Fiber Cells

The nuclear fiber cells are shorter and irregular in shape. The membrane is intact and gap junctions are numerous. The cell is densely packed with granular cytoplasm, and organelles are absent. The surface morphology is less ordered than in the cortical fiber cells with few balls and sockets but many surface ridges and/or mounds.

"Spherical structures" consisting of multiple (2–10) unit membranes are present in the nucleus of some adult rabbits (see later Figures 48,49; ref. 63). Their globular surface ultrastructure is similar to that of normal nuclear fiber cells. Globules seen in the cataractous lens do not have multiple unit membranes at their surface (63). There is no significant difference between the cortical and nuclear fiber cells with respect to the morphology and presumably function of the membrane gap junctions.

2.6 THE CYTOSKELETON OF LENS CELLS

2.6.1 Epithelial Cells

The cytoskeleton of the epithelial cell consists of microtubles, intermediate filaments (about 12 nm diameter), and actin filaments (7 nm diameter) (Figures 24–28). Rafferty and Goossens (33) described the distribution of cytoplasmic filaments (5–11 nm in diameter) in five species. In the spherical

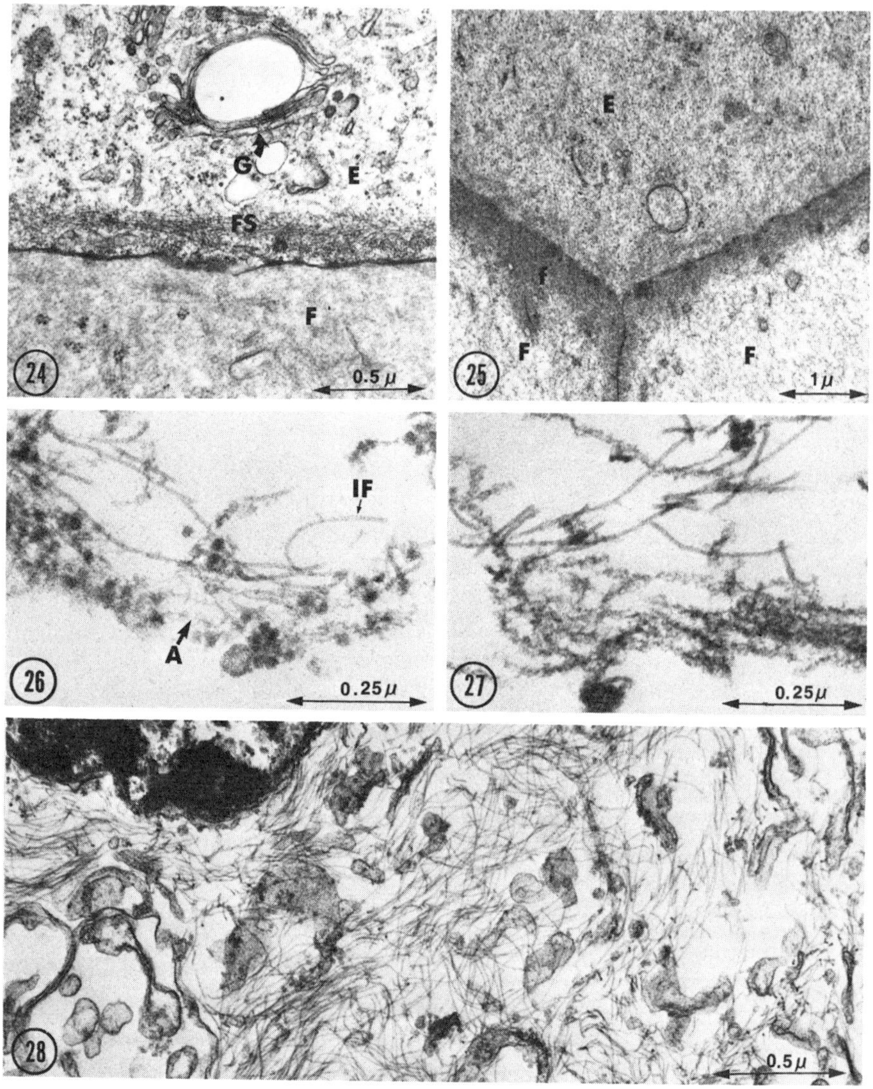

Figure 24 A filamentous sheath is present in the chick lens epithelial cell (E) parallel to the fiber cell (F) junction (×30,750) (64). G, Golgi apparatus. Reprinted by permission from Academic Press Inc. (London) Ltd.

Figure 25 Epithelial cell (E)–lens fiber (F) junction of juvenile squirrel lens showing a heavy concentration of filaments (f) at the fiber plasma membrane. A less dense array is present on the epithelial side of the junction (×11,875) (33). Reprinted by permission from Academic Press Inc. (London) Ltd.

Figure 26 Intermediate filaments (IF) and actin filaments are seen in the filamentous sheath of a glycerinated chick lens epithelial cell (×80,000) (64). Reprinted by permission from Academic Press Inc. (London) Ltd.

lenses of mice and rats, filaments are grouped in perinuclear bundles, in cell processes, and throughout the subcortical cytoplasm. In the anteriorly flattened accommodating lenses of man, squirrel, and frog, the filaments are scattered in the cytoplasm. Although the perinuclear sheath and subcortical bundles are absent, the filaments are accumulated on either side of the epithelial–fiber cell junction (Figure 25).

Studies of glycerinated lenses revealed actin and intermediate filaments in the epithelial cells of human, rabbit, bovine, and chick lenses (64–66). A sheath of intermediate and actin filaments lies parallel to the apical surface of the chick and rabbit lens epithelial cell membrane at the epithelial–fiber cell junction (Figure 24). Intermediate filaments are found throughout the cytoplasm as well as perinuclear in position.

Actin and tubulin have been identified by immunofluorescence using antibodies to actin (51, 67), and by their reaction with heavy meromyosin (Figures 26, 27) (64, 65). Cytochalasins B and D, which disrupt actin microfilaments, inhibit differentiation and elongation of cultured rat lens epithelial cells and the production of γ-crystallin (51).

A complex system of microfilaments and microtubules is present in long-term monolayer cultures of bovine epithelial cells (67). The microfilaments (including actin) form arrays in contact with the plasma membrane, bundles of fibers connected particularly to junctional complexes between adjacent cells, and thick bundles of stress fibers running longitudinally throughout the cell. Microtubules are aligned along the long axis of the cell. Colchicine, which disrupts microtubules, prevents the elongation of cultured embryonic chick epithelial cells. Insulin, however, promotes cell elongation and the alignment of microtubules (50).

2.6.2 Fiber Cells

Microtubules are present only in the superficial fiber cells (3). In the adult lens intermediate filaments are numerous in the superficial cortical fiber cells, decrease in number in the deeper fibers, and are absent from the nuclear fiber cells. However, in the newborn human lens, and in the chick lens at hatching, intermediate filaments are present in cortical and nuclear fiber cells (64). Since filaments are present in the superficial fiber cells of the adult lens, their absence from the nuclear cells must be due to degradation with age. Intermediate filaments are present in fish, frog, bird, and mammalian lenses (68) (Figures 29–31).

Figure 27 Only the actin filaments seen in Figure 26 are decorated by heavy meromysin (×80,000) (64). Reprinted by permission from Academic Press Inc. (London) Ltd.

Figure 28 Intermediate filaments are prominent in the epithelium of a 80-year-old male (×31,500).

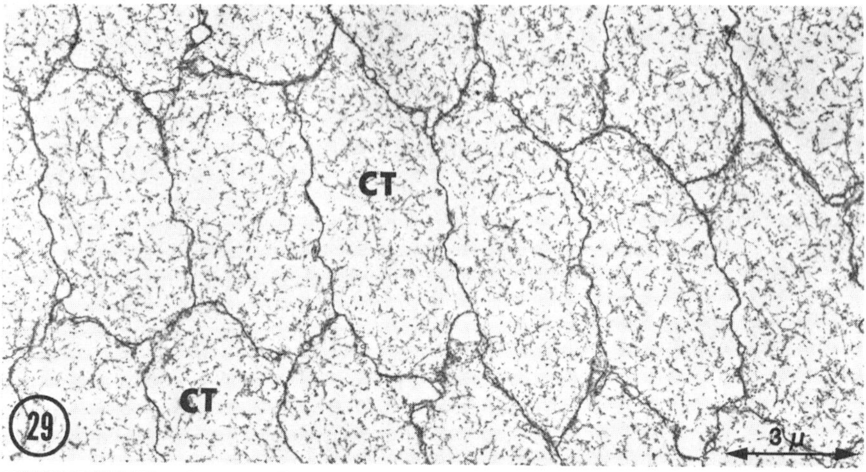

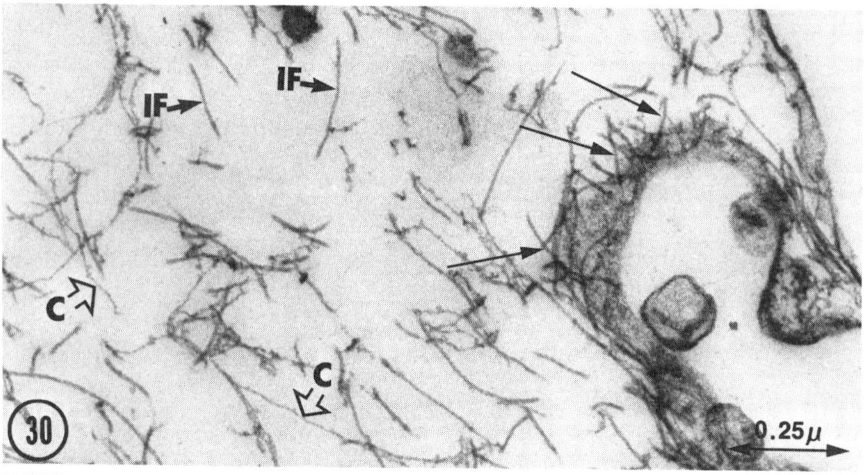

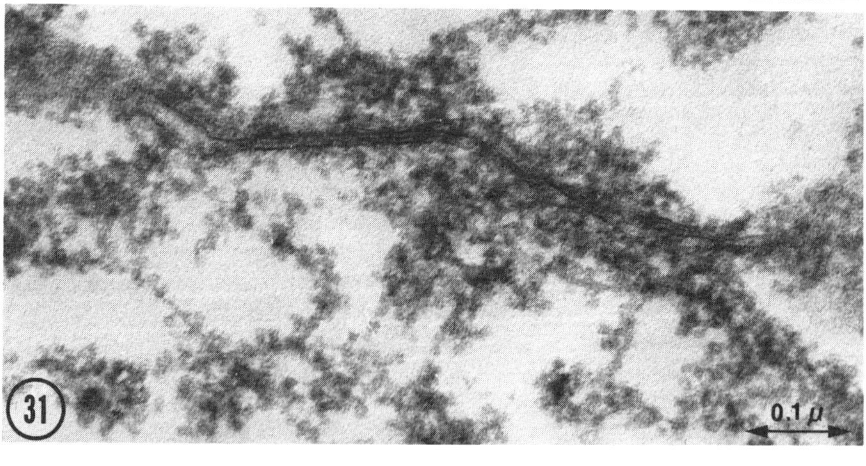

Figure 29 Glycerinated mature rabbit fiber cells, showing the cytoskeleton (CT) (×6125).

Rafferty and Goossens (33) found that filaments (5–11 nm in diameter) are sparse in the fiber cells of mice and rats, and only a loose array is found at the epithelial–fiber cell junction. However, in the squirrel and infant human lens, a heavy accumulation of filaments is present on either side of the epithelial–fiber cell junction, especially on the cytoplasmic side of fibers (compare Figure 25). These filaments form a three-dimensional lattice and some are associated at one end with the membrane. The lattice is less dense along the sides of the fiber cells not abutting on the epithelium. In deeper anterior fibers the filaments are located parallel to the membrane.

Another element of the fiber cell cytoskeleton, represented by beaded filaments, was described by Wanko and Gavin (36), Resnik et al. (69), Lasser and Balasz (70), and Maisel and Perry (71). In chick and frog lenses these structures, referred to as chains, consist of globular particles 12 to 15 nm in diameter, arranged along a filamentous backbone (7 nm in diameter) (Figures 32–34). In the mammalian lens the globular particles measured 15 to 20 nm in diameter (72), similar to that of α-crystallin (73). The chains appear as widely dispersed elongated elements in cortical fiber cells, but are aggregated in nuclear fiber cells (64, 65) (Figure 31). The chains, which are not seen in epithelial cells, are best observed in sections of glycerinated fiber cells or by negative stain of lens homogenates. Although the size of the filament backbone corresponds to that of actin, positive morphological identification has not yet been made. Nevertheless it is the likely candidate for the actin molecule that has been found biochemically (51, 110).

The lens cytoskeleton thus shows regional variations. Microtubules, actin, and intermediate filaments are prevalent in the epithelial cells but chains are lacking. In the mature cortical fiber cells chains are abundant, whereas microtubules are absent and intermediate filaments are decreased in number. Nuclear fiber cells contain only aggregated chains.

The interaction between the plasma membrane and the cytoskeleton has been demonstrated morphologically. The tightness of this association is attested to by the need for agents such as 8 *M* urea for dissociating the membrane from the cytoskeleton. Lasser and Balasz (70) described the filaments and chains as forming a continuous intracellular matrix of the fiber cell. The matrix is water-insoluble, and represents the urea-soluble fraction of the fiber as distinct from the water-soluble protein (crystallins), and the urea-insoluble (membrane) fractions (55, 70, 74–76). The intermediate filaments re-form upon removal of urea but the chains do not (77).

Figure 30 Glycerinated mature cortical fiber cells of rabbit lens showing intermediate filaments (IF) and chains (C). The long arrows point to the attachment of the filaments to the plasma membrane (×65,000) (66). Reprinted by permission from S. Karger AG, Basel.

Figure 31 Glycerinated rabbit nuclear fiber cell showing the absence of filaments and the clumping of the chains. (×138,750) (66). Reprinted by permission from S. Karger AG, Basel.

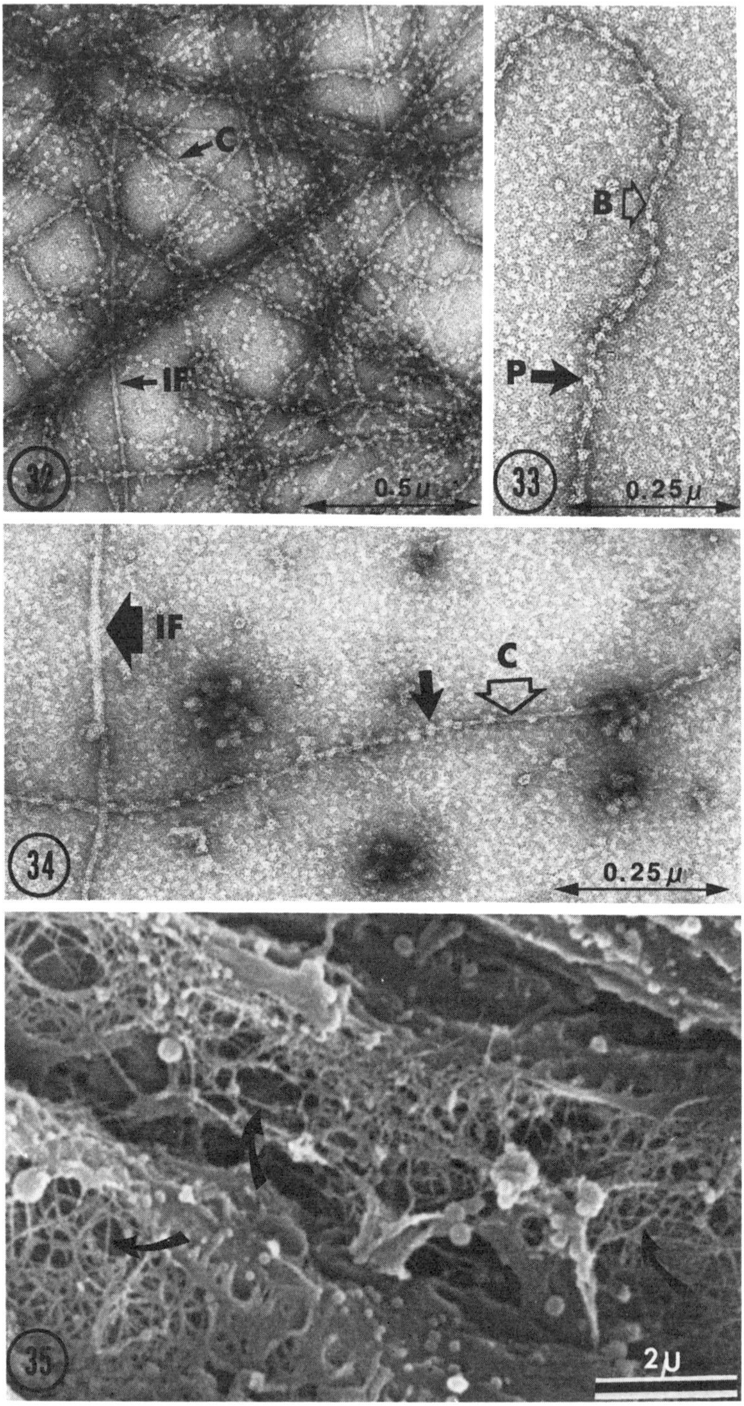
C
IF
32
0.5 μ
B
P
33
0.25 μ
IF
C
34
0.25 μ
35
2 μ

The role of the cytoplasmic filaments remains to be elucidated. Rafferty and Goossens (33) correlated the pattern of filament distribution with the gross shape of the lens and its accommodative ability. They proposed that the sheaths and thick bundles of filaments in the nonaccommodating spherical lens (e.g., mouse lens) impart structural support to the epithelium to maintain the spherical lens shape. The loose filament arrangement in accommodating lenses may be compatible with lens deformation, and the lattice arrangement along the fiber membrane may provide a contractile force or elasticity to return the flattened anterior lens surface to the accommodated state in conjunction with the elasticity of the lens capsule. The increased filament accumulation in the older human lenses may impose a resistance to change in lens shape and thus in part account for the loss of accommodation with age.

The proteinaceous chains of particles may serve to stabilize the organization of some lens crystallins in a nonrandom form that minimizes disorder in the fiber cell. Indeed their clumping or aggregation is unique to the aged nuclear fiber cell.

It appears that the interior of the fiber cell is occupied by a three-dimensional filamentous network (Figure 35), the interstices of which are occupied by the crystallins. This network, attached to the cell membrane, imparts an internal architecture to the fiber cell that minimizes light scattering, stabilizes the order and distribution of proteins, and maintains the shape of the fiber and the lens.

2.7 JUNCTIONAL APPARATUS OF LENS FIBER CELLS

Scanning electron microscopy (SEM) has greatly extended our knowledge of the complex system of interdigitations between fiber cells. Intercellular attachments were first observed by transmission electron microscopy (TEM) and described by Cohen (18, 19), Wanko and Gavin (35, 36), and Gaspar (78). Leeson (28) combined TEM and freeze-fracture to demonstrate

---

Figure 32 Negative stain of intermediate filaments and chains of chick lens fiber cells (×47,000). Courtesy of M. M. Perry, University of Edinburgh, Scotland; see also ref. 108.

Figure 33 The particles (P) attached to the chick lens chain can obscure the filament backbone (B) (×95,000) (71). Reprinted by permission from Academic Press Inc. (London) Ltd.

Figure 34 The filamentous backbone of a chick lens chain (C) is contrasted with that of an intermediate filament (IF). (×95,000) (71). Reprinted by permission from Academic Press Inc. (London) Ltd.

Figure 35 Lens fiber cytoskeleton (SEM) of cataractous nude mouse cortical fiber cells (×95,000) (109). Reprinted by permission from Academic Press Inc. (London) Ltd. A similar pattern is seen with normal chick lens fiber cells.

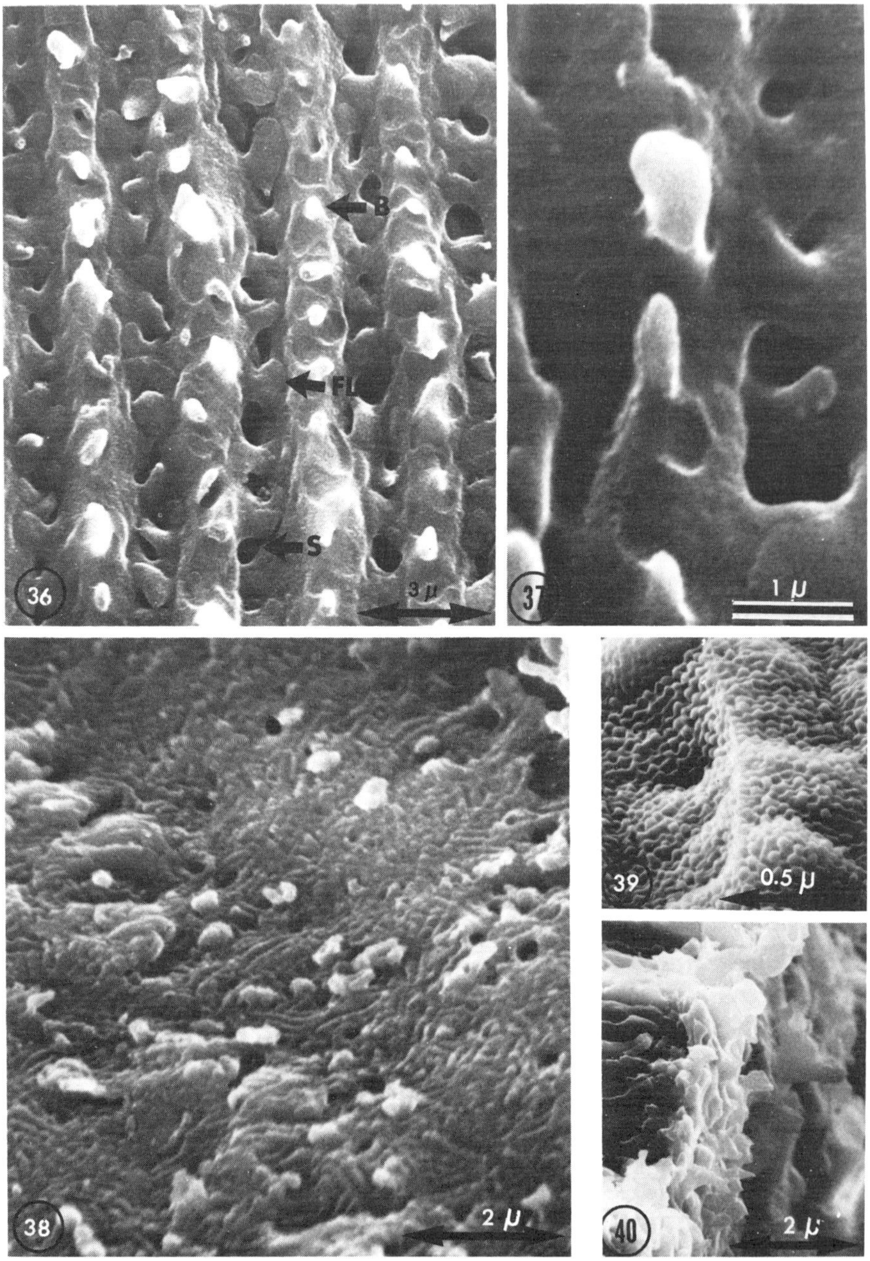

Figure 36 SEM of chick lens cortical fiber cells showing balls (B) and sockets (S). Interdigitating flaps (FL) extend from the angles at the junction of the short and long sides. (×6000).

Figure 37 A ball is shown arising from the apex of a chick lens fiber cell (×16,000).

the interdigitating processes of lens fiber cells, whereas Kuwabara (27) was the first to correlate the TEM appearance of lens fibers with the SEM pattern. SEM studies of the lens have been described for man (3, 79), monkey (3, 20), rat (57, 80, 81), mouse (58, 82), and rabbit (59).

The junctional complexes include balls (knobs) and sockets, ridges and grooves (tongues and grooves), mounds (globules), and pits. The ball consists of a short stalk which expands to form a spherical head embedded in a complementary depression in the adjacent fiber (Figures 22, 36, and 37). In the adult rhesus monkey the diameter of the ball is about 1 μm and the depth of the socket about 1.5 μm (3). The ridge consists of a linear evagination of the membrane about 0.1 μm high, 0.1 to 0.15 μm wide, and 1 to 2 μm long. The ridges interdigitate with others on adjacent cells. Mounds (globules) are small elevations about 0.2 μm in diameter in the albino mouse and rat lens (58).

Ball-and-socket joints first appear in the fibers of the bow region, and in the elongated fiber of the adult Macaque monkey cortex they are present over the whole of the cell surface (83). In the nuclear fiber cells the ball-and-socket joints are mainly found at or near the angles of the cell, whereas the long flat surfaces are covered with many ridges (tongues) 83). Some ridges are grouped into parallel clusters about 0.1 μm apart in the rhesus monkey lens (3) (compare Figure 38). Thus in the primate lens there is a decrease in ball-and-socket joints proceeding from the cortex to the nucleus, coincident with an increase in the ridges and grooves. At the suture planes the fiber cells have irregular edges, ball-and-socket joints are absent but ridges and grooves are present (3) (see Figure 40).

Ball-and-socket joints are also present at the angles of the hexagonal cortical and adult nuclear fiber cells of the mouse lens. Pronounced fingerlike projections from the apexes of the anterior cortical fiber cells are not found in the posterior fiber cells (82). The surfaces of the adult nuclear fiber cells are covered with many mounds and depressions, and the surfaces of fetal nuclear fibers are covered with prominent ridges and grooves. However, Sakuragawa et al. (58) found ridges and grooves on the surface of the mouse deep cortical fibers, and mounds and pits on the nuclear fiber cells. Ridge formation begins in the second postnatal week, and in older mice fibers covered with ridges and mounds occupy the major part of the lens (58).

Hollenberg et al. (57) reported that the superficial cortical fibers of the

---

Figure 38 SEM of an adult nuclear fiber cell of human showing prominent surface ridges. (×10,500).

Figure 39 SEM of adult rabbit embryonic nuclear fiber cell showing numerous surface globules (×3800) (59). Reprinted by permission from S. Karger AG, Basel.

Figure 40 SEM showing irregular interdigitations of rabbit fiber cells at a suture plane (×850) (59). Reprinted by permission from S. Karger AG, Basel.

rat followed a straight course from the region of one pole to the other. In contrast, the deeper cortical fibers displayed a pronounced undulating course in the equatorial and surrounding areas. A similar wave pattern was also observed by Hansson (80), Leeson (28), Farnsworth et al. (81), and Nelson and Rafferty (82). Hollenberg et al (57) observed ball-and-socket joints along one irregular line down the center of the long fiber surface. Interdigitating processes were noted at each of the angles but were not further characterized.

In the rabbit lens there is a progressive accumulation of mounds (globules) on the surface of the deep fiber cells (59) (Figure 39). The superficial fiber cells have many interdigitations (undefined) at the angles of the hexagon and balls and sockets along the long sides. The deeper fibers have more prominent interdigitations at the angles, but no balls and sockets on their long sides.

Reported differences of the description of the junctional apparatus are probably due to species variations, as well as inconsistency in identification. Thus ball-and-socket joints are present as rows at the angles of the cortical and adult nuclear fiber cells of the mouse lens (82), but in the rat they form a row down the center of the long surface of the cortical fiber cell (57). In the rabbit they are identified only on the wide sides of the superficial cortical fiber cells. On the other hand, in the primate lens they are noted over the whole surface of the cortical fiber, and concentrated at or near the six angles in the nuclear fiber cells (83). However, in all these species interdigitations of one form or another are present at all the angles. In the chicken lens the angular junctions resemble flaplike projections that interdigitate with one another (111). In this species large ball-and-socket joints are found at the apexes of the fibers (Figure 36), and numerous smaller ones on the long surfaces of the cortical fiber cells (Figure 41).

Ridges and grooves appear in increasing number on the surface of the older fibers. Furthermore, in the rat, mouse, and rabbit lens many mounds (globules) are found on the surface of the nuclear fiber cells. Such mounds were not seen in the monkey, human, or chick lens.

Figure 41 SEM of the long side of 3-month-old chick cortical fiber cell (×3625) (60). The numerous gap junction (GJ) placques are seen. Note numerous small balls (B) and sockets (S) on the surface, and the lateral flaps (FL). (60) Reprinted by permission from Academic Press Inc. (London) Ltd.

Figure 42 Freeze-fracture analysis of chick cortical fiber cell membrane showing several gap junctions (60). Reprinted by permission from Academic Press Inc. (London) Ltd.

Figure 43 The labeled gap junctions A–E of Figure 42 are matched with surface plaques of comparable areas seen by SEM (× 12,125) (60). Reprinted by permission from Academic Press Inc. (London) Ltd.

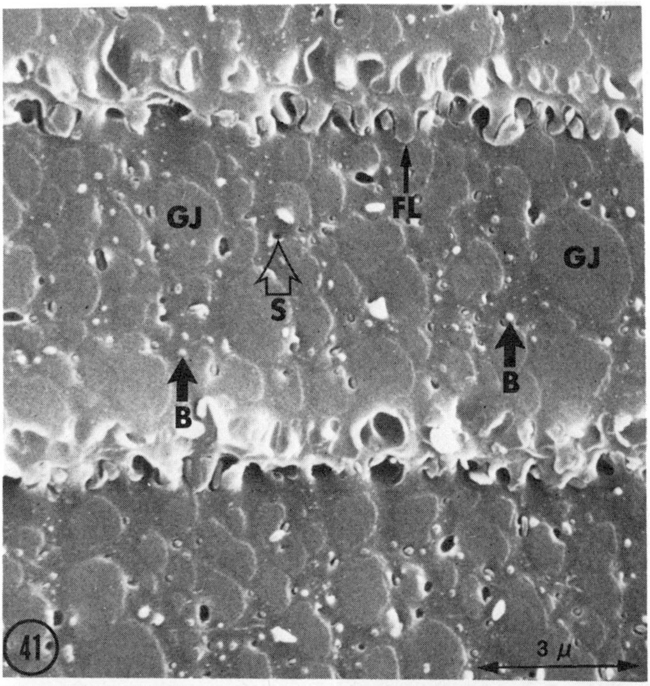
GJ
FL
GJ
S
B
B
41
3 μ

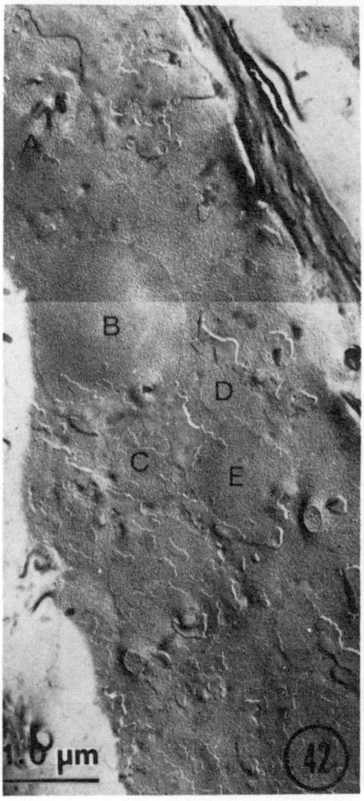
A
B
D
C
E
1.0 μm
42

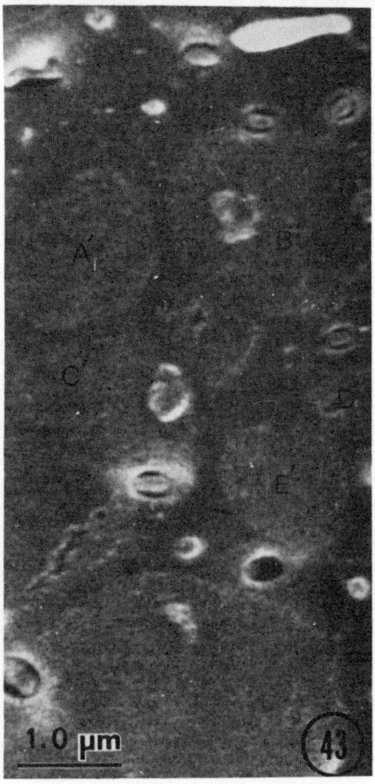
A′
B′
C′
D′
E′
1.0 μm
43

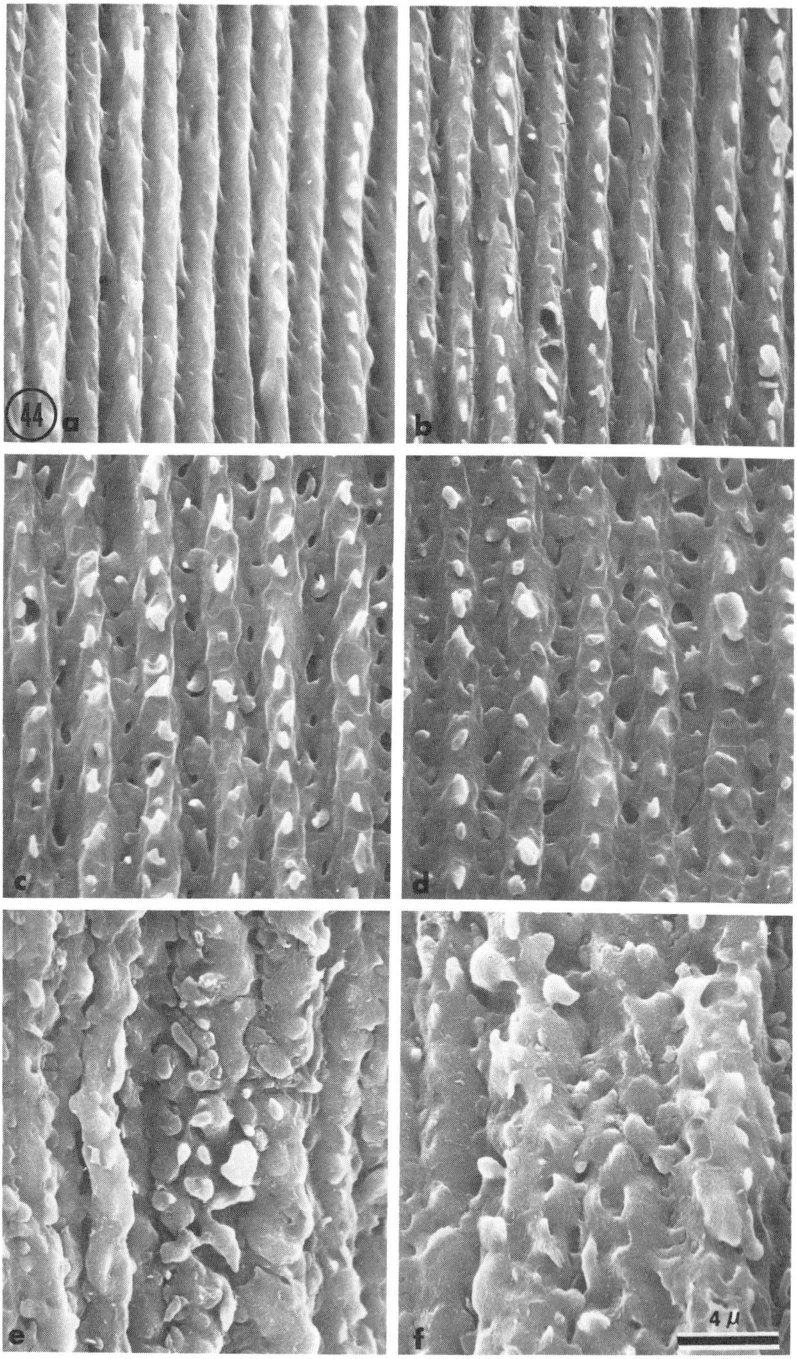
44
a
b
c
d
e
f
4 μ

The progressive elaboration of interdigitations during maturation of the chicken fiber cell is seen in Figure 44. The surfaces of the nuclear fiber cells (primary and secondary) are irregular, but show no ridges or mounds as observed in mammalian lenses. SEM studies of the development of the chick lens clearly show that aging of the mature fiber cell is associated with the progressive disorganization of the junctional apparatus. Furthermore, Kuszak et al (60) showed that it is possible to detect gap junction areas on the cell surface using SEM (Figures 42 and 43).

Irrespective of the species variations, these studies show that the surface morphology in a given location is fairly uniform, and that it changes in different layers of the lens. There is evidence that the surface morphology of a fiber cell undergoes some change as it passes from the posterior to the anterior region of the lens.

The elaborate junctional apparatus serves to maintain the shape of the lens during focusing movements and minimizes cell slippage. Kuwabara (3) suggests that cellular shape changes mainly in the superficial zone, and only a little intercellular gliding occurs during focusing. According to Hollenberg et al (57), the shape of the interdigitating processes appears well-adapted for intrinsic movements of a rotatory kind, whereas Dickson and Crock (83) indicate that the ball and socket joints bind fibers firmly together, and the ridges and grooves prevent lateral sliding movements.

These observations point out the need for further studies of the morphology and function of the fiber cell interdigitations.

2.8 MORPHOLOGICAL CHANGES ASSOCIATED WITH AGING OF THE LENS

The morphology of the aging lens is of particular interest with respect to the cataractous process. Gorthy (24) analyzed these changes in the rat lens and compared them with observations of older and senile cataractous human lenses by Kobayashi and Suzuki (84), Seland (77), Dilley et al. (85), Brini et al. (86), Phillipson (87), and Huang and Kuwabara (88).

The following cytological alterations appear to be common to the aging rat and human lens. In the epithelium there is an overall reduction in cellular and nuclear height, frequently accompanied by karyopycnosis (24). Mitochondria are swollen, and dense epithelial cells with elaborate endoplasmic reticulum are frequent. Huang and Kuwabara (88) also observed a

---

Figure 44 SEM of fiber cell maturation in the 3-month-old chick lens (×3375). *(a–d)* Progressively internalized cortical fiber cells; *(e)* secondary nuclear fiber cells; *(f)* primary fiber cells. Note the elaboration of interdigitating processes with cortical fiber cell maturation, and their disorganization in the nuclear region *(e)*.

decrease in size of the bow arrangement of the cell nuclei at the equator of the human lens. The bow cells become irregular in size and shape and some show distended microorganelles although the background cytoplasm is normal.

The surface of the cortical fiber cells becomes more irregular with age, and there is an increase in fiber cell projections into the postequatorial capsule. Such projections may penetrate up to two-thirds of the capsular thickness (11). Accumulations of atypical mitochondria and an increase in the number of vesicular profiles are observed in the polar regions of cortical fiber cells in the older rat (24). Isolated focuses of fiber cell swelling and disintegration are common in the sutural area and adjoining subepithelial region in the rat lens. Such focuses show multilamellar bodies and swollen twisted fibers. Cortical globular bodies (89) and multilamellar inclusions are common in the older human lens (Figures 45 to 47), as well as intercellular clefts (84). These features probably arise from the disruption of fiber cells.

According to Huang and Kuwabara (88) epithelial cell differentiation slows down considerably before the lens fiber cells degenerate in the cataractous process. Gorthy (24) suggests that the swelling and degeneration of fibers might be caused by hydrolases, secreted by the Golgi–lysosomal system of the epithelium.

2.9 SOME BIOCHEMICAL AND FUNCTIONAL CORRELATES OF LENS MORPHOLOGY

The morphology of the epithelium reflects its metabolically active state. In addition to general cellular proteins the epithelium cell synthesizes the collagen and glycoprotein of the capsule as well as some lens crystallins. The epithelium is the major lenticular site for oxidative phosphorylation. The cation pump (Na-K-ATPase) is located on the lateral plasma membrane of the epithelial cells, and in the anterior cortex (91). The large surface membrane area due to the many interdigitations appears to contribute significantly to transport activity at the anterior lens (3). The epithelium is likely to be electrically coupled to the fibers via a pathway through the plane of the epithelium, connecting to lens fibers in the bow region (97). Gap junctions between epithelial cells and the underlying fibers may provide a pathway linking epithelial cells directly to the underlying fiber cells (61).

Cell elongation in the bow region is specifically associated with the synthesis of γ-crystallin in the bovine lens (93), the enlargement of the metabolically active nucleus and nucleoli, an increase in ribosomal population, and an increase in plasma membrane, especially in the form of gap junction and the associated 26K intrinsic membrane polypeptide. Further maturation of the fiber cell results in a loss of the nucleus, a significant drop in the ribosomal population and its sequestration, a loss of the

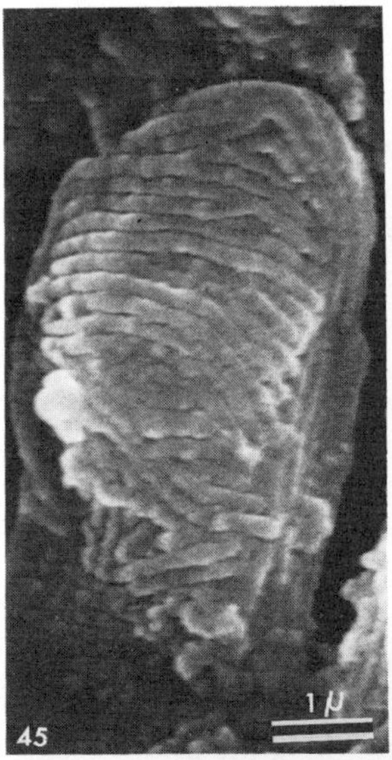

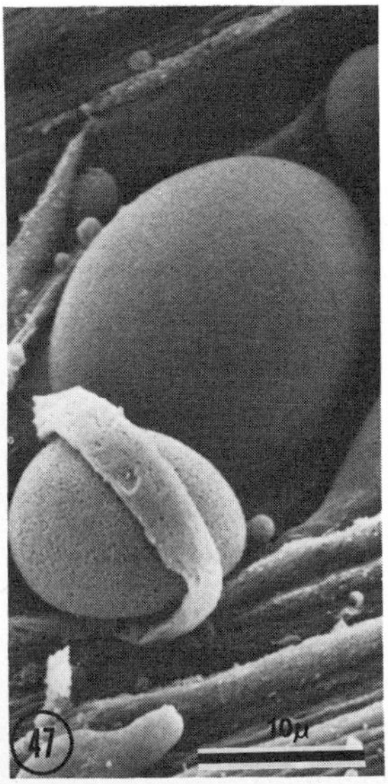

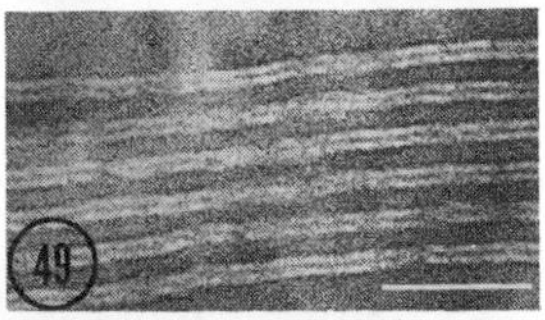

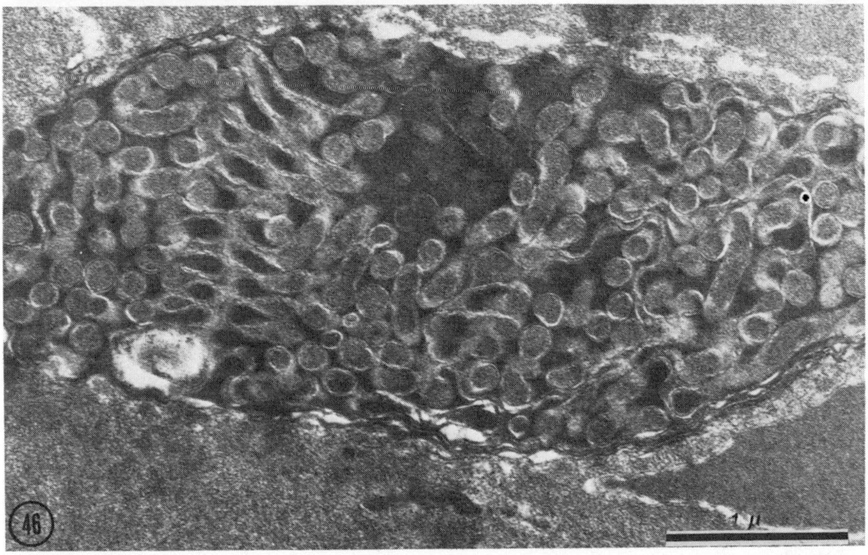

Figure 45 SEM of a multilamellar body in a senile human cataractous lens (×11,000).
Figure 46 The multilamellar body consists of multiple membrane enclosed tubes (×24,900).
Figure 47 SEM of globule from a senile cataractous lens (×1590).
Figure 48 "Sperical structure" in adult rabbit lens nucleus showing ridge and globular surface patterns (×14,750) (63). Reprinted by permission from S. Karger AG, Basel.
Figure 49 Portion of multiple unit membranes of a "spherical structure" (×32,000) (63). Reprinted by permission from S. Karger AG, Basel.

endoplasmic reticulum, and a shift to long-lived high molecular weight messenger RNA. Qualitative and quantitative changes in the crystallin population of the fiber cells become apparent while protein synthesis occurs in the cortical fibers and little or no synthesis occurs in the nuclear fiber cells (94 review) (95, 96). The loss of cellular organelles with fiber cell maturation minimizes light scattering.

The nearly thousandfold increase in plasma membrane associated with the formation of a mature fiber cell (73) is associated with changes in lipid composition (97), the elaboration of enormous numbers of gap junction (60), and the accumulation of the 26K polypeptide to nearly 60% of the total membrane protein (75). The cytoskeleton in the mature fiber also differs from that of the epithelial cell. There is a loss of microtubules, a decrease in intermediate filaments and actin, and the appearance of proteinaceous chains. In the nucleus of the adult lens only aggregated chains remain. These changes are reflected biochemically in the polypeptide composition of the urea-soluble fractions (USF) of cortical and nuclear fiber cells (98), (see Figures 50 and 51). Specifically the 53K polypeptide of the intermediate filament (98–100) is readily apparent in the cortical USF but not seen in the nuclear USF. Other high molecular weight polypeptides present in the cortical USF but absent from the nuclear USF are also likely components of the cytoskeleton (112). The cytoskeleton of the mature cortical fiber cell appears to have a specific interaction with the plasma membrane. The densely granular cytoplasm of the nuclear fiber cells is probably related to the known heavy accumulation of high molecular weight protein consisting mainly of α-crystallin (102).

The topographical organization of the macromolecular constituents of the fiber cells is relevant to the understanding of lens transparency. Optical considerations for maintaining zero total birefringence and hence complete transparency of the lens requires some special, nonrandom organization of the structural proteins that possess intrinsic birefringence (103). The presence of an elaborate cytoskeleton may provide the required morphological organization. The cytoskeleton, through a contractile force exerted on the fiber cell plasma membrane, may also serve to maintain and alter the shape of the lens fiber cell during accommodation.

The large number of coupled gap junctions in the fiber cell plasma membrane suggests that the fibers exist in a state that facilitates the movement of water, ions, and other small metabolites. This aspect of lens morphology is discussed further by Benedetti et al. (Chapter 4).

Figure 50 5% polyacrylamide–1% SDS gel of bovine cortical fiber water-soluble protein (A); urea-soluble protein (B, E); isolated intermediate filament pellet (C); and plasma membrane (D). Note the selective increase of the 53K polypeptide in the filament pellet preparation and the absence of the 53K and the 41K polypeptide (actin) from the nuclear urea-soluble fraction (F).

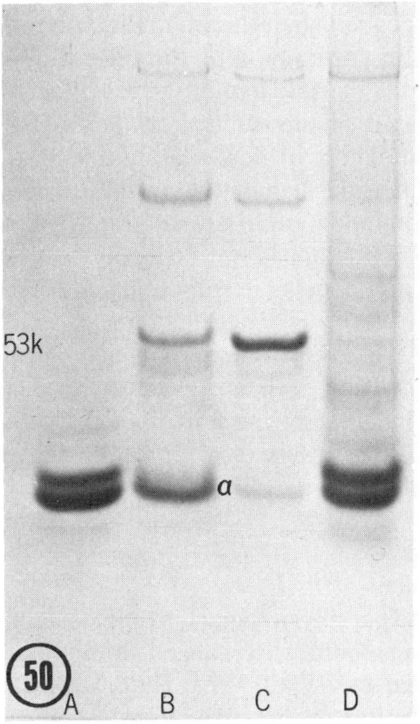

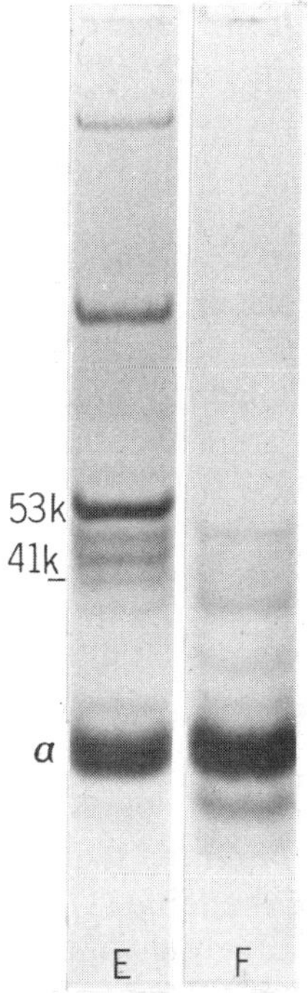

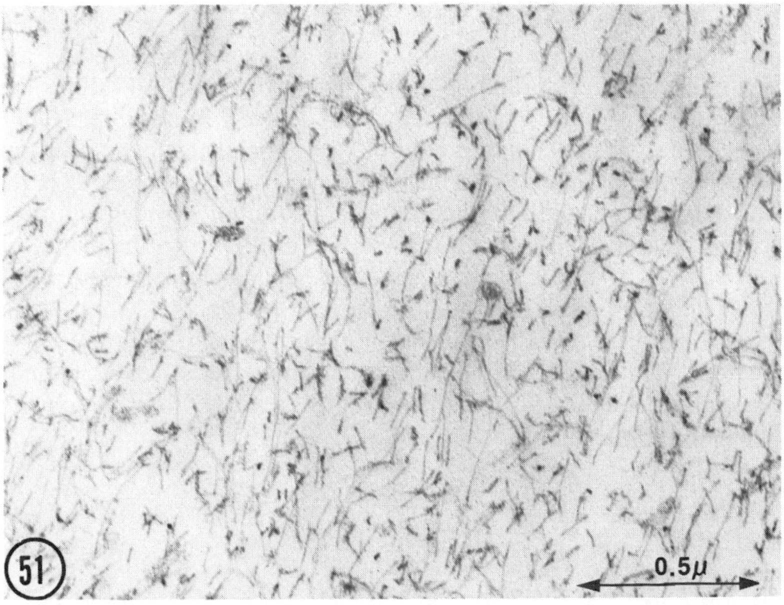

Figure 51 Isolated bovine intermediate filament preparation, seen by transmission electron microscopy (×43,500).

It is generally accepted that accommodation, which involves an increase in curvature of the lens surfaces, is accomplished through relaxation of the suspensory apparatus of the lens upon contraction of the ciliary muscle, and the separate elastic properties of the capsule and of the lens fiber mass (7, 12, 104). Brown (7) believes that it is the nucleus that is active in changing lens shape, expanding, and molding the cortex around it. Rafferty and Goossens (33) suggest that cytoplasmic filaments may be involved in providing an elastic force during accommodation. The elaborate junctional apparatus of the fiber cell minimizes cell slippage but probably allows for some intercellular movement. The precise cellular forces involved in accommodation are as yet unresolved.

It is of interest that fiber cells have the ability to undergo some degree of repair after injury (31) or a cataractous insult such as a galactose diet (105). This indicates a degree of plasticity inherent in the lens fiber cells that is not readily apparent from their morphology or low rate of protein synthesis.

2.10 CONCLUDING REMARKS

Cohen (19) observed that three principal physiological problems confront the student of the lens. (1) How are metabolites exchanged throughout the depth of a relatively thick and avascular tissue? (2) What factors contribute to transparency? (3) What structural factors are involved in maintaining the structural integrity of the lens despite normal physiological deformities? The essential morphological elements were described in his paper and in earlier studies. The interpretation of the morphological observations is now more refined as a result of the advances in cell biology. To facilitate exchange of metabolites the lens fiber cell membrane acquires an enormous number of gap junctions. A sophisticated junctional apparatus serves to link the lens cells into a functional unit no doubt through appropriate surface molecules. This apparatus probably allows for some intrinsic rotatory movement during the movements of accommodation. The presence of cytoskeleton of filaments and organized protein chains provides an internal order to the fiber cell that may help to (1) maintain cell integrity, (2) provide "contractile" force during lens movement, and (3) facilitate transparency of the lens. Despite this progress each of the questions posed by Cohen (19) requires careful further analysis.

REFERENCES

1 M. J. Hogan, J. A. Alvarado, and J. Wedell, in *Histology of the Human Eye*, Saunders, Philadelphia, (1975).

2 A. J. Coulombre, *Invest. Ophthalmol.*, **4,** 411 (1965).

3 T. Kuwabara, *Exp. Eye Res.*, **20,** 427 (1975).

4 C. Hanna and H. C. Keatts, *Exp. Eye Res.,* **7,** 244 (1968).

5 B. V. Worgul, G. R. Merriam, A. Szechter, and B. D. Srinivason, *Arch. Ophthalmol.,* **94,** 996 (1976).

6 A. Beyer-Mears, P. N. Farnsworth, S-C Fu, and C-K Yeh, *Exp. Eye Res.,* **27,** 627 (1978).

7 N. Brown, *Exp. Eye Res.,* **15,** 441 (1973).

8 J. Nordmann, G. Mack, and G. Mack, *Ophthalmol Res.,* **6,** 216 (1974).

9 Z. Dische and E. Borenfreund, *Am. J. Ophthalmol.,* **38,** 165 (1954).

10 R. C. Hughes, M. Laurent, M. Longchampt, and J. E. Courtois, *Eur. J. Biochem.,* **52,** 143 (1975).

11. J. H. Seland, *Acta Ophthalmol.,* **52,** 688 (1974).

12 R. F. Fisher and J. Wakely, *Proc. R. Soc. London B,* **193,** 335 (1976).

13 A. J. Dark, B. W. Streeten, and D. Jones, *Arch. Ophthalmol.,* **82,** 815 (1969).

14 P. N. Farnsworth, J. A. Mauriello, P. Burke-Gadomski, T. Kulyk, and A. A. Cinotti, *Invest. Ophthalmol.,* **15,** 36 (1976).

15 B. W. Streeten, *Invest. Ophthalmol.,* **16,** 364 (1977).

16 P. N. Farnsworth and P. Burke, *Exp. Eye Res.,* **25,** 563 (1977).

17 A. Brini and A. Porte, *Bull. Soc. Ophthalmol.,* **5,** 340 (1961).

18 A. I. Cohen, *Am. J. Anat.,* **103,** 219 (1958).

19 A. I. Cohen, *Invest. Ophthalmol.,* **4,** 433 (1965).

20 D. H. Dickson and G. W. Crock, *Invest. Ophthalmol.,* **11,** *809 (1972).*

21 T. Futagami, *Acta Soc. Ophthalmol. Jap.,* **66,** 1166 (1962).

22 W. C. Gorthy, *Exp. Eye Res.,* **7,** 394 (1968).

23 W. C. Gorthy, M. R. Snavely, and N. D. Berrong, *Exp. Eye Res.,* **12,** 112 (1971).

24 W. C. Gorthy, *Exp. Eye Res.,* **27,** 301 (1978).

25 C. V. Harding, J. R. Reddan, N. J. Unakar, and M. Bagchi, *Int. Rev. Cytol.,* **31,** 215 (1971).

26 T. Kuwabara, *Arch Ophthalmol.,* **79,** 189 (1968).

27 T. Kuwabara, Surface structure of the eye tissues, *Proc. III Annual Scanning Electron Microscope Symposium, Chicago,* 1970, p. 189.

28 T. S. Leeson, *Exp. Eye Res.,* **11,** 78 (1971).

29 A. Porte, A. Brini, and M. E. Stoeckel, *Ann. Ophthalmol.,* **7,** 623 (1975).

30 N. S. Rafferty and E. A. Esson, *J. Ultrastruc. Res.,* **46,** 239 (1974).

31 N. S. Rafferty and W. Goossens, *Am. J. Anat.,* **142,** 177 (1975).

32 N. S. Rafferty and W. Goossens, *Am. J. Anat.* **148,** 385 (1977).

33 N. S. Rafferty and W. Goossens, *Exp. Eye Res.,* **26,** 177 (1978).

34 T. Wanko and M. A. Gavin, *Arch. Ophthalmol.,* **60,** 868 (1958).

35 T. Wanko and M. A. Gavin, *J. Biophys. Biochem. Cytol.,* **6,** 97 (1959).

36 T. Wanko and M. A. Gavin, "Cell Surfaces in the Crystalline Lens," in G. K. Smelser, Ed., *The Structure of the Eye,* Academic Press, New York, 1961, p. 221.

37 N. R. Willis, M. J. Hollenberg, and C. R. Braekvelt, *Can. J. Ophthalmol.,* **4,** 307 (1969).

38 C. V. Harding, A. Donn, and B. D. Srinivasan. *Exp. Cell Res.*, **18,** 582 (1959).
39 A. Weinsieder, H. Rothstein, and D. Drebert, *Cytobiologie,* **7,** 406 (1973).
40 H. Rothstein and B. Worgul, *Ophthalmol. Res.*, **5,** 151 (1973).
41 L. van Sallmann, *Arch. Ophthalmol.*, **47,** 305 (1952).
41a B. V. Worgul, "Anatomy and Congenital Anomalies of the Lens," in T. Duane and F. Jakobiec, Eds., *Biomedical Foundation of Ophthalmology,* Harper and Row, in press.
42 S. P. Modak and S. W. Perdue, *Exp. Cell Res.*, **59,** 43 (1970).
43 S. P. Modak, V. R. R. Uppuluri, D. W. Appleby, A. M. Therwath, and W. E. Lever, "Growth Kinetics of Epithelial and Fiber Cells in Developing Chick Lens," in Y. Courtois and F. Regnault, Eds., *Biology of the Epithelial Lens Cells,* Publ. Inserm, Paris, 1976, p. 105.
44 A. A. Swanson, J. Peter, and K. W. Raley, *Exp. Eye Res.*, **6,** 351 (1967).
45 V. R. Reddy and N. J. Unakar, *Exp. Eye Res.*, **17,** 405 (1973).
46 K. Ashghar, B. G. Reddy and G. Krishna, *J. Histochem. Cytochem.*, **23,** 774 (1975).
47 P. N. Farnsworth, P. Burke-Gadomski, T. Kulyk, J. A. Mauriello, and A. A. Cinotti, *Exp. Eye Res.*, **22,** 615 (1976).
48 M. Palva and A. Palkama, *Exp. Eye Res.*, **22,** 229 (1976).
49 C. V. Harding, D. Harding, V. Peters, T. Kuwabara, J. Reddan, N. Unakar, M. Bagchi, T. Schnur and S. Gordon, *Exp. Eye Res.*, **16,** 1 (1973).
50 J. Piatigorsky, *Ann. N.Y. Acad. Sci.*, **253,** 333 (1975).
51 G. Y. Mousa and J. R. Trevithick, *Dev. Biol.*, **60,** 14 (1977).
52 B. V. Worgul and H. Rothstein, *Medikon,* **6,** 5 (1977).
53 T. Kuwabara and M. Imaizumi, *Invest. Ophthalmol.*, **13,** 973 (1974).
54 B. D. Srinivasan and T. Iwamoto, *Exp. Eye Res.*, **16,** 9 (1973).
54a B. V. Worgul, T. Iwamoto, and G. R. Merriam, *Ophthalmol. Res.*, **9,** 388 (1977).
55 J. Alcalá, N. Lieska, and H. Maisel, *Exp. Eye Res.*, **21,** 581 (1975).
56 P. Waggoner and H. Maisel, *Exp. Eye Res.*, **27,** 151 (1978).
57 M. J. Hollenberg, J. P. H. Wyse, and B. J. Lewis, *Cell Tissue Res.*, **167,** 425 (1976).
58 M. Sakuragawa, T. Kuwabara, J. H. Kinoshita, and H. N. Fukui, *Exp. Eye Res.*, **21,** 381 (1975).
59 C. V. Harding, S. Susan, and H. Murphy, *Ophthal. Res.*, **8,** 443 (1976).
60 J. Kuszak, H. Maisel, and C. V. Harding, *Exp. Eye Res.*, **27,** 495 (1978).
61 G. Duncan, "Comparative Physiology of the Lens Membranes," in M. Davson and L. T. Graham, Jr., Eds., *The Eye,* Vol. 5, Academic Press, New York, 1974, p. 357.
62 T. Yorio and R. J. Bentley, *Exp. Eye Res.*, **23,** 601 (1976).
63 C. V. Harding, S. Susan, R. S. Jampel, and E. Cohen, *Ophthal. Res.*, **10,** 7 (1978).
64 R. H. Bradley, M. Ireland, and H. Maisel, *Exp. Eye Res.*, **28,** 441 (1979).
65 R. H. Bradley, M. Ireland, and H. Maisel, *Acta Ophthalmol.*, **57,** 461 (1979).

66 M. Ireland, H. Maisel, and R. H. Bradley. *Ophthalmol. Res.*, **10**, 231 (1978).

67 M. O. Longchampt, M. Laurent, Y. Courtois, P. Trenchev, and R. C. Hughes, *Exp. Eye Res.*, **23**, 505 (1976).

68 H. Maisel, *Experientia*, **33**, 525 (1977).

69 R. A. Resnik, T. Wanko, and M. A. Gavin, *J. Biophys. Biochem. Cytol.*, **7**, 403 (1960).

70 A. Lasser and E. A. Balasz, *Exp. Eye Res.*, **13**, 292 (1972).

71 H. Maisel and M. M. Perry, *Exp. Eye Res.*, **14**, 7 (1972).

72 R. Bradley and H. Maisel, *Experientia*, **34**, 470 (1978).

73 E. L. Benedetti, I. Dunia, C. J. Bentzel, A. J. M. Vermorken, M. Kibbelaar, and H. Bloemendal, *Biochim. Biophys. Acta*, **457**, 353 (1976).

74 z. Dische, *Invest. Ophthalmol.*, **4**, 759 (1965).

75 H. Maisel, J. Alcalá, and N. Lieska, *Doc. Ophthalmol.*, **8**, 121 (1976).

76 H. Maisel, M. Perry, J. Alcalá, and P. R. Waggoner. *Ophthalmol. Res.*, **8**: 55 (1976).

77 H. Maisel, *Exp. Eye Res.*, **25**, 595 (1977).

78 S. Gaspar, *Cesk. Morfol.*, **12**, 117 (1964).

79 T. Matsuto. *Acta Soc. Ophthalmol. Jap.*, **77**, 853 (1973).

80 H. A. Hansson, *Z. Zellforsch.*, **107**, 187 (1970).

81 P. N. Farnsworth, S. C. Fu, P. A. Burke, and I. Bahia, *Invest. Ophthalmol.*, **13**, 274 (1974).

82 K. J. Nelson and N. S. Rafferty, *Exp. Eye Res.*, **22**, 335 (1976).

83 D. H. Dickson and G. W. Crock, "Fine Structure of Primate Lens Fibers," in J. G. Bellows, Ed., *Cataract and Abnormalities of the Lens*, Grune and Stratton, New York, (1975), p. 49.

84 Y. Kobayashi and T. Suzuki, "The Aging Lens: Ultrastructural Changes in Cataract," in J. G. Bellows, Ed., *Cataract and Abnormalities of the Lens*, Grune and Stratton, New York, (1975), p. 313.

85 K. J. Dilley, A. J. Bron, and J. O. Habgood, *Exp. Eye Res.*, **22**, 155 (1976).

86 A. Brini, A. Porte, and M. E. Stoecke, *Bull, Soc. Fr. Ophthalmol.*, **76**, 193 (1963).

87 B. Phillipson, *Exp. Eye Res.*, **16**, 29 (1973).

88 F. L. Huang and T. Kuwabara, "Early Cytological Changes in Senile Cataract," *Abstr. Assoc. Res. Vision Ophthalmol.*, 1977, p. 143.

89 M. O. Creighton, J. R. Trevithick, G. Y. Mousa, D. H. Perry, A. J. McKinna, C. Dyson, H. Maisel, and R. Bradley, *Can. J. Ophthalmol.*, **13**, 166 (1978).

90 R. van Heyningen, "The Lens: Metabolism and Cataract," in H. Davson, Ed., *The Eye*, 2nd ed., Academic Press, 1969, p. 381.

91 M. C. Neville, C. A. Paterson, and P. M. Hamilton, *Exp. Eye Res.*, **27**, 637 (1978).

92 D. A. Goodenough, *Invest. Ophthalmol.*, **18**, 1104 (1979).

93 J. Papaconstantinou, *Science*, **156**, 338 (1967).

94 J. J. Harding and K. J. Dilley, *Exp. Eye Res.*, **22**, 1 (1976).

95 J. Genis-Galvez, H. Maisel, and J. Castro, *Exp. Eye Res.*, **71**, 593 (1968).

96 H. Bloemendal, *Science,* **197,** 127 (1977).
97 P. Zelenka, *Exp. Eye Res.,* **26,** 267 (1978).
98 H. Maisel, J. Alcala, N. Lieska, and N. Rafferty, *Ophthal. Res.,* **9,** 147 (1977).
99 H. Maisel, N. Lieska, and R. Bradley, *Experientia,* **34,** 352 (1978).
100 H. Maisel and N. S. Rasmussen, *Ophthal. Res.,* **10,** 241 (1979).
101 G. Y. Mousa and J. R. Trevithick, *Invest. Ophthalmol.* Suppl., 165 (1978).
102 D. Roy and A. Spector, *Exp. Eye Res.,* **22,** 273 (1976).
103 F. A. Bettelheim and T. J. Y. Wang, *Exp. Eye Res.,* **25,** 613 (1977).
104 R. F. Fisher, *J. Physiol. London.,* **228,** *765 (1973).*
105 N. J. Unakar, C. Genyea, J. R. Reddan, and V. N. Reddy, *Exp. Eye Res.,* **26,** 123 (1978).
106 A. Pirie and R. van Heyningen, in *Biochemistry of the Eye,* Charles C Thomas, Springfield, Ill., 1956.
107 N. S. Rafferty, *Anat. Rec.,* **173,** 225 (1972).
108 M. M. Perry. *Exp. Eye Res.,* **22,** 125 (1976).
109 L. D. Hazlett and R. H. Bradley, *Exp. Eye Res.,* 26, 207 (1978).
110 M. A. Kibbelaar, F. C. S. Ramaekers, P. J. Ringens, A. M. E. Selten-Versteegen, L. G. Poels, P. H. K. Jap, A. L. van Rossum, T. E. W. Feltkamp, and H. Bloemendal, *Nature,* **285,** 506 (1980).
111 J. Kuszak, J. Alcalá, and H. Maisel, *Am. J. Anat.* (in press).
112 S. Nasser, R. Bradley, J. Alcalá, and H. Maisel, *Exp. Eye Res.,* 30, 109 (1980).

CHAPTER

3

Cytoskeletal and Contractile Structures in Lens Cell Differentiation

Frans C. S. Ramaekers and Hans Bloemendal

Department of Biochemistry, University of Nijmegen, Nijmegen, The Netherlands

3.1 LENS CELL DIFFERENTIATION

3.1.1 Why Study Lens Cell Differentiation?

One of the major problems that investigators of the mechanism(s) of cellular differentiation are confronted with is: "How does a developing cell acquire its specific biochemical characteristics and how are these linked to morphological development?" (1). During the process of differentiation the morphological diversification and specialization of a cell is paralleled by the synthesis of specific enzymes and structural proteins. An example of such a highly specialized cell is the lens cell, which is characterized by the synthesis of a group of highly specific structural proteins, the crystallins (compare Chapter 1).

The process of lenticular differentiation is easy to recognize visually by the transition of the lens epithelium to the strongly elongated lens fibers. The fiber cells remain functional on the level of protein biosynthesis for a relatively long period, which makes them very suitable not only for studies on the mechanism of protein synthesis *per se,* but also for its regulation during cellular differentiation. A number of excellent reviews dealing with the biochemistry of lens cell differentiation have appeared in the literature (1–6).

Investigations by Benedetti et al. (7; see also Chapter 4) have shown that during lens cell elongation the amount of gap junctions between the cells increases dramatically. This renders the lens a suitable organ for the study of these junctions and membrane differentiation in general. Moreover, the lens offers a pure population of mitotically active cells, at least in the germinative region, which allows the culturing of lens cells and the study of lens cell differentiation *in vitro*. Moreover, recent investigations, many of which are still unpublished, indicate important functions of the cytoskeletal and contractile elements in lens cell differentiation. For this reason, in Section 3.2 of this chapter we outline these structures in more detail and their role in lens cell elongation is discussed.

3.1.2 Morphological Aspects of Lens Cell Differentiation

The lens of adult vertebrate species is surrounded by an external noncellular capsule and can be divided into four major regions: (*1*) A monolayer of epithelial cells, located at the lateral side of the lens, which in their turn can be separated into two distinct regions, namely, the epithelial cells of the central region, which have a low rate of mitosis, and the germinative zone that is composed of cells with a high mitotic rate. These latter cells elongate at (*2*) the so-called equatorial zone to form (*3*) the cortical or peripheral fibers. It is in this region that most of the apparent morphological changes associated with lens cell differentiation take place. The nuclear part of the lens is composed of fiber cells formed during the fetal and embryonic growth of the lens. Newly formed fiber cells are continuously laid onto this core throughout the postnatal life of the animal. Lens fibers are long tube-like structures that have a hexahedral shape and are, as already mentioned, the result of a process of terminal differentiation.

During lens cell differentiation several changes occur in the cuboidal epithelial cells. In the zone of elongation where the epithelial cells commence the process of fiber cell formation, the cell volume increases, nuclei and nucleoli enlarge, and the number of polyribosomes increases, but the cells no longer divide. In the completed fiber cells most of the intracellular organelles have disappeared with the exception of polyribosomes, the plasma membranes (7), and the cytoskeleton (see Chapter 4).

3.1.3 Protein Biosynthesis during Lens Cell Differentiation

From the morphological data presented in Section 3.1.2, two biochemical features of lens fibers are obvious. First, since nuclei have disappeared after fibrogenesis, DNA and RNA synthesis cease. On the other hand, the increase in the number of polyribosomes adjacent to the enlarged nuclei in the elongation zone suggests a high rate of RNA synthesis and/or protein biosynthesis in this region (1). Second, because of the tremendous enlargement of the cell surface a great amount of plasma membrane constituents has to be newly synthesized.

3.1.3.1 Stabilization of RNA

From experiments in which the effect of actinomycin D on RNA biosynthesis was studied, Papaconstantinou (1) concluded that the synthesis of crystallins in bovine lens epithelium is directed by short-lived mRNA, whereas in the fiber cells long-lived mRNAs are operative. How stabilization of mRNA is accomplished in the fiber cells remains to be answered, but the phenomenon clearly exists since protein biosynthesis is still possible in lens fibers and appreciable amounts of active mRNA can be isolated from the lenticular cortex region (see Chapter 5). Some further evidence concerning protein biosynthesis during lens cell differentiation can be derived from studies dealing with the fate of ribosomal structures. Electron microscopic observations reveal an increase in the amount of ribosomes during elongation, possibly required for the overall increase in protein synthesis and necessary for the increase in cell volume. After the fiber cells have formed the ribosomal population decreases drastically.

Piatigorsky (2) used complementary DNA (cDNA) for the detection of crystallin mRNAs in detailed studies concerning the stability of mRNA in the different stages of lens cell differentiation. Embryonic chick lenses showed an accumulation of δ-crystallin mRNA during elongation, resulting in a preferential stimulation of δ-crystallin biosynthesis (8). These data suggest that the increase in the synthesis of a specific protein is not the result of a more effective utilization of a fixed mRNA pool, but is simply the result of an increase in its mRNA synthesis. Whether or not these latter conclusions drawn for embryonic lenses are also true for postnatal lenses remains to be established.

3.1.3.2 Synthesis of Specific Crystallins as a Result of Lens Cell Differentiation

Several authors have reviewed the sequential changes that parallel lenticular cell differentiation both during embryogenesis and in the postnatal lens. The specific appearance of crystallins during embryogenesis of the lens and the formation of fiber cells is reviewed by van Doorenmaalen in Chapter 10. Lens epithelial cells from chicken embryos, rat, or calf in culture have been shown to produce the specific proteins only when they form "lentoid bodies," which contain fiberlike structures (see Section 3.1.4). These proteins, therefore, can also be used as typical *in vitro* differentiation markers.

Early studies by Papaconstantinou and co-workers (reviewed in ref. 1) in postpartum bovine lenses revealed that the synthesis of γ-crystallin is restricted to lens fibers and is absent in the epithelium. Slingsby and Croft (9) showed a gradual replacement of the four members of the low molecular weight crystallin family from γ_{IV} to β_S. Furthermore, Delcour and Papaconstantinou (10, 11), Vermorken et al. (12, 13), and Bloemendal and Vermorken (14) reported a change in the stoichiometry of the αA_2 and αB_2 chains during fiber formation. The amount of αB_2 synthesized in epithelial cells is three times more than that of αA_2. In cells isolated from the elongation zone the synthesis of αA_2- and β-crystallins becomes abundant, indicating that the synthesis of these polypeptides accompanies differentiation. According to van Kamp et al. (15, 16) two embryonic α-crystallin subunits are lost as lens development proceeds. In the chick lens the high amount of δ-crystallin decreases upon aging and is replaced by α- and β-crystallins (17–19). Changes in the distribution of the LDH isoenzymes during bovine lens differentiation have been reported by Papaconstantinou (1).

The question of whether the synthesis of specific gene products during elongation is the trigger for differentiation or a result of the morphological changes also remains to be elucidated.

3.1.3.3 Membrane Differentiation

The structure and biochemistry of the lenticular plasma membranes are described in Chapter 4. Here we want only to mention that the dramatic increase in cellular surface during lens cell differentiation is accompanied by the incorporation of a special intrinsic membrane protein exclusively observed in fiber cells (20). This membrane protein has been designated MP26 by Bloemendal et al. (21) because of its apparent molecular weight of 26,000 and main intrinsic polypeptide (MIP) by Broekhuyse et al. (22).

Since MP26 has been shown to occur in fractions enriched in gap junctions (7), it is assumed that the appearance of MP26 in lens fiber membranes is not only an indication for lens membrane differentiation but primarily for gap junction assembly. Polyribosomes, specifically synthesizing MP26, are attached to microfilaments as judged from their release from the plasma membrane–cytoskeleton complex by DNAse I (23–25). This enzyme has been shown to depolymerize actin filaments specifically (26).

3.1.4 Lens Cell Cultures for Differentiation Studies *in Vitro*

Lens cell differentiation appears to involve a sequence of morphological and biochemical events. These processes have been studied intensively in the intact lens, but yet the molecular mechanism underlying the morphological changes that become manifest during differentiation is poorly understood. It would therefore be of great help if the processes normally going on *in vivo* could be maintained in lens cell cultures.

In the following sections we review the various attempts that have been made by several investigators to develop such an *in vitro* differentiating system and the results emerging from these efforts (27–77).

3.1.4.1 Criteria for *in Vitro* Differentiation

Creighton et al. (53) summarized the criteria that are required for a lens cell system in order to be suitable for differentiation studies in culture:

1 Differentiation in culture should be recognizable by comparable morphological changes also observed *in situ*.
2 Specific protein markers of lenticular differentiation should be absent in primary epithelial cells and should arise during the course of the morphological changes.
3 It should be possible to keep the epithelial cells in culture in an "undifferentiated" state and to induce differentiation, as is achieved, for example, in the embryonic lens by some component(s) from the retina (78).
4 Clones of epithelial cells should show the characteristics of freshly explanted epithelial cells.

If such an *in vitro* system is available the effect of external factors such as hormones, cyclic nucleotides, or inhibitors of protein or RNA synthesis, can be studied fruitfully. Moreover, RNA and protein biosynthesis during lens cell differentiation can be followed easily by incorporation studies of radioactive precursors. Another advantage of such a system is the possibility of investigating the role of cytoskeletal and contractile structures in lens cell differentiation by adding specific microtubule- or microfilament-disrupting drugs to the cell culture medium.

3.1.4.2 Differentiation Studies in Primary Lens Cultures and Explants

Embryological and anatomical features of the crystallin lens make the organ a unique system to isolate a homogeneous line of epithelial cells. If the epithelium is carefully excised by peeling off the lens capsule from the center of the anterior region, the culture is completely free of cells of other origin. Differentiated fiber cells that contaminate the epithelial cells do not disturb the culture since they have lost their mitotic activity.

The first attempts to culture lens epithelial cells were made by Kirby et al. in 1929 (27) and almost 20 years later by Mann (28). However, the lens epithelial cells could not be maintained for more than a few days in culture by these investigators and many others thereafter (29–35). Moreover, these workers failed in obtaining differentiation in the monolayer cultures. Many efforts have been made since then to find conditions that could stimulate growth and differentiation of the epithelial lens cells brought into culture.

The first succesful results concerning lens epithelial cell differentiation *in vitro* were achieved with explants from embryonic chick lens (33, 34, 36). These lens epithelia, still attached to the capsules, were cultured in media either containing or depleted of fetal calf serum. Addition of serum or protein supplement appeared to be crucial in initiating or maintaining the differentiated state. Pre-culturing the explants in serum-free medium resulted in a drastic decrease of their differentiating capability under the influence of serum. In this case the cells were stimulated to proliferation and migration (36). Piatigorsky and co-workers used this system to study the role of protein synthesis and microtubules in chick lens differentiation (for a review see ref. 49 and Section 3.4.2.3). These workers also found that insulin, next to fetal calf serum, can induce elongation of explanted chick embryo epithelial lens cells. Other factors, such as cyclic AMP, GMP, nerve growth factor, or hydrocortisone, failed to induce differentiation (49). In view of these findings the embryonic chick lens system is very satisfactory for the study of lens cell differentiation. Besides an inducible elongation [the cell length increases minimally 25-fold (41)], an increase of δ-crystallin synthesis up to 40 to 50% of the newly synthesized protein is also observed during elongation (cf. ref. 42). Results by Milstone and Piatigorsky (60) show that δ-crystallin synthesis is stimulated initially by an increase in translational efficiency and secondarily by an increase in the amount of δ-crystallin mRNA. Furthermore, age-dependent changes in the ratio of synthesis of δ-crystallin polypeptides seem to occur upon culturing the cells in the absence of the vitreous body (75). Moreover, a factor has been demonstrated in the vitreous humor which promotes fiber differentiation (71).

The first monolayer cultures of chick lens epithelial cells were described by Okado et al. (39). These investigators showed that the flattened epithelial cells retain their ability to differentiate and thus to form lens fibers which accumulate δ-crystallin in so-called "lentoid bodies." Single-plated cells proliferate to produce these spherical bodies showing lens fibers upon electron microscopic examination (43). In addition, appropriately absorbed lens-specific antisera reacted strongly with the lentoid bodies. Subcultured cells were also shown to form these three-dimensional dense aggregates (39).

Hamada et al. (61, 70) showed the formation of "lentoid bodies" containing lens fibers in monolayers of cultured rat lens epithelia. The presence of γ-crystallin, a presumed marker protein for a mature lens fiber, was confirmed by means of the fluorescent antibody technique. Again, *in vitro* differentiation was also seen in clones originating from a single cell. In these cells the synthesis of α-crystallin rapidly ceased after culturing, but was detected after some time in the lentoid bodies and in a few adjacent epithelial cells.

Another type of differentiation of rat lens epithelial cells has been described by Creighton et al. (44, 53) and by Mousa and Trevithick (54). These authors show that rat lens cells cultured by plating cell suspensions differentiate as concluded from the elongation of single cells. These cells either elongate spontaneously or can be induced to differentiate by dibutyryl cAMP or caffeine (44). Upon elongation synthesis of γ-crystallin was detected by means of immunofluorescence. Furthermore, the authors suggest on the basis of experiments using cytochalasin B and D that a proper organization of actin-containing microfilaments is required for rat lens cell elongation and the production of γ-crystallin (see also Section 3.1.3.2).

Primary cultures of bovine epithelial lens cells still growing on the capsule synthesize only one of the two α-crystallin subunits, namely, αB_2(12–14; see Figure 1). The synthesis of this polypeptide decreases dramatically when the cells grow off the capsule onto the bottom of the Petri dish. These results suggest a possible role of the capsule in the enhancement of crystallin synthesis in lens cell culture. A similar result for the capability of collagenlike material to enhance the differentiated state of a tissue was recently reported by Yasuda for the pigmented cells of the retina (79).

Lentoid bodies can be observed in primary cultures 2 to 3 weeks after confluence (see Figure 2). These structures seem to synthesize α- and β-crystallinlike polypeptides as concluded from SDS gel electrophoretic analysis (Figure 1, *l*, *m*). Lentoid bodies in long-term cultures have been described to synthesize capsulelike material (57).

Primary cultures from human epithelial lens cells have also been described (62, 69, 72). In these cultures piles of elongated cells with typical profiles of mature lens fibers (lentoid bodies) are observed too. γ-Crystallin, undetectable in the lens epithelium, is observed in these differentiated

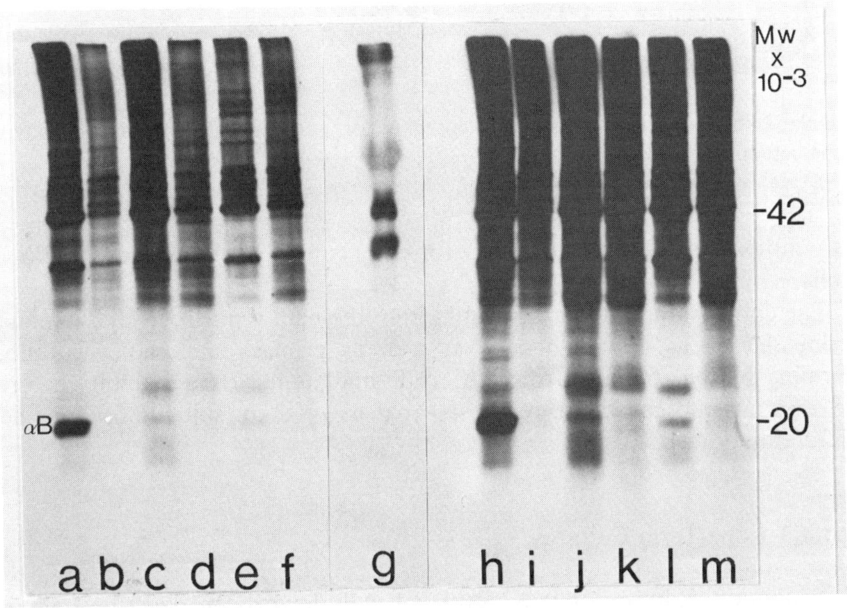

Figure 1 Newly synthesized proteins in primary bovine lens cells in culture. (*a*) Soluble fraction of cells growing on the capsule. (*b*) Insoluble fraction of cells growing on the capsule. (*c*) Soluble fraction of cells growing on plastic. (*d*) Insoluble fraction of cells on plastic. (*e*) Soluble fraction of dense cell aggregates (lentoid bodies). (*f*) Insoluble fraction of lentoid bodies. (*g*) Labeled proteins in the culture medium. (*h–m*) As *a–f*, but with longer exposure of the autoradiograph.

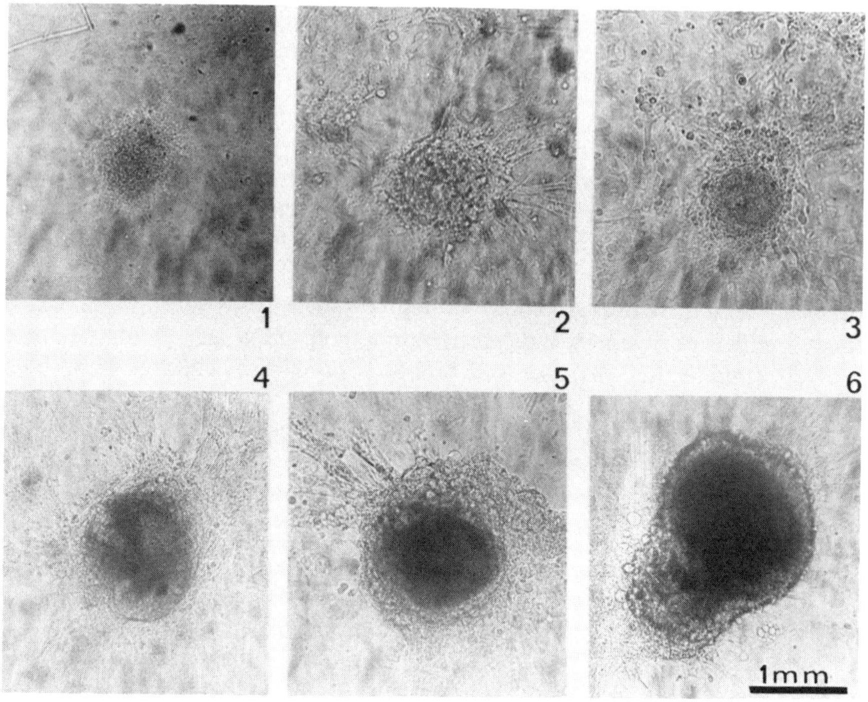

Figure 2 Six stages of dense cell aggregate formations (lentoid bodies) in bovine lens cell cultures.

structures (80). This was the case for embryonic as well as very old donor cells.

The conclusion that can be drawn from the data reported is that explants from chick lens epithelial cells as well as primary monolayer cultures, forming lentoid bodies, from rat, calf, and human lens exhibit features (elongation and crystallin synthesis) that make them suitable for lens cell differentiation studies *in vitro*.

3.1.4.3 Long-Term Cultures

Long-term cultures of lens epithelial cells have been described almost exclusively for bovine lens cells. The first reports by van Venrooij et al. (45, 46) showed that although these highly proliferative cells seemed to synthesize some β-crystallin, as concluded from immunological experiments, they produced neither α- nor γ-crystallin. However, young cuboidal bovine epithelial cells in long-term culture elongate either spontaneously in a time-dependent process after a certain number of transfers (20 or 30 passages) (45, 51, 52, 66, 76), under the influence of dexamethasone (46), or after addition of a retinal extract (63). If the hormone or the extract is omitted the induced elongation is abolished. The *in vitro* elongation of lens cells after incubation with retinal extract is an interesting feature. Several authors (see also Chapter 10) reported that in lens embryogenesis the formation of the lens placode and subsequent elongation of the cells are induced by the neural retina (78).

The elongated cells formed from the polygonal ones are rearranged into whorls of spindlelike cells and exhibit many intracytoplasmic fibrils, which have been identified as microfilament bundles (76, 77, 81). The crucial question concerning this process of elongation in bovine lens epithelium monolayer cultures is whether or not this phenomenon reflects morphologically the *in vivo* differentiation process. As already mentioned, the cells stop synthesizing crystallins as soon as they lose contact with the capsule. In fact there are no indications for the synthesis of specific differentiation markers on elongation *in vitro*. Glässer et al. (56) state that fiber formation in the "lentoid bodies" does not reflect differentiation *in vitro*. However, the cells continue to synthesize collagen type IV, which has been shown to be identical to the collagen occurring in the capsular basement membrane (57, 64). Furthermore, Courtois et al. (65) report that when bovine lens cells are transformed and injected into nude mice, they form tumors that contain elongated cells, resembling lens fibers (55, 65). In these structures α-crystallin could be detected by means of immunofluorescence and immunodiffusion techniques. The latter results show that transformation of bovine lens cells growing *in vitro* does not affect their ability to differentiate. A continuous cell line from rat lens epithelium has been described, which was suggested to offer a convenient model for the study of a stepwise regulation of cytodifferentiation (74).

The resynthesis of α-crystallin in the tumors of bovine lens cells suggests the requirement of a three-dimensional arrangement of the cells in differentiation, because the same phenomenon is also seen in lentoid bodies. Moreover, the necessity for primary lens epithelial cells to adhere to capsular material in order to enable the synthesis of α-crystallin supports this idea (see also ref. 82).

The fact that lens cells in monolayer can elongate, either upon aging or as a result of induction with dexamethasone or retinal extract, means that crystallin synthesis is not a prerequisite for the morphological feature of elongation. The synthesis of crystallins in monolayer culture seems to depend on a proper spatial arrangement of the cells. For example, Creighton et al. (53) point out that cell-to-cell contact is the main factor inducing differentiation accompanied by crystallin synthesis in newborn rat lens epithelium *in vitro*. The reversibility of elongation, induced by dexamethasone or retinal extract, seems to indicate that this phenomenon is not interrelated with differentiation. One should, however, keep in mind that differentiated nuclei-containing fiber cells have been shown to dedifferentiate into motile lens epitheliumlike cells (59).

It would be premature to conclude here that elongation of bovine lens cells in long-term monolayer cultures reflects true differentiation. Further studies in this field should be undertaken in order to establish whether or not the markers of differentiation *in vitro* are identical to those *in vivo*. Not only the crystallins should be considered in this respect, but also membrane and/or cytoskeletal proteins. In Section 3.2 we show that for the study of the role of cytoskeletal and contractile structures in lens cell elongation the bovine monolayer culture is very useful. At any rate it should be kept in mind that cells devoid of their natural environment frequently adapt to the *in vitro* conditions (66, 73).

3.1.4.4 Transdifferentiation of Ocular Tissues into Lens Cells

When differentiated cells divide, their progeny usually show a similar differentiated state, which means that cytodifferentiation results in stabilized cells. Furthermore, the differentiated cells may continue along a pathway that is characteristic of that cell type, for example, myotube formation from myoblasts or lens fiber formation from lens epithelial cells. There are, however, some striking exceptions to the rule that differentiated cells are stable and cannot change into other cell types. These exceptions, which may be of great interest in the study of differentiation in general, include such ocular tissues as the pigmented epithelium of the dorsal and ventral iris epithelium (83, 84), the pigmented cells of the retina (85), the neural retina (86), and cornea. These tissues can, under proper conditions, transform (transdifferentiate) into lens fiberlike cells, a process that has been reviewed by Clayton (82). Although all the ocular tissues mentioned above in principle originate from the ectodermal layer of the early embryonic life, they have gone through several stages of cellular differentiation (see Chapter 10). Therefore, crucial questions that arise from this model deal with the regulatory processes in the synthesis of tissue-specific mRNA and/or proteins.

Iris. One of the classic examples of the transdifferentiational capacity of ocular tissue was observed first in adult newts. After lentectomy, the dorsal epithelium of the iris regenerates a lens in these animals. This phenomenon is called Wolffian regeneration and has been reviewed by several authors (87–90). The ability of cells from the dorsal iris to transform into lens fiber cells either *in vivo* or *in vitro*, as concluded from the appearance of lentoid bodies and the occurrence of α-, β-, and γ-crystallins in these structures, depends on the presence of the neural retina (91, 92) or other neural tissue

(93). However, formation of "lentoid bodies" can be obtained from both the dorsal and ventral iris (the latter tissue normally is not able to transdifferentiate) when *N*-methyl-*N*-nitrosoguanidine is added to the cell culture (85, 94).

Cornea. Upon lentectomy of *Xenopus laevis* the cornea regenerates a lens (95–97). Also in culture this phenomenon can be observed, as was concluded from the formation of lentoids, synthesizing some of the crystallins present in the lens of *Xenopus* (98, 99).

Pigmented Epithelial Cells. Such cells from chick embryo retina, newt, and quail can form crystallin-synthesizing cells (84, 85). Yasuda (79) has shown that this kind of transdifferentiation can be inhibited in pigmented retina cells from chick when grown on collagen-coated substrates. This result suggests for the *in situ* situation a stabilizing effect of the basal lamina on the differentiated state of the pigmented retina and an inhibition of transdifferentiation.

Neural Retina. The most extensively studied ocular tissue known to transdifferentiate into lens cells, but also into pigmented epithelium, is the neural retina (82, 100). Although it was originally suggested by Okada (86) and Araki and Okada (101) that no single-cell clone in chick embryo neural retina cultures gives rise to both pigment and lens cells, later a multipotential nature of embryonic neural retina cells was proposed by the same author (102). Culture medium conditions can affect the selection between the two pathways of transdifferentiation (101, 103). Results by Thomson et al. (104, 105) show that the amount of lens-specific mRNA increases in transdifferentiating neural retina cells, forming lentoid bodies. Furthermore, Clayton et al. found a relationship between crystallin mRNA expression in retina cells and their capacity to redifferentiate into lens cells (106). These results provide evidence for a transcriptional control of crystallin synthesis upon culturing. A very remarkable phenomenon observed independently and at about the same time by de Pomerai and Clayton (107) and Nomura and Okada (108) is the effect of the embryonic stage on transdifferentiation of chick neural retina cells. In contrast to iris in newt, the transdifferentiation capacity of the former tissue is age-dependent. The latest developmental stage at which differentiation of cells took place was in 18-day-old embryos. A gradual decrease in transdifferentiation capacity was seen in going from very young embryonic stages (3½ days) up to this stage. Also the accumulation of δ-crystallin is much more pronounced in cultures of the early stages of neural retina embryogenesis.

Speculations dealing with and trying to explain the phenomenon of transdifferentiation of ocular tissues have, until now, mainly taken into account the regulatory mechanisms of the synthesis of crystallins. A role of intermediate-sized filaments in transdifferentiation deserves consideration also. As we shall see in Section 3.2.2.2, bovine and murine cornea contain only prekeratin as tonofilament protein, whereas in the mammalian lens only vimentin is observed as intermediate-sized filament constituent. Transdifferentiation of corneal into lenticular tissue, as observed in *Xenopus laevis,* therefore possibly also involves a switch in the synthesis of the proteins building up the intermediate-sized filaments. In this respect it is also interesting to note that the epitheliumlike layers covering the iris and ciliary muscles invariably contain vimentin filaments (77, 109). These cells may represent, together with lens and neural retina cells, lines of early ectodermal differentiation, characterized by the absence of prekeratin. It should be kept in mind, however, when dealing with the problem of transdifferentiation, that many, if not all, cells brought into culture after a certain lag time start to develop vimentin containing intermediate-sized filaments. Moreover, the presence of vimentin demonstrated by immunofluorescence with the aid of monospecific antibodies in hamster embryo neural retina (Figure 3) may possibly explain the age-dependent transdifferentiation of the neural retina into lens cells as observed in chick embryo tissue.

We believe that studies on the development of the intermediate-sized filaments during embryogenesis will give valuable information concerning transdifferentiation and differentiation *per se*.

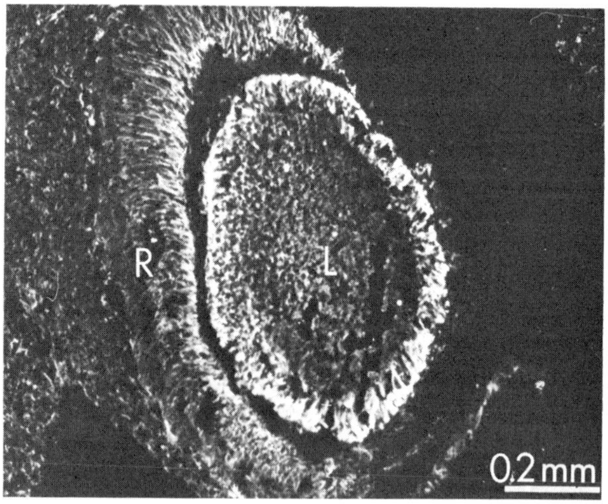

Figure 3 Immunofluorescence micrograph of a freeze section from hamster embryo eye tissue after reaction with antibodies to murine vimentin (R, retina; L, lens).

3.2 CYTOSKELETAL AND CONTRACTILE ELEMENTS AND THEIR ROLE IN LENS CELL DIFFERENTIATION

3.2.1 Introduction

In Chapters 2 and 4 some aspects of the lenticular cytoskeleton have been dealt with. Here we describe in somewhat more detail the structure and function of the cytoskeleton during lens cell differentiation.

Three types of cytoskeletal and contractile structures have been shown to occur in eukaryotic nonmuscle cells (110–116). They can be classified by their filament diameter and by immunological and biochemical properties:

1 Microfilaments (5 nm), containing the two cytoplasmic forms of actin (β- and γ-actin), which assemble into bundles frequently called stress fibers. The latter structures may be associated with myosin, α-actinin, and tropomyosin.

2 Microtubules (25 nm), which consist of α- and β-tubulin.

3 Intermediate-sized filaments or tonofilaments (7–11 nm). Although these filaments have for a rather long time been assumed to represent a single class of cytoskeletal fibers (see refs. 117–130), recent investigations by Franke and co-workers (131) have shown, mainly on the basis of immunological and biochemical data, that intermediate filaments can be divided into at least five subgroups (132–135). The subdivision could be made after the application of specific antisera in immunofluorescence studies. The following subgroups were observed:

   a Tonofilamentlike filaments characteristic of epithelial cells and containing the prekeratin (cytokeratin) polypeptides ranging in molecular weight between 46 and 68 kilodalton.

   b Intermediate-sized filaments characteristic of cells of the mesenchymal type. These filaments can be detected by using antibodies directed against a polypeptide with an approximate molecular weight of 57,000 for which the name vimentin or decamin has been proposed. They also occur in cultured cells (136).

   c Intermediate-sized filaments characteristic of muscle tissue which can be indentified with sera directed against desmin (skeletin).

   d Neurofilaments occurring only in cells of neural origin.

   e Glial filaments, which are specific to astroglia in the central nervous system (137).

Possible functions of these cytoskeletal and contractile structures in eukaryotic nonmuscle cells have been the object of many studies. It is now generally accepted that microfilaments are involved in the realization of cell shape and cell dynamics. Processes such as cell contraction and elongation, exocytosis, membrane ruffling, and movement on a substrate find their molecular basis in the interaction of continuously aggregating and disaggregating microfilament bundles with the plasma membrane (138). Lazarides et al. (139, 140) showed, by using antibodies directed against α-actinin isolated from Z-lines of muscle cells, that this protein occurs in the microfilament adhesion sites of nonmuscle cells. It was concluded that any part of the cell membrane can function as an attachment site for actin filaments (140).

Although microfilament bundles (stress fibers) appear to be highly organized structures, they can disaggregate rapidly into an amorphous meshwork, mainly observed in rounded cells that do not adhere strongly to a substratum. This meshwork is commonly located just below the cell membrane and is also seen in areas of membrane ruffling in motile cells. The two states of microfilament aggregates should be considered as being in equilibrium and interconvertible. Bundles of microfilaments give the cell structural support, and the microfilaments building up the meshwork are responsible, at least partly, for cellular motility processes. Interaction of actin with the protein profilin has been suggested to be responsible for the regulation of micro-filament polymerization (141).

As a rule microtubules exist as singlets in the cytoplasm and play a role both in cell division and in the determination of cell shape. They are involved in motility processes that are based on either microtubule assembly or tubular sliding mechanisms. It has been proposed (142) that microtubules are mainly involved in cell elongation, whereas microfilaments are responsible for cytoplasmic contraction. The growth of axons, for example, depends on the elongation of microtubules, since the disassembly of these filaments by mitotic drugs blocks axonal extension and promotes axonal retraction *in vitro* (143–145). However, Dustin (115) recently concluded that motion results from an interaction between actin and myosin and that microtubule poisons modify cell motility indirectly, by destroying the microtubules that maintain the general structure and polarity of the cell.

The function of microfilaments and/or microtubules in the process of lens cell differentiation, which becomes manifest by the elongation phenomenon, has been the subject of study and speculation. In the following sections these studies are reviewed and we provide some evidence suggesting a possible role of microfilaments in lens cell differentiation, both *in vivo* and *in vitro*.

3.2.2 Cytoskeletal and Contractile Structures in Lens Cells

Since in Chapter 4 electron microscopic aspects of the lens cytoskeleton will be dealt with exhaustively we restrict ourselves here mainly to the immunofluorescence and biochemical characterization of these structures in lens cells.

3.2.2.1 Introduction

The presence and biochemical nature of lenticular microfilaments have been documented by several investigators (7, 54, 81, 146–152). The criteria used are those of morphological comparison with other tissues (147), immunological methods (81, 146), chemical parameters (149), and the heavy meromyosin reaction (150). Microfilaments in lens fiber cells are covered with globular particles identified as α-crystallin aggregates and ribosomes (23; see also Chapter 4). An attempt has been made to correlate distribution of actin filaments in lenses of various animals to their accommodative capability (147, 153). Further functions ascribed to these microfilaments in lens concern epithelial wound healing (154), maintenance of cell structure (76), and cell elongation during differentiation (54). Moreover, Mousa et al. have found that disorganization of actin filaments by cytochalasin D gives rise to a cortical cataract (155). Myosin has been shown to occur in ocular nonmuscle tissues by Drenckhahn and Gröschel-Stewart (146). Microtubules have been shown to occur mainly in the lens epithelial cells and in the newly formed cortical fibers (49, 81, 150, 151, 156). Piatigorsky and coworkers (49) reported a function in lens differentiation for microtubules (see Section 3.2.4). According to Farnsworth et al. (156a) microtubules are the only surviving structural elements in mature human lens fibers.

The nature of the lenticular intermediate-sized filaments has until now been a subject of speculation and confusion (149–152, 157–161). For chick lens Maisel et al. (157) show a decrease in the abundance of the filaments in going from epithelium to the nucleus, where no intermediate-sized filaments can be observed. The protein building these filaments occurs in the urea-soluble fraction of the lens (152; see also refs. 162–164) and was described to have a molecular weight slightly higher than those of both tubulin and desmin (165). In the following sections we show that the protein subunit of the intermediate-sized filaments in both bovine and murine lens cells can be identified with vimentin immunologically and biochemically. Intermediate-sized filaments in many other tissues of mesenchymal origin and in cultured cells have been shown to contain vimentin as the major constituent (140, 166).

3.2.2.2 The Cytoskeleton of the Lens

Actin. Since microfilaments can be obtained by co-isolation with highly purified lens plasma membranes (see Chapter 4) one may ask how the actin-containing structures are localized *in situ*. To answer this question, the indirect immunofluorescence technique (IFT) (167) was applied to unfixed cryostat sections of lens tissue, obtained from various species (Figure 4). A specific fluorescence at the membranes was preferentially observed when the sections were incubated with an antiserum directed against purified calf

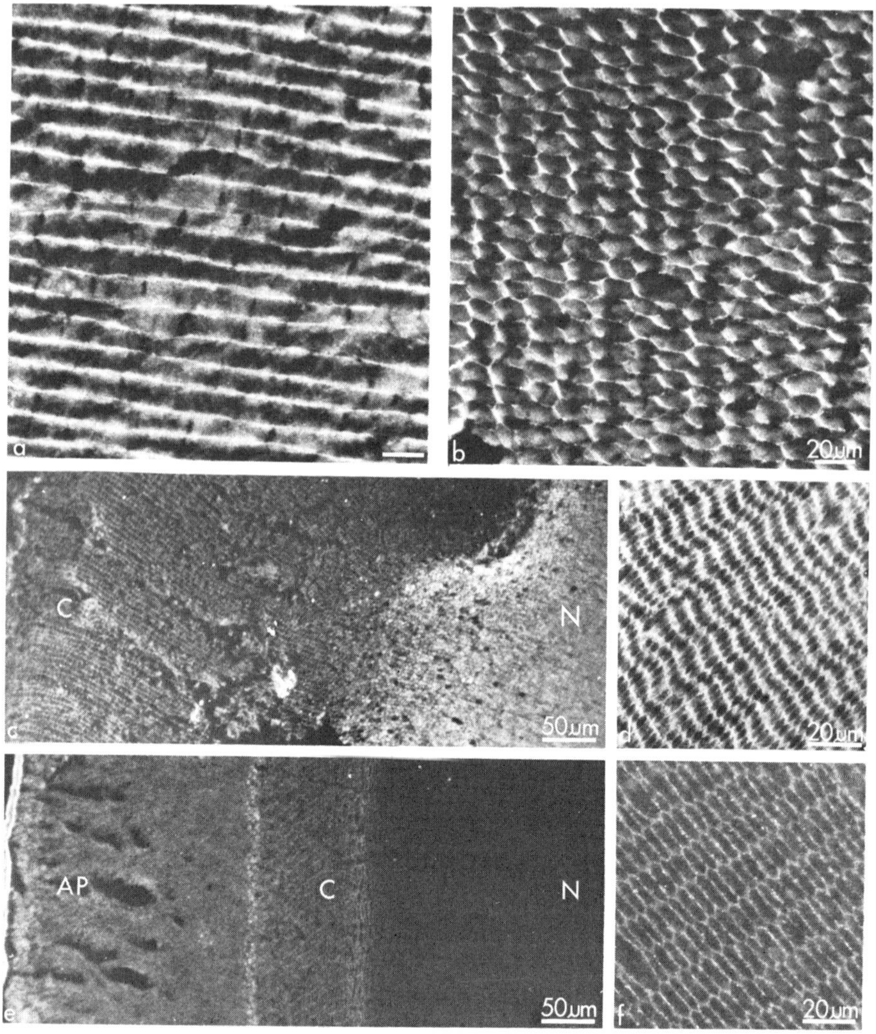

Figure 4 Immunofluorescence micrographs of freeze sections from lenses of three species after reaction with antibodies to bovine lens actin. (*a, b*) Bovine lens; (*c*) overview of the rat lens (C, cortex; N, nucleus); (*d*) detail of the rat cortex; (*e*) overview of the pigeon lens (AP, annular pad; C, cortex; N, nucleus); (*f*) detail of the pigeon cortex. Courtesy of Drs. M. Kibbelaar, L. Poels, and P. Jap, University of Nijmegen.

lens actin (see Figure 4*a*, *b*) (149). An interrelationship seems to exist between the accommodative power of the lens and its actin distribution (cf. ref. 147). In the rat lens, which is assumed to lack accommodative power, fluorescence is rather intensive as compared to the fluorescence of lenses of the other species studied. In rat the highest concentration of actin is observed in the rigid lens body (nucleus) (Figure 4*c*), where a strong but diffuse fluorescence can be seen in the cytoplasm, whereas at the lens periphery the weak fluorescence is restricted to the membrane region (Figure 4*d*). A species with a high accommodative power is the pigeon. A pigeon accommodates by bulging forward the rather soft lens body through a ring of scleral ossicles. The peripheral layer (annular pad) of the pigeon lens is somewhat more rigid, a condition required to ensure proper accommodation. Immunofluorescence reveals a relatively low concentration of actin compared to the less accommodating calf and rat (see Figure 4*e*). The immunofluorescence is mainly localized in the peripheral region and highest at the membranes (Figure 4*f*).

Intermediate-Sized Filaments. Immunofluorescence microscopy on frozen sections of calf and mouse lenses using antibodies directed against vimentin shows a strong reaction in the epithelioid cell layer underneath the capsule and, to a significantly lesser degree, in lens fiber cells (Figure 5*a–c*). Surprisingly, lens cells are in contrast to cornea epithelium stained neither with antibodies directed against prekeratin (Figure 5*d*, *e*), nor with antibodies against desmin. The latter antibodies show, in some regions, only weak spotty fluorescence along the inner side of the capsule.

Electron microscopy of ultrathin sections through tissue derived from calf and mouse lens reveals intermediate-sized filaments in the cytoplasm of the outermost cell layer, commonly designated lens epithelium (Figure 6). In these cells the 7- to 11-nm-thick intermediate-sized filaments represent the most abundant cytoplasmic structure, exceeding even membranes of endoplasmic reticulum and polyribosomes (Figures 6*b*, *c*). These filaments often appear to be arranged in bundles, which show a minimal interspace of about 8 nm (Figure 6*c*). The bundle arrays are frequently arranged at certain angles to give rise in sections to the "herringbone" patterns decribed by Rafferty and Esson (168), which these authors interpreted as 5-nm microfilaments. In centripetal direction of the lens the density of intermediate-sized filaments in the cytoplasm of the fiber cells gradually decreases, whereas in the innermost parts of the lenses practically no intermediate-sized filaments can be observed. Neither in the "epithelial" cell layer nor in the fiber cell can desmosomes, tonofibrils or other intermediate-sized tonofilaments of the prekeratin type be observed.

3.2.2.3 Biosynthesis of Lenticular Cytoskeletal and Contractile Proteins

The biosynthesis of the lenticular cytoskeletal and contractile constituents occurs mainly on "free" polyribosomes that can easily be isolated from the cortical fibers (169). When these polyribosomes are translated in a heterologous cell-free system (e.g., a reticulocyte lysate), they mainly direct the synthesis of the α-, β-, and γ-crystallin polypeptides (see also Chapter 5). However, next to these products the cytoskeletal and intrinsic or extrinsic

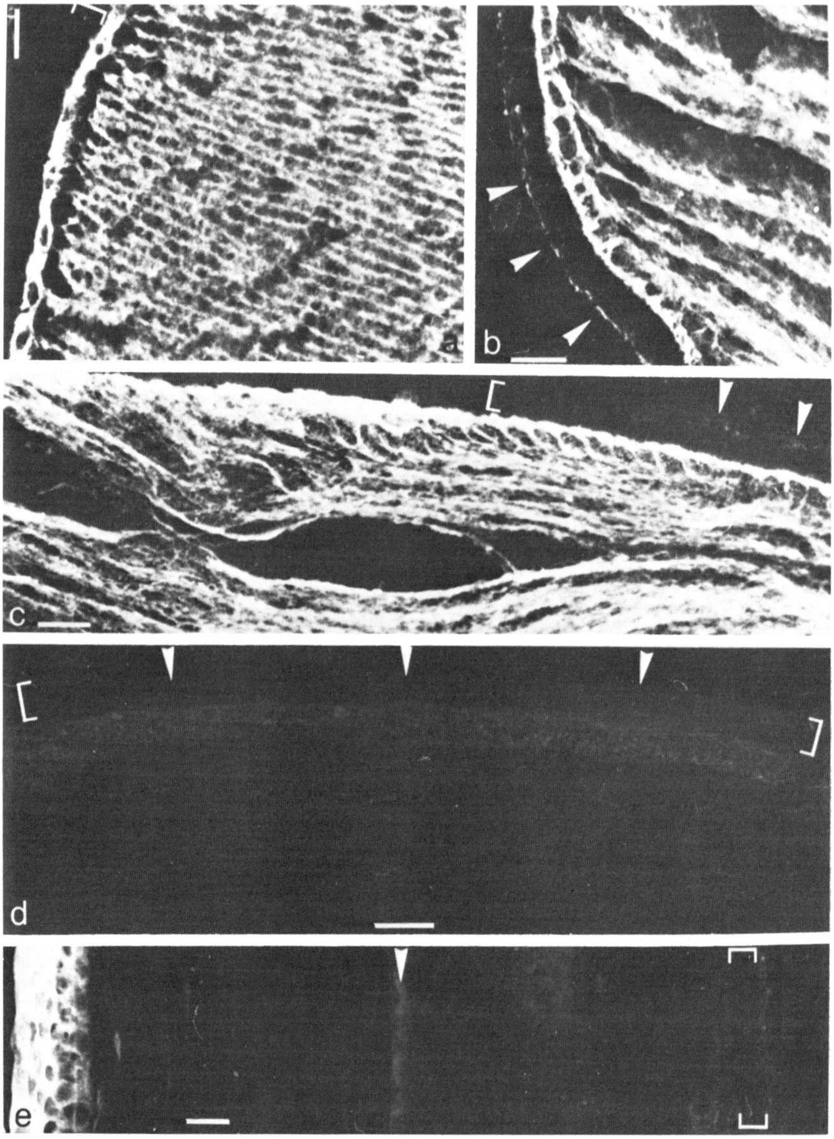

Figure 5 Immunofluorescence microscopy on frozen sections of murine lenses and eyes, respectively, using antibodies to vimentin (*a–c*) and prekeratin (*d, e*). Note intense positive reaction of outer lens cell layer and lens fibers with antibodies to vimentin (*a–c*, different orientations of section; *c* is from the equatorial region of the lens where the epithelioid cells elongate and are transformed into lens fiber cells) but not with prekeratin antibodies (*d, e*). Arrowheads and brackets denote contours and position of the lens capsule. Note in (*e*) the positive staining of the cornea (at the outer left margin) with prekeratin antibodies, and in contrast the absence of staining by prekeratin antibodies of cells of corneal stroma, corneal "endothelium" (denoted by arrowhead in *c*), iris, lens capsule (bracket), and lens tissue (at the right margin). Bars denote 30 μm.

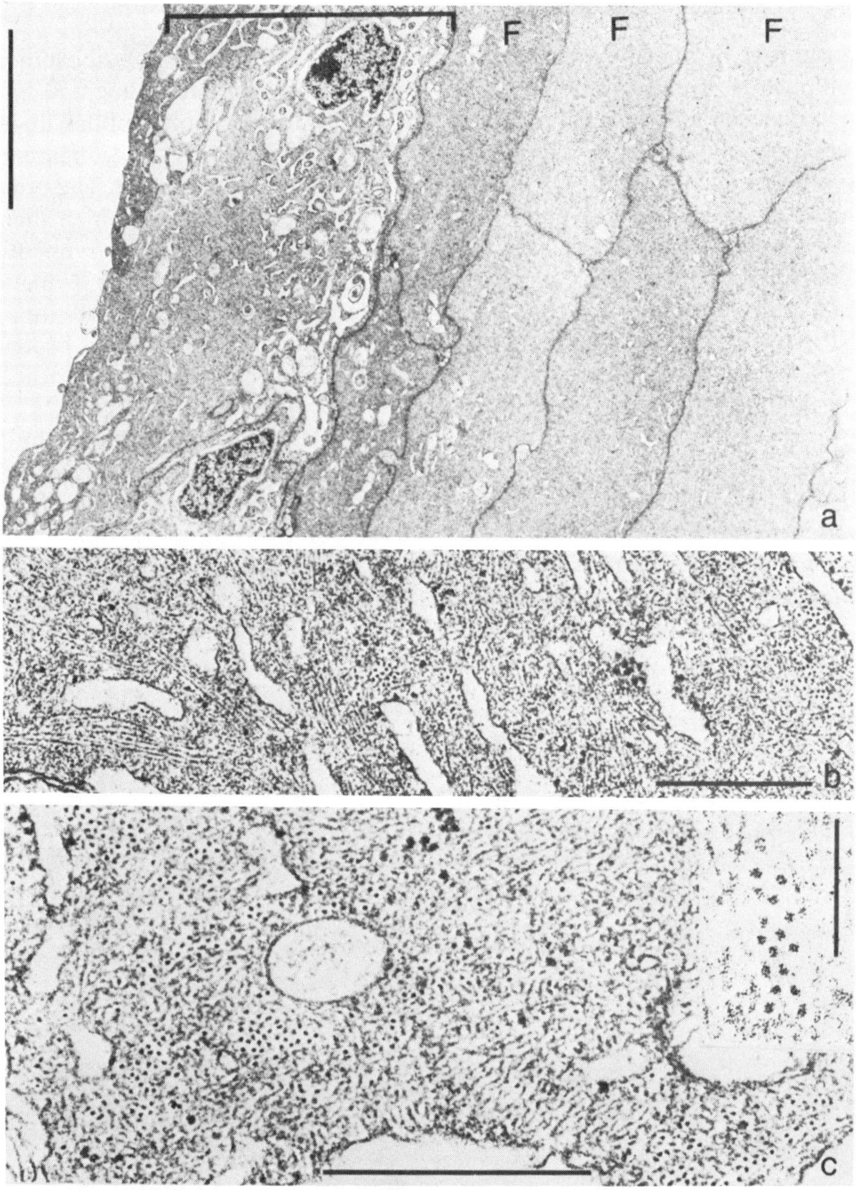

Figure 6 Electron micrographs of sections through the outer cell layers of lenses of bovine eyes, including fiber cells (F) as well as the epithelioid outermost cell layer, characterized by relatively rich endoplasmic reticulum and nuclei (bracket in *a*). At higher magnification of such sections (*b, c*) the very high density of intermediate-sized filaments is apparent. These filaments tend to form relatively loosely packed bundles running in diverse directions (*b, c*). Cross-sections through such bundles often reveal their unstained, apparently hollow core (insert in *c*). Bars denote 5 μm (*a*), 0.5 μm (*b, c*), and 0.1 μm (insert in *c*).

membrane proteins (except MP26) are synthesized by these polyribosomes, although to a much lesser extent. Figure 7 shows an autoradiograph of a two-dimensional separation (170) of the newly formed polypeptides under direction of calf lens polyribosomes. The main cytoskeletal component synthesized is vimentin, followed by actin, α-actinin, and tubulin. The order seems to be in good agreement with the relative amounts in which these constituents occur in the native fiber cells. Like β- and γ-actin, vimentin also exhibits two isoelectric variants as described earlier for Chinese hamster ovary cells and chicken nonmuscle cells (171, 172). In addition, a hitherto unidentified 47,000 d component is consistently found. This protein presumably represents a cytoskeletal constituent, as has also been suggested for the erythrocyte system (173).

In order to get an idea of the size of the mRNAs directing the synthesis of the cytoskeletal and contractile polypeptides, polyribosomal RNA was selected on the basis of its poly A content. The poly A-containing mRNA was separated by linear sucrose gradient centrifugation, fractions were collected and incubated in a reticulocyte cell-free system, and the products were analyzed by one-dimensional SDS gel electrophoresis. Figure 8 shows that an mRNA peak in the 16 to 18 S area codes for the synthesis of the cytoskeletal and contractile proteins actin, tubulin, and vimentin. Also

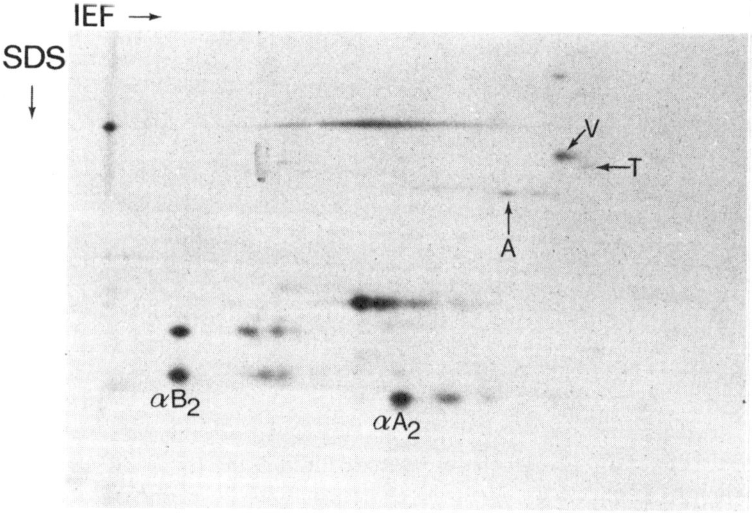

Figure 7 Autoradiograph of a two-dimensional gel electrophoretic pattern of newly synthesized proteins in a reticulocyte cell-free system under the direction of lens fiber polyribosomes (A, actin; V, vimentin; T, tubulin; αA_2 and αB_2, subunits of α-crystallin).

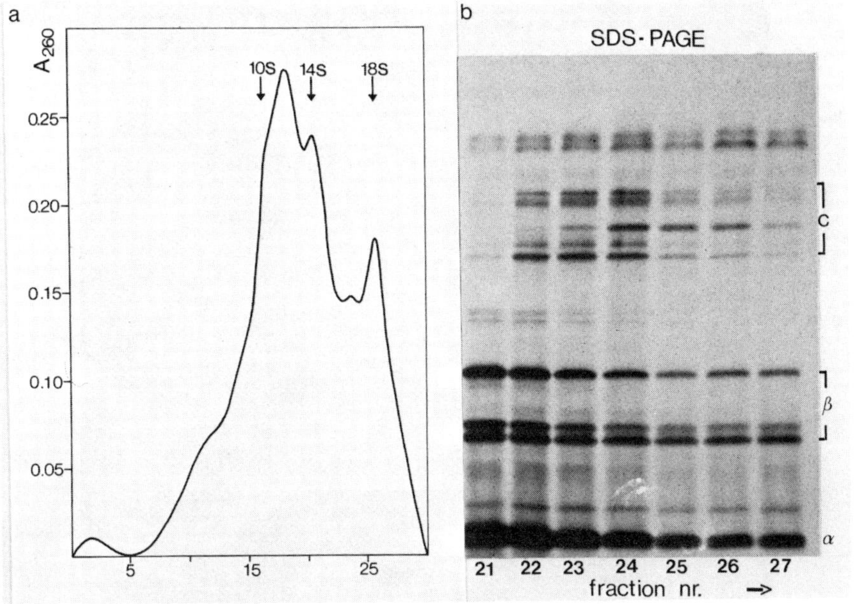

Figure 8 (*a*) Sucrose gradient centrifugation profile of poly A-containing RNA from bovine lens fibers. (*b*) SDS–polyacrylamide gel electrophoretic pattern of newly synthesized products of some fraction from (*a*) when added to a reticulocyte cell-free system, (α, α-crystallins, β, β-crystallins, c, cytoskeletal proteins). Courtesy of H. Dodemont, University of Nijmegen.

some of the unidentified proteins of the plasma membrane–cytoskeleton complex are synthesized by mRNAs from this region of the sucrose gradient.

3.2.3 The Cytoskeleton in Lens Cell Cultures

3.2.3.1 Localization of the Cytoskeletal and Contractile Structures in Cultured Bovine Lens Cells

Cultures derived from the outer monolayer of lens cells were described in Section 3.1.4. Under culture conditions routinely used in our laboratory the cells growing in monolayers maintain a polygonal epithelioid shape, at least up to passage 20. Thereafter, an increasing number of cells tends to elongate whereas the division rate gradually decreases. Usually cell cultures do not grow for more than 40 to 45 passages, and soon after culturing the synthesis of crystallins is no longer detected (45, 174). Cell cultures,

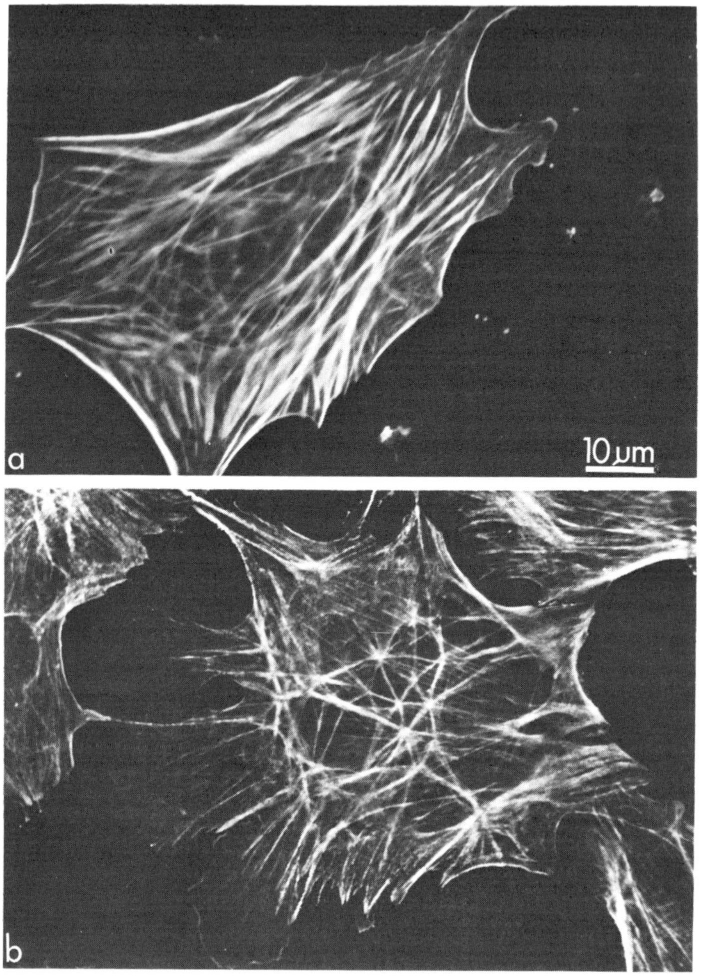

Figure 9 Cultured bovine eye lens cells stained with antibodies against actin (for explanation see text).

however, continually produce procollagen type I (77) and seem to secrete collagen type I, apparently in addition to type IV collagen (cf. ref. 64). Production of type I procollagen and collagen in these cells is remarkable since this type of collagen is not formed in cells of intact lenses as shown by immunofluorescence microscopy of lenticular cells found in adult mammals (unpublished) and in chick embryos in later stages of development (175). The cells also produce and secrete a protein of similar molecular weight and preparative behavior as fibronectin.

A considerable part of the cellular actin can be localized by immunofluorescence microscopy using antibodies against actin in extended fiberlike structures (Figure 9). These findings are in agreement with observations by Lonchampt et al. (81). In cells that were transferred up to 20 times, a rather random pattern of microfilament bundles is seen (Figure 9*a*). In cells at higher passage number, which are more elongated, parallel arrays of stress fibers are more frequent and often run along the whole length of the cell. Occasionally polygonal nets (''geodesic domes'') are observed (Figure 9*b*), which are similar to those described in other cell types grown in culture (176, 177). On electron micrographs of such cells, these network arrays appear to consist of focally centered microfilament bundles similar to the ''starlike arrangements'' described in such cultures (81) and other epithelioid cell cultures (see e.g., ref. 177). A correlation between the different microfilament organizations and lens cell elongation is presented in Section 3.2.5.2.

Antibodies directed against myosin (Figure 10), α-actinin (Figure 11), and tropomyosin (not shown) also decorate the actin-containing fibrils, indicating that these proteins form part of the microfilament bundles, including typical stress fibers. Myosin and especially α-actinin often appear periodically associated with the microfilament bundles. Upon closer examination of the α-actinin fluorescence also some patched fluorescence near the cell membrane can be seen (Figure 11*b*).

Antibodies directed against bovine brain tubulin give a positive staining

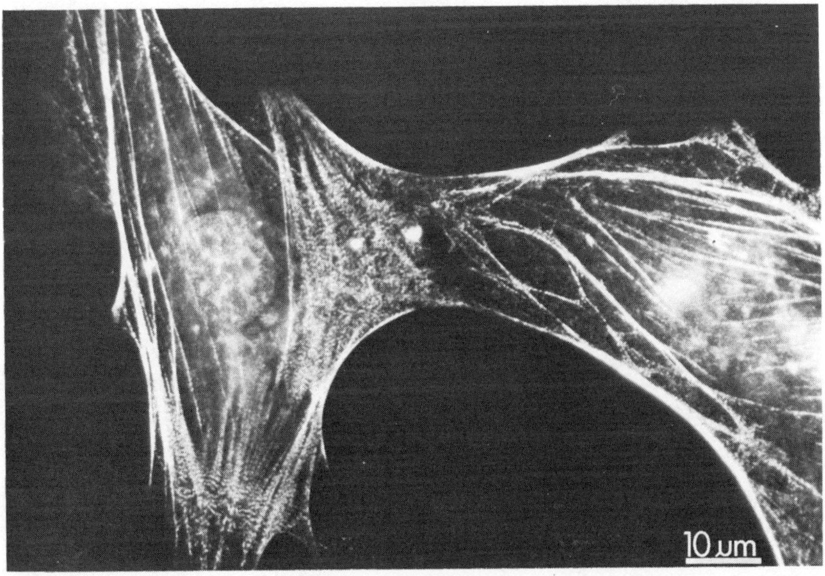

Figure 10 Cultured bovine lens cells stained with antibodies against myosin.

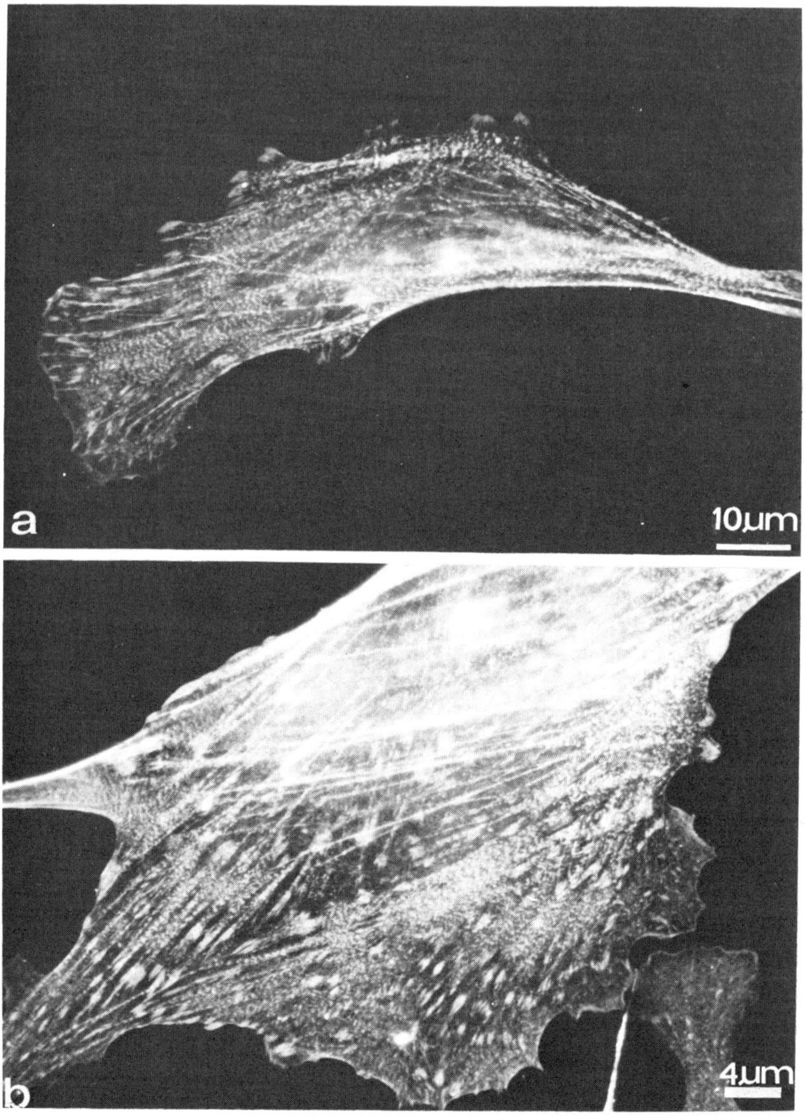

Figure 11 Cultured bovine lens cells stained with antibodies against α-actinin (for explanation see text).

in cultured bovine cells. The fluorescence micrographs (Figure 12) display microtubules similar to those normally seen with this technique in other cell types in culture (e.g., 112, 178, 179) but appear different from the thick longitudinally oriented, almost stress fiberlike arrays reported to be stained by tubulin antibodies in cultured bovine lens cells (81). Fibril staining by the tubulin antibodies disappears after treatment of the cells with colcemid.

In order to establish the type of intermediate-sized filaments present in lens cells antibodies against vimentin, desmin, and prekeratin were raised. Of the antibodies tested only that directed against vimentin shows positive fibril staining (Figure 13*a*).

Strong vimentin fluorescence localized in extended fiber arrays is seen in all cells. Upon treatment with colcemid the vimentin filaments aggregate into one large juxtanuclear, strongly stained mass (Figure 13*b*). The cells are not significantly stained by antibodies against desmin or by antibodies directed against prekeratin (77, 109). The same result is obtained with lens cells at early and late passages (from passage 1 to passage 39) as well as with cells still adhering to the lens capsule, which also show strong staining with antibodies directed against bovine lens α-crystallin. It seems, therefore, that intermediate-sized filaments in lens cells are exclusively of the vimentin type and that these cells do not contain filaments of the cytokeratin type, although cytokeratin-type intermediate filaments are, as a rule, observed in cells of epithelial origin.

3.2.3.2 Electron Microscopy of Lens-Forming Cells Grown in Culture

Some ultrastructural aspects of bovine lens cells grown *in vitro* have been reported earlier (e.g., see ref. 81). Survey micrographs of sections parallel or slightly oblique to the substratum illustrate the abundance of the different filamentous elements (compare Figure 14):

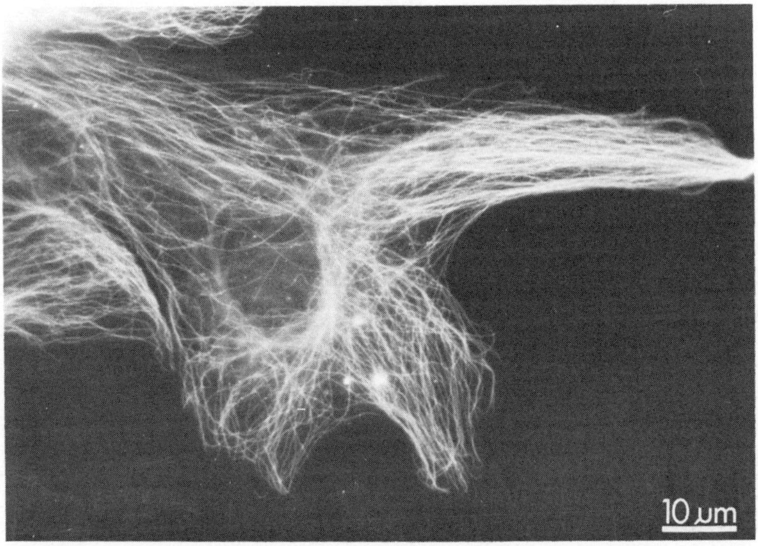

Figure 12 A cultured bovine lens cell stained with antibodies against tubulin.

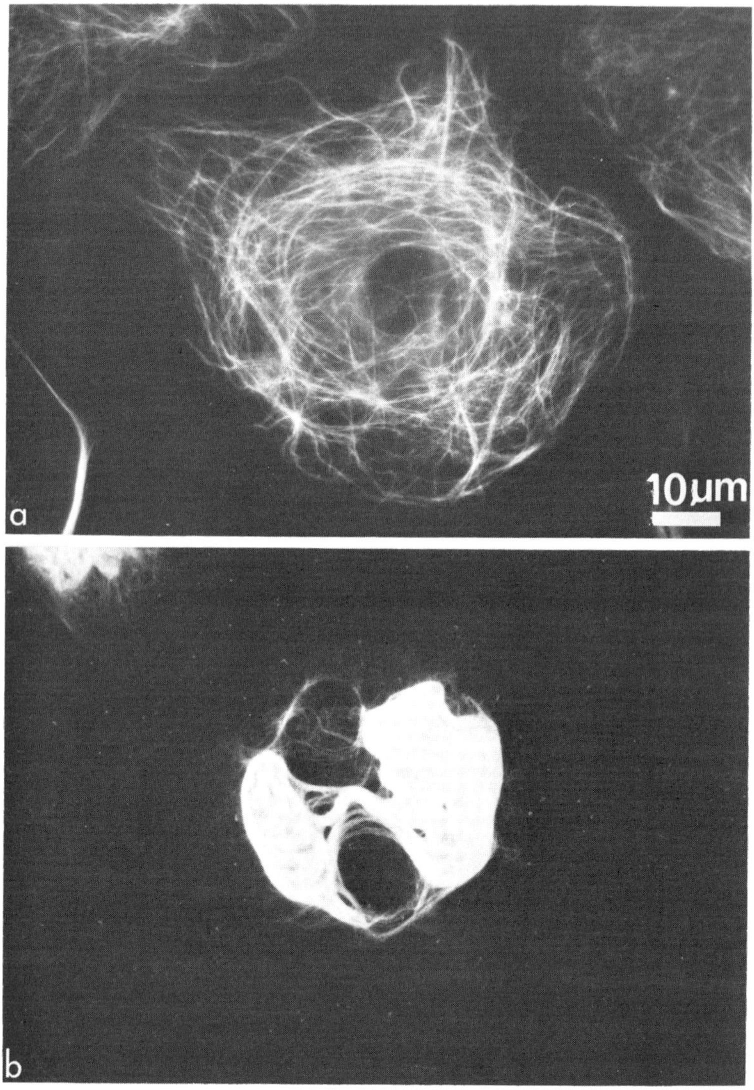

Figure 13 Cultured bovine lens cells, (*a*) normally grown and (*b*) colchicine-treated, visualized by immunofluorescence microscopy using antibodies to vimentin.

(1) microfilaments, mostly arranged in massive cablelike bundles; (2) microtubules, in general running as individual tubules in various directions but displaying some focal and radial orientation in the pericentriolar region; (3) intermediate-sized (7–11 nm) filaments occurring as individual filaments running in all directions as well as in relatively loosely packed bundles (Figure 14*a*).

In certain regions of the cytoplasm the intermediate-sized filaments of the vimentin-type are the most common structures (Figure 14*a*). In the cells parallel alignments of individual microtubules and intermediate-sized filaments seem to be maintained by short lateral linker structures 2 to 5 nm wide and ranging from 10 to 30 nm long (Figure 14*b–d*; cf. refs. 127, 180, 181). A number of bundles of intermediate-sized filaments, with and without included microtubules, are arranged almost perpendicularly to the microfilament bundles (Figure 14*a*). Prolonged treatment of the cells with antimicrotubular drugs such as colcemid results in the disappearance of microtubules, concomitantly with the disappearance of fibrillar staining by tubulin antibodies, and a simultaneous formation of vimentin coils around the nucleus.

The epithelioid lens-forming cells grown *in vitro* as monolayer sheets are connected by gap junctions and a special type of junction that resembles the *fascia adhaerens* type of junctions (for review see ref. 182). This particular form of junction, which has been described in various cells in culture (183), predominates and occasionally occupies more than half of the total cell-to-cell boundary encountered (Figure 15*a*). Intermediate-sized filaments occur in close proximity to the intercellular boundary but appear to be arranged randomly and are never organized in tonofibrillike bundles attached to the junctions (Figure 15*b*). On the other hand, the prevailing junctions (the *fascia adhaerens*-like ones) represent terminal anchorage sites of microfilament bundles, whereas a densely stained web of fine interfilamentous material appears at the membrane anchoring site (Figure 15*b*).

3.2.3.3 Analysis of Cytoskeletal Proteins from Cultured Lens Cells

Whole cells and cytoskeleton fractions are characterized by the predominance of actin and vimentin (Figure 16). Cytoskeletal preparations obtained by extraction of cultured bovine lens cells in buffers containing high salt concentrations and Triton X-100 consist primarily of intermediate-sized filaments and some tangles of microfilamentous structures, similar to the appearance of such preparations derived from a variety of other cultured cells (e.g., 131, 184–186) and to the intermediate-sized filaments described in the "water-insoluble fraction" of lens homogenates (23, 158, 160, 161, 187). Analysis of such preparations by SDS–polyacrylamide gel electrophoresis (Figure 16*a*) shows in addition to some residual actin the predominance of a polypeptide with an apparent molecular weight of 57,000 that comigrates with murine and human vimentin. Moreover, some minor high molecular weight polypeptide bands with mobilities similar to those of fibronectin and collagen I are sometimes noticed. Polypeptides co-migrating with avian and mammalian desmin and with any of the known polypeptides of authentic epidermal and tonofilamentous prekeratin are not seen in the

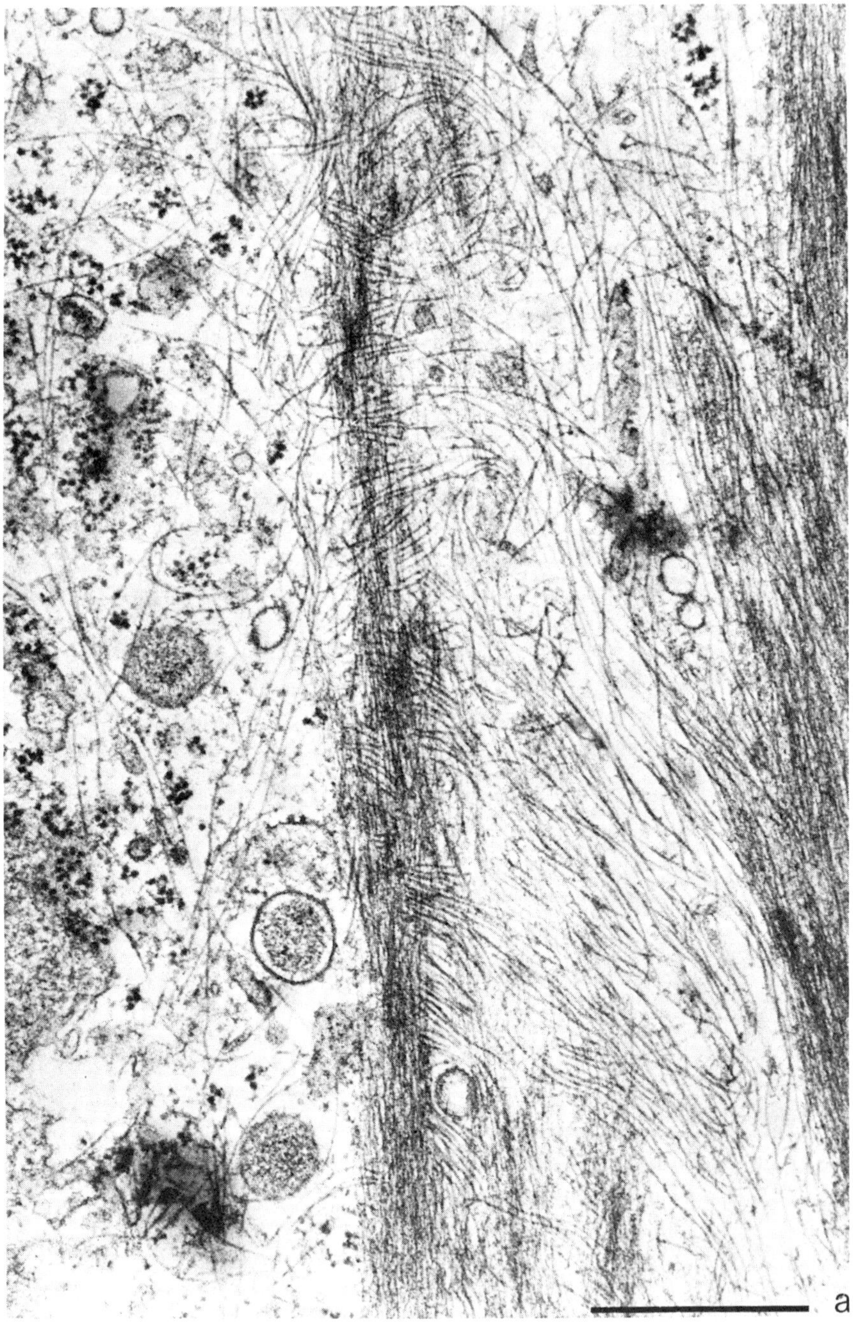
a

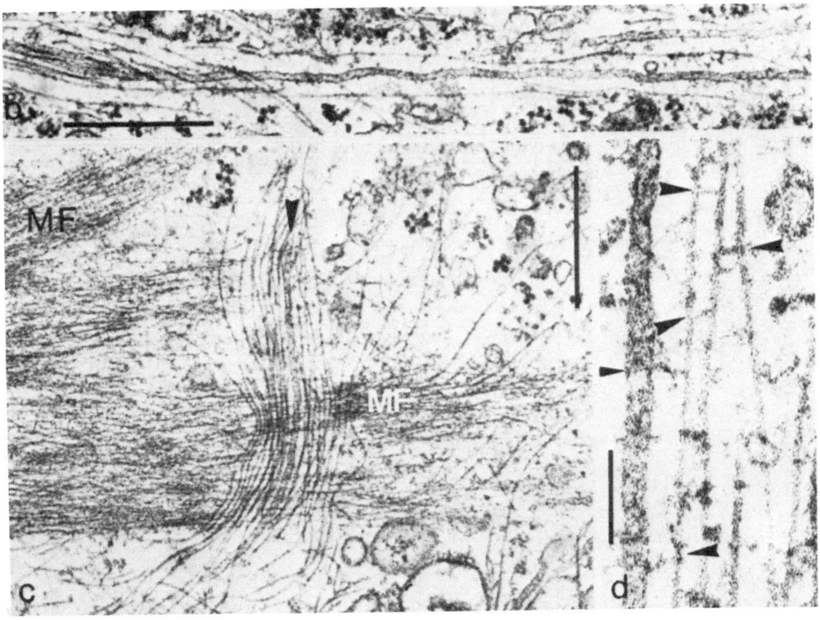

Figure 14 Electron microscopic details of organization of intermediate-sized filaments of the vimentin type in cultured bovine lens cells, showing (*a*) the high density of such filaments and their apparent irregular display, (*b*) extended parallel alignments with individual microtubules, (*c*) groups of parallel microtubules (arrowhead) and intermediate filaments running perpendicularly to microfilament (MF) bundles, and (*d*) lateral bridges between individual vimentin filaments and between microtubules and vimentin filaments (some are denoted by arrowheads). Bars denote 0.5 μm (*a–c*) and 0.1 μm (*d*).

gel electrophoresis pattern of the proteins of cytoskeletons of cultured bovine lens cells. Upon examination by two-dimensional gel electrophoresis patterns the only major protein identified as an intermediate-sized filament component again is vimentin (Figure 16*b*), which has an apparent isoelectric pH of 5.5. Furthermore, a minor, slightly more acidic isoelectric variant is observed (cf. ref. 171, 172). Actin, exclusively of the β and γ type, is the second most frequently occurring protein.

3.2.4 Microtubules and Lens Cell Differentiation

Much of the evidence that microtubules play a role in cell shape changes and other cellular motility processes has come from the use of antimitotic drugs, such as colchicine or vinblastine, that disrupt the microtubule structure. Although it is likely that microtubules may have important

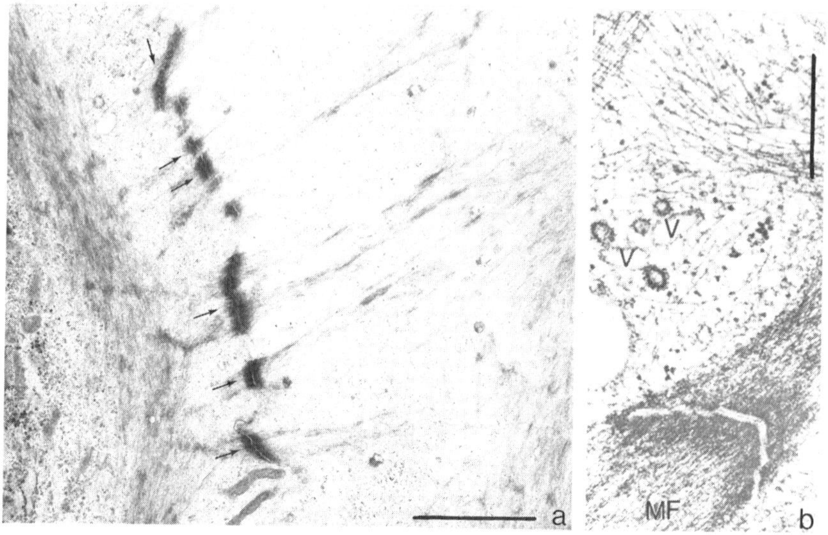

Figure 15 (*a*) Survey electron micrograph showing the most frequent type of intercellular junction found in cultured bovine lens cells (arrows). Bar denotes 5 μm. Courtesy G. Mungyer and P. Jap, University of Nijmegen. (*b*) Higher magnification. Note that intermediate-sized filaments occur in the cell periphery, in close vicinity to the junctions, but apparently are not attached to them. In contrast, bundles of filaments of microfilamentlike diameters (MF) are terminally attached to junctions of this type. V, Smooth-surfaced vesicles. Bar denotes 0.5 μm.

functions in these processes, the observed effects of the drugs should be interpreted with great care. It should be kept in mind that in some cases the applied treatments affect not only the microtubules but also other cellular structures (49, 115).

The elongation of lens cells has been correlated by several authors with the appearance of longitudinally oriented microtubules. Byers and Porter (188) were the first investigators to report this phenomenon in the embryonic chick lens rudiment. They suggested a causal relationship between microtubule formation and lens cell elongation (188). Thereafter a number of papers dealing with the same topic appeared in the literature (144, 189–205). In particular, the group of Piatigorsky (for a review see ref. 49) has intensively studied the role of microtubules in the differentiation of chick lens epithelium. The tissue used by these investigators for the study of elongation consisted of square pieces of anterior chick embryo lens epithelium still attached to the capsule (33). The cells, which are initially cubically shaped, elongate under defined growth medium conditions such as the addition of fetal calf serum or insulin (206, 207). The ultrastructural changes that occur in these elongating epithelia have been shown to resemble the changes that take place during fibrogenesis in mammals (156, 208–211), chicks (43, 188, 212), amphibians (213–215), and squids (216).

Microtubules appear randomly arranged in the cubical cells. In contrast, elongating cells exhibit numerous longitudinally oriented microtubules, mainly located along surface membranes. This longitudinal arrangement in columnar chick lens epithelial cells suggests a relationship between microtubules and lens cell elongation. Additional evidence for a role of microtubules in lens cell differentiation was obtained from experiments using either colchicine or vinblastine. Elongation was inhibited by these drugs, but cells already elongated retained their shape after incubation with colchicine. Pearce and Zwaan (198) showed that in chick lens placode cells also the elongated shape is maintained after destruction of the microtubules. These results suggest an involvement of microtubules in the process of elongation rather than in the maintenance of the shape of closely opposed cells. However, recent findings by Beebe et al. (217), who used nocodazole to disrupt microtubules, provide strong evidence that elongation of embryonic chick lens epithelial cells can occur in the absence of microtubules. Lens cells treated with the latter drug completely lacked cytoplasmic microtu-

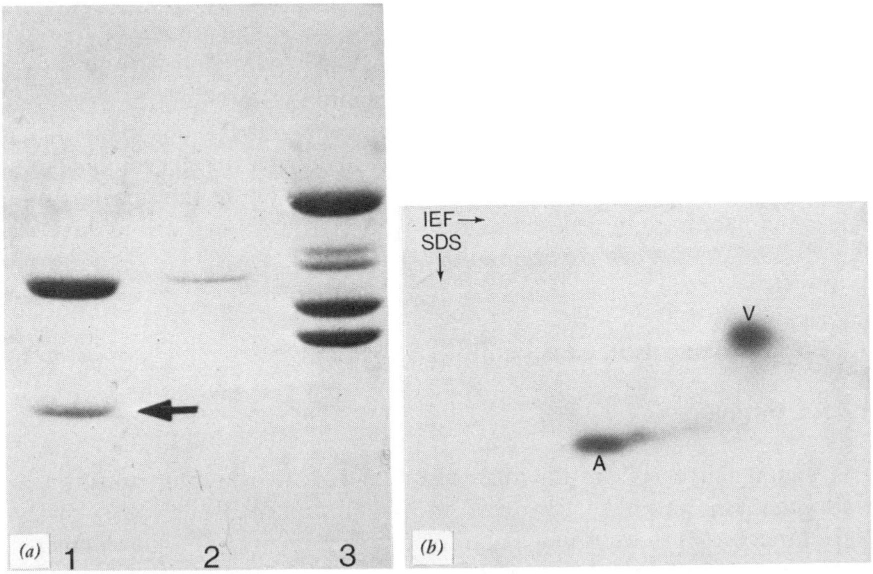

Figure 16 Gel electrophoretic analyses of cytoskeletal proteins. (*a*) Comparison of a cytoskeleton preparation from bovine lens cell culture (lane 1) with purified murine vimentin from 3T3 cells (lane 2) and prekeratin from bovine snout epidermis (lane 3) (arrow indicates actin). (*b*) Two-dimensional gel pattern of cytoskeletal proteins from high salt- and dertergent-extracted cultured bovine lens cells (V, vimentin; A, actin).

bules whereas the typical aggregates of intermediate-sized filaments were similar to those described in Section 3.2.3.1. Also in other cells treated with anti-microtubule agents (218) these structures were seen. Despite these effects nocodazole did not prevent lens cell elongation induced by fetal calf serum, although it blocked mitosis. Furthermore, Beebe et al. (217) concluded from experiments in which extremely low concentrations of colchicine (10^{-10}–10^{-8} M) were used that an increase in cell volume is responsible for lens cell elongation. The concentrations mentioned still (partially) inhibit elongation but do not dissociate microtubules. That an increase in cell volume might be responsible for lens cell elongation was proposed in 1973 by Zwaan and Hendrix (219). A localized crowding of cells as a result of continued cell division was thought to be the driving force for the differentiation process during lens cell embryogenesis. Piatigorsky et al. (40) showed that the number of microtubules in elongating lens cells was not impressive. Moreover, electron micrographs revealed that the more elongated cultured lens cells did not contain appreciably more microtubules than the shorter cells. These workers studied the biosynthesis of tubulin during the elongation process. It seemed that inhibition of both RNA and/or protein biosynthesis (220) did not prevent the initial rapid phase of lens cell elongation induced by fetal calf serum. Furthermore, it was found that the rate of tubulin synthesis during the elongation process was not increased. It was concluded that, since the total cumulative length of the microtubules per cell appeared to be conserved, a reorganization of microtubules was responsible for lens cell elongation. However, the relatively low amount of microtubules in chick lens cells and the observation that colchicine inhibits elongation not by direct action on the microtubules but by its involvement in cell volume regulation (217) led to the consideration that other factors (e.g., microfilaments and cell junctions) are involved in the process of lens cell differentiation.

3.2.5 Microfilaments in Lens Cell Differentiation

3.2.5.1 Introduction

A major active role of microfilaments in lens invaginations during lens embryogenesis has been proposed by Wrenn and Wessells (221). Zwaan and Hendrix (219) ascribed a more passive function of microfilaments during invagination, namely, anchoring the cell junctions. To our knowledge the first indication that microfilaments are involved in lens cell elongation was published by Piatigorsky et al. in 1972 (40). The authors described some preliminary tests showing that next to colchicine, cytochalasin B, which interferes with the contractile function of microfilaments (222, 223), also inhibits cell elongation in cultured chick lens epithelium. On the basis of this and former experiments they proposed a possible cooperation of microtubules and microfilaments in lens cell elongation. However, the role of microfilaments has not been investigated as intensively as the role of microtubules. Lonchampt et al. (81) demonstrated the presence of microfilaments in bovine lens cells by using the indirect immunofluorescence technique. These structures have been shown to be attached to lens plasma membranes if isolated under proper conditions (Chapter 4).

Mousa and Trevithick (54) were able to show an involvement of microfilaments in the *in vitro* differentiation process of rat lens cells (see also ref. 155). They found a diffuse, nonfibrillar form of actin in round undifferentiated cells which changed to the appearance of longitudinal microfilaments on differentiation. Furthermore, the effect of cytochalasins B and D was different on both kinds of cells. The microfilament-disrupting drugs had no effect on undifferentiated cells, whereas differentiated cells attained an arborized shape. Similar effects were described for the action of cytochalasin B on bovine lens cells by Ramaekers et al. (76) (see also Section 3.2.5.4). Furthermore these two drugs were shown to inhibit morphological differentiation (elongation) and the production of γ-crystallin, a presumed marker of terminal cellular differentiation in rat lenses (53).

A study on changes in lenticular actin distribution during differentiation in rat lens has been published by Mousa and Trevithick (148). Using one- and two-dimensional gel electrophoretic techniques the authors showed that the amount of actin in rat lens decreases from about 11% of the total amount of protein in lens epithelial cells to about 1% in the cortical fiber cells. These authors also claim that the lens nucleus lacks actin, but contains a high amount of plasma membranes. One may question whether this nuclear membrane fraction contains some actin, not detectable by the electrophoretic experiments of Mousa and Trevithick (148), but which can unequivocally be visualized by immunofluorescence (see Section 3.2.2). This discrepancy presumably can be explained further by the fact that in the experiments of Mousa and Trevithick 17% of the nuclear fiber proteins were not dissolved in the lysis buffer. Moreover, the ratio between the two species of cytoplasmic actin, β- and γ-actin, changes in going from epithelium to the nucleus. In the epithelium the ratio β-actin/γ-actin was 2.2 but in the mature cortical lens fibers this ratio had decreased to 1.2. On the basis of these results the investigators conclude that the relatively high level of actin in the lens epithelial cells is probably required for a role in the maintenance of cell shape, growth, migration, and differentiation. The low amount of actin in the cortical fibers might be required for lens movements during accommodation; the structural support of these cells is thought to result from the very high amounts of crystallins in these cells.

In the next sections evidence is provided for a role of microfilaments in the *in vivo* and *in vitro* elongation (differentiation) process in bovine lens cells. The experimental approach is based on the idea that the accumulation of microfilaments as a result of assembly of globular actin (G-actin) into filamentous actin (F-actin) is a major factor of the morphological features of lens cell differentiation.

3.2.5.2 Elongation of Bovine Lens Cells *in Vitro* and the Reorganization of Microfilament Bundle Distribution

Bovine epithelial lens cells brought in culture retain their cubical shape at least during the first 15 passages. Occasionally some elongated cells are observed. After about 20 transfers the cells start to change their appearance spontaneously (52), resulting in elongated cells. Morphologically this phenomenon has its counterpart in the *in vivo* differentiation. Elongation of cultured bovine lens cells can also be induced by dexamethasone (46) or by an extract of the bovine retina (63). In the phase-contrast microscope these elongated cells exhibit pronounced intracellular filamentous organizations running parallel to the direction of elongation and crossing the whole length of the cell. These filaments can be visualized by the indirect immunofluorescence technique using either an auto-immuneserum from rabbit (76) or highly specific antibodies directed against actin (77). In this way these filaments are identified as stress fibers.

When young cubical cells are incubated with specific antibodies directed against actin a rather random distribution of the microfilaments (microfilament bundles) is observed (see Figure 17*a*). In elongated cells, in cultures of higher passage number (30–40 transfers) strong arrays of parallel stress fibers occur, often running across the whole length of the cell in the direction of elongation (see Figure 17*b*). Also young cells that have been incubated with retinal extracts as described by Arruti and Courtois (63) elongate within 48 hours. Here again a drastic reorganization of stress fibers in the direction of elongation is observed (compare Figures 17*c, d*). Furthermore a relative increase in the abundance of microfilament bundles is seen on the fluorescence micrographs. An optimal cellular density is needed to induce the morphological change, which may lead to the conclusion that cell-to-cell contact is necessary. To show that the immunofluorescence observed is specifically due to actin-containing fibers, cells were treated with cytochalasin B. Figure 17*f* shows that after incubation of the cells with this microfilament-disrupting drug, only a diffuse actin fluorescence is observed. From these results it is concluded that the morphological differentiation of bovine lens cells in culture resulting in elongated cells is accompanied by a reorganization of microfilament bundles. In cells of intermediate age (as reflected by the number of passages), which do not yet show pronounced elongation, a structure consisting of regular polygonal actin networks is observed (Figure 17*e*). These structures, which have been described by Lazarides (140) in fibroblasts and designated "geodesic

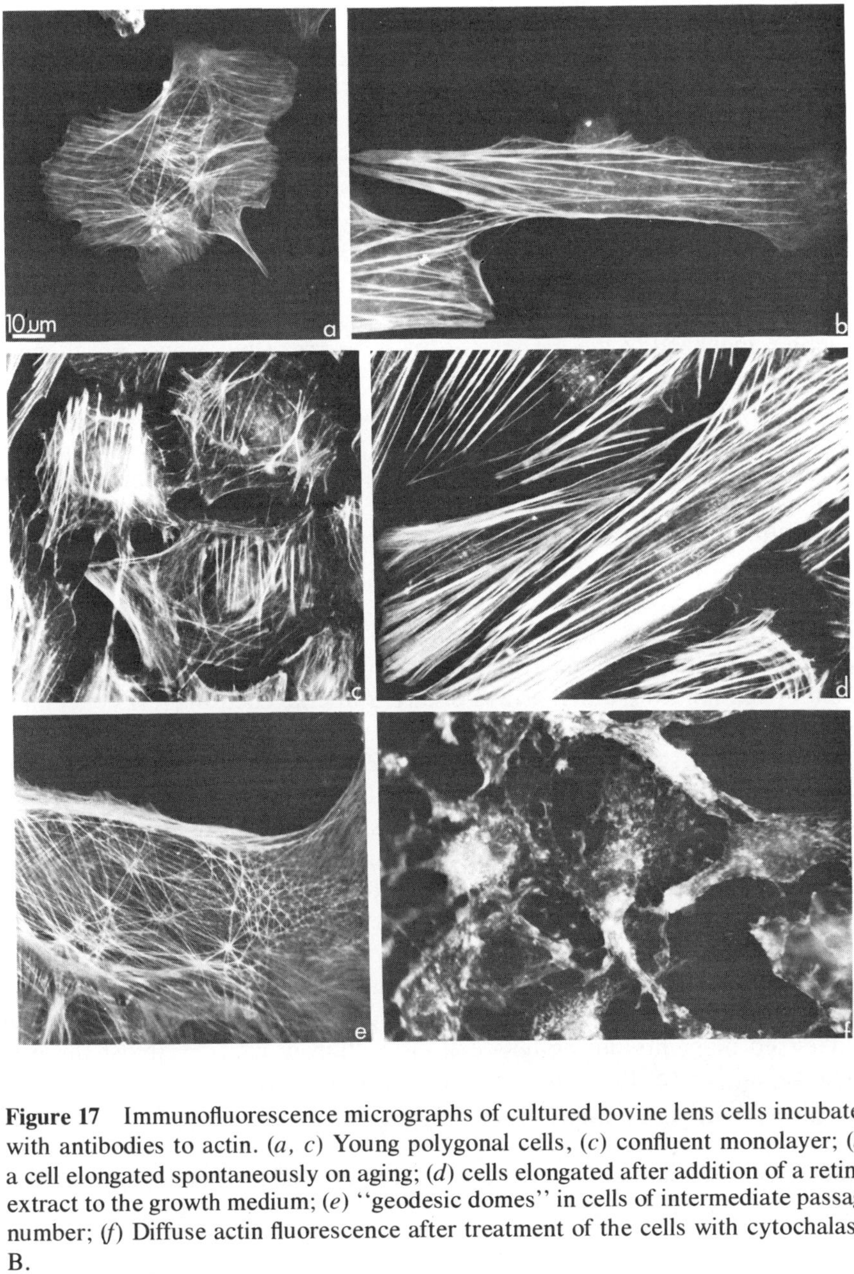

Figure 17 Immunofluorescence micrographs of cultured bovine lens cells incubated with antibodies to actin. (*a, c*) Young polygonal cells, (*c*) confluent monolayer; (*b*) a cell elongated spontaneously on aging; (*d*) cells elongated after addition of a retinal extract to the growth medium; (*e*) "geodesic domes" in cells of intermediate passage number; (*f*) Diffuse actin fluorescence after treatment of the cells with cytochalasin B.

Table 1 Distribution of "Geodesic Domes" in Elongating Cultured Bovine Lens Cells

Passage number	% Cells containing "geodesic domes"	Relative shape of "geodesic domes"
1	16	Medium
10	23	Large
18	5	Small
34	5	Very small

domes," encompass the entire area above and around the nucleus and have α-actinin localized at the vertices of the network. The vertices are thought to act as organization centers for actin filament bundles resulting in or correlated with the elongation and flattening of fibroblasts. This temporary actin organization in fibroblasts is an intermediate structure between the diffuse meshwork state in rounded cells (e.g., seen after trypsinization) and the linear actin bundles in the fully flattened cells.

In cultured bovine lens cells the networks are also present after flattening of the cells in established cells (Figure 17*e*) and should thus not be considered as a temporary structure in cells of intermediate passage number. Thus during subculturing these "geodesic dome"-like structures occur mainly in young cells that have not yet elongated and do not yet exhibit pronounced parallel arrays of stress fibers. Their highest concentration and greatest size is observed in cells that were subcultured 10 to 15 times. Further subculturing results in decreased amounts of polygonal networks and a decrease in their size (see Table 1). These results show that a transient organization of actin networks occurs in established flattened bovine lens cells in culture and represents an intermediate structure between the random distribution of actin fibers in polygonal cells and strongly organized parallel arrays of stress fibers in elongated cells.

Antibodies directed against tubulin and vimentin were also tested on both cubical and elongated lens cells in culture (unpublished). The results showed that real reorganization occurred neither for microtubules nor for intermediate-sized filaments during lens cell elongation *in vitro*. At most there is some adjustment of the latter structures to the changing cell shape.

3.2.5.3 Decreased Tubulin and Crystallin Synthesis In Elongating Bovine Lens Cells in Culture

Arruti and Courtois (63) reported that protein biosynthesis seems to be necessary for the elongation of cultured bovine lens cells under the direction of a retinal extract. Furthermore these authors claimed that protein biosynthesis is a requirement for the maintenance of the elongated shape. This was concluded from the reversal to the polygonal shape of previously elongated cells if treated with cycloheximide. This latter result is surprising, since Piatigorsky (49) showed that chick lens epithelial explants are independent of protein biosynthesis for their elongation induced by insulin or under certain growth medium conditions. When cubical and elongated bovine lens cells in monolayer cultures are compared with respect to their protein biosynthetic activity (see Figure 18), it is obvious that the synthesis of both crystallin and tubulin is undetectable in the elongated cells. This phenomenon is observed after only 2 to 3 passages, long before the cells start to elongate. However, immunofluorescence shows the presence of microtubules also in elongated cells, which means that a low level of tubulin synthesis must occur. From these results we conclude that neither tubulin nor crystallin synthesis is a prerequisite for bovine lens cell elongation *in vitro*. This does not exclude a role of microtubules in this process, since the assembly of these structures may be independent of tubulin synthesis but may merely represent an equilibrium between a soluble and a cytoskeletal form of tubulin. If the observation made by Arruti and Courtois (63) that elongation depends on protein synthesis holds true for the calf lens system also, one has to conclude that microfilaments rather than microtubules play a major role in lenticular cell elongation *in vitro*.

3.2.5.4 Studies using Cytochalasin B and Colchicine

Mousa and Trevithick (54) have reported a differential effect of cytochalasin B and D on cubical and elongated rat lens cells in culture. Young cells that have not yet elongated are less sensitive to treatment with these drugs than are the elongated ones, which readily arborize. The cytochalasins inhibit elongation in cultures from rat lens epithelial cells (54), calf lens cultures (63), and chick lens epithelium explants (40).

The effect of cytochalasin B on calf lens epithelial cells in culture is shown in Figure 19. In this figure three stages of *in vitro* elongating lens cells are depicted, which all exhibit the typcial arborized shape (224) when incubated with the drug (10 μg/ml growth medium). However, the effect of cytochalasin B seems to be more drastic in older cultures with elongated cells than in the younger ones containing the cubical cells. Surprisingly the differential effect of cytochalasin B on calf lens cells becomes more pronounced when the drug is withdrawn after several hours of incubation. The young cells that were formerly of polygonal shape regain their original form within 3 hours whereas the older, originally elongated cells do not restore their shape (even after maintenance for 24 hours in normal serum). Cells of intermediate age show only a partial recovery after 24 hours. These results provide evidence for an involvement of microfilaments in the structure of lens cells in culture and suggest an important role of actin-containing stress fibers in elongated cells *in vitro*. Furthermore, there seems

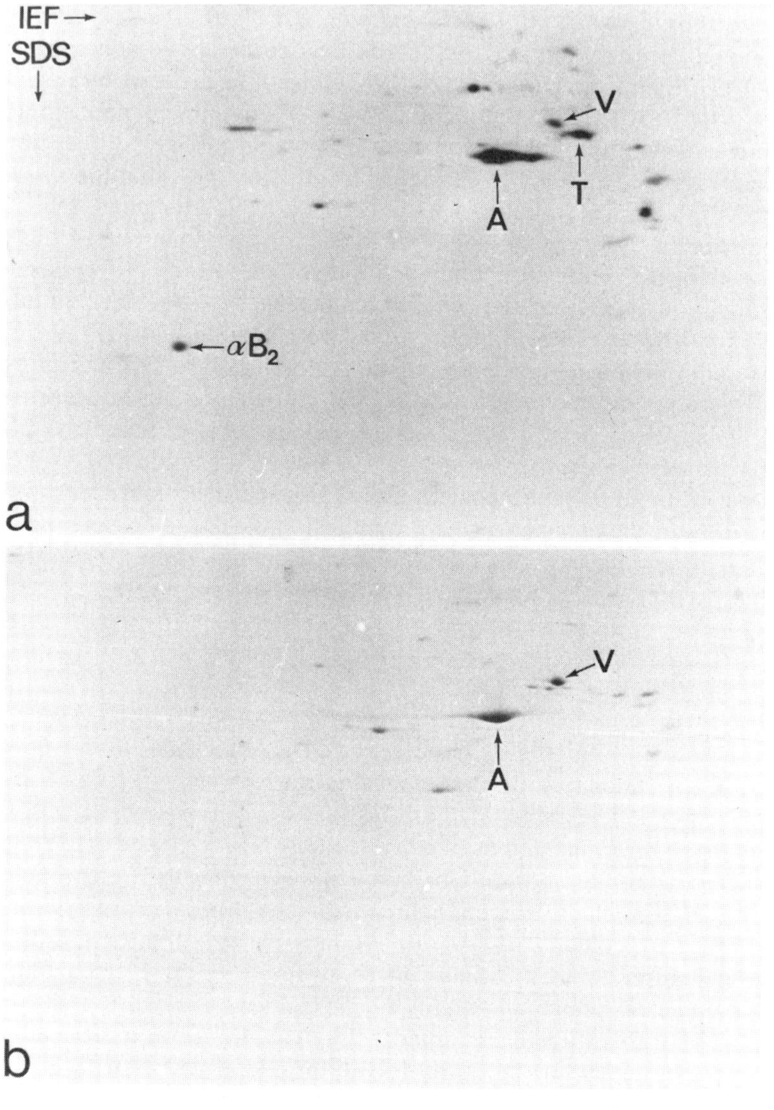

Figure 18 Autoradiographs of two-dimensional gel electrophoretic patterns of newly synthesized polypeptides in (*a*) primary bovine lens cell culture and (*b*) older elongated cells (A, actin; T, tubulin; V, vimentin).

to be a differential irreversible effect of cytochalasin B on the microfilaments in senescent, elongated cells, which structures may occur less frequently in young cells or alternatively may not be functional in the maintenance of their cell shape.

The effects of cytochalasin B and colchicine on the process of bovine lens cell elongation *in vitro* under the direction of a retinal extract were also examined. Since Arruti and Courtois (63) showed an inhibitory effect of colchicine (6×10^{-5} to 6×10^{-9} M) on the elongation of adult bovine epithelial lens cells and van Venrooij et al. (46) demonstrated that cytochalasin B (5 μg/ml medium) does not prevent elongation of calf lens cells under the direction of dexamethasone, one would expect microtubules to

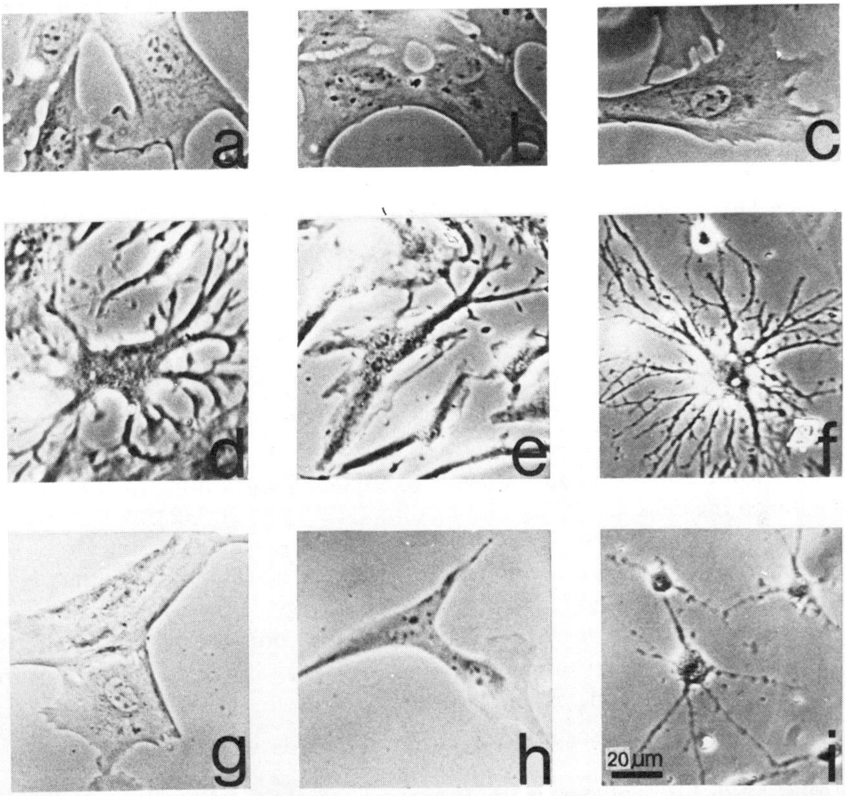

Figure 19 Treatment of cultured bovine epithlial lens cells with cytochalasin B (CB). (*a–c*) Untreated cells that have been subcultured 2, 10, and 30 times, respectively; (*d–f*) the same cells treated with CB; (*g–i*) same cells 24 hours after fresh medium, depleted of CB, has been added.

Table 2 The Effect of Colchicine and Cytochalasin B on the Elongation of Bovine Lens Cells in Culture under the Direction of Retinal Extracts (63)

Retinal extract (μl/5 ml medium)	Colchicine (m*M*)	Elongation	Cell shape changes
—	—	—	—
—	6.25×10^{-9}	—	—
—	6.25×10^{-7}	—	—
200	—	+++	Elongation
200	6.25×10^{-9}	++	Elongation
200	6.25×10^{-7}	++	Elongation
200	6.25×10^{-6}	—	Arborization
200	6.25×10^{-4}	—	Arborization
	Cytochalasin B (m*M*)		
—	—	—	—
—	2×10^{-9}	—	—
—	2×10^{-6}	—	—
300	—	+++	Elongation
300	2×10^{-7}	++	Elongation
300	2×10^{-6}	++	Elongation
300	2×10^{-5}	—	Arborization
300	2×10^{-4}	—	Arborization
300	2×10^{-3}	—	Arborization

have a role in the process of lens cell elongation *in vitro* as also proposed for chick lens. Careful re-examination, however, has shown that concentrations of both cytochalasin B and colchicine that change the shape of normal nonelongating cells also inhibit elongation (see Table 2). On the other hand, as soon as the drug concentrations are so low that they do not affect the cell shape, they also do not prevent elongation. This shows that both microtubules and microfilaments may be involved in elongation and/or maintenance of cell shape. Therefore, taking into consideration all the available data, we want to propose, in agreement with Dustin (115), a more static role for microtubules and a rather active role for microfilaments.

3.2.5.5 Microfilament Assembly during *in Vitro* and *in Vivo* Lens Cell Elongation

As can be concluded from the foregoing sections microfilaments seem to play an important role in the elongation and/or maintenance of the cell shape of calf lens cells in monolayer culture. From the immunofluorescence

Table 3 Distribution of F- and G-Actin in Elongating Cultured Bovine Lens Cells

Passage no.	G-Actin (%)	F-Actin (%)
6	61	39
22	48	52
37	26	74

studies described in Section 3.2.5.2 it seems that elongation can be correlated with a relative increase in the amount of filamentous actin. In order to get an impression of the molecular basis of this process, the ratio between globular actin (G-actin) and filamentous actin (F-actin) was estimated in both young, cubical cells and elongated cells in culture. The method used for these estimations was first described by Blikstad et al. (225) and makes use of specific inhibition of DNAse I by G-actin. The amount of F-actin is calculated by subtracting the amount of G-actin (as estimated in the absence of guanidine hydrochloride) from the total amount of actin in a cell, estimated in the presence of guanidine hydrochloride. From the results summarized in Table 3 it becomes clear that there is a shift from G-actin to F-actin on lens cell elongation. The relative amount of total actin per cell, however, does not change very dramatically if compared to the total amount of protein per cell (results not shown).

One may ask how these results can be related to *in vivo* differentiation. As already mentioned, Mousa and Trevithick (148) showed, by means of isoelectric focusing, that in going from epithelium to nucleus the relative amount of actin decreases and the ratio between β- and γ-actin changes. However, no results were presented concerning the ratio between G- and F-actin in these parts of the lens.

Using the method of Blikstad et al. (225), we estimated the F/G-actin ratios in calf lens cells (3-month-old animals) from several parts of the lens. These include the germinal zone of the anterior epithelium, the epithelial cells in the elongation zone, the elongated fiber cells in the superficial cortical region, and older fiber cells in deeper layers surrounding the nucleus. Table 4 shows that again there is an increase in the relative amount of filamentous actin in going from epithelium to fibers. Especially, a drastic increase in F-actin from 22 to 40% is seen upon comparing epithelial cells before elongation and fiber cells in the elongation zone. From these results it can be concluded that microfilament assembly accompanies lens cell differentiation (elongation) and that this phenomenon represents a common feature of both *in vivo* and *in vitro* differentiation. Despite the loss of ability of calf lens cells in culture to synthesize crystallins, those cultures can serve as models for the study of microfilament assembly during the process of elongation.

Table 4 Distribution of Filamentous Actin in the Bovine Lens

Site	F-Actin (%)
Lens epithelium	13 ± 7
Pre-elongation and elongation zones	22 ± 7
Young fibers	40 ± 7
Old fibers	46 ± 5

3.2.5.6 A Role of β-Actin and Myosin in *in Vitro* Lens Cell Elongation

Closer examination of cytoskeletal preparations from young cubical calf lens epithelial cells and elongated cells reveals some interesting aspects concerning the molecular processes involved in microfilament bundle assembly during elongation *in vitro*. Newly synthesized proteins incorporated into the cytoskeletal fractions, isolated as described (77), were analyzed by two-dimensional gel electrophoresis (10) followed by autoradiography (see Figure 20). A clear change in the intensity of vimentin relative to actin can be observed going from polygonal to elongating cells. This observation is paralleled by the finding *in vivo* where a similar change was seen in electron micrographs of epithelial and fiber cells (see Section 3.2.2). Upon elongation incorporation of myosin synthesized *de novo* gradually increases. At the same time the incorporated β-actin, which is virtually absent in the cytoskeletal preparation from young cells, becomes more pronounced. In older preparations the hitherto unindentified 47-kilo dalton protein is found. Its intensity increases on aging of the cells. These findings again demonstrate the strong involvement of specific contractile proteins in the process of lens cell elongation. It seems as if particularly β-actin and myosin are responsible for the formation of the parallel arrays of stress fibers described in the immunofluorescence studies. The differential effect of the cytochalasins may therefore tentatively be ascribed to an action of these drugs on β-actin and myosin.

3.3 CONCLUDING REMARKS

One of the definitions that has been proposed for cell differentiation is related to the expression of specific gene products (226). For instance, hemoglobin for the red blood cell, γ-crystallin for lens fibers or α-actin and myosin for myotubules. However, together with the production of these

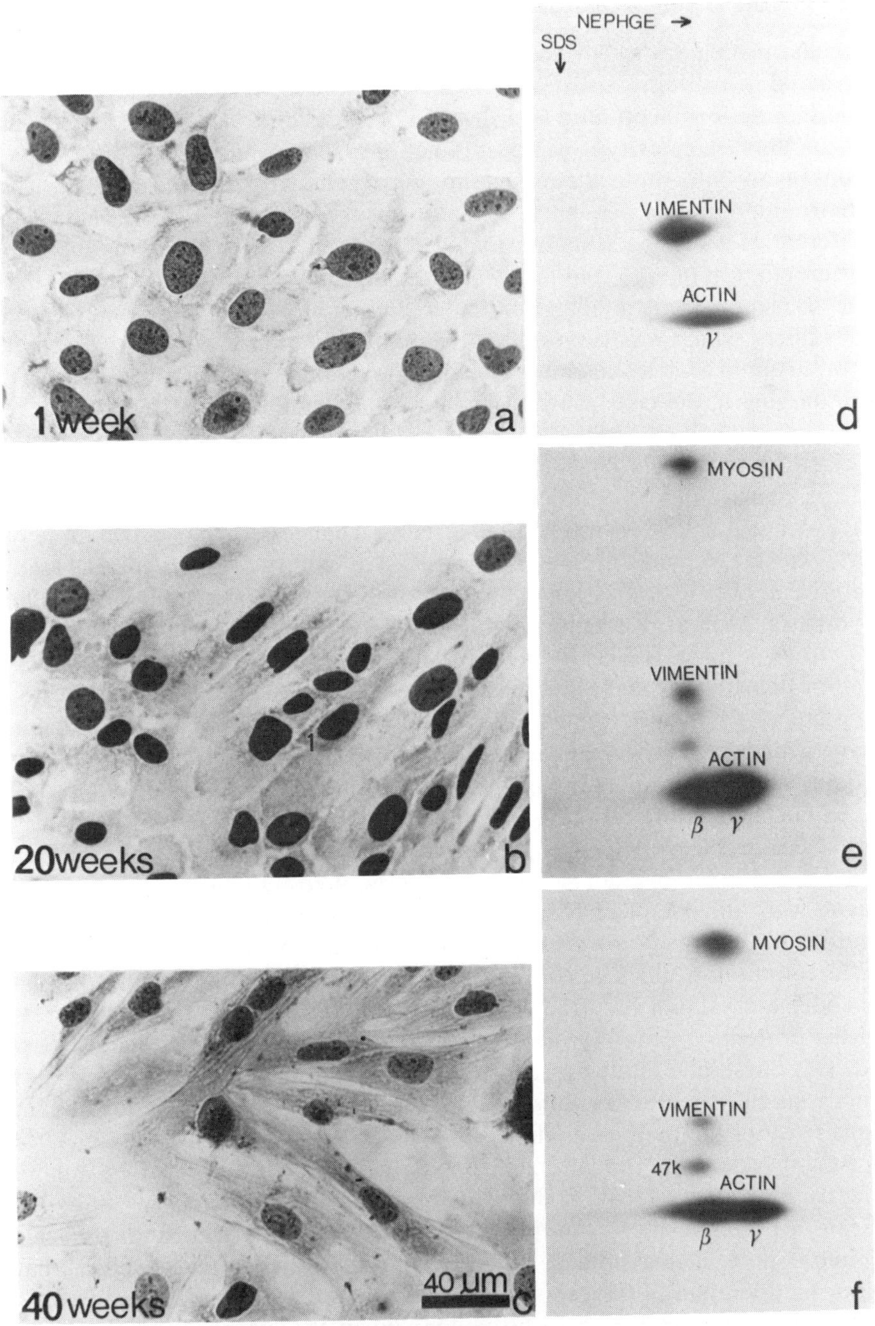

Figure 20 Incorporation of β-actin and myosin into stress fibers of elongating lens cells. (*a–c*) Bovine epithelial lens cells of different age; (*d–f*) autoradiographs of 2D gel electrophoretic patterns of the cytoskeletons derived from the cells shown in (*a–c*).

specific proteins, the biosynthesis of many other cell constituents may be involved in cell differentiation, in particular those that maintain cell shape and interactions according to a designed principle of tissue differentiation. Taking this complexity in mind, it is not easy to compare tissue differentiation in an integral cellular system with cells growing *in vitro*, where environmental and feedback mechanisms of regulation are completely different. This is the situation that we are facing when studying lens cell differentiation *in vivo* and *in vitro*. Actually, if we want to mimic *in vitro* the transition from dividing lenticular epithelial cells into the anucleated lens fibers which represent a good example of terminal differentiation, we have to admit that at least the major specific markers, namely the crystallins, are missing in the cells growing under artificial conditions. However, if we take into consideration that terminal differentiation *in vivo* is also characterized by a specific change of the cell shape leading to fiber elongation (227), then we may conclude that this parameter can be studied *in vitro*. In fact, we have demonstrated that striking changes in the distribution of cytoskeletal elements are detectable either in culture or in the intact organ. *In vivo* the formation of lens fibers is paralleled by considerable crystallin synthesis. But as we have mentioned, elongated epithelial lens cells if cultured on glass or plastic have lost the latter capacity, at least as far as calf or hamster lens cells are concerned. Attempts to switch on the arrested crystallin synthesis by the addition of various agents to the culture medium have generally failed. Courtois et al. (65), however, reported the synthesis of α-crystallin in tumors induced by transformed bovine lens cells.

At this moment it cannot be excluded that a certain amount of "silent" crystallin messenger RNA is present in the elongated calf or hamster lens cells in culture. This can rather easily be verified by hybridization experiments with the use of cDNA probes for the corresponding mRNA's. Such studies have already successfully been undertaken in examining terminal differentiation in chick lens (8). In that case accumulation of mRNA that encodes δ-crystallin occurred at the same time as the onset of the synthesis of the protein. Apparently messenger- and protein-synthesis are coordinated. Of course one is always confronted with the difficulty whether or not small quantities of crystallin mRNA's escape detection due to lack of sensitivity of the method applied.

At any rate, experiments carried out in our laboratory and by others clearly indicate that calf or hamster crystallin synthesis *per se* is by no means a prerequesite for morphological differentiation (elongation). In our opinion one of the advantages of studying differentiation in lens cells is that cloned populations of these cells are available. Although in some cases crystallin synthesis seems to be uncoupled in the established cultures, a valuable alternative for these biochemical markers of differentiation *in vivo* may be found in the polypeptides that constitute the cytoskeleton. Another important feature of fiber differentiation concerns the biosynthesis of a specific membrane protein MP26 (25) and the assembly of lens plasma membrane junctions (227). The question as to whether this protein synthesized on polyribosomes attached to the cytoskeleton is committed to a certain program of differentiation is one of the numerous questions that has urgently to be answered.

From a number of data presented in this chapter it is strongly suggested that the assembly of microfilaments is one of the main driving forces of lens cell elongation, albeit the underlying regulatory processes are not yet fully understood. In our opinion, in the studies of lens cell differentiation and terminal differentiation in general, the role of the cytoskeleton and its constituent proteins should not be underestimated. No doubt investigations on the gene organization and expression of the lenticular cytoskeleton will deepen our insight in regulatory mechanisms that control differentiation.

ACKNOWLEDGMENTS

The authors are grateful to Drs W. Franke, K. Weber, and M. Osborn for kind gifts of antisera against cytoskeletal proteins.

REFERENCES

1 J. Papaconstantinou, *Science,* **156,** 338 (1967).

2 J. Piatigorsky, *Dev. Biol.,* **40,** f 21 (1974).

3 R. M. Clayton, G. Eguchi, D. E. S. Truman, M. M. Perry, J. Jacob, and O. P. Flint, *J. Embryol. Exp. Morphol.,* **35,** 1 (1976).

4 R. M. Clayton, *Curr. Topics Dev. Biol.,* **5,** 115 (1970).

5 J. L. Coulombre and A. J. Coulombre, *Science,* **142,** 1489 (1963).

6 Y. Courtois and R. Regnault, (Eds.), *Biology of the Epithelial Lens Cells,* INSERM, Vol. 60, 1976.

7 E. L. Benedetti, I. Dunia, C. J. Bentzel, A. J. M. Vermorken, M. Kibbelaar, and H. Bloemendal, *Biochim. Biophys. Acta,* **457,** 353 (1976).

8 L. M. Milstone, P. Zelenka, and J. Piatigorsky, *Dev. Biol.,* **48,** 197 (1976).

9 C. Slingsby and L. R. Croft, *Exp. Eye Res.,* **17,** 369 (1973).

10 J. Delcour and J. Papaconstantinou, *J. Biol. Chem.,* **247,** 3289 (1972).

11 J. Delcour and J. Papaconstantinou, *Biochem. Biophys. Res. Commun.,* **31,** 257 (1974).

12 A. J. M. Vermorken, J. M. H. C. Hilderink, W. J. M. van de Ven, and H. Bloemendal, *J. Cell Biol.,* **76,** 175 (1978).

13 A. J. M. Vermorken and H. Bloemendal, *Nature,* **271,** 779 (1978).

14 H. Bloemendal and A. J. M. Vermorken, *Interdiscipl. Topics Gerontol.,* **12,** 41 (1978).

15 G. J. van Kamp, H. A. J. Struyker-Boudier, and H. J. Hoenders, *Comp. Biochem. Physiol.,* **49B,** 445 (1974).

16 G. J. van Kamp, F. S. M. van Kleef, and H. J. Hoenders, *Biochim. Biophys. Acta,* **342,** 89 (1974).

17 M. Rabaey, *Exp. Eye Res.,* **1,** 310 (1962).

18 D. E. S. Truman, A. G. Brown, and J. C. Campbell, *Exp. Eye Res.,* **13,** 58 (1972).

19 J. M. Génis-Gálvez, *Cell Diff.,* **4,** 49 (1975).

20 A. J. M. Vermorken, J. M. H. C. Hilderink, I. Dunia, E. L. Benedetti, and H. Bloemendal, *FEBS Lett.* **83,** 301 (1977).

21 H. Bloemendal, A. J. M. Vermorken, M. Kibbelaar, I. Dunia, and E. L. Benedetti, *Exp. Eye Res.,* **24,** 413 (1977).

22 R. M. Broekhuyse, E. D. Kuhlmann, and H. J. Winkens, *Exp. Eye Res.,* **29,** 303 (1979).

23 H. Bloemendal, M. A. Kibbelaar, F. C. S. Ramaekers, A. M. E. Selten-Versteegen, I. Dunia, and E. L. Benedetti, in: H. Peeters, Ed., *Prot. Biol. Fluids, Proc. 26th Colloq., Brussels,* Pergamon Press, Oxford, 1979, p. 499.

24 H. Bloemendal, E. L. Benedetti, F. C. S. Ramaekers, I. Dunia, M. A. Kibbelaar, and A. J. M. Vermorken, *Mol. Biol. Rep.,* **5,** 99 (1979).

25 F. C. S. Ramaekers, A. M. E. Selten-Versteegen, I. Dunia, E. L. Benedetti, and H. Bloemendal, *Proc. Natl. Acad. Sci. U.S.* **77,** 725 (1980).

26 E. Lazarides and U. Lindberg, *Proc. Natl. Acad. Sci. U.S.,* **71,** 4742 (1974).

27 D. B. Kirby, K. Estey, and F. Tabor, *Arch. Ophthalmol.,* **1,** 358 (1929).

28 I. Mann, *Brit. J. Ophthalmol.,* **30,** 591 (1948).

29 G. Mamo and P. J. Leinfelder, *Am. Med. Assoc. Arch. Ophthalmol.,* **57,** 417 (1958).

30 J. Van der Veen and C. F. A. Heyen, *Nature,* **183,** 1137 (1959).

31 A. Bryan, P. J. Leinfelder, and M. Meltzer, *Am. Med. Assoc. Arch. Ophthalmol.,* **63,** 108 (1960).

32 S. Tamura, *Jap. J. Ophthalmol.,* **9,** 177 (1965).

33 G. W. Philpott and A. J. Coulombre, *Exp. Cell Res.,* **38,** 635 (1965).

34 G. W. Philpott and A. J. Coulombre, *Exp. Cell Res.,* **52,** 140 (1968).

35 A. L. Shapiro, I. M. Seigel, M. D. Scharff, and E. Robbins, *Invest. Ophthalmol.,* **8,** 393 (1969).

36 G. W. Philpott, *Exp. Cell Res.,* **59,** 57 (1970).

37 J. R. Reddan, M. W. Crotty, and C. V. Harding, *Exp. Eye Res.,* **9,** 165 (1970).

38 C. V. Harding, J. R. Reddan, N. J. Unaker, and M. Bagchi, *Int. Rev. Cytol.,* **31,** 215 (1971).

39 T. S. Okada, G. Eguchi, and M. Takeichi, *Dev. Growth Differ.,* **13,** 323 (1971).

40 J. Piatigorsky, H. DeF. Webster, and M. Wollberg, *J. Cell Biol.,* **55,** 82 (1972).

41 J. Piatigorsky, S. S. Rothschild, and L. M. Milstone, *Dev. Biol.,* **34,** 334 (1973).

42 H. Watanabe, and I. Kawakami, *Dev. Growth Differ.,* **15,** 101 (1973).

43 T. S. Okada, G. Eguchi, and M. Takeichi, *Dev. Biol.,* **34,** 321 (1973).

44 M. O. Creighton, and J. R. Trevithick, *Nature,* **249,** 767 (1974).

45 W. J. van Venrooij, A. A. Groeneveld, H. Bloemendal, and E. L. Benedetti, *Exp. Eye Res.,* **18,** 517 (1974).

46 W. J. van Venrooij, A. A. Groeneveld, H. Bloemendal, and E. L. Benedetti, *Exp. Eye Res.,* **18,** 527 (1974).

47 M. Iwig and D. Glässer, *Acta Biol. Med. Ger.,* **34,** 987 (1975).

48 M. Iwig and D. Glässer, *Acta Biol. Med. Ger.,* **34,** 997 (1975).

49 J. Piatigorsky, *Ann. N.Y. Acad. Sci.,* **253,** 333 (1975).

50 M. D. Hollenberg, *Arch. Biochem. Biophys.,* **171,** 371 (1975).

51 G. Mungyer and P. H. K. Jap, *Verh. Anat. Ges.,* **65,** 469 (1971).

52 G. Mungyer and P. H. K. Jap, *Verh Anat. Ges.,* **69,** 673 (1975).

53 M. O. Creighton, G. Y. Mousa, and J. R. Trevithick, *Differentiation,* **6,** 155 (1976).

54 G. Y. Mousa and J. R. Trevithick, *Dev. Biol.,* **60,** 14 (1977).

55 J. Tassin, L. Simonneau, and Y. Courtois, *INSERM,* **60,** 145 (1976).

56 D. Glässer, M. Iwig, S. Ansorge, and C. Fischer, *Ophthal. Res.,* **8,** 283 (1976).

57 M. Iwig, G. Geyer, I. Marquardt, and D. Glässer, *Acta Biol Med. Ger.,* **35,** 413 (1976).

58 D. C. Beebe and J. Piatigorsky, *Exp. Eye Res.,* **22,** 277 (1976).

59 M. Iwig and D. Glässer, in O. Hockwin, Ed., *Progress of Lens Biochemistry Research,* Documenta Ophtalmologica Proceedings Series Vol. 8, Dr. W. Junk B. V. Publishers, The Hague, 1976, p. 47.

60 L. M. Milstone and J. Piatigorsky, *Exp. Cell Res.,* **105,** 9 (1977).

61 Y. Hamada and T. S. Okada, *Dev. Growth Differ.,* **19,** 265 (1977).

62 Y. Hamada and T. S. Okada, *Exp. Eye Res.,* **26,** 91 (1978).

63 C. Arruti and Y. Courtois, *Exp. Cell Res.,* **117,** 283 (1978).

64 M. Laurent, M-O. Lonchampt, F., Regnault, J. Tassin, and Y. Courtois, *Exp. Cell Res.,* **15,** 127 (1978).

65 Y. Courtois, L. Simmonneau, J. Tassin, M. V. Laurent, and E. Malaise, *Differentiation,* **10,** 23 (1978).

66 G. Mungyer and P. H. K. Jap, *Interdiscipl. Topics Gerontol.,* **12,** 13 (1978).

67 A. Victoria-Ihler, *Interdiscipl. Topics Gerontol.,* **12,** 34 (1978).

68 H. Rink, *Interdiscipl. Topics Gerontol.,* **12,** 24 (1978).

69 Y. Courtois, M. F. Counis, M. Laurent, K. Simonneau, and J. Treton, *Interdiscipl. Topics Gerontol.,* **12,** 2 (1978).

70 Y. Hamada, K. Watanabe, H. Aoyama, and T. S. Okada, *Dev. Growth Differ.,* **21,** 205 (1979).

71 D. C. Beebe, D. E. Feagans, and H. A. Jebens, *Proc. Natl. Acad. Sci. U.S.,* **77,** 490 (1980).

72 J. Tassin, E. Malaise, and Y. Courtois, *Exp. Cell Res.,* **123,** 388 (1979).

73 D. Glässer, W. Rattke, and M. Iwig, *Exp. Cell Res.,* **122,** 281 (1979).

74 G. G. Miller, D. G. Blair, E. Hunter, G. Y. Mousa, and J. R. Trevithick, *Dev., Growth Differ.,* **21,** 19 (1979).

75 T. Shinohara and J. Piatigorsky, *Dev. Biol.*, **72,** 385 (1979).

76 F. C. S. Ramaekers, M. W. A. C. Hukkelhoven, A. Groeneveld, and H. Bloemendal, *Ophthal. Res.*, **11,** 283 (1980).

77 F. C. S. Ramaekers, M. Osborn, E. Schmid, K. Weber, H. Bloemendal, and W. W. Franke, *Exp. Cell Res.*, **127,** 309 (1980).

78 A. J. Coulombre, in R. L. De Haan, Ed., *Organogenesis,* New York, Holt, Rinehart and Winston, New York, 1965, p. 219.

79 K. Yasuda, *Dev. Biol.*, **68,** 618 (1979).

80 M. Perry, and Y. Courtois, *Exp. Eye Res.*, **28,** 327 (1979).

81 M. O. Lonchampt, M. Laurent, Y. Courtois, P. Trenchev, and R. C. Hughes, *Exp. Eye Res.*, **23,** 505 (1976).

82 R. M. Clayton in B. I. Lord, C. S. Potten, and R. J. Cole, Eds., *Stem Cells and Tissue Homeostasis.* British Society for Cell Biology, Symposium 2, Cambridge University Press, 1978, p. 115.

83 G. Eguchi, S-I. Abe, and K. Watanabe, *Proc. Natl. Acad. Sci. U.S.*, **71,** 5052 (1974).

84 G. Eguchi, in *Embryogenesis in Mammals, Ciba Symp.*, **40,** 241 (1976).

85 G. Eguchi and T. S. Okada, *Proc. Natl. Acad. Sci. U.S.*, **70,** 1495 (1973).

86 T. S. Okada, *Dev. Growth Differ.*, **19,** 47 (1977).

87 R. W. Reyer, *Symp. Soc. Study Dev. Growth,* **20,** 211 (1962).

88 L. S. Stone, *Invest. Ophthalmol.*, **4,** 420 (1965).

89 D. Scheib, *Ergeb. Anat. Entwicklungsges.*, **38,** 47 (1965).

90 T. Yamada, *Curr. Topics Dev. Biol.*, **2,** 247 (1976).

91 T. Yamada, T. H. Reese, and D. S. McDevitt, *Differentiation,* **1,** 65 (1973).

92 G. Eguchi, *Embryologia,* **9,** 246 (1967).

93 J. A. Powell and N. Segil, *Dev. Biol.*, **52,** 128 (1976).

94 G. Eguchi and K. Watanabe, *J. Embryol. Exp. Morphol.*, **30,** 63 (1973).

95 J. C. Campbell and D. E. S. Truman, *Exp. Eye Res.*, **25,** 99 (1977).

96 G. J. Freeman, *Exp. Zool.*, **154,** 39 (1963).

97 J. C. Campbell, *J. Embryol. Exp. Morphol.*, **13,** 171 (1965).

98 J. C. Campbell and K. W. Jones, *Dev. Biol.*, **17,** 1 (1968).

99 S. A. Brahma and D. S. McDevitt, *J. Embryol. Exp. Morphol.*, **32,** 783 (1974).

100 Y. Itch, *Dev. Biol.*, **54,** 157 (1976).

101 M. Araki and T. S. Okada, *Dev. Growth Differ,* **20,** 71 (1978).

102 T. S. Okada, K. Yasuda, M. Araki, and G. Eguchi, *Dev. Biol.*, **68,** 600 (1979).

103 R. M. Clayton, D. de Pomerai, and D. J. Pritchard, *Dev. Growth Differ.*, **19,** 319 (1977).

104 I. Thomson, C. E. Wilkinson, J. F. Jackson, D. I. de Pomerai, R. M. Clayton, D. E. S. Truman, and R. Williamson, *Dev. Biol.*, **65,** 372 (1978).

105 I. Thomson, D. I. de Pomerai, F. F. Jackson, and R. M. Clayton, *Exp. Cell Res.*, **122,** 73 (1979).

106 R. M. Clayton, I. Thomson, and D. I. de Pomerai, *Nature,* **282,** 628 (1979).

107 D. I. de Pomerai and R. M. Clayton, *J. Embryol. Exp. Morphol.*, **47,** 179 (1978).

108 R. Nomura and T. S. Okada, *Dev. Growth Differ.*, **21,** 161 (1979).

109 T-T. Sun, C. Shih, and H. Green, *Proc. Natl. Acad. Sci. U.S.*, **76,** 2813 (1979).

110 E. D. Korn, *Proc. Natl. Acad. Sci. U.S.*, **75,** 588 (1978).

111 M. Clarke and J. A. Spudich, *Ann. Rev. Biochem.*, **46,** 797 (1977).

112 M. Osborn, T. Born, H. J. Koitsch, and K. Weber, *Cell,* **14,** 477 (1978).

113 E. Lazarides and J. P. Revel, *Sci. Am.*, **240,** 88 (1979).

114 E. Lazarides, *Nature,* **283,** 249 (1980).

115 P. Dustin, *Microtubules,* Springer-Verlag, Berlin, 1978.

116 R. Goldman, T. Pollard, and J. Rosenbaum, Eds., *Cell Motility Book A–C,* Vol. 3, Cold Spring Harbor Conferences on Cell Proliferation, Cold Spring Harbor Laboratory, 1976.

117 H. Ishikawa, R. Bischoff, and H. Holtzer, *J. Cell Biol.*, **38,** 538 (1968).

118 W. W. Franke, *Protoplasma,* **73,** 263 (1971).

119 R. D. Goldman, *J. Cell Biol.*, **51,** 752 (1971).

120 H. Holtzer, S. Fellini, N. Rubinstein, J. Chi, and K. Strahs, in R. D. Goldman, T. Pollard, and J. Rosenbaum, Eds., *Cell Motility,* Cold Spring Harbor Lab., Cold Spring Harbor, N.Y., 1976, p. 823.

121 E. Lazarides and B. D. Hubbard, *Proc. Natl. Acad. Sci. U.S.*, **73,** 4344 (1976).

122 M. L. Shelanski, S-H. Yen, and V. M. Lee, R. D. Goldman, T. Pollard, and J. Rosenbaum, Eds., *Cell Motility,* Cold Spring Harbor Lab., Cold Spring Harbor, N.Y., 1976, p. 1007.

123 S-H. Yen, D. Dahl, M. Schachner, and M. L. Shelanski, *Proc. Natl. Acad. Sci. U.S.*, **73,** 529 (1976).

124 S. H. Blose, M. L. Shelanski, and S. Chacko, *Proc. Natl. Acad. Sci. U.S.*, **74,** 662 (1977).

125 M. Osborn, W. W. Franke, and K. Weber, *Proc. Natl. Acad. Sci. U.S.*, **74,** 2490 (1977).

126 J. M. Starger and R. D. Goldman, *Proc. Natl. Acad. Sci. U.S.*, **74,** 2422 (1977).

127 W. W. Franke, C. Grund, M. Osborn, and K. Weber, *Cytobiologie,* **17,** 365 (1978).

128 W. E. Gordon, A. Bushnell, and K. Burridge, *Cell,* **13,** 249 (1978).

129 R. O. Hynes and A. T. Destree, *Cell,* **13,** 151 (1978).

130 J. V. Small and J. E. Celis, *J. Cell Sci.*, **31,** 393 (1978).

131 W. W. Franke, E. Schmid, K. Weber, and M. Osborn, *Exp. Cell Res.*, **118,** 95 (1979).

132 E. Lazarides, *Exp. Cell Res.*, **112;** 265 (1978).

133 W. W. Franke, K. Weber, M. Osborn, E. Schmid, and E. Freudenstein, *Exp. Cell Res.*, **116,** 429 (1978).

134 W. W. Franke, E. Schmid, M. Osborn, and K. Weber, *Proc. Natl. Acad. Sci. U.S.*, **75,** 5034 (1978).

135 P. F. Davison, B. S. Hong, and P. Cooke, *Exp. Cell. Res.*, **109,** 471 (1977).

136 W. W. Franke, E. Schmid, S. Winter, M. Osborn, and K. Weber, *Exp. Cell Res.*, **123,** 25 (1979).

137 D. C. Rueger, J. S. Huston, D. Dahl, and A. Bignami, *J. Mol. Biol.,* **135,** 53 (1979).

138 C. Cohen, *TIBS,* **4,** 73 (1979).

139 E. Lazarides and K. Burridge, *Cell,* **6,** 289 (1975).

140 E. Lazarides, *J. Cell Biol.,* **68,** 202 (1976).

141 L. Carlsson, L. E. Nyström, I. Sundkvist, F. Markey, and U. Lindberg, *J. Mol. Biol.,* **115,** 465 (1977).

142 B. Burnside, *J. Supramol. Struct.,* **227,** 257 (1976).

143 M. P. Daniels, *J. Cell Biol.,* **53,** 164 (1972).

144 K. M. Yamada, B. S. Spooner, and N. K. Wessells, *Proc. Natl. Acad. Sci. U.S.,* **66,** 1206 (1970).

145 P. C. Marchisio, M. Osborn, and K. Weber, *Brain Res.,* **155,** 229 (1978).

146 D. Drenckhahn and U. Gröschel-Stewart, *Cell Tiss. Res.,* **181,** 493 (1977).

147 N. S. Rafferty and W. Goossens, *Exp. Eye Res.,* **26,** 177 (1978).

148 G. Y. Mousa and J. R. Trevithick, *Exp. Eye Res.,* **29,** 71 (1979).

149 M. Kibbelaar, A. M. E. Selten-Versteegen, I. Dunia, E. L. Benedetti, and H. Bloemendal, *Eur. J. Biochem.,* **95,** 543 (1979).

150 R. H. Bradley, M. E. Ireland, and H. Maisel, *Exp. Eye Res.,* **28,** 441 (1979).

151 R. H. Bradley, M. E. Ireland, and H. Maisel, *Acta Ophthalmol.,* **57,** 461 (1979).

152 S. Nasser, R. Bradley, J. Alcalá, and H. Maisel, *Exp. Eye Res.,* **30,** 109 (1980).

153 M. A. Kibbelaar, F. C. S. Ramaekers, P. J. Ringens, A. M. E. Selten-Versteegen, L. G. Poels, P. H. K. Jap, A. L. Van Rossum, T. E. W. Feltkamp, and H. Bloemendal, *Nature,* **285,** 506 (1980).

154 N. S. Rafferty, *Anat. Rec.,* **186,** 115 (1976).

155 G. Y. Mousa, M. O. Creighton, and J. R. Trevithick, *Exp. Eye Res.,* **29,** 379 (1979).

156 T. Kuwabara, *Arch. Ophthalmol.,* **79,** 189 (1968).

156a P. N. Farnsworth, S. E. Shyne, S. J. Caputo, A. U. Fosano, and A. Spector, *Exp. Eye Res.,* **30,** 611 (1980).

157 H. Maisel, J. Alcalá, N. Lieska, and N. Rafferty, *Ophthal. Res.,* **9,** 147 (1979).

158 H. Maisel, N. Lieska, and R. Bradley, *Experientia,* **34,** 352 (1978).

159 M. Ireland, H. Maisel, and R. Bradley, *Ophthal. Res.,* **10,** 231 (1978).

160 H. Maisel, *Exp. Eye Res.,* **25,** 595 (1977).

161 H. Maisel and M. M. Perry, *Exp. Eye Res.,* **14,** 7 (1972).

162 H. Maisel and N. S. Rasmussen, *Ophthal. Res.,* **10,** 241 (1978).

163 M. A. Kibbelaar and H. Bloemendal, *Exp. Eye Res.,* **29,** 679 (1979).

164 M. A. Kibbelaar, Biochemical and Ultrastructural Features of the Plasma Membrane–Cytoskeleton Complex in Eye Lens Fibers (Actin in the Lens), Ph.D. Thesis, University of Nijmegen, 1979.

165 W. R. Lo and H. Maisel, *Opthal. Res.,* **11,** 129 (1979).

166 W. W. Franke, E. Schmid, M. Osborn, and K. Weber, *J. Cell Biol.,* **81,** 570 (1979).

167 T. E. W. Feltkamp and A. L. van Rossum, *Clin. Exp. Immunol.,* **3,** 1 (1968).

168 N. S. Rafferty and E. A. Esson, *J. Ultrastruct. Res.,* **46,** 239 (1974).

169 H. Bloemendal, J. Schoenmakers, A. Zweers, R. Matze, and E. L. Benedetti, *Biochim. Biophys. Acta,* **123,** 217 (1966).

170 P. H. O'Farrell, *J. Biol. Chem.,* **250,** 4007 (1975).

171 F. Cabral and M. M. Gottesman, *J. Biol. Chem.,* **254,** 6203 (1979).

172 C. M. O'Connor, D. R. Balzer, and E. Lazarides, *Proc. Natl. Acad. Sci. U.S.,* **76,** 819 (1979).

173 E. Ungewickell, P. M. Bennett, R. Calvert, V. Ohanian, and W. B. Gratzer, *Nature,* **280,** 811 (1979).

174 A. J. M. Vermorken, A. A. Groeneveld, J. M. H. C. Hilderink, R. de Waal, and H. Bloemendal, *Mol. Biol. Rep.,* **3,** 371 (1977).

175 K. von der Mark, H. von der Mark, R. Timpl, and R. L. Trelstad, *Dev. Biol.* **59,** 75 (1977).

176 E. Lazarides, *J. Cell Biol.,* **65,** 549 (1975).

177 P. C. Rathke, M. Osborn, and K. Weber, *Eur. J. Cell Biol.,* **19,** 40 (1979).

178 M. Osborn, and K. Weber, *Exp. Cell Res.,* **103,** 331 (1976).

179. K. Weber, R. Pollack, and T. Bibring, *Proc. Natl. Acad. Sci. U.S.,* **72,** 459 (1975).

180 M. Schliwa and U. Euteneuer, *Exp. Cell Res.,* **122,** 93 (1979).

181 E. Wang and R. D. Goldman, *J. Cell Biol.,* **79,** 708 (1978).

182 L. A. Stachelin, *Int. Rev. Cytol.,* **39,** 191 (1974).

183 H. Zerban and W. W. Franke, *Cytobiologie,* **18,** 360 (1978).

184 J. M. Starger, W. E. Brown, A. E. Goldman, and R. D. Goldman, *J. Cell Biol.,* **78,** 93 (1978).

185 M. Osborn and K. Weber, *Eur. J. Cell Biol.,* **20,** 78 (1979).

186 W. W. Franke, E. Schmid, M. Osborn, and K. Weber, *Cytobiologie,* **17,** 392 (1978).

187 A. Lasser and E. A. Balazs, *Exp. Eye Res.,* **13,** 292 (1972).

188 B. C. Byers and K. R. Porter, *Proc. Natl. Acad. Sci. U.S.,* **52,** 1091 (1964).

189 J. Overton, *J. Cell Biol.,* **29,** 293 (1966).

190 M. M. Perry and C. H. Waddington, *J. Embryol. Exp. Morphol.,* **15,** 317 (1966).

191 D. A. Fischman, *J. Cell Biol.,* **32,** 557 (1967).

192 L. G. Tilney, *Dev. Biol.,* **2** (suppl.), 63 (1968).

193 L. G. Tilney and J. R. Gibbins, *J. Cell Biol.,* **41,** 227 (1969).

194 D. Schubert, S. Humphreys, C. Baroni, and M. Cohn, *Proc. Natl. Acad. Sci. U.S.,* **64,** 316 (1969).

195 J. B. Olmsted, K. Carlson, R. Klebe, F. Ruddle, and J. Rosenbaum, *Proc. Natl. Acad. Sci. U.S.,* **65,** 129 (1970).

196 T. E. Schroeder, *J. Embryol. Exp. Morphol.,* **23,** 427 (1970).

197 N. W. Seeds, A. G. Gilman, T. Amano, and M. W. Nirenberg, *Proc. Natl. Acad. Sci. U.S.,* **66,** 160 (1970).

198 T. L. Pearce and J. Zwaan, *J. Embryol. Exp. Morphol.,* **23,** 491 (1970).

199 K. M. Yamada, B. S. Spooner, and N. K. Wessells, *J. Cell Biol.*, **49**, 614 (1971).

200 K. M. Yamada and N. K. Wessells, *Exp. Cell Res.*, **66**, 346 (1971).

201 B. Burnside, *Dev. Biol.*, **26**, 416 (1971).

202 M. A. Handel and L. E. Roth, *Dev. Biol.*, **25**, 78 (1971).

203 P. Karfunkel, *Dev. Biol.*, **25**, 30 (1971).

204 B. Burnside, *Am. Zool.*, **13**, 989 (1973).

205 J. B. Olmsted and G. G. Borisy, *Ann. Rev. Biochem.*, **42**, 507 (1973).

206 J. Piatigorsky, H. DeF. Webster, and S. P. Craig, *Dev. Biol.*, **27**, 176 (1972).

207 J. Piatigorsky, *Dev. Biol.*, **30**, 214 (1973).

208 T. Wanko and M. A. Gavin, *J. Biophys. Biochem. Cytol.*, **6**, 97 (1959).

209 A. I. Cohen, *Am. J. Anat.*, **103**, 219 (1958).

210 A. I. Cohen, *Dev. Biol.*, **3**, 297 (1961).

211 M. Rabaey and A. Lagasse, in J. W. Rohen, Ed., *Eye Structure II*, Schattauer, Stuttgart, 1965, p. 317.

212 A. Porte, M. E. Stoeckel, and A. Brini, *Arch. Ophthalmol.*, **28**, 259 (1968).

213 G. Eguchi, *Bull. Mar. Biol. Stn. Asamushi Tohoku Univ.* **11**, 223 (1963).

214 G. Eguchi, *Embryologia*, **8**, 247 (1964).

215 J. Overton, *J. Cell Biol.*, **24**, 211 (1965).

216 J. M. Arnold, *J. Ultrastruct. Res.*, **14**, 534 (1966).

217 D. C. Beebe, D. E. Feagans, E. J. Blanchette-Mackie, and M. E. Nau, *Science*, **206**, 836 (1979).

218 J. Hoebeke, G. van Kijen, and M. de Brabander, *Biochim. Biophys. Res. Commun.*, **69**, 319 (1976).

219 J. Zwaan and R. W. Hendrix, *Am. Zool.*, **13**, 1039 (1973).

220 S. P. Craig and J. Piatigorsky, *Biochim. Biophys. Acta*, **299**, 642 (1973).

221 J. T. Wrenn and K. Wessells, *J. Exp. Zool.*, **171**, 359 (1969).

222 D. Bray, *Nature*, **282**, 671 (1979).

223 S. W. Tanenbaum, Ed., *Cytochalasins: Biochemical and Cell Biological Aspects*, North-Holland, Amsterdam, 1978.

224 G. C. Godman and A. F. Miranda, in S. W. Tanenbaum, Ed., *Cytochalasins: Biochemical and Cell Biological Aspects* North-Holland, Amsterdam, 1978, p. 227.

225 I. Blikstad, F. Markey, L. Carlsson, T. Persson, and U. Lindberg, *Cell*, **15**, 935 (1978).

226 F. Jacob and J. Monod, in M. Locke, Ed., *Cytodifferentiation and Macromolecular Synthesis* Academic Press, New York, 1963, p. 30.

227 E. L. Benedetti, I. Dunia, and H. Bloemendal, *Proc. Natl. Acad. Sci. U.S.*, **71**, 5073 (1974).

CHAPTER

4

Lenticular Plasma Membranes and Cytoskeleton

E. L. Benedetti and I. Dunia

Institut de Recherche en Biologie Moléculaire, C.N.R.S., Université Paris, Paris, France

Frans C. S. Ramaekers and M. A. Kibbelaar

Department of Biochemistry, University of Nijmegen, Nijmegen, The Netherlands

4.1 INTRODUCTION

The plasmalemma in eye lens has some unique structural and developmental features as compared to cell membranes in other cellular systems. In fact during the transition from epithelium into fiber (see Chapters 2 and 3) most of the other membranous organelles, including endoplasmic reticulum, disappear. Conversely, during this process of terminal differentiation, which occurs at the lens cortical region, a dramatic assembly of newly formed plasma membranes takes place, and extensive junctional domains are settled. In the absence of other intracellular organella, the plasma membrane in lens fibers has the role of providing to the cytoskeleton unique sites of nucleation and attachment. Hence the plasma membrane participates in the maintenance of the lens shape and supplies the means for the control of the mechanical properties of the lenticular mass. In the lens fiber system where the endoplasmic reticulum has been deleted, the polyribosomes engaged in the active protein biosynthesis of the set of crystallins have direct association with both the plasma membrane and cytoskeleton.

The aim of this chapter is to review the experimental evidence that has been provided on lenticular plasma membrane structure and function. We are aware that the reader of the following pages may find the selection of the topics to a certain extent incomplete and arbitrary. We must admit that, although the study of the lens is certainly very attractive, it is also rather distressing because of the overwhelming number of structural and functional features that are comprised in such a small and only apparently "simple" suborgan.

4.2 SOME BIOCHEMICAL FEATURES OF EYE LENS PLASMA MEMBRANES

As described in Chapter 1, about 35% of the eye lens weight is due to protein, whereas the remaining 65% is virtually water. It is remarkable that the water-soluble lens proteins—the crystallins—contribute more than 90% of the total lens protein. The remaining 10% of the lens protein in the water-insoluble fraction contains the total bulk of the lens plasma membranes (1). The plasma membrane architecture results from the association of two main components, namely, a lipid phase primarily occurring as a bilayer and associated or intercalated protein entities (compare Section 4.2.2).

4.2.1 Preparation of Plasma Membranes

We have developed a method for the preparation of lens membranes both from epithelium and from fibers, which is based mainly on the classical technique of purification of plasma membranes (2). If the isolation concerns only lens fiber plasma membranes, then the capsule together with the epithelium are removed, and the cortical region of the lens is selected and pooled. The tissue is gently homogenized at low temperature in 1 m*M* sodium bicarbonate containing 1 m*M* calcium chloride. Upon low-speed centrifugation, a plasma membrane fraction is obtained under circumstances that avoid the stabilization of other cytoplasmic constituents such as cytoskeleton and polyribosomes. The definition of the "purity" of the plasma membrane-rich fraction is based on electron microscopic observations on thin sections, negative staining, and freeze-fracturing. The membrane fraction purified on sucrose density gradients consists mainly of large fragments of membranes displaying the classical triple layer aspect of the unit membrane (Figure 1). Plasma membranes of two adjacent lens fibers appear as differentiated membrane segments, and both plasma membranes merge together forming junctional domains (Figure 1*a*, arrow) of pentalayered appearence. Isolated membrane fragments originating from the anterior and posterior sutures, where *in situ* extensive fiber interdigitation is found, have the appearence of vesicular membrane profiles of rather various sizes.

4.2.2 The Proteins

The study of the lens membrane protein composition, although having started only recently, has already yielded an impressive and exponentially growing amount of data. Some of the membrane protein constituents can be eluted by water and/or by EDTA-containing buffers. Other polypeptides can be progressively solubilized by urea and/or by ionic or nonionic

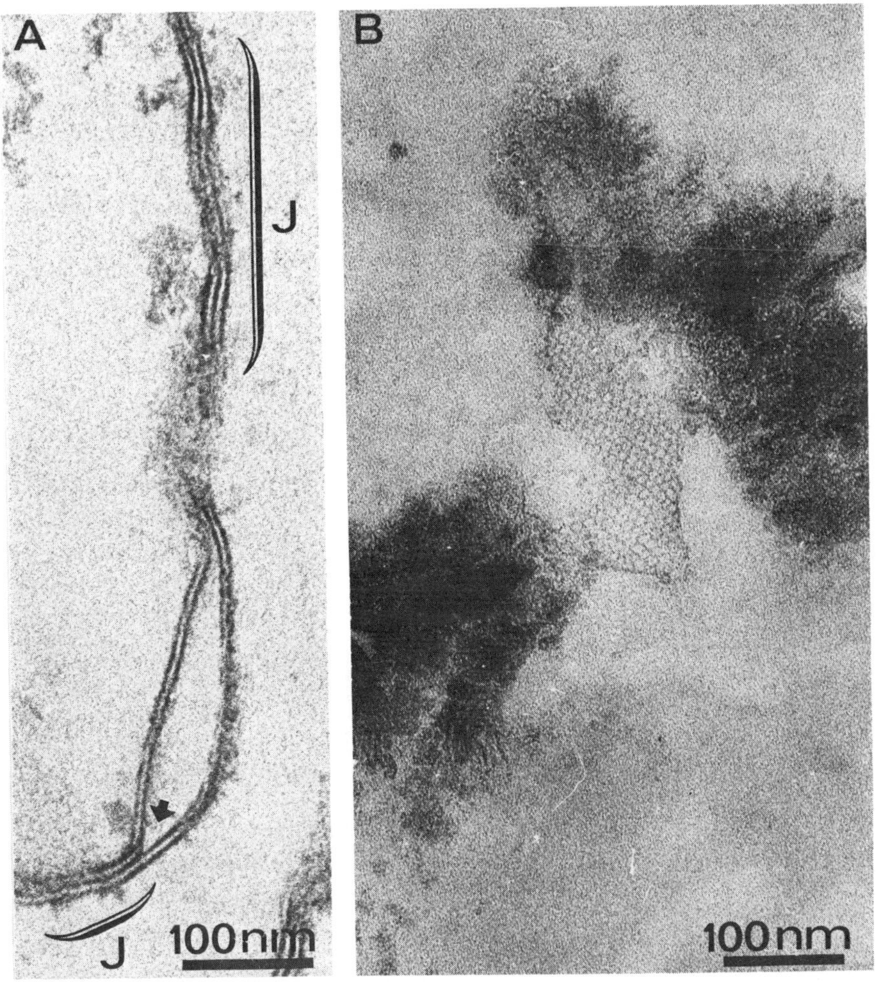

Figure 1 (*a*) Isolated lens fiber plasma membrane (cf. ref. 2). Two adjacent membrane elements displaying triple-layered structure merge (arrow) into junctional domains (*J*). Thin section stained with uranyl acetate and lead citrate. (*b*) Lens elongation zone impregnated with lanthanum nitrate. In the thin section the electron dense marker filling the intercellular spaces has outlined the gap junctional region. Note the two-dimensional geometric pattern of repeating subunits with a center-to-center space of 9 nm.

detergents (1). "Intrinsic" membrane proteins can also be extracted by chloroform, methanol, and butanol, according to methods applied to the solubilization of proteolipids (3–8). These differential solubilization properties and the complex features of the electrophoretic profiles suggest the existence of multiple associations between membrane protein and other lens constituents. Hence the definition of the protein components, which could be identified either as "intrinsic" or "peripheral" to the plasma membrane backbone, is not easily apparent and essentially operational. Even more complex protein profiles are obtained if the plasma membranes are isolated under preserving condition(s) that stabilize(s) the labile and perhaps transient association between the membrane bilayer and motility protein of the cytoskeleton (9).

The isolated plasma membrane of the cortical fibers comprises more than 50% of protein (2); thus, when membranes are submitted to electrophoretic analysis in polyacrylamide gels containing SDS, a number of components are detected. Unfortunately, none of these polypeptides has been studied in terms of a possible association with enzymatic or receptor activity. Thus the main interest has been focused on a rather tedious determination of the molecular weight of the lens membrane proteins and on a few other molecular features. Only for a limited number of polypeptides do some data suggest their possible role in the specialized domain formation in lens membranes (1, 10).

Early studies of Dische et al. (11) demonstrated that a membrane fraction could be isolated if the water-insoluble "albuminoid" was extracted by guanidine hydrochloride or urea. Ever since, a major trend of protein analysis of lens membranes has been accomplished on the intact lens and its water-insoluble fraction, in both cases thoroughly extracted by highly concentrated urea or guanidine hydrochloride. Obviously these treatments affect the integrity of the plasma membrane's morphological and biochemical features since inherent constituents of the plasma membrane can be extracted, and in turn the conformation and solubility properties of even intrinsic membrane components might be changed (12).

The characteristic protein pattern of isolated lens fiber plasma membranes is shown in Figure 2. In sodium dodecylsulfate gel electrophoresis a multitude of polypeptide bands are seen varying in molecular weight roughly between 18,000 and 100,000 dalton and even higher (1, 3, 10, 13). A significant fact is that crystallin polypeptides—especially α-crystallin subunits—remain detectable in the lens membrane protein pattern (approximately 18,000 to 31,000 dalton) (1, 10). The crystallin polypeptides, which are not removed by repeated washing in buffer of low ionic strength, even resist mild proteolytic digestion applied to the isolated plasma membranes (cf. ref. 2). When the isolated membranes are repeatedly washed with 6 *M* urea, the crystallin polypeptides are partially solubilized; however, α-crystallin (Figure 3) is still detectable (4–13). This finding is consistent with the observation of Broekhuyse and Kuhlmann (3) and of Alcalá and Maisel (13), who have also observed crystallin polypeptides in the urea-extracted membrane. An interesting molecular model has recently been purported by Spector et al. (14) and Garner et al. (15). The crystallin polypeptides could be membrane-bound by means of strong interaction with another membrane protein of 43,000 dalton. The latter constituent, according to these authors, differs from actin and may form water-insoluble aggregates with crystallin

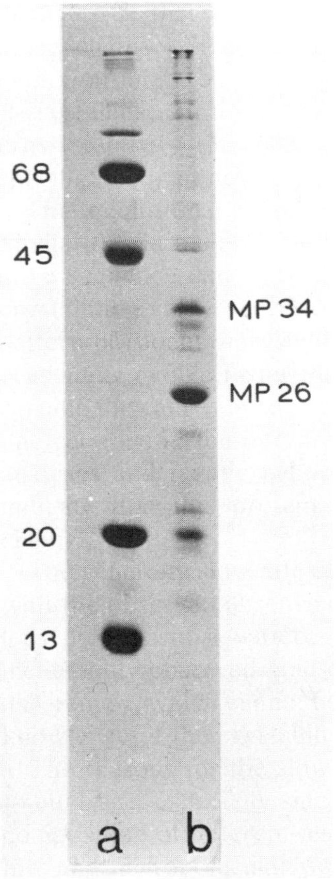

Figure 2 SDS–polyacrylamide gel electrophoresis of lens fiber membrane proteins showing the two major bands MP34 and MP26 (*b*). For comparison, the pattern of the marker polypeptides is shown (*a*). The molecular weights of cytochrome C, αA-crystallin, ovalbumin, and bovine serum albumin are given in kilodalton.

and higher molecular weight proteins, in particular during cataract formation.

In an early report on the protein profile of isolated lens fiber membranes we were mainly impressed by a major membrane component forming a rather diffuse band of a molecular weight estimated according to the method of Shapiro et al. (15a, cf. ref. 2) to be approximately 38,000 dalton, being probably a doublet. Further studies on the protein pattern of isolated lens fiber membranes and junctional subfractions using Laemmli's method demonstrated that two major polypeptides, one of 34,000 dalton and one of 26,000 dalton, are characteristic of the lens membrane (10). We must admit that the obtained profiles were inconsistent and that the number of bands and their presumptive molecular weights were different in various experiments, in particular when enzymatic digestion or detergent solubilization were applied (10). Moreover, it appeared that the 26,000-dalton component is particularly heat-sensitive, and its conversion to high molecular weight

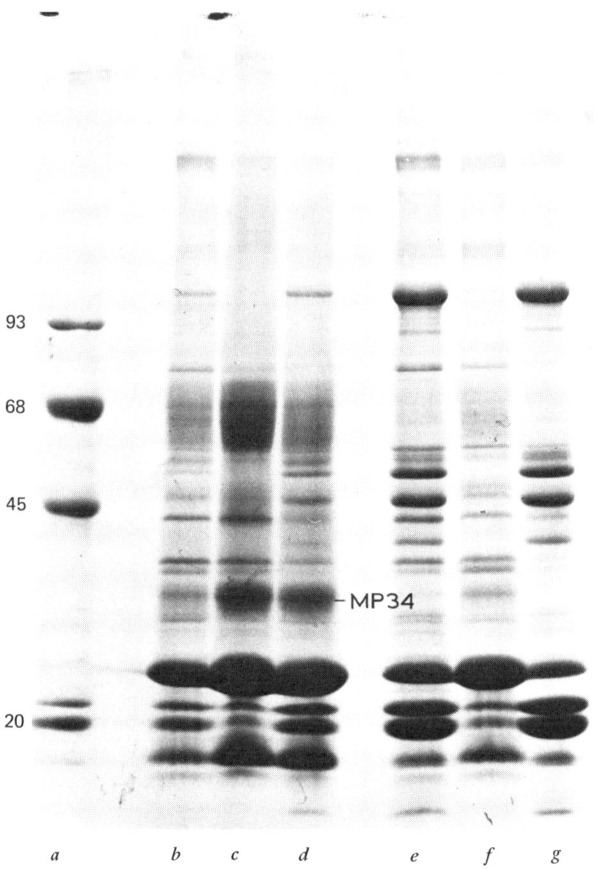

Figure 3 Protein profile of different lens plasma membrane fractions. (*a*) Markers (for comparison): αA-crystallin, 20,000; ovalbumin, 45,000; bovine serum albumin, 68,000; phosphorylase A, 93,000. (*b*) Lens plasma membranes isolated in bicarbonate at densities 1.20 to 1.18 and 1.18 to 1.16 g/cm^3. (*c*) As (*b*), but washed with 6 *M* urea. (*d*) As (*b*), but washed with 1% DOC. (*e*) Lens plasma membranes isolated in bicarbonate at density 1.22 to 1.20 g/cm^3. (*f*) As (*e*), but washed with 6 *M* urea. (*g*) As (*e*), but washed with 1% DOC.

aggregates may easily result from overheating and inappropriate ionic strength conditions during solubilization (16, 17). If the isolation procedure and solubilization in reducing conditions are well standardized and optimized, then consistently the 34,000- and 26,000-dalton components are found as membrane-associated constituents (18). Yet, discrepancy and contradictory interpretation cast some doubt on the intrinsic or peripheral nature of these two major components. To avoid misleading terminology in the identification of lens membrane proteins and to facilitate discussion, we proposed a simple nomenclature based on the only well-defined parameter, namely the apparent molecular weight of each individual polypeptide (18).

4.2.2.1 MP26

Experiments in SDS–polyacrylamide gels have shown that the major component of water- and urea-insoluble lens membranes is a subunit of 26,000 dalton (MP26). This MP26 can easily be extracted in chloroform, methanol, and butanol, as found by Broekhuyse et al. (7), who called this component the "main intrinsic polypeptide" (MIP) of bovine lens plasma membranes. This predominant component comprises more than 50% of the protein recovered from the membranes (1). It is noteworthy that the solubility in organic solvents of the MP26 is much more effective if the membrane fractions are obtained by a previous and repeated urea extraction (5, 7). This interesting observation might reflect the occurrence of architectural changes of "urea-insoluble" membrane components induced by the chaotropic influence that enhances the exposition of the protein to the organic solvents and/or its strongly bound lipid environment. In a similar manner, urea treatment favors the solubility of MP26 in 1% Triton X100 and in 0.1 *M* dodecyltrimethylammonium (3, 5, 7). A similar, if not identical, polypeptide of 26,000 dalton has been detected in human lens by Roy et al. (19). This major component and another one of 22,000 dalton are found in membranes originating from the lens cortex as well as from the nucleus. However, the latter component becomes more prominent in the nuclear region where no protein synthesis can be detected. Hence the 22,000-dalton component presumably is produced by post-translational processing of the major MP26. A similar assumption has been postulated previously for the observation that in stored calf lens fiber membrane preparations the MP26 diminishes, while a smaller component MP22 is found in the protein profile (cf. ref. 1). Although most of the experimental evidence is consistent with the ubiquitous nature of the MP26—no matter whether the lens membranes are derived from different animal sources—studies with immunodiffusion point to the existence of a specific antigenic property of this membrane protein. The antiserum directed against chick lens fiber MP26 precipitates the corresponding polypeptide isolated from turkey lens fiber, but not the MP26 of bovine lens plasma membranes (13).

The complex nature of the MP26 is even more obvious if the protein profile of the isolated membranes from lens epithelium is compared with that of fiber membranes. Vermorken et al. (20) have demonstrated that in the lens plasma membranes from epithelium the MP26 is absent. This interesting observation, which has been further substantiated by the study of Broekhuyse et al. (7), points to the existence of the specific synthesis of MP26 during lens fiber differentiation (21). Comparing the molecular weights estimated in sodium dodecylsulfate gel electrophoresis, Alcalá et al. (22) suggested that MP26 could be the predominant polypeptide of β-crystallin (βB_p). Conversely, Broekhuyse et al. (4, 18) and Kibbelaar and Bloemendal (23) reported an amino acid composition of MP26 that was quite different from that of the βB_p-crystallin chain. Additional chemical data provided by Kibbelaar and Bloemendal (23) have confirmed this latter conclusion and extended our knowledge on the molecular nature of the MP26.

All the experimental information that we have so far provided indicates that MP26 is a major fiber membrane consitituent, amphipatic in nature, and associated with the hydrophobic lipid core. Evidence has been provided that at least in calf lens fiber membranes the MP26 is a glycoprotein. Conversely, the results of gas chromatographic analysis by Broekhuyse and Kuhlmann (3, 5) clearly show that MP26 does not comprise carbohydrates.

MP26 was purified by preparative gel electrophoresis and analyzed by analytical gel electrophoresis. The estimated purity was more than 95%. The amino acid composition was compared to the amino acid composition reported by Broekhuyse et al. (7) and to the amino acid composition of βB_p (24) (compare Chapter 1). This comparison shows that identical MP26 preparations are obtained by different purification procedures such as chloroform-methanol extraction and sodium dodecylsulfate preparative gel electrophoresis. Comparison of the amino acid compositions of βB_p and MP26 also reveals a quite different degree of polarity of the two polypeptides. MP26 meets the criterion which is currently found for intrinsic membrane proteins. An additional proof for the dissimilarity of MP26 and βB_p is provided by the peptide maps of both proteins shown in Figure 4.

4.2.2.2 MP34

Serious difficulties are encountered if one tries to define the type of association of the MP34 with the membrane backbone. This protein (about 5% on membrane protein basis) is defined by Bouman et al. (6) as an "extrinsic" lens membrane protein, different from the crystallins and bound to the membrane by calcium ions since it can be eluted from the urea-insoluble membrane fraction by EDTA-containing buffer. From a careful reading of the elegant analysis of Bouman et al. it appears, however, that the MP34 or EDTA-extractable protein (EEP), although it has been called "extrinsic," is almost urea-insoluble (cf. also Figure 21 in Chapter 1). Furthermore, EEP can be extracted in a "pure" form only when the membranes have extensively been treated with urea. A rather strong affinity between α-crystallin and MP34 seems also to exist, and it cannot completely be ruled out that crystallin "stabilizes" the MP34 in the intact membrane. This kind of supramolecular assembly could easily be affected by the urea extraction that changes the overall membrane architecture so that the MP34 could in turn be released by chelating agents. Consistent with the amphipatic nature of the MP34 is the amino acid composition (6) showing the rough

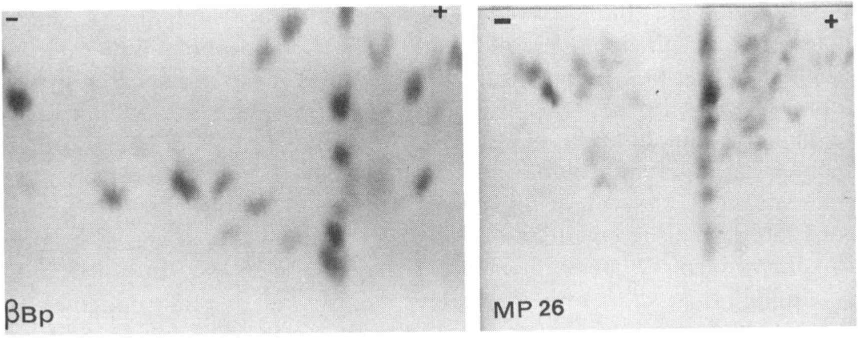

Figure 4 Tryptic peptide maps of MP26 and βB_P. (The finger print of βB_P was kindly provided by Huub Driessen.)

percentage of hydrophobicity. Even if it is not as high as that estimated from MP26, it is certainly not negligible. This molecular feature could reinforce the assumption that MP34, which in our experiments appears to be comprised in the protein profile of a DOC-insoluble membrane fraction, is at least partially in contact with the hydrophobic lipid core. Unfortunately, so far, the membrane-bound MP34 has not yet been analyzed by verifying its accessibility to protease. Although no experimental evidence has been provided, the EDTA-extractable protein has been considered to be exposed and/or associated to the cytoplasmic side of the lens plasma membrane (5). In our early reports on the protein pattern of isolated lens plasma membranes (2) and junctions (10) the component, migrating as a band of 34,000 dalton, appeared as a doublet. More recently, Bouman et al. (6) have shown that the protein constituent that can be extracted by EDTA-solution contains indeed two "subunits" of 32,000 and 35,000 dalton, respectively. It is noteworthy that by two-dimensional electrophoresis and by immunoelectrofocusing, Bouman et al. could show that the two subunits of the bovine EEP membrane component were antigenically different. A common antigenic determinant seems, however, to exist between EEP and α-crystallin polypeptides. A sound biochemical and immunological analysis of the MP34 from lens membranes of different species has not yet been provided. We know, however, that the MP34 is common to both isolated epithelial plasma membranes and lens fiber plasmalemma, in contrast to MP26, which is absent in the protein profile of the epithelium (20).

4.2.3 The Lipids

The plasma membranes isolated in bicarbonate solution from the cortical region of the calf lens contain about 1.25 mg of total lipid per mg of protein. As much as 46% of the lipids are phospholipids, and a considerable amount of cholesterol is also present (2). Urea-extracted plasma membranes at the cortical region are characterized by a weight ratio protein/lipid and a molar cholesterol/phospholipid ratio of about 1 : 1. The phospholipid distribution is similar to that of the nonextracted membranes (i.e., 21.5 sphingomyelin, 29 phosphatidylcholine, 11 phosphatidylserine, 34 phosphatidylethanolamine mol/100 mol lipid P) (3, 25–28). A study of the lipid and protein composition of urea-insoluble membranes isolated from different lens regions indicates that the cholesterol/phospholipid molar ratio increases remarkably from the equator to the cortex (from 0.5 to 0.8) and even more (1.6) if "nucleous" membranes are analyzed (3, 26–28). In parallel the protein/lipid ratio shifts to higher values due to a relative phosphoglyceride loss toward deeper lens layers. Furthermore, an interesting and accurate analysis by Broekhuyse and Veerkamp (25) on the phospholipid composition of the lens membrane indicates that during the process of differentiation of the epithelium into fibers higher lipid metabolic activity, as can be monitored by ^{32}P incorporation, is characteristic of the epithelium and outer cortex, while the lens nucleus has no activity as far as phosphate turnover of phospholipids is concerned. Furthermore, fiber cell differentiation is accompanied by a change in the relative lipid composition in so far as the phosphatidylcholine and phosphatidylinositol percentage decreases, whereas a remarkable increase of phosphatidylserine and sphingomyelin is found. The lens fiber membrane sphingomyelin has some features in common with the class of phospholipids found in myelin. In both cases most of the fatty acids have a very high C24 : 1/C24 : 0 ratio (25–28). Hence restriction in selected areas of the lens plasma membrane of the free lateral mobility of protein could depend on local accumulation of more highly saturated fatty acids in the chains of membrane phospholipids. Restriction could also come from the high concentration of cholesterol, which is thought to impair the mobility of phospholipid fatty acid chains, thus provoking an "intermediate fluid condition" rather close to the gel phase. These circumstances would favor the segregation of either preexisting or newly inserted proteins from the rather rigid lipid domain and enhance protein-to-protein interaction.

4.2.4 Carbohydrates and Glycoproteins

Isolated lens fiber plasma membranes (2) are also characterized by the presence of carbohydrates as minor components (total sugar content 2.7 to 4.0%) according to Broekhuyse et al. (3–7). A major portion (73%) of sialic acid (25.6 nmol/mg protein) is liberated from the membrane by neuraminidase. After mild proteolytic treatment, neuraminidase action releases almost 93% of terminal sialic acid. Fucose, glucose, mannose, and *N*-acetylgalactosamine are present in relatively lower amounts than galactose and *N*-acetylglucosamine (3, 5). Since, at least for neutral sugars and hexosamine, similar values are found in bicarbonate-isolated (2) and urea-extracted fiber plasma membranes (3, 5), one might assume that these carbohydrates are mainly associated with the membrane intrinsic glycoprotein. Although the lipid composition of the lens membrane has been the object of several investigations, to our knowledge the presence of gangliosides has not yet been taken into consideration. Hence it remains to be established whether lipid-associated sugars might eventually be lens membrane constituents. Glycoproteins have been described since the early observation of Dische et al. (11). More recently, Alcalá et al. (22) have studied by gel electrophoresis in SDS the protein pattern of urea-extracted bovine lens membranes and found only one component of 68,000 dalton that could be stained by periodic acid-Schiff reaction.

4.3 DOMAIN FEATURES IN EYE LENS EPITHELIUM AND FIBERS

The eye lens is morphologically asymmetric, since it is provided with a cuboïdal epithelial monolayer associated only to the lens anterior side (see Chapter 2). On its posterior side the fiber cells are directly in contact with the posterior capsule. The asymmetrical organization of the cellular elements is presumably relevant to the lens function and its electrophysiological features in that the lens acts as a controlling pathway between two different compartments.

In vision we may ascribe to the lens the unique function of forming an image by refracting and focusing the light onto the retina in a controlled manner (29). Thus the lenticular body must be provided with sophisticated mechanisms controlling the ionic balance and the water content of the "internal milieu" in order to maintain the solubility properties of the crystallins and to ensure lens transparency.

4.3.1 The Vectorial Organization of the Lens

Several interesting postulates emerge from a recent and outstanding review by Rae (30) on lens electrophysiology. Consistent with the morphological asymmetry of the lens, the sites for active transport of ions might also be asymmetrically associated with the epithelial domain and perhaps with the anterior cortical fibers (31). On the other hand, the posterior lens fibers would have a merely passive role (30). Additional information on the nature of the ionic fluxes would also imply the existence of an active mechanism pumping sodium ions out through the anterior surface. The maintenance of the lenticular water balance could then rely on a Na^+ pump asymmetrically located only on the anterior epithelial surface actively extruding Na^+ from the inside of the lens to the outside.

Hence this active transport system in the lens would have some features in common with other types of epithelium involved in ion "secretion." For instance, in the salt gland of elasmobranch and marine catfish the ion-secreting epithelium usually transports ions from the inside blood luminal domain to the outside (1). In contrast with other types of epithelial cells, those involved in ion secretion are not provided with typical occluding junctions forming a continuous belt around each mural cell (1). It is remarkable that also the lens epithelium is lacking these kinds of junctional complexes that usually control the paracellular routes of permeation (1). This conclusion, reached only in freeze-fracture experiments, receives further support from the elegant study of Rae and Stacy (32), who have shown that in the lens both lanthanum and Procion yellow readily penetrate between the anterior epithelial cells, because of the absence of a "tight" barrier to solute movement around the cells. In whole lens incubation, Rae and Stacy (32) have also found that the tracers penetrate into the extracellular space around the surface cortical fibers. Conversely, lanthanum or Procion yellow do not easily fill up the space between deeper fibers and even less the lenticular core. These interesting observations might shed some light on the mechanism controlling the transepithelial potential. This mechanism, implying both the action of an active cellular transport and spatial and electrical properties of the lenticular intercellular spaces, may prevent (or at least restrain) backflow of cations in a direction opposite to the active transport (1).

4.3.2 The Lens Extracellular Domain

The exact values for translenticular electrical resistance have not yet been fully established. By using different circuit models, contradictory results are obtained (30). It is, nevertheless, most probable that the electrical resistance values range between 2000 and 1000 Ω/cm^2 (30). Even if we take the lower value into consideration, still the lens system cannot be classified among the "leaky" epithelia (33, 34). Rather, it may be compared to an intermediate tight type of mural epithelium. In the absence of the morphological features of the tight junctions, the considerable electrical resistance across the lens could not be maintained if the structural and electrical properties of the lens extracellular spaces would not play a role similar to that of the *zonulae occludens* by imposing a restriction to the free movement of water and ions. It should be stressed that in the lens the intercellular spaces are relatively narrow and folded (see Chapter 2). The extracellular space in the lens is equivalent to 7 to 12% of the water content of the tissue (30, 35). It is remarkable that 80% of this space is comprised in the lens cellular capsule (35). Moreover, the lens fibers are very tightly packed, and the minimal extracellular space is even more reduced where the close opposite plasma membranes merge and differentiate into an extensive junctional domain (20 Å gap). These junctions, morphologically differing from the *zonulae occludentes*, may control not only the horizontal intracellular transit but also the overall translenticular permeability (1). In agreement with our assumption, Rae and Stacy (32) have also suggested that a barrier to water and ion-free movement could reside in the extracellular space, where types of junctional complexes other than the "tight" type could simply limit the ratio of diffusion. To this view, further credit has also been given by the interesting observation of Ginzberg and Gilula concerning the junctional organization in the sensory epithelium of chicken otocyst (36).

4.3.3 Cell-to-Cell Pathways

The maintenance of an electrochemical gradient across the entire lens can be interpreted in a framework of a single-cell syncitial model. In this model the source and means of energy production would lie in the epithelium and perhaps in the outer cortex (31, 37). Through an extensive communicating network the energy would be provided both for ion transport and for keeping the required ion and water balance in the internal milieu of each individual cell (1). Indeed the potential difference (PD) of the lens measured with the insertion of glass microelectrodes has been shown to be independent of the position of the electrode within the lens, as the lenticular body was isopotential (cf. ref. 30). This phenomenon probably indicates that the cell membrane of each individual fiber has domains of low electrical resistance. Along this line experimental evidence has been provided showing that the cellular elements—epithelial monolayer and fibers—are connected by very extensive junctional complexes.

Measurement of the electrical coupling, however, between individual lens fibers with the insertion of microelectrodes is not very easily apparent, since the fibers are indeed several millimeters long and 8 to 10 μm wide, but only 2 to 5 μm thick. The cellular elements are really tightly packed (cf. ref. 35). In spite of all these difficulties, it has been possible to demonstrate in frog and rat lenses that Procion yellow (injected by using glass microelectrodes and iontophoretic techniques) did not remain within a single fiber but moved easily from one fiber to another, while no extracellular leakness was detected (29). Rather extensive electrical coupling was also directly measured by placing two electrodes into two close neighboring fibers and recording the induced potential difference (30).

4.3.4 The Junctional Complexes Found in Lens Epithelium and Fibers

The electrophysiological observations that we have so far summarized provide compelling evidence for electrical coupling between the lens cellular elements, in particular the fibers. The communicating pathway also allows the passage of a tracer substance of relatively higher molecular size (Procion yellow) than the small ions carrying the electrical current. A number of electron microscopic investigations have revealed the presence of junctional complexes connecting the lens cellular elements (cf. ref. 1). In thin sections of intact lens or isolated plasma membranes the junctional complexes appear as a domain of close apposition of two adjacent plasma membranes (compare Figure 1*a*). There the two outer leaflets do not merge in a medial electron dense tract, but are rather separated by a 2-nm gap. Incidentally, on fortunate tangential sections of junctions impregnated by lanthanum, the subunit lattice organization of the intracellular gap is delineated (Figure 1*b*). Surprising, however, is the absence of any subunit pattern on the negatively stained isolated lens fiber junctions. Even upon raising the temperature of staining up to 37°C, as we originally proposed for a better visualization of isolated rat liver gap junctions (38), calf lens fiber junctions never displayed any geometrical array of repeating subunits.

However, the presence of lens junctional domains can be easily identified by studying both epithelium and fibers by the freeze-fracture technique (1, 39). Likewise in other tissues (40–45), the lens gap junctions possess arrays of 9-nm intramembranous particles, which after cleavage remain associated with the protoplasmic fracture face (PF). The external fracture face (EF) is characterized by rows of small pits (Figure 5). The results obtained both on thin-sectioned material and on freeze-fracture replicas indicate that junctional complexes are localized at the lateral surface of the epithelial cell and between the basal surface of the epithelial monolayer and the cortical fibers. Furthermore, in all cortical fibers, very extensive junctional domains are also present. Remarkably, the organization of the gap junctions found in the epithelium differs from that of the junctional domains connecting lens fibers. We have observed (1, 39) that the epithelial monolayer comprises small domains characterized by closely and geometrically packed arrays of intramembranous particles on the PF and of pits on the EF (Figure 6). In contrast, the lens fiber junctions are pleomorphic and constantly lack any geometrical packing of the particulate entities (Figure 5).

Electron microscopic observations have pointed to the fact that the membrane specialized domains assumed to be involved in short-range cell-to-cell communication have a common feature: the presence of paired intrammembranous protein oligomers that extend and merge together across the narrowed, 2.0-nm wide intermembranous gap (46, 47). However, gap junction in various tissues may be characterized by architectural pleomorphism, which could be considered suggestive for junctional diversity of the junctional competence (cf. ref. 48).

The packing and the number of the transmembrane oligomers, the size of individual gap junctions, and the total areas per cell-to-cell communicating interface may vary considerably (cf. ref. 48). Several studies have also tried to correlate the existence of reversible structural changes of the communicating pathways with the functional uncoupling of gap junctions (41–45).

It has been claimed that when, by various experimental means, the junctional resistance is increased, the transmembrane oligomers form more

Figure 5 Freeze-fracture replica of lens fibers. Extensive junctional domains connect the fibers. Arrows indicate the intercellular space.

closely and geometrically packed arrays in the plane of the junctional membrane (''crystallization'') (41); moreover, a considerable reduction of the lattice constant has also been estimated (42). Transition from an open to a closed state has also been correlated with the enlargement of the gap width by a factor of 1.6 and intramembranous particle ''contraction,'' resulting in the partial or complete lack of continuity with the connecting oligomers of adjacent plasma membrane (43). Goodenough (44) and Goodenough et al. (45) have postulated that the lens fiber gap junctions, unlike other communicating junctions; do not show in a high resistance state any structural transition (''crystallization'') of the transmembrane oligomers.

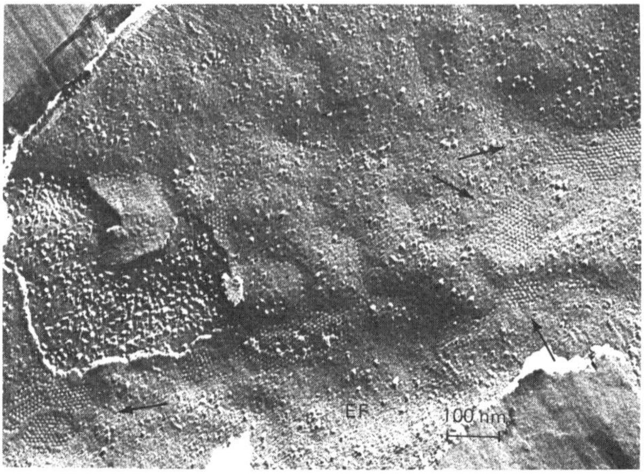

Figure 6 Freeze-fracture replica of lens epithelium. Several junctional domains (arrows) are represented by hexagonally packed arrays of pits on the EF face.

The implication of this interesting assumption is that lens fiber gap junction could be insensitive to the regulatory mechanism that switches the junctional domain organization from the low- to the high-resistance state. Conversely, the gap junctions coupling the lens epithelial cells, which are seen in freeze-fracture replicas of fixed material as small domains of geometrically packed intramembranous particles, could be sensitive to conditions that modulate the junctional permeability. However, the accurate study of Peracchia (42) casts some doubts on the existence of nonregulated low-resistance pathways connecting lens fibers. According to Peracchia the exposure of isolated lens fiber gap junctions to 5×10^{-7}M (or greater) [Ca^{2+}] results in irreversible aggregation of particles and pits tightly into regularly hexagonal arrays (42). Our electron microscopic observations on isolated lens fiber plasma membrane exposed to high concentration of Mg^{2+} show that the junction fracture face is characterized by tightly packed hexagonal arrays of particles and pits (Figure 7). Our study as well as Peracchia's results obtained in rather different experimental conditions than that developed by Goodenough et al. (45), do not, however, rule out the possibility that not only structural but also functional features might characterize, respectively, the gap junctions found in the lens epithelium and those connecting the fibers. This assumption also gains some credit from the following data concerning developmental properties and chemical features of these two types of lenticular junctions. We have found that the epithelial layer (1) possesses a remarkable number of "internalized" annular junctions (Figures 8 and 9). Internalization of gap junctions has been described in a

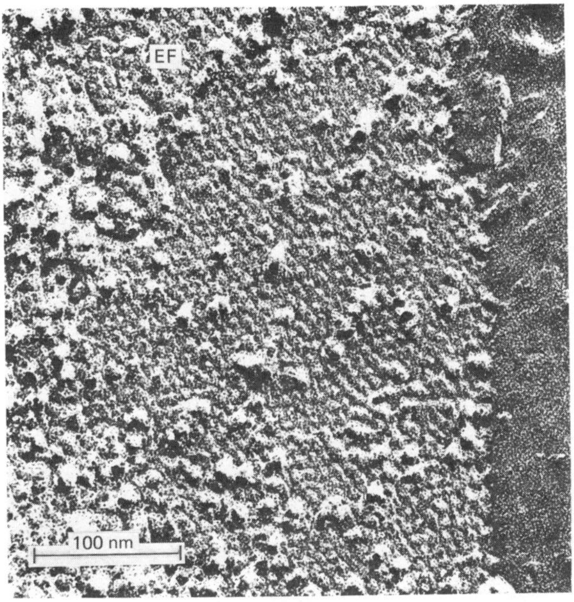

Figure 7 Freeze-fracture of isolated lens fiber plasma membrane in TKM. Note the hexagonally packed array on pits of the EF face.

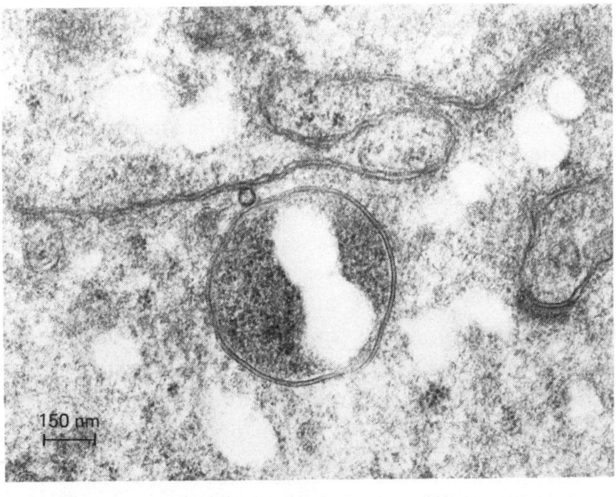

Figure 8 Thin section of lens epithelium and internalized gap junction comprising a round-shaped cytoplasmic domain.

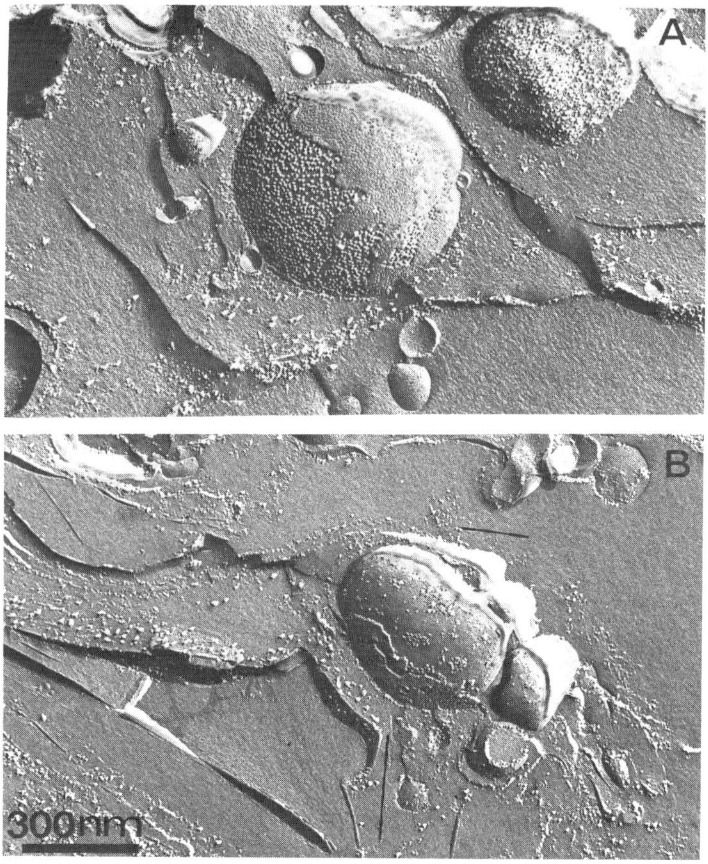

Figure 9 Annular internalized gap junctions visualized in the replicas of freeze-fractured epithelium close to the elongation zone. Note the deteriorated organization of both particulated and pitted entities of the fracture faces (*b*).

variety of normal and pathological conditions (49, 50). Moreover, internalization of junctional complexes has been observed as a consequence of cell dissociation obtained by protease action. We have recently studied in a mouse embryonal carcinoma cell line the effect of Fab fragments against the surface antigen of the EC F9 cells (51). The action of this ligand upon the cell surface is followed by a number of events that result in the deletion of the junctional complexes. The latter process is characterized by a progressive internalization and formation of annular gap junctions.

It is tempting to postulate that during the process of terminal differentiation of the epithelium into lens fibers, the junctional complexes and perhaps other cell surface adhesion properties of the former cell disappear, while in the fibers pleomorphic junctional domains are formed by recruiting novel precursors. Conversely, the preexisting epithelial junctions most probably do not act as nucleation sites for the apposition of new junctional material (39). The process of junctional assembly at the elongation zone is, in fact, characterized by the convergence of 9-nm particles, forming linear arrays of streams in the two matching plasma membrane domains (39). The linear rows have been described as either forming an irregular particulate network or as merging in larger aggregates in potential areas for cell-to-cell close contact (Figure 10).

An elegant study by Ginzberg and Gilula (36) indeed shows that during

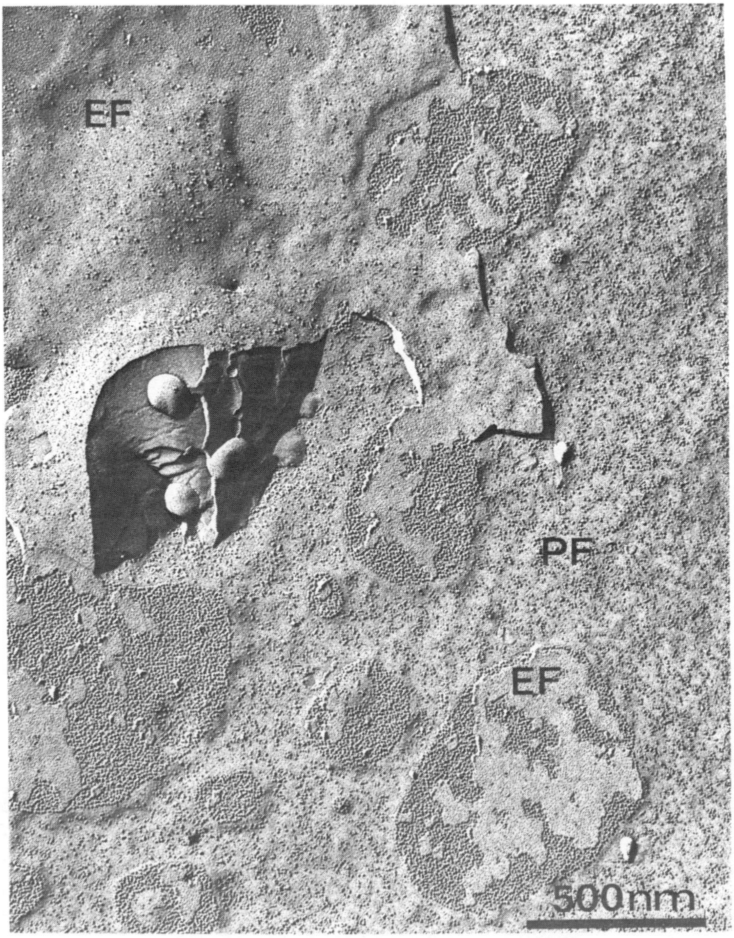

Figure 10 Replica of freeze-fractured elongating zone. Multiple clusters of 9 nm intramembranous particles (PF) characterize the multicentric assembly of gap junctional elements. Pitted fracture face (EF) are also visible.

differentiation of the chicken otocyst sensory epithelium a sequential internalization and intracellular breakdown of gap junctions occur as sensory cells differentiate. In contrast, a separate population of very extensive gap junction domains is formed between differentiating support cells.

It seems relevant here to add some data about the postulated chemical nature of the lens junctions. At least, the lens fiber junctions have in common with liver gap junctions the characteristic of being almost insoluble in sodium deoxycholate (DOC). Hence the isolation of the lens fiber junctions can also be obtained by detergent solubilization followed by equilibrium density grandient centrifugation (10) (Figure 11). When sodium dodecylsulfate gel electrophoresis is applied to the detergent-insoluble junctional fraction of lens fibers, two polypeptides (MP34 and MP26) characterize the protein pattern. It is also noteworthy that the MP26 is abundant in the urea-insoluble membrane fraction that comprises mostly junctional fragments (1). Our study, still in progress (I. Dunia, C. Quemard, and E. L. Benedetti unpublished), on the effect of various types of detergents and chaotropic agents on lens fiber junctions indicates, however, that the solubility properties of the junctional domain(s) may vary according to the choice of the surfactants and perturbants.

A recent report by Henderson et al. (52) indicates that a 26,000-dalton polypeptide is the major component of the isolated liver gap junctions. A minor 21,000-mol. wt. subunit is also detected, being most probably a degradation product of the larger polypeptide. If we take into consideration the estimated molecular weight of the identified polypeptides, then we may conclude that liver and lens fiber junctions have at least one major component of 26,000 dalton in common. We have also found in the fiber plasma membranes a similar polypeptide of 22,000 dalton; which may be the product of either a post-translational processing or, more likely, proteolitic degradation (18). Comparison of the amino acid analysis of both lens MP26 and the major junctional polypeptide from liver indicates that the overall polarity of the two proteins is similar. Hence both can be defined as "intrinsic" membrane proteins. The conclusion that the MP26 is a lens fiber junctional constituent is also consistent with the results of the remarkable study of Tanaka et al. (53) on the structural and chemical alterations that characterize the plasma membrane in Nakano mouse lens. The 26,000-dalton polypeptide is the major component of the plasma membrane before the development of the cataract and when the fibers are connected by extensive junctional domains. During cataract formation, the progressive removal of gap junctions is positively associated with the deletion of the 26,000-dalton polypeptide (53).

It is, however, puzzling that in the lens epithelium, where geometrically packed arrays of transmembrane oligomers indicate the existence of typical gap junctions, polyacrylamide gel electrophoretic experiments have failed to demonstrate the presence of the MP26 that may be considered as a

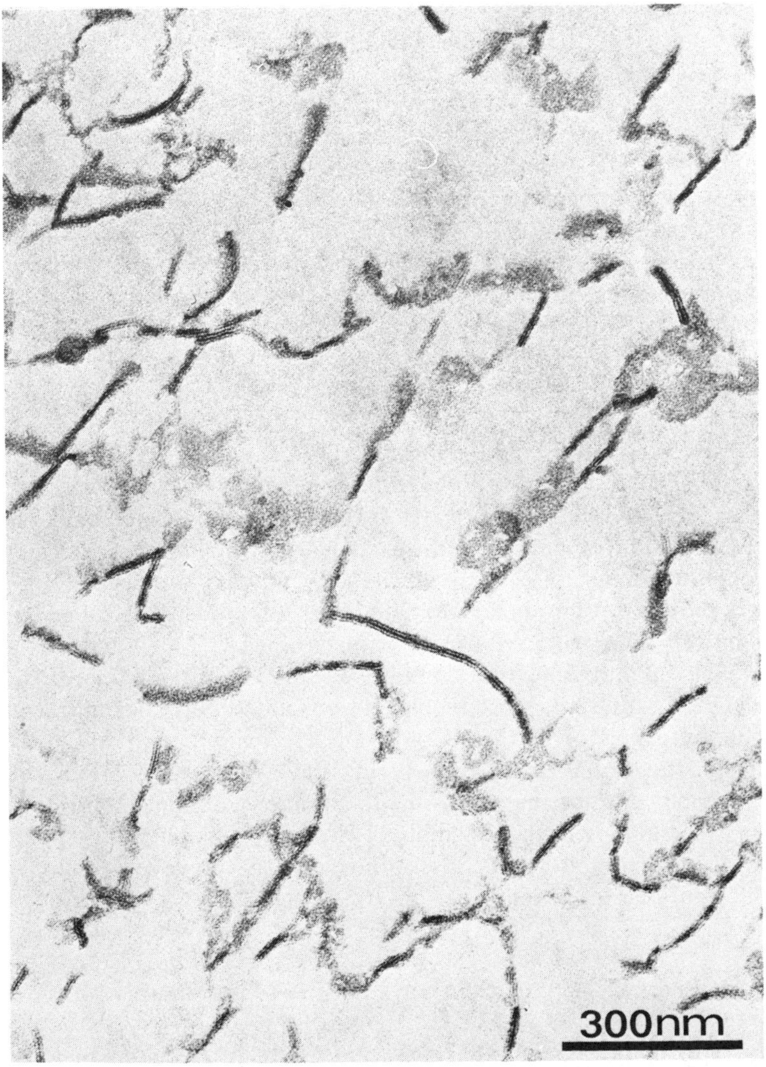

Figure 11 Isolated lens fiber gap junctions obtained by detergent solubilization followed by equilibrium density gradient centrifugation. Thin section stained with uranyl acetate and lead citrate.

universal gap junctional chemical marker. Conversely, the protein profile of isolated lens epithelium plasma membrane is characterized by the presence of the MP34 (20). The absence of the MP26 in the lens epithelium has also been corroborated by the observations of Broekhuyse et al. (7). This interesting study demonstrates by immunofluorescence technique that antiserum raised against the 26,000-dalton polypeptide does not react with the lens epithelial structures (7). Thus it is conceivable that the structural diversity shown by the epithelial gap junctions is also associated with some intrinsic difference in their protein composition if the latter feature is compared with that of the lens fiber junctions. This condition indicates that qualitative difference in protein composition, structural organization, and assembly may exist even within one presumed single junctional type. These variations seem to be inherent markers, at least in the lens, at two different stages of the differentiation program of one unique cell type. Immunoelectron microscopy of isolated epithelial plasma membranes incubated with specific antibodies directed against the MP34 and MP26 combined with ferritin-labeling will probably reinforce the latter assumption. These techniques applied to the fiber plasma membrane fraction indicate that specific antibodies directed against the MP26 do react with the junctional domains (Figure 12).

We know very little about the formation of the junctions in the epithelium. Conversely, some clues have recently emerged concerning the assembly process of the lens fiber junctional domains. In the lens region of elongation, the accumulation of newly synthesized plasma membranes and junctions parallels the increase of cell surface areas by at least a thousandfold. There the incorporation of radioactive precursors of lipid and protein is most abundant (cf. ref. 1). Polyribosomes isolated by DOC from the lens cortex are capable of directing the synthesis of MP34, but not of MP26, in a reticulocyte lysate (20). More recently, we isolated polyribosomes from the lens fiber cytoskeleton by the action of DNAse I and found that this fraction tested in a heterologous system directs the synthesis of MP26 (54, 55). Hence we may assume that the formation of fiber junctions in the elongation zone is associated with the assembly of two major membrane polypeptides, that is, the MP34 and MP26, where biosynthesis, however, is directed by two different classes of lens fiber polyribosomes.

4.3.5 The Presence of "Square Arrays" in Lens Fiber Plasma Membrane

This type of intramembrane particulate domain is found in lens fiber plasma membrane sites different from the junctions, which in some cases could be identified in replicas where extensive plasma membrane intradigitation is found (anterior and posterior sutures). The square array consists of rectangularly packed rows of particles on the fracture face P and of a linear series of grooves on the EF. The center-to-center distance is approximately 7 nm. It is remarkable that the square array organization is retained following homogenization and purification of the lens fiber plasma membrane and even after solubilization with detergent. In fact, the detergent-insoluble membrane fraction comprises the gap junctions and orthogonal particulate aggregates that appear to be indistinguishable from those of square arrays that are found in the intact lens fiber plasma membrane (Figures 13–15). This type of geometrically packed intramembranous par-

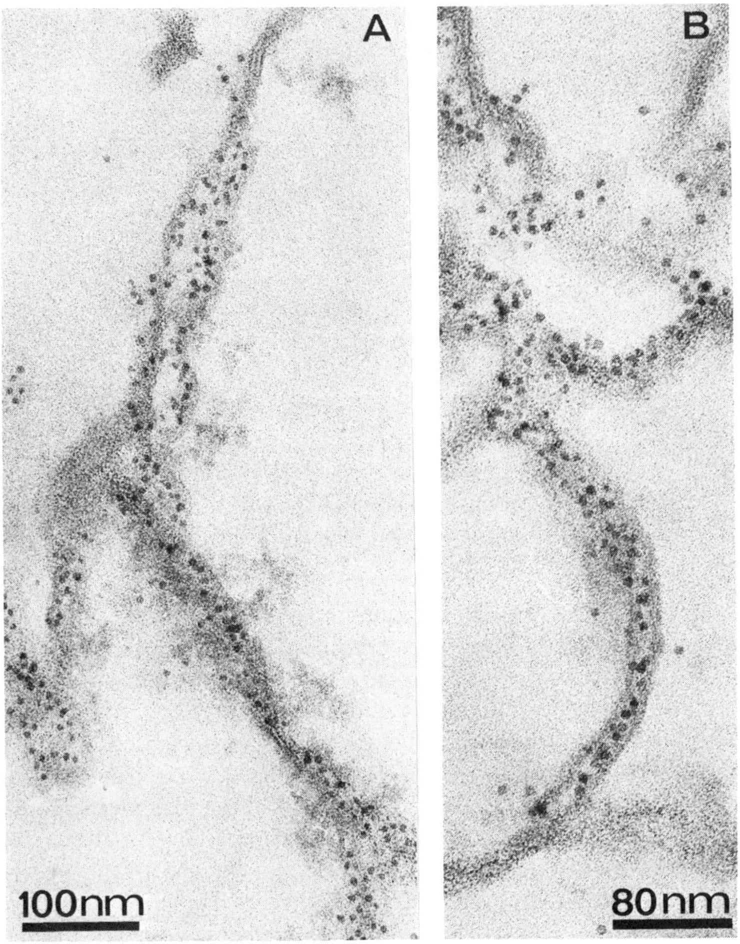

Figure 12 Isolated lens fiber plasma membranes extracted with 6 *M* urea gathered at the interface of 1.16 to 1.14 g/cm^3 in a discontinuous sucrose gradient (see ref. 2). (*A*) The isolated membranes resuspended in phosphate buffer-saline (PBS) have been incubated (overnight at 4°C) with a rabbit serum containing antibodies directed against MP26. After washing in PBS, the membranes were incubated for 1 hour with goat ferritin-labeled IGg directed against rabbit serum. The immunoprecipitate, visualized by ferritin molecules, appears asymmetrically associated with the outer leaflet of the plasma membrane. (*B*) Images suggesting that the incubation with antibodies against MP26 "unzips" the median gap region.

ticle array, distinct from the junctional domains, is found as local differentiation in plasma membranes of several types of tissues (56–61). But only in a few instances could a possible correlation between the presence of particulate arrays and membrane transport junction(s) be investigated. That is the case of toad bladder membranes, where the water movement is

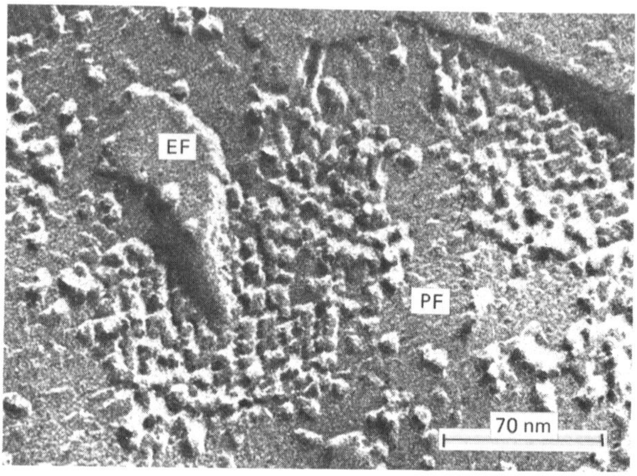

Figure 13 Freeze-fracture replica of lens fibers. The PF face is characterized by the presence of square arrays of particulates entities with a repeating distance of 7.0 nm. Details of the EF face are presented in Figure 14.

mediated by vasopressin stimulation and cAMP level. In the bladder luminal membrane exposed to different concentration of vasopressin, the number of sites and the area occupied by square arrays of particles are parallel to the level of water permeability that could be measured (57, 59, 61).

In other interesting observations, both the increase in sodium transepithelial transport and increased water permeability were porportional to the extension of the square particulate array found in the plasma membrane of frog urinary bladder epithelium exposed to oxytocin (56–58). At present we cannot conclude as to the junctional role of the "square array" present in the lens fiber plasma membrane. We can only speculate that as in other tissues they might represent membrane domains with increased permeability. In contrast to the gap junctions, which ensure rapid horizontal communication between one fiber and another, the "square array" of transmembrane particulate entities could be the site of preferential and controlled permeability between the intracellular milieu and the narrow lens extracellular domain.

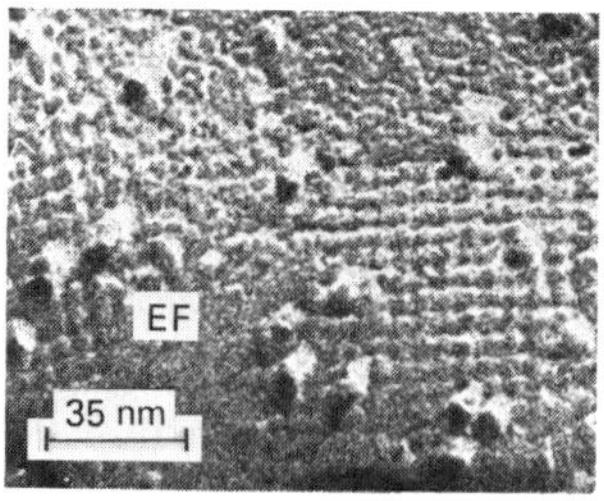

Figure 14 Freeze-fracture replica of lens fibers and square arrays of grooves on the EF face.

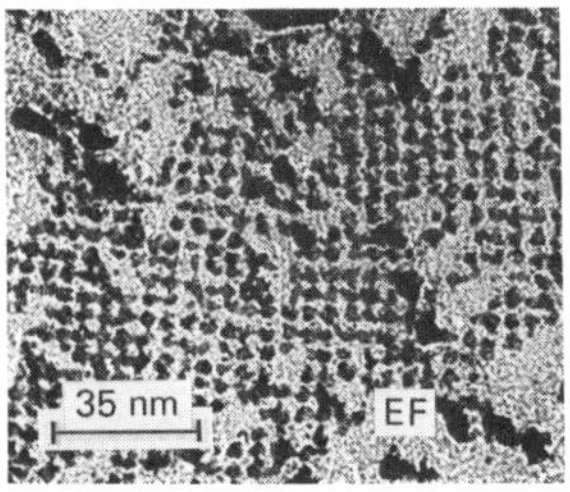

Figure 15 Freeze-fracture replica of isolated plasma membrane thoroughly extracted with 6 *M* urea. The orthogonal pattern is resistant to the chaotropic agent. Similar freeze-fracture features are found in membrane fractions insoluble either in deosycholate or in a mixture of 6 *M* urea and 5% Triton X100.

Since the detergent-insoluble fraction comprises a consistent amount of membrane fragments displaying square array organization together with remaining gap junctions, the existence of a structural transition between these two types of membrane domain architecture cannot completely be ruled out. Whether this transition is allusive of a functional switch operating *in vivo* requires further investigation. The fact that square arrays of intramembranous particulate entities are found in intact lens fiber plasma membranes casts some doubt on the assumption that the presence of the orthogonal pattern is merely the effect of junctional deterioration and/or uncoupled state.

4.4 STRUCTURAL AND BIOCHEMICAL DEFINITION OF THE PLASMA MEMBRANE-CYTOSKELETON COMPLEX

More specifically we attempt in this section to explore the existence of structural and biochemical relationships between the lens fiber plasma membrane and the motility protein that, also in the lens, may display various molecular forms. It is not surprising that, like many other (if not all) nonmuscle cells, the lens cell possesses a well-developed and elaborated cytoskeleton (see Chapters 2 and 3). In particular, electron microscopic evidence has been provided indicating that the lens fiber cytoplasmic matrix comprises both thin (7-nm) filaments and intermediate-sized (10-nm) filaments (9). In agreement with the morphological evidence, actin polypeptides and a putative protein subunit of the 10-nm filaments have been thoroughly purified and characterized from both water-soluble and water-insoluble lens fractions (62). In that respect the lens cytoskeleton constituents show, as do other nonmuscle cell systems, the property of being in a rather transient and dynamic state resulting from the combination of both a water-soluble and a highly polymerized insoluble configuration at physiological temperature. In contrast to the stable sarcomeric design principle, the assembly and disassembly of the motility protein in nonmuscle cells are regulated by several cofactors, either favoring filament formation or preventing this process (63). In the lens this control mechanism has not been elucidated yet.

4.4.1 Isolation of the Plasma Membrane-Cytoskeleton Complex

The isolation of plasma membranes from lens fibers involves several steps, including homogenization in 1 m*M* sodium bicarbonate and differential gradient centrifugation, the whole procedure being usually performed at low temperature (2). To obtain the "purest" membrane fraction, digestion by proteases is eventually applied. Furthermore, many studies on lens membranes have been carried out on the urea-insoluble lens fraction. Obviously, under all these conditions the labile structural organization of the cytoskeleton and its association with plasma membranes are severely damaged. In fact, the stability of this complex greatly depends on proper ionic and temperature conditions. To obtain the plasma membrane-cytoskeleton complex we have applied an isolation procedure analogous to the former (2) with the exception that the temperature was kept at 15°C and that the homogenization and gradient centrifugation were carried out in a medium containing 40 m*M* KC1, 1 m*M* $MgCl_2$, and 1 m*M* sodium bicarbonate (9, 54, 62). Eventually, for preservation of the ribosomal structure, the Mg^{2+} concentration was raised to 10 m*M* $MgCl_2$.

A striking morphological difference characterizes the membrane fractions that have been isolated in sodium bicarbonate compared to those obtained under stabilizing conditions for cytoskeletal elements. Whereas in the former fractions the cytoplasmic surface of the plasma membranes is practically devoid of any filamentous or stranded structure, the membrane fractions isolated under the latter conditions are characterized by close association of filamentous structures to the inner site of the membrane sheets (Figure 16).

In thin sections membrane profiles, displaying a "ghostlike" appearance, contain two different types of filamentous structures. Examination at suitably high magnification of thin sections of isolated fractions reveals a considerable amount of thin filaments (about 5 to 7 nm in diameter) with end-on attachment to the inner cytoplasmic surface of the plasma membranes. Some of these interconnected filaments have a smooth surface; instead, multiple particulate entities ranging in diameter between 10 to 20 nm are associated with the fibrillar arrays (Figure 17). Some rows of particles are connected by filamentous material whose diameter ranges down to dimensions difficult to measure accurately in thin sections (Figure 17*a*). The other type of cytoskeletal elements is represented by intermediate-sized 10-nm filaments that are "clean" and smooth (Figures 18 and 19).

The one-dimensional SDS gel electrophoretic pattern of the proteins that constitute the plasma membrane-cytoskeleton complex (Figure 20, and for comparison Figures 2 and 3) shows that a polypeptide of 57,000 mol. wt. comigrating with vimentin (see Chapter 3) is one of the main constituents of the complex. Actin is also a predominent polypeptide. α-Crystallin, presumably α-actinin, and some minor components in the 47,000 to 55,000

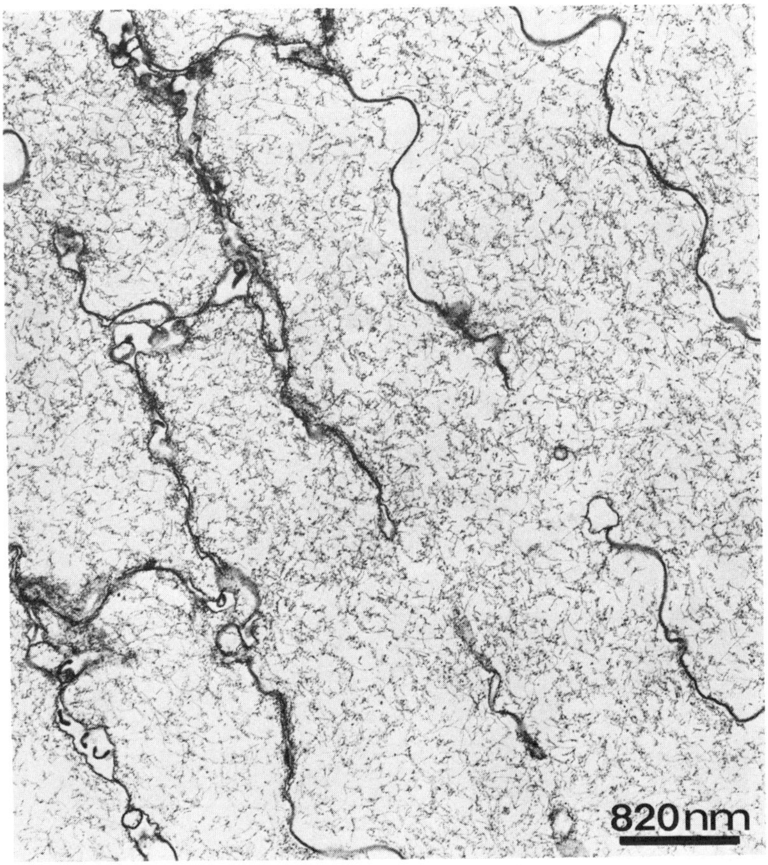

Figure 16 Thin section stained with uranyl acetate and lead citrate of isolated lens fiber plasma membrane-cytoskeleton complex. Note the "ghostlike" appearance of the membrane profiles comprising cytoskeletal elements.

mol. wt. region are seen. The nature of the high molecular weight polypeptides is not yet easily apparent. The component with an apparent molecular weight of 250,000 does not comigrate with muscle myosin (unpublished results). However, the presence of myosin in the lens has been claimed (64).

Two-dimensional gel electrophoresis (Figure 21) indicates that the cytoskeletal constituents have acidic isoelectric points. Next to β- and γ-actin (not visibly separated in this gel), vimentin exhibits a slightly more acidic isoelectric variant as described for chinese hamster ovary cells and chicken nonmuscle cells (65, 66). The α-actinin polypeptide has an isoelectric point comparable to that of vimentin.

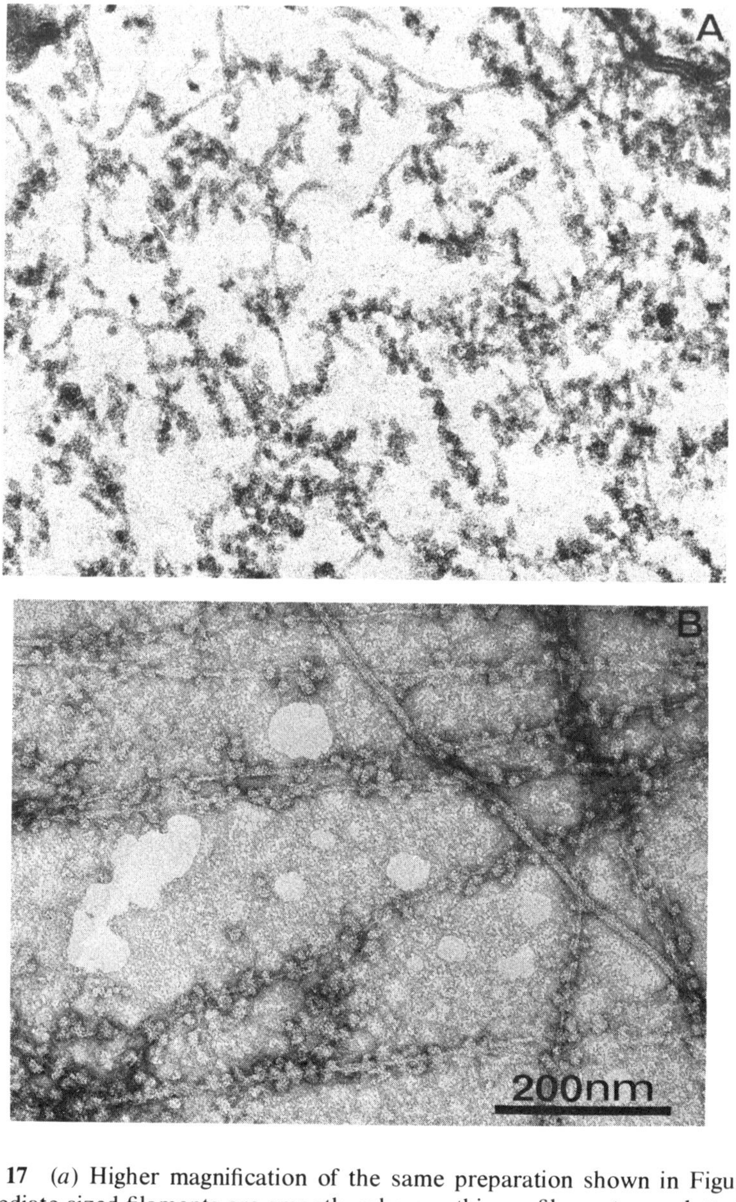

Figure 17 (*a*) Higher magnification of the same preparation shown in Figure 16. Intermediate-sized filaments are smooth, whereas thinner filaments are dotted with particulated entities of 10 to 20 nm. (*b*) Isolated lens fiber cytoskeleton negatively stained by uranyl acetate. The thin filaments (7 nm) are "decorated" by globular subunits. Note the smooth appearance of the intermediate-sized filaments (10 nm).

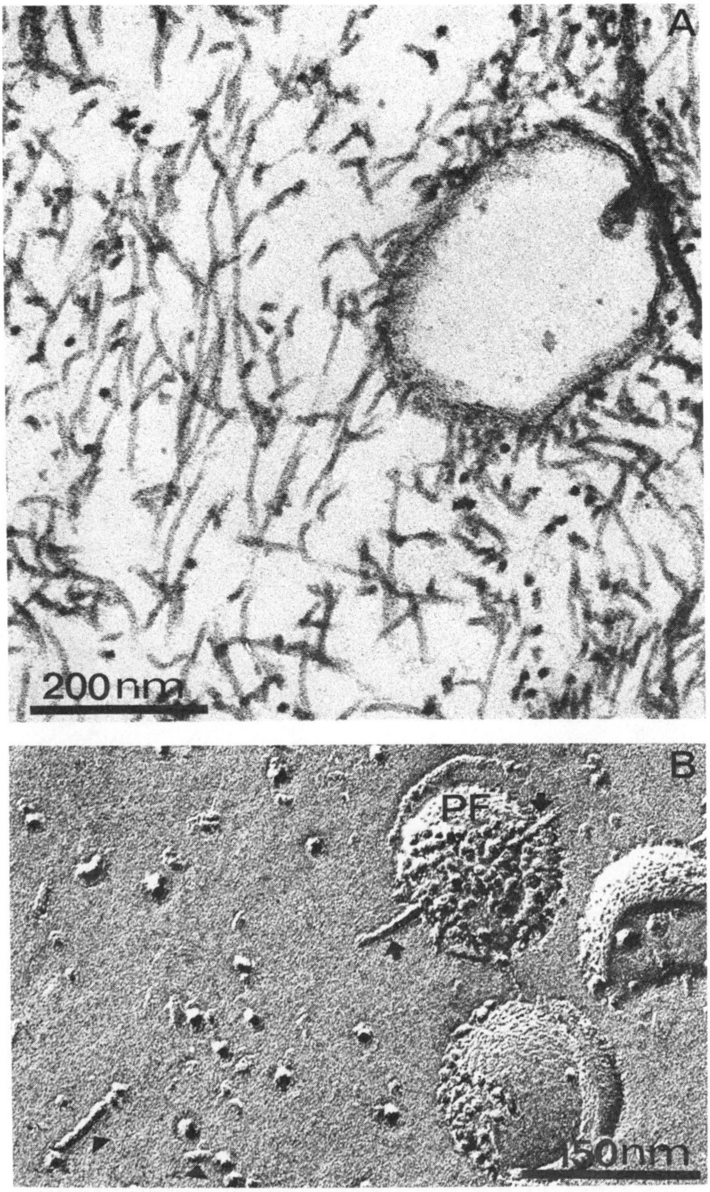

Figure 18 (*A*) Isolated lens fiber plasma membrane-cytoskeleton complex. Note the end-on attachment of the intermediate-sized filaments to the membrane profile. Thin section stained with uranyl acetate and lead citrate. (*B*) Freeze-fracture replica of lens fiber plasma membrane-cytoskeleton complex. The arrows point to intermediate-sized filaments that appear to be associated, with their terminal segment, to the hydrophobic protoplasmic fracture face (PF). Intermediate-sized filaments (arrow head) are visualized in cross or longitudinal fracture.

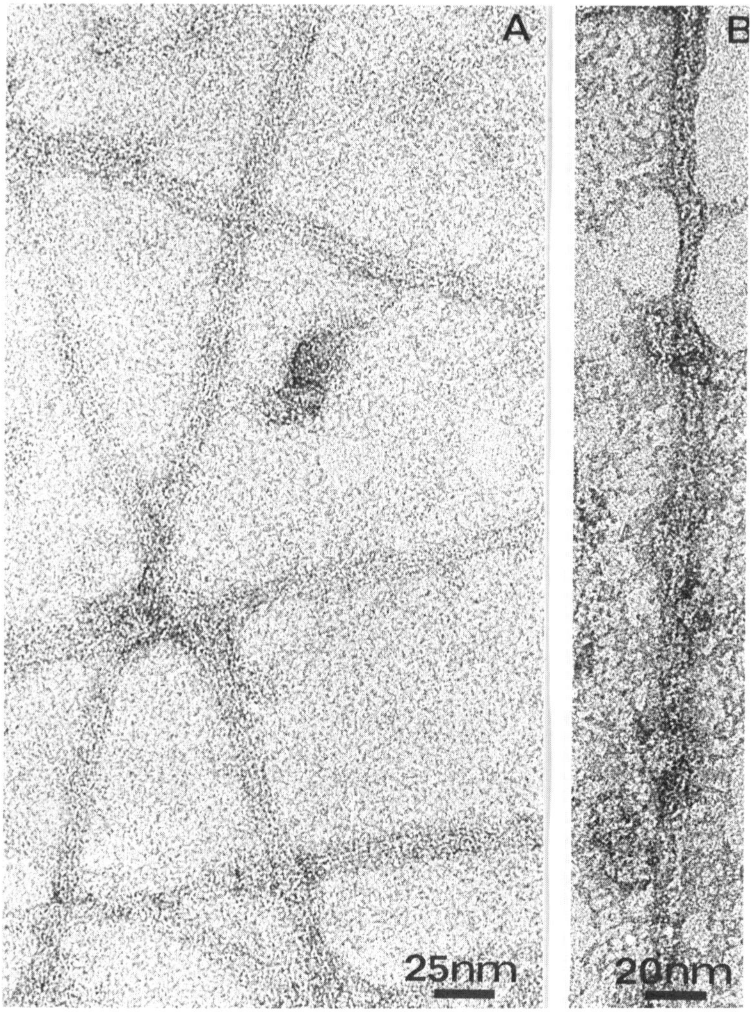

Figure 19 (*a*) Lens fiber cytoskeleton intermediate filaments (10 nm). Note the protofibrillar organization outlined by the negative staining with uranyl acetate. (*b*) An intermediate filament after the action of DNAse I incubation. Irregular-sized segments and disassembly are visible along the filament length.

Figure 21 Two-dimensional gel electrophoretic analysis of lenticular cytoskeletal proteins; actin (A) and vimentin (V). Markers as in Figure 20.

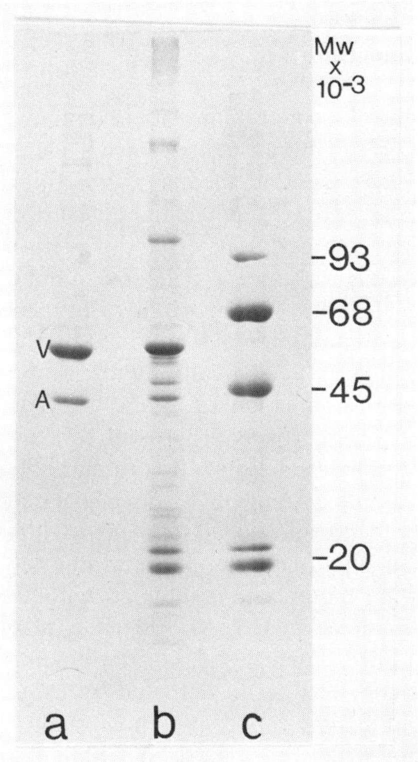

Figure 20 SDS gel electrophoretic analysis of polypeptides that make up the lenticular plasma membrane-cytoskeleton complex. (*a*) Markers from bovine lens cell culture; actin (A) and vimentin (V). (*b*) Plasma membrane-cytoskeleton complex from bovine lens fibers. (*c*) Marker proteins, (α-crystallin 20,000 D; ovalbumin, 45,000 D; bovine serum albumin, 68,000 D; and phosphorylase A, 93,000 D).

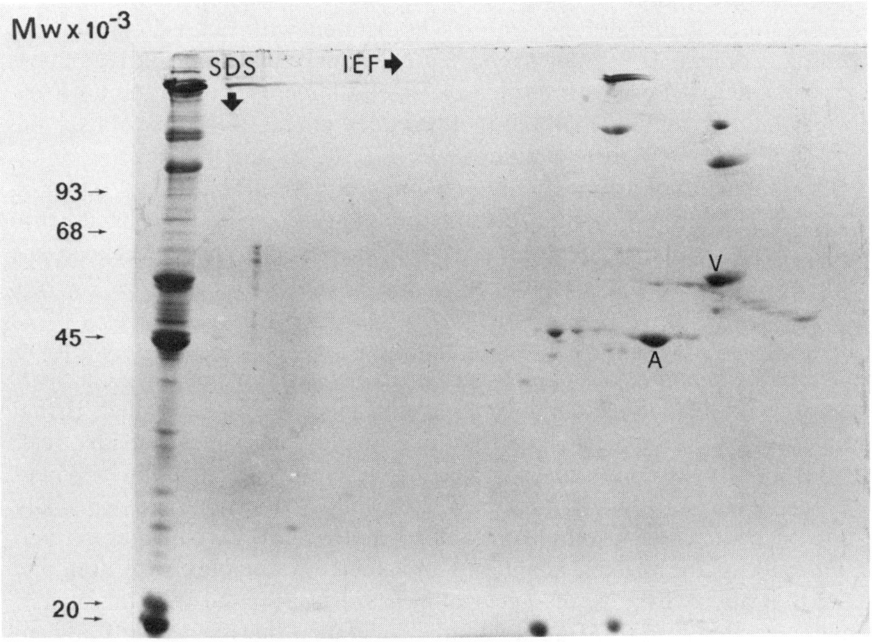

4.4.2 The Nature of the Beaded Microfilaments

Further information on the nature of the thin microfilaments is obtained by the application of the negative staining technique to the plasma membrane-cytoskeleton complexes (Figure 22). By this means, 7-nm thick filaments, when not marked by the globular entities, appear to consist of a double row of subunits arranged in a helical configuration. From this feature we may assume that 7-nm lens fiber filaments have a common subunit organization as actin filaments visualized in muscle or nonmuscle cell types. Further support for the actin nature of the thin filaments is provided by the observation that the latter class of filaments specifically binds rabbit antibodies directed against purified lens actin (62) (Figures 23 and 24). The binding is visualized in transmission electron microscopy by using ferritin-labeled antibodies directed against rabbit immunoglobulin. It is remarkable that antiactin ferritin-labeled antibody complexes appear to be associated even to plasma membrane domains where no end-on attachment of filamentous structure is observed (Figure 25). This feature might reinforce the assumption that actin polypeptides may be membrane-associated without displaying a polymerized filamentous organization (63, 67, 68). It is also noteworthy that in contrast to the results presented in ref. 69, the binding of rabbit immunoglobulin raised against calf lens actin, detected by using peroxidase-labeled Fab fragments directed against rabbit immunoglobulins, occurs on the totality of the filamentous structure comprised within the fiber ghost. No preferential accumulation of the immunoperoxidase complex is found at the cytoplasmic surface of the membranes (Figures 26 and 27).

The nature of the globular entities associated with actin filaments is a puzzling problem. It has been reported previously that in chick lens fibers the cytoskeleton, besides intermediate-sized filaments, comprises an irregular beaded filamentous material that has been characterized by Maisel and Perry (70), Perry (71), and Bradley et al. (72) as "chains" consisting of 12- to 15-nm particles arranged on a filamentous backbone 7 to 9 nm in diameter. The latter structure, presumably consisting of polymerized actin, was not consistently "decorated" by heavy meromyosin, because the globular material most probably occupied specific binding sites (72; see Chapter 2).

From the interesting observations mentioned before, however, the identification of the nature of the globular particules still remains to be clarified. At least two alternatives come to mind, the most obvious of which is that a molecular class of particulated entities display similar, if not identical, structural features to the macromolecular assembly of crystallin polypeptides, especially of α-crystallin (see Chapter 2). In addition, and especially to demonstrate water-insoluble crystallin fractions, we have further extracted the bulk of the plasma membrane-cytoskeleton complex with urea and found that about 2 to 5% of α-crystallin is still associated with this water-insoluble material (73). More qualitative evidence that crystallin polypep-

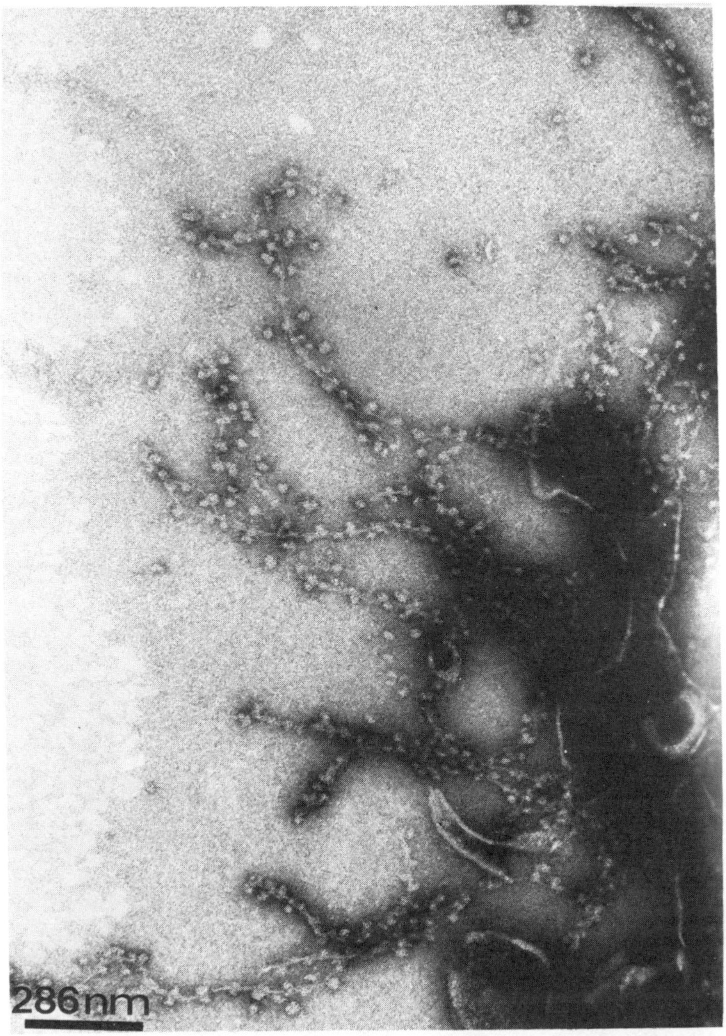

Figure 22 Isolated lens fiber plasma membrane-cytoskeleton complex negatively stained with uranyl acetate. Double-stranded 7-nm filaments are associated with the membrane sheets. Several particulated entities decorated the filaments.

tides could well be associated with microfilaments was also obtained by showing that actin is consistently isolated together with α-crystallin after urea extraction. It is also remarkable that if the water-insoluble lens fraction was extracted by 5 m*M*, TES buffer [2 tris (hydroxymethyl)methyl-2-aminoethane sulfonic acid] (51), both actin and a considerable amount of α-crystallin were detected in the dodecylsulfate gel electrophoretic pattern of

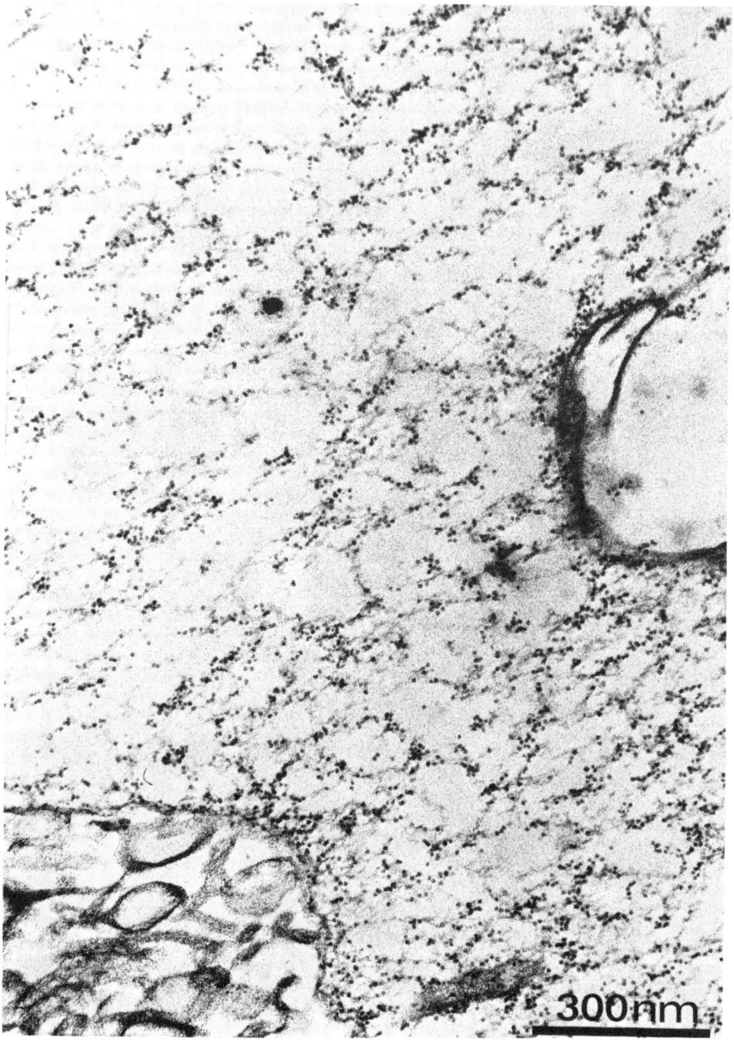

Figure 23 Isolated lens fiber plasma membrane-cytoskeleton complex incubated with rabbit antibodies against calf lens purified actin. The immunocomplex associated with the cytoskeletal elements is revealed by ferritin-labeled antibodies directed against rabbit immunoglobulin. The ferritin-labeled antibody forms rows along the thin filaments. Thin section stained with lead citrate.

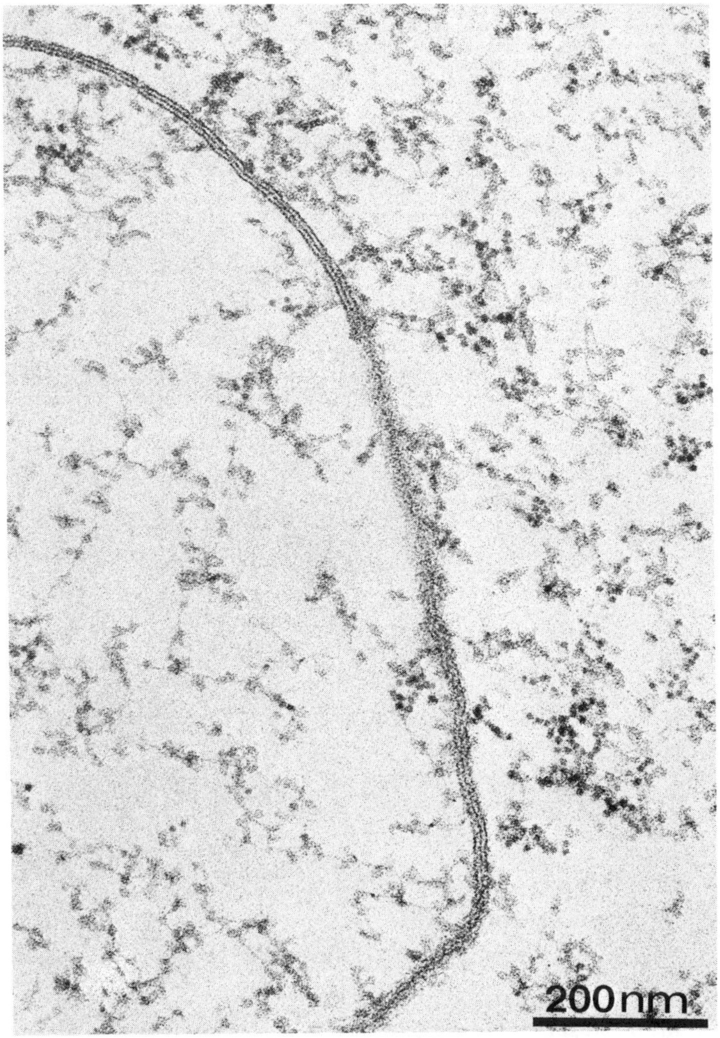

Figure 24 The same preparation as Figure 12, with the exception that the incubation was made with rabbit antibodies against calf lens purified α-crystallin. Note the existence of two classes of particulate entities associated with thin filaments. One, most probably the ribosomes, does not react with the anti-α-crystallin; the other, crystallin globular structures, is marked with ferritin molecules.

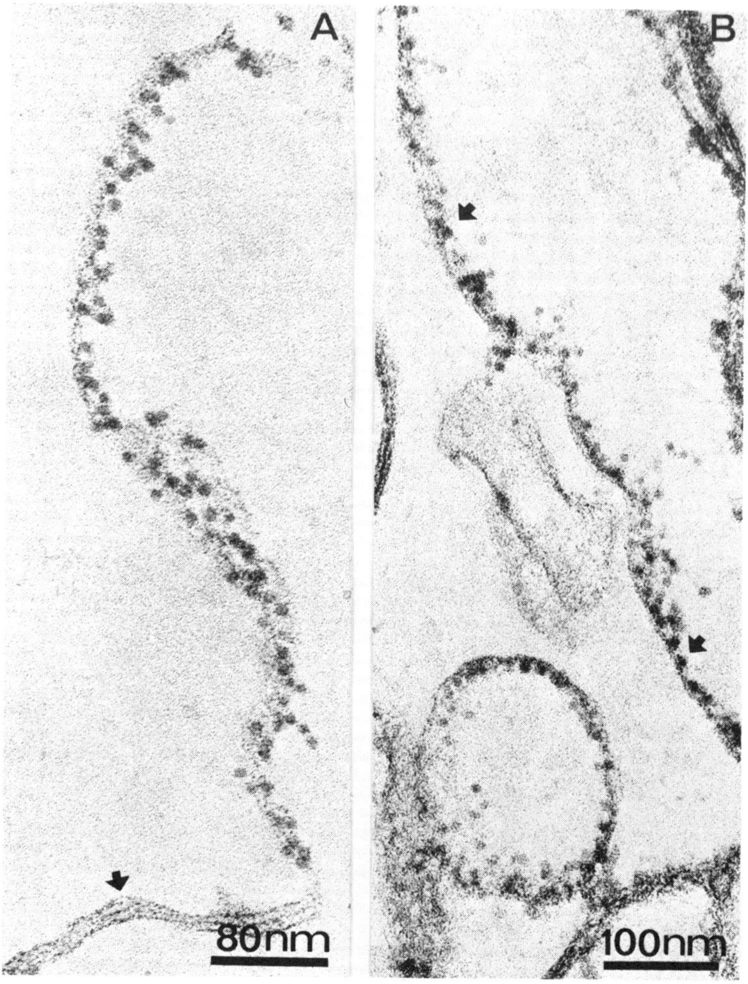

Figure 25 (*a*) Isolated lens fiber plasma membrane under conditions unfavorable to preserving the cytoskeleton and incubated with antibodies against lens actin as indicated in Figure 12. The ferritin-labeled immunocomplex is asymmetrically associated with the membrane element but not with the gap junction (arrow). (*b*) Same preparation with the exception that the antiserum was directed against α-actinin. The latter constituent is apparently asymmetrically associated with the cytoplasmic site of the plasma membrane (arrow).

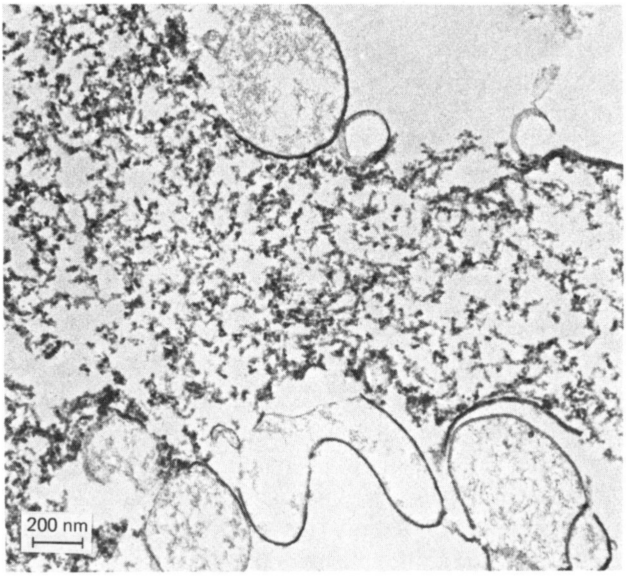

Figure 26 Isolated plasma membrane-cytoskeleton complex from lens fiber. The fraction has been incubated with rabbit antiserum raised against calf lens actin. After repeated washing, the fraction has been incubated in peroxidase-labelled Fab fragments against rabbit immunoglobulins, was treated with diaminobenzidine, and then processed for thin sectioning. The immunoperoxidase complex is found to be associated with the totality of the filamentous material (actin filaments) comprised within the membrane profiles.

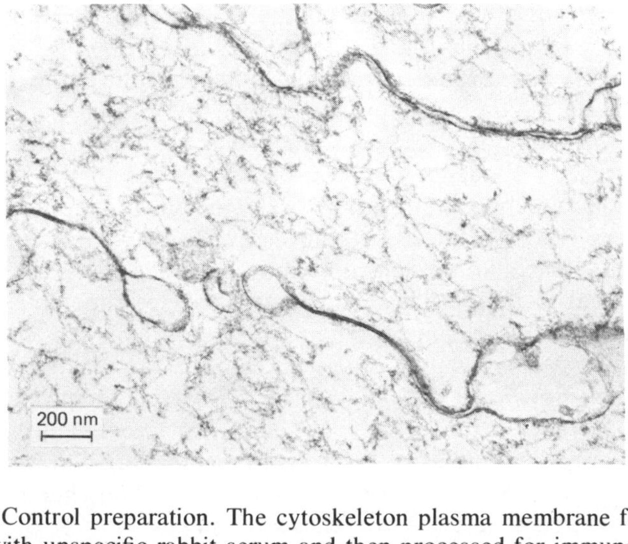

Figure 27 Control preparation. The cytoskeleton plasma membrane fraction was incubated with unspecific rabbit serum and then processed for immunoperoxidase reaction. Note the absence of immunoprecipitate, whereas the actin filaments are easily detectable.

the eluted material. Other results also indicate the existence of some interaction between actin and crystallin, since both components are retained on a Sephadex-DNAse I affinity column and are eluted together only by high concentrations of guanidine hydrochloride (62). Alternatively, we may assume that at least part of the globular entities associated with the thin microfilaments in the lens fibers consists of ribosomal structures (9). Electron microscopy is expected to provide the illustration of whether polyribosomes are associated with the filamentous consitituents of the lens cortex cytoskeleton. Unfortunately, in this system it is not easy to identify morphologically which particulate entities are indeed ribosomes rather than crystallin assemblies. We must admit that the evidence for the ribosomal nature of the particulate entities associated with actin filaments is essentially indirect. During the isolation of the plasma membrane fraction from the cortical region of the calf lens using discontinuous gradient centrifugation, we observed (2) that the heavy plasma membrane fraction gathering at the interface between densities 1.16 and 1.18 g/cm^3 was bearing giant ribosomal clusters, strongly attached to the plasma membrane via a protease-sensitive link. These ribosomal clusters have a close resemblance to the giant ''polyribosomes'' depicted in the early observations of Spector and Travis (74). This latter supramolecular assembly, active in amino acid incorporation *in vitro*, was found to be heterogeneous in nature in so far as, besides RNA, it comprised large amounts of crystallin polypeptides (75). Furthermore, the giant assemblies were RNAse-resistant, but highly trypsin-sensitive (75). Electron microscopic observations clearly indicated that as a result of careful tryptic digestion breakdown of the giant structures occurred and the end product was a class of polyribosomes consisting of a maximum of nine monomers per strand (75). In addition to these earlier observations, more recently stereo high-voltage electron microscopy of whole cultured cell lines (76,77) has given a strong indication that the three-dimensional network forming the cytoskeleton comprises polyribosomes in association with filamentous components (''microtrabecular lattice'' of the cytoplasmic ground substance). Evidence has also been provided that in HeLa cells, after membrane solubilization, the cytoskeleton retains the bulk of active polyribosomes (78). All these facts strongly support the assumption that the structural organization of lens polyribosomes relies, to some extent, on their attachment to the plasma membrane-cytoskeleton complex. This led us to incubate the plasma membrane-cytoskeleton complex isolated from lens fibers with DNAse I rather than with trypsin. The result indicates that polyribosomes showing *in vitro* uptake of radioactive amino acids can be released by DNAse I and further purified on a sucrose density gradient. Actually, from earlier studies of Lazarides and Lindberg (79) it was known that actin binds to DNAse I. Further observations of Hitchcock et al. (80) have demonstrated that DNAse I causes depolymerization of F-actin to form a 1 : 1 complex. This process is slowed down but not prevented if tropomyosin and troponin are associated with F-actin. Conversely, in the absence of ATP heavy meromyosin completely stabilizes F-actin against DNAse I action. It is important to note that other factor(s) and not only these regulatory proteins might prevent or slow down the depolymerization of F-actin by DNAse I, or in a more general sense "stabilize" the microfilaments (81). DNAse I-resistant macromolecular actin has been found in red blood cells where it is stabilized by a protease-sensitive component (81). Even in the lens plasma membrane-cytoskeleton complex, DNAse I-resistant microfilaments are found. SDS gel electrophoresis experiments show that this fraction still contains a considerable amount of actin. Our knowledge of the regulatory protein that in the lens may control actin polymerization is too poor to allow further discussion of this interesting problem. It would not be surprising if crystallin polypeptides should play a role in the stabilization of the microfilaments. Whatever will be the answer to this hypothesis, it is not unreasonable to conclude that in the plasma membrane-cytoskeleton complex there exists at least a class of F-actin-associated giant strand of polyribosomes that can be disassembled by incubation with protease-free DNAse I (Figures 28 and 29). We have already discussed in Section 4.2.2 that the synthesis of MP34 occurred upon addition of "free" lens polyribosomes to a heterologous cell-free system, whereas MP26 was absent among the translation products (20, 45). The routine procedure in which polyribosomes were prepared consisted of DOC treatment of the total lens cortex homogenate followed by centrifugation at 15,000 × *g* as described earlier (75). Conversely, when the polyribosomes released from the plasma membrane-cytoskeleton complex by DNAse I incubation are added to a reticulocyte lysate, it appears that they are capable of directing the synthesis of the missing MP26 polypeptide (9, 54, 55).

A question, which cannot easily be answered, concerns the nature of the link between F-actin and polyribosomes. After DNAse I incubation of the plasma membrane-cytoskeleton complex, the 10,000 × *g* supernatant contains large clusters of polyribosomes linked by strands displaying the features of partially depolymerized actin filaments (Figures 28 and 29; cf. Figure 30). From these images it is, however, impossible to determine whether the attachment to the actin filaments is made by the small or the large ribosomal subunit or by the messenger strand. Penman et al. (82) claim that most of the cellular polyribosomes are linked to the skeletal framework via mRNA. At least in the lens fiber system the missing link remains to be elucidated.

The complexity of the lens fiber plasma membrane-cytoskeleton intersection is further demonstrated by the solubilization experiments in 0.6 *M* KI (9). For this purpose, the isolated membrane fraction was resuspended in 0.6 *M* KI and stirred for 1 hour at 4°C. Centrifugation at 100,000 × *g* for 1 hour was applied, and the obtained pellet was further extracted in 0.6 *M* KI, resuspended, and kept while stirring for 12 hours at 4°C. This centrifugation step was repeated, and the final pellet as well as the supernatant

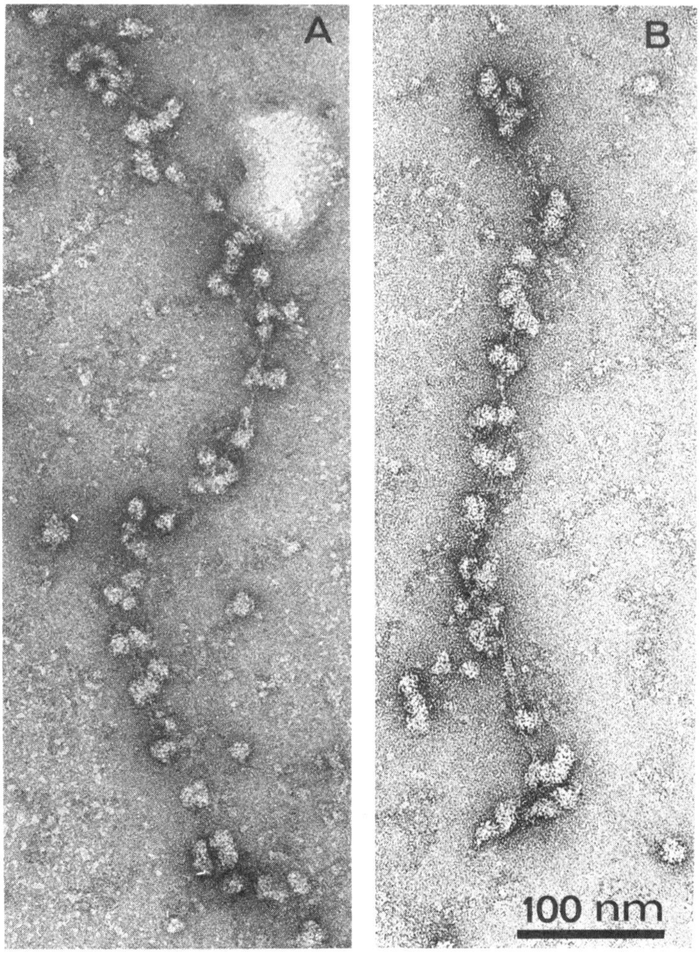

Figure 28 Giant arrays of particulate entities linked together by thin cytoskeletal filaments partially degraded by the incubation (1 hour) in protease-free DNAse I (0.2 mg/mg of protein). Negative staining with uranyl acetate.

were both analyzed by polyacrylamide gel electrophoresis in SDS and by electron microscopy. In another experiment the pellet obtained after the second extraction in 0.6 *M* KI was resuspended and centrifuged through a discontinuous sucrose gradient (9). The plasma membrane-cytoskeleton complex was also extracted with TES-buffer (62), and the insoluble material was obtained with the same centrifugation steps as described for the 0.6 *M* KI extraction. Electron microscopic observation on the KI- or TES-insoluble fraction shows that even after prolonged extraction, thin microfilaments, heavily "decorated" by globular entities—most likely macromolecular crystallins—are still detectalbe in close association with membrane fragments (Figure 31). On the other hand, SDS gel electrophoresis indicates that polypeptides comigrating with actin are present both in the soluble supernatant and in the sedimentable material where microfilaments are found by electron microscopy.

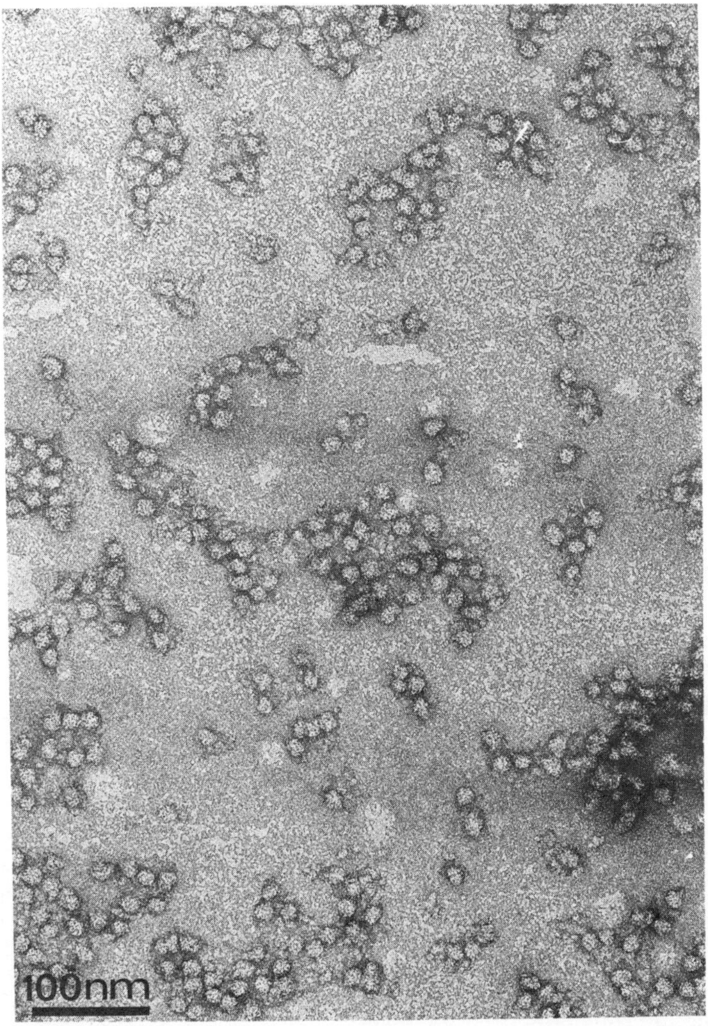

Figure 29 Polyribosomes released by DNAse I incubation, purified through a discontinuous sucrose density gradient. The polyribosomes are collected in the pell. Negative staining with uranyl acetate.

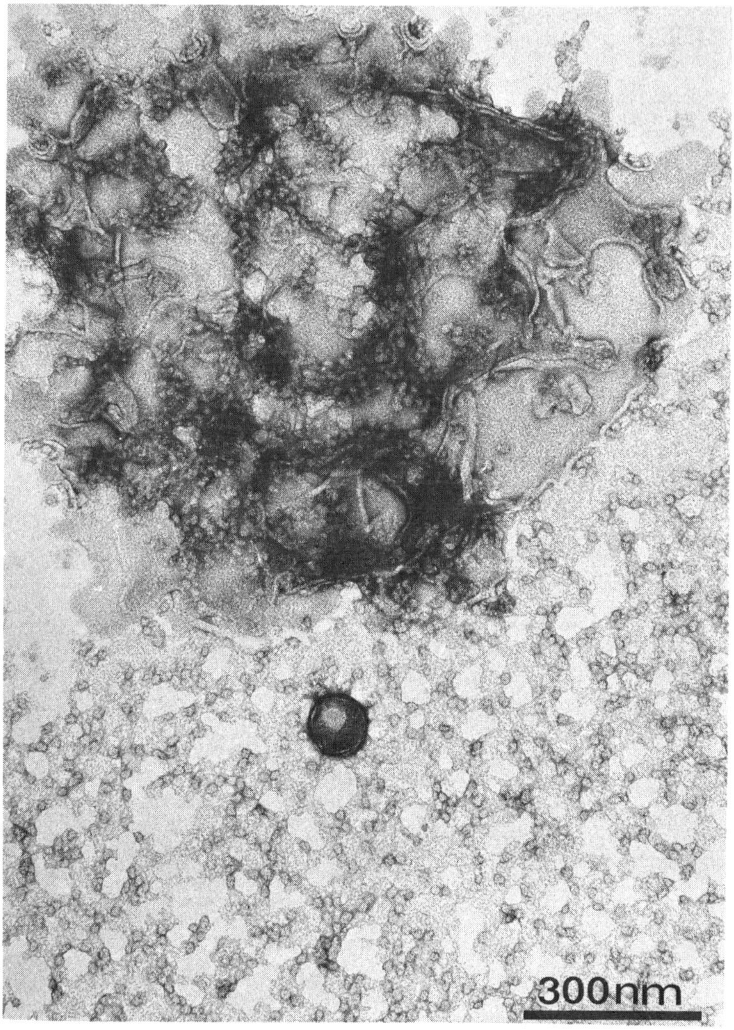

Figure 30 Membrane-bound polyribosomes gathered at the interface between 1.5–2.0 M sucrose layer of the discontinuous gradient as in Figure 16. Membrane-bound polyribosomes are found, which may be directly attached to the plasma membrane via a DNAse-resistant link.

4.4.3 The Intermediate-Sized Filaments

Electron microscopic observations clearly show that the other type of filamentous structure found in the isolated plasma membrane-cytoskeleton complex has features in common with the intermediate-sized (10-nm) filaments that have been described in a variety of nonmuscle cells as well as in smooth muscle (83–89). The membrane fraction enriched by a high amount of intermediate filaments comprises a major component that in SDS gel electrophretic patterns is found in the 55,000 to 57,000-dalton region (Figures 32 and 33) (1, 9, 54–62). The problem of the subunit composition of intermediate-sized filaments is extremely complex if one compares the subunit features of this cytoskeletal constituent present in various types of

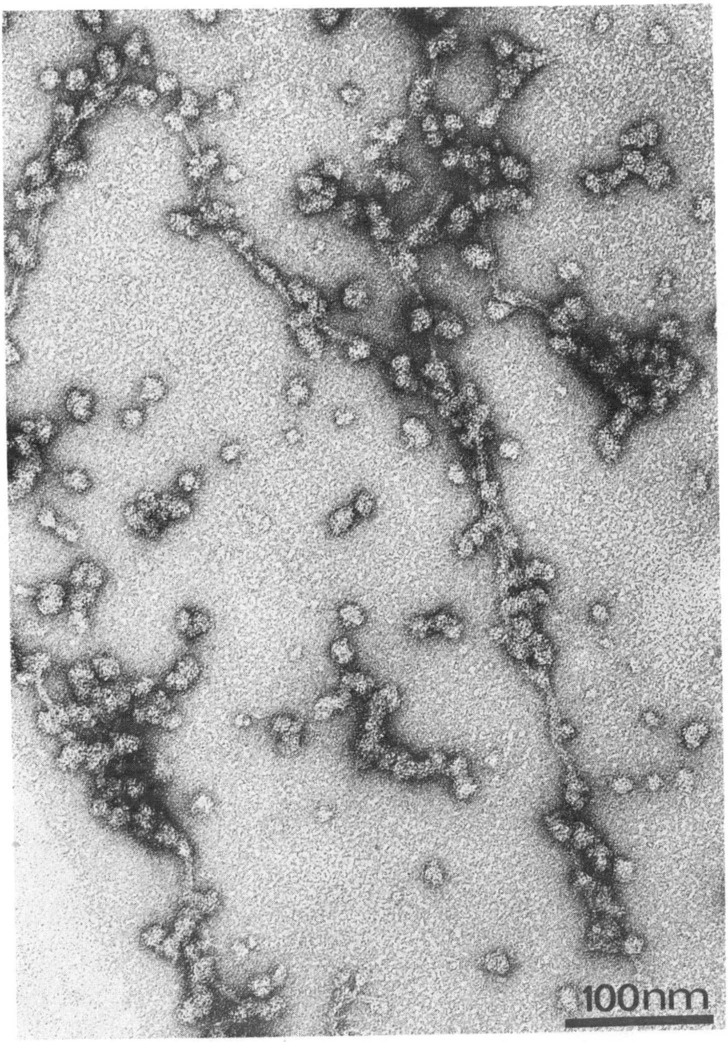

Figure 31 Giant assemblies of crystallin molecules and ribosomes associated by actinlike filaments that resist 0.6 *M* KI solubilization. Negative staining with uranyl acetate.

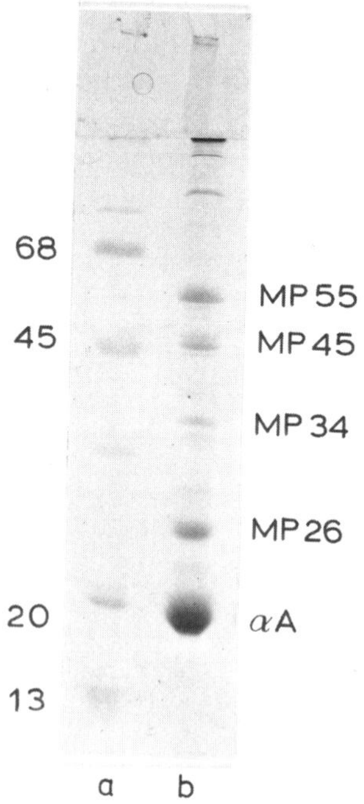

Figure 32 SDS gel electrophoresis of isolated lens fiber plasma membrane-cytoskeleton complex. (*a*) Marker proteins: cytochrome C, 13,000; αA-crystallin, 20,000; ovalbumin, 45,000; and bovine serum albumin, 68,000 (for comparison). (*b*) Two major polypeptides are visualized in the 45,000- and 55,000-dalton region, in addition to some crystallins and the membrane proteins MP26 and MP34.

cells (83–93). This class of polypeptides originally identified in smooth muscle as desmin or skeletin is, in fact, heterogeneous, and isoelectric variants can be indentified (63, 85–87, 90–94).

In higher vertebrate cells of mesenchymal origin, which contain intermediate-sized filaments, a major polypeptide—called vimentin—with an apparent molecular weight of 55,000 to 58,000 has been identified (86) (see Chapter 3). In a number of immunological and biochemical properties vimentin-containing filaments are distinct from the intermediate filaments found in several epithelial types that comprise prekeratinlike proteins or cytokeratins (95). From these interesting studies it emerges that the terms "desmin" and "skeletin" should be applied only to the identification of the subunits of intermediate filaments found in smooth muscle cells (85, 90). On the other hand, neurofilaments, although having morphological aspects in common with the other types of intermediate-sized filaments, have a polypeptide composition that differs in many respects from vimentin, cytokeratin, and desmin (cf. ref. 86). Thus one might assume that the subunit composition of intermediate filaments could be considered a taxonomic character that allows the identification of the cell origin.

Our study still in progress in association with B. Whalen (unpublished) has already shown that the intermediate filaments found in the plasma membrane-cytoskeleton complex of calf lens fibers comprise vimentinlike polypeptides, which in principle should indentify vertebrate cells of mesenchymal origin. This result is surprising, because the lens fibers have an epithelial origin, and the entire lens is generated from ectodermal cells. It is also noteworthy that vimentin-containing intermediate filaments have been identified in elongating bovine lens cells in culture.

Further studies, in particular those leading to the knowledge of the primary structure of the different types of intermediate-sized filament subunits, are needed to establish whether these polypeptides are evolutionarily related and have a taxonomic character. Coexistence of desmin and vimentin subunits has already been described in muscle as well as in nonmuscle cells (92).

Another intriguing result concerns the effect of protease-free DNAse I, which induces morphological alterations of the lens intermediate-sized filaments (Figure 34), although the 57,000-dalton component seems to remain mainly associated with the DNAse I-insoluble fraction of the plasma membrane-cytoskeleton complex (Figures 18–20). Further doubts about the

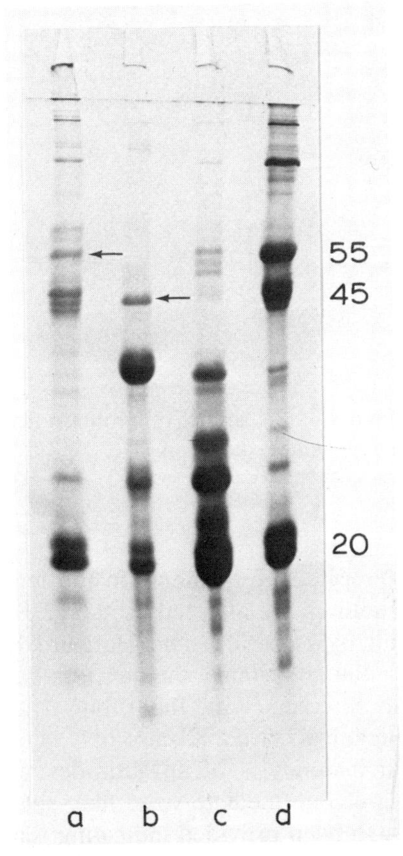

Figure 33 SDS gel electrophoretic pattern of lens fiber plasma membrane-cytoskeleton complex after the action of DNAse I. (*a*) Preparation collected from a sucrose density gradient at density 1.20 to 1. Apparently, the 55,000-dalton component (vimentin, arrow) remains mainly associated with the DNAse-insoluble pellet. (*b*) In the supernatant a polypeptide comigrating with actin is found (arrow). For comparison, the water-soluble crystallins are shown (*c*) as well as the urea-soluble lens proteins (*d*). Molecular weights are given in kilodalton.

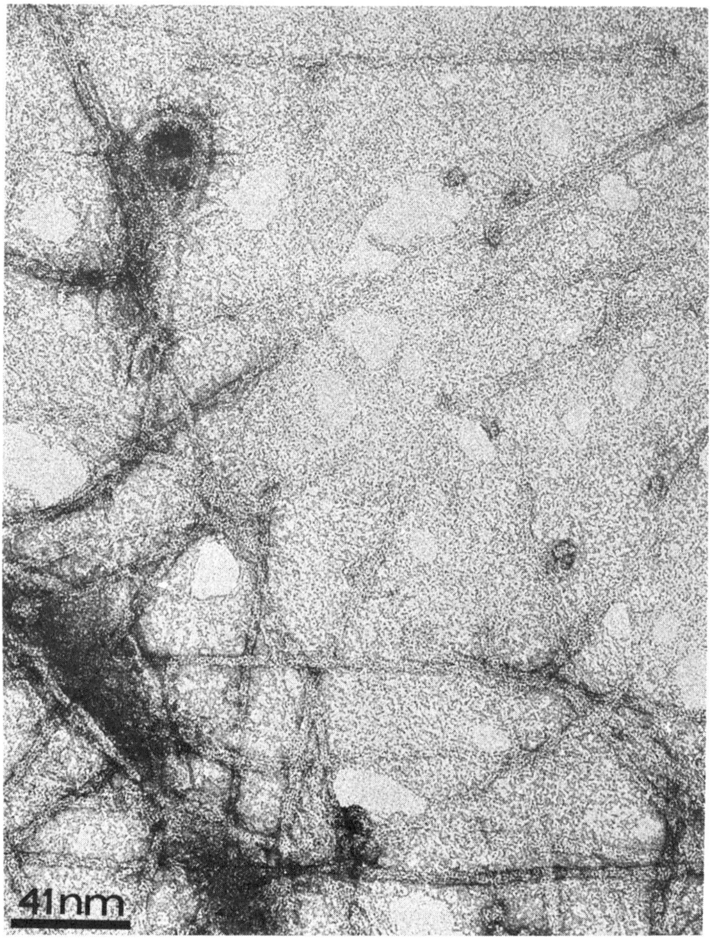

Figure 34 Intermediate filaments disassembled by the action of protease-free DNAse I. The filaments are swollen, fragmented, and reduced in size. Negative staining with uranyl acetate.

sole and unique vimentin nature of the intermediate filaments in the lens are cast by the observation that these structures are affected by 0.6 *M* KI treatment, as has also been demonstrated by Lazarides and Hubbard in smooth muscle (96). With regard to other cellular systems, the question has already been raised as to whether actin interacts with the subunits of intermediate filaments. For instance, in chicken gizzard a KI-insoluble actin fraction copurifies with desmin (97), and by dialysis of an actin-desmin solution a filamentous network, consisting of intermediate-sized filaments, is observed. Experimental evidence has also been provided indicating that in a variety of cell types intermediate filaments might contain a core of F-actin (98, 99). The same study also shows that when applied to glycerinated myosin-extracted cells, DNAse I induces a progressive breakdown in all filaments including the intermediate-sized filaments (99). All these data lead to the notion that actin might be involved in a structural interaction (cf. ref. 100) with other polypeptides such as vimentin and desmin forming the intermediate filaments.

4.5 CONCLUDING REMARKS

The vertebrate eye lens has repeatedly been described as an efficient tissue for the elucidation of a great number of cellular and molecular biological problems, in particular those concerning cell differentiation. These assumptions are certainly true if the student of the eye lens keeps in mind that this suborgan is among the most intriguing biological systems. In fact, our observations raised more unsolved questions than allowed firm conclusions.

We have outlined the quite unusual regulatory mechanism that controls the extracellular and intracellular ion and water content in lens. Further electrophysiological analysis must shed more light on the active transport, in particular the secretory character of the sodium fluxes. Yet, the molecular mechanism controlling the junctional assembly and removal is far from being elucidated. A funtional role must be found for the particulate square arrays. Characterization of the subunit composition of the communicating junctions has only just been started, and several questions concern the relationship that eventually might exist between the MP26 and the MP34 during junctional removal and assembly at the elongation zone.

Although the experimental evidence tends to indicate the existence of a close association between crystallin subunits and other components of the "water-insoluble" lens fraction, the information is still missing about the role that the crystallins play in the stabilization of the membrane and cytoskeletal constituents. The latter association is well established from a morphological point of view, but still we do not know what the other proteins are that operate in the end-on attachment of actin and intermediate filaments. Preliminary results using immunocytochemistry are consistent with the presence of α-actinin on the membrane cytoplasmic inner surface (Figure 25b). This protein has been considered in other tissues as binding protein for microfilament attachment to plasma membranes (101, 102). The observation that both actin and vimentin may occur in the lens either in a water-soluble or highly polymerized and insoluble form anticipates the questions about other protein cofactors that regulate the assembly of the cytoskeleton.

It is remarkable that lens fiber polyribosomes in a cellular system, where the endoplasmic reticulum is almost completely deleted, are associated with the plasma membranes either directly or via intimate relation with the actin filaments. Thus the plasma membrane-cytoskeleton complex in eye lens fibers is the site of protein biosynthesis. The molecular mechanism controlling transfer and insertion of the newly synthesized polypeptides at the plasma membrane level requires further study.

One of the most striking observations made by us is that an actin-associated unique class of polyribosomes directs in a heterologous system the translation of a specific intrinsic membrane protein (MP26), whereas another class of polyribosomes is involved in the maintenance of the pool of other lens proteins, including crystallins.

ACKNOWLEDGMENTS

Our work was supported by grants from DGRST (78. 703 71) and INSERM (A 650 5075). It is a pleasure to thank Mrs. C. Emmanuel for typing the manuscript, and our sincere appreciation is extended to Miss M. C. Potjens for her help in preparing this chapter. The serum directed against α-actinin was a precious gift of Dr. Louvard of the European Molecular Biology Laboratory at Heidelberg.

REFERENCES

1 E. L. Benedetti, I. Dunia, C. J. Bentzel, A. J. M. Vermorken, M, Kibbelaar, and H. Bloemendal, *Biochim. Biophys. Acta,* **457,** 353 (1976).

2 H. Bloemendal, A. Zweers, F. Vermorken, I. Dunia, and E. L. Benedetti, *Cell Diff.,* **1,** 91 (1972).

3 R. M. Broekhuyse and E. D. Kuhlmann, *Exp. Cell Res.,* **19,** 297 (1974).

4 R. M. Broekhuyse, E. D. Kuhlmann, and A. L. H. Stols, *Exp. Eye Res.,* **23,** 365 (1976).

5 R. M. Broekhuyse and E. D. Kuhlmann, *Exp. Eye Res.,* **26,** 305 (1978).

6 A. A. Bouman, A. L. M. de Leeuw, E. F. J. Tolhuizen, and R. M. Broekhuyse, *Exp. Eye Res.,* **29,** 83 (1979).

7 R. M. Broekhuyse, E. D. Kuhlmann, and H. J. Winkens, *Exp. Eye Res.,* **29,** 303 (1979).

8 R. M. Broekhuyse and E. D. Kuhlmann, *Exp. Eye Res.,* **28,** 615 (1979).

9 E. L. Benedetti, I. Dunia, J. Cartaud, T. Hatae, C. Favard-Sereno, C. Bentzel, M. Kibbelaar, and H. Bloemendal, in: J. Dumont and J. Nunez, Eds., *Hormones and Cell Regulation,* Vol. 2, North Holland, 1978, p. 305.

10 I. Dunia, C. Sen-Ghosh, E. L. Benedetti, A. Zweers, and H. Bloemendal, *FEBS Lett.,* **45,** 139 (1974).

11 Z. Dische, M. A. Hairstone, and O. Zelmenis, *Protides Biol. Fluids, Proc. Colloq.,* **15,** 123 (1969).

12 H. Maisel, *Exp. Eye Res.,* **24,** 417 (1977).

13 J. Alcalá and H. Maisel, *Exp. Eye Res.*, **26,** 219 (1978).

14 A. Spector, M. H. Garner, W. H. Garner, D. Roy, P. Farnsworth, and S. Shyne, *Science,* **204,** 1323 (1979).

15 W. H. Garner, M. H. Garner, and A. Spector, *Exp. Eye Res.*, **29,** 257 (1979).

15a. A. L. Shapiro, A. L. Viñuela, and J. V. Maizel, *Biochim. Biophys. Res. Comm.*, **28,** 815 (1967).

16 M. M. Wong, N. P. Robertson, and J. Horwitz, *Biochim Biophys. Res. Comm.*, **84,** 158 (1978).

17 J. Horwitz, N. P. Robertson, M. M. Wong, J. S. Zigler, and J. H. Kinoshita, *Exp. Eye Res.*, **28,** 359 (1979).

18 H. Bloemendal, A. J. M. Vermorken, M. Kibbelaar, I. Dunia, and E. L. Benedetti, *Exp. Eye Res.*, **24,** 413 (1977).

19 D. Roy, A. Spector, and P. N. Farnsworth, *Exp. Eye Res.*, **28,** 353 (1979).

20 A. J. M. Vermorken, J. M. H. C. Hilderink, I. Dunia, E. L. Benedetti, and H. Bloemendal, *FEBS Lett.*, **83,** 301 (1977).

21 H. Bloemendal, *Ophthal. Res.*, **11,** 243 (1979).

22 J. Alcalá, N. Lieska, and H. Maisel, *Exp. Eye Res.*, **21,** 581 (1975).

23 M. A. Kibbelaar and H. Bloemendal, *Exp. Eye Res.*, **291,** 679 (1979).

24 P. Herbrink and H. Bloemendal, *Biochim. Biophys. Acta,* **336,** 370 (1974).

25 R. M. Broekhuyse and J. H. Veerkamp, *Biochim. Biophys. Acta,* **152,** 316 (1968).

26 R. M. Broekhuyse, *Biochim Biophys. Acta,* **187,** 354 (1969).

27 R. M. Broekhuyse, H. Roelfzema, M. E. Breimer, and K. A. Karlsson, *Exp. Eye Res.*, **19,** 477 (1979).

28 H. Roelfzema, R. M. Broekhuyse, and J. H. Veerkamp, *Exp. Eye Res.*, **23,** 409 (1976).

29 H. Bloemendal, *Science,* **197,** 127 (1977).

30 J. L. Rae, in: J. A. Zadunaisky and H. Davson, Eds., *Current Topics in Eye Research,* Vol. 1, Academic Press, New York, 1979, p. 37.

31 M. C. Neville, C. A. Paterson, and P. M. Hamilton, *Exp. Eye Res.*, **27,** 637 (1978).

32 J. L. Rae and T. Stacey, *Exp. Eye Res.*, **28,** 1 (1979).

33 E. Frömter and J. M. Diamond, *Nature,* **235,** 9 (1972).

34 Ph. Claude and D. A. Goodenough, *J. Cell Biol.*, **58,** 390 (1973).

35 N. A. Delamere and C. A. Paterson, *J. Gen. Physiol.* **71,** 581 (1978).

36 R. D. Ginzberg and N. B. Gilula, *Dev. Biol.*, **68,** 110 (1979).

37 S. L. Bonting "Membrane and Ion Transport," in: E. E. Bittar, Ed., Vol. 1, Wiley London, 1970, p. 357.

38 E. L. Benedetti and P. Emmelot, *J. Cell Biol.*, **26,** 299 (1965).

39 E. L. Benedetti, I. Dunia, and H. Bloemendal, *Proc. Natl. Acad. Sci. U.S.*, **71,** 5073 (1974).

40 W. J. Larsen, *Tissue and Cell,* **9,** 373 (1977).

41 M. V. L. Bennett and D. A. Goodenogh, *Neurosci. Res. Prog. Bull.*, **16,** 377 (1978).

42 C. Peracchia, *Nature (London)* **271,** 66g (1978).

43 G. Dahl and G. Isenberg, *J. Membr. Biol.,* **53,** 63, (1980).

44 D. A. Goodenough, *Invest. Ophthalmol., Vis. Sci.,* **18,** 1104 (1979).

45 D. A. Goodenough, J. S. B. Dick, II, and J. E. Lyons, *J. Cell Biol.,* **86,** 576 (1980).

46 P. N. T. Unwin and G. Zampighi, *Nature,* **283,** 545 (1979).

47 G. Zampighi and P. N. T. Unwin, *J. Mol. Biol.,* **135,** 451 (1979).

48 E. L. Benedetti and I. Dunia, in: G. Balian, M. Chabre, and Ph. Devaux, Eds., *Biophysics of Membrane and Cell Communication,* North-Holland, Amsterdam, 1980.

49 R. Azarnia and W. J. Larsen, in: W. C. DeMello, Ed., *Intercellular Communication and Cancer,* Plenum Press, New York, 1977, p. 145.

50 A. Amsterdam and J. D. Jamieson, *J. Cell Biol.,* **63,** 1037 (1974).

51 I. Dunia, J. F. Nicolas, H. Jakob, E. L. Benedetti, and F. Jacob, *Proc. Natl. Acad. Sci. U.S.,* **76,** 3387 (1979).

52 D. Henderson, H. Eibl, and K. Weber, *J. Mol. Biol.,* **132,** 193 (1979).

53 M. Takana, P. Russel, S. Smith, S. Uga, T. Kuwabara, and J. H. Kinoshita, *Invest. Ophthalmol. Vis. Sci.,* **19,** 619 (1980).

54 H. Bloemendal, E. L. Benedetti, F. C. Ramaekers, I. Dunia, M. A. Kibbelaar, and A. J. M. Vermorken, *Molec. Biol. Rep.,* **5,** 99 (1979).

55 F. C. S. Ramaekers, A. M. E. Selten-Versteegen, H. Bloemendal, I. Dunia, and E. L. Benedetti, *Proc. Natl. Acad. Sci. U.S.,* **77,** 725 (1980).

56 J. Chevalier, J. Bourguet, and J. S. Hugon, *Cell Tissue Res.,* **152,** 129 (1974).

57 W. A. Kachadorian, J. B. Wade, and U. A. Discala, *Science,* **190,** 67, (1975).

58 J. Bourguet, J. Chevalier, and J. S. Hugon, *Biophys. J.,* **16,** 627 (1976).

59 W. A. Kachadorian, J. B. Wade, C. C. Uiterwyk, and V. A. Discala, *J. Mem. Biol.,* **30,** 381 (1877).

60 J. B. Wade, W. A. Kachadorian, and U. A. Discala, *Ann. J. Physiol.,* **232**(1), F77 (1977).

61 J. B. Wade, *J. Mem. Biol.* **40,** 281 (1978).

62 M. Kibbelaar, A. E. Selten-Versteegen, H. Bloemendal, I. Dunia, and E. L. Benedetti, *Eur. J. Biochem.,* **95,** 543 (1979).

63 M. Clarke and J. A. Spudich, *Ann. Rev. Biochem.,* **46,** 797 (1977).

64 E. Ungewickell, P. M. Bennett, R. Calvert, V. Ohanian, and W. B. Gratzer, *Nature,* **280,** 811 (1979).

65 F. Cabral and M. M. Gottesman, *J. Biol. Chem.,* **254,** 6203 (1979).

66 C. M. O'Connor, D. R. Balzer, and E. Lazarides, *Proc. Natl. Acad. Sci. U.S.,* **76,** 819 (1979).

67 G. Babbiani, C. Chaponnier, A. Zumbe, and P. Vassalli, *Nature,* **269,** 697 (1977).

68 K. G. Sundqvist, P. Otteskog, and T. Ege, *Nature,* **274,** 915 (1978).

69 M. A. Kibbelaar, F. C. S. Ramaekers, P. J. Ringens, A. M. E. Selten-Versteegen, L. G. Poels, P. H. K. Jap, A. L. Van Rossum, T. E. W. Feltkamp, and H. Bloemendal, *Nature,* **285,** 506 (1980).

70 H. Maisel and M. M. Perry, *Exp. Eye Res.*, **14,** 7 (1972).

71 M. M. Perry, *Exp. Eye Res.*, **22,** 125 (1976).

72 R. H. Bradley, M. Ireland, and H. Maisel, *Exp. Eye Res.*, **28,** 441 (1979).

73 H. Bloemendal, M. A. Kibbelaar, F. C. S. Ramaekers, A. M. E. Selten-Versteegen, I. Dunia, and E. L. Benedetti, in: H. Peeters, Ed., *Prot. Biol. Fluids, Proc. 26th Colloq., Brussels,* Pergamon Press, Oxford, 1978, pp. 499–505.

74 A. Spector and D. Travis, *J. Biol. Chem.*, **241,** 1294 (1966).

75 E. L. Benedetti, A. Zweers, and H. Bloemendal, *Biochem. J.*, **108,** 765 (1968).

76 J. J. Wolosewick and K. R. Porter, *Am. J. Anat.*, **147,** 303 (1976).

77 J. J. Wolosewick and K. R. Porter, *J. Cell Biol.*, **82,** 114 (1979).

78 R. Lenk, L. Ramson, Y. Kaufmann, and S. Penman, *Cell,* **10,** 67 (1977).

79 E. Lazarides and U. Lindberg, *Proc. Natl. Acad. Sci. U.S.*, **71,** 4742 (1974).

80 S. E. Hitchcock, L. Carlsson, and U. Lindberg, *Cell,* **7,** 531 (1976).

81 M. P. Sheetz, *J. Cell Biol.*, **81,** 266 (1979).

82 S. Penman, R. Lenk, A. Ben-Ze'ev, F. Solomon, A. Duer, and S. Farmer, in: *Abs. Cold Spring Harbor Meet. The Cytoskeleton,* 1979, p. 159.

83 J. M. Starger, W. E. Brown, A. E. Goldman, and R. D. Goldman, *J. Cell Biol.*, **78,** 93 (1978).

84 M. Osborn, *Mol Biol. Report,* **5,** 132 (1979).

85 J. V. Small and A. Sobieszek, *J. Cell Sci.*, **23,** 234 (1977).

86 W. W. Franke, E. Schmid, S. Winter, M. Osborn, and K. Weber *Exp. Cell Res.*, **123,** 25 (1979).

87 S. A. Fellini, G. S. Bennett, Y. Toyama, and H. Holtzer, *Differentiation,* **12,** 59 (1978).

88 G. S. Bennett, S. A. Fellini, and H. Holtzer, *Differentiation,* **12,** 71 (1978).

89 E. D. Frank, G. P. Tuszynski, and L. Warren, in: *Abs. Gold Spring Harbor Meet. The Cytoskeleton,* 1979, p. 130.

90 R. G. Whalen, G. S. Butler-Browne, and F. Gros, in: *Abs. Cold Spring Harbor Meet. The Cytoskeletal and Contractile Networks of Nonmuscle Cells,* 1978, p. 53.

91 J. F. Perdue, *J. Cell Biol.*, **58,** 265 (1973).

92 D. L. Gard, P. B. Bell, and E. Lazarides, in: *Abs. Cold Spring Harbor Meet. The Cytoskeleton,* 1979, p. 95.

93 H. Holtzer, S. Fellini, N. Rubinstein, J. Chi, and K. Strahs, in: R. Goldman, T. Pollard, and J. Rosenbaum, Eds. *Cell Motility,* Cold Spring Harbor Laboratory, 1976, 823.

94 G. S. Bennett, S. A. Fellini, J. M. Croop, J. J. Otto, J. Bryan, and H. Holtzer, *Proc. Natl. Acad. Sci. U.S.*, **75,** 4364 (1978).

95 W. W. Franke, B. Appelhans, E. Schmid, C. Freudenstein, M. Osborn, and K. Weber, *Eur. J. Cell Biol.*, **19,** 255 (1979).

96 E. Lazarides and B. D. Hubbard, *Proc. Natl. Acad. Sci. U.S.*, **73,** 4344 (1976).

97 B. D. Hubbard and E. Lazarides, *J. Cell Biol.*, **80,** 166 (1979).

98 I. K. Buckley, T. R. Raju, and M. Stewart, *J. Cell Biol.*, **78,** 644 (1978).

99 I. K. Buckley, M. Stewart, D. O. Irving, T. R. Raju, and W. E. Gordon, in: *Abs. Cold Spring Harbor Meet. The Cytoskeleton,* 1979, p. 135.

100 E. Lazarides, B. D. Hubbard, and B. L. Granger in: S. Hatano, H. Ishikawa, and H. Sato, Eds. *Cell Motility: Molecules and Organization,* University Park Press, Tokyo, 1979.

101 B. Geiger and S. J. Singer, *Cell,* **16,** 213 (1979).

102 S. W. Craig and J. V. Pardo, *J. Cell Biol.,* **80,** 203 (1979).

CHAPTER 5

Biosynthesis of Lens Crystallins

Hans Bloemendal

Department of Biochemistry, University of Nijmegen, Nijmegen, The Netherlands

5.1 INTRODUCTION

The complicated pathway of protein biosynthesis can be summarized in a simple schematic representation:

$$\text{DNA} \rightarrow \text{RNA} \rightarrow \text{protein}$$

This means that messenger RNA (mRNA) is transcribed from the gene and transfers the genetic information encoded in DNA to electron-dense cell particles, the ribosomes, for the synthesis of polypeptide chains. The biosynthetic process *per se* can be subdivided into three major steps, initiation, elongation, and termination, which in some cases are followed by postsynthetic modification and eventually by assembly of the newly synthesized chains into a polymeric protein. Although it is beyond the scope of this chapter to give a detailed description of protein biosynthesis, it has to be mentioned here that a few fundamental aspects of the biosynthetic process and of the post-translational modifications in eukaryotes could be approached experimentally using lens tissue as the model system.

This chapter describes briefly the isolation and in detail the properties of some major components of the translational machinery in lens tissue. The major intent is to give a systematic survey rather than a chronological summary of the data available on lens protein biosynthesis.

The major constituents of the lens are the crystallins, the only proteins that are believed to be organ-specific. Lens fiber differentiation is accompanied on the molecular level by the synthesis of these crystallins. Although *de novo* biosynthesis of crystallins has been studied incidentally in the intact organ, most investigations were carried out *in vitro*.

Upon incubation of isolated polyribosomes or messenger RNA in a cell-free system the newly synthesized products can be analyzed by electrophoresis on sodium dodecylsulfate–polyacrylamide gels followed by scintillation autoradiography (1). As far as the lens cell-free system *per se* is concerned, examination of the newly synthesized products and their assembly might sometimes be hampered by the presence of large amounts of endogenous crystallins. This complication can be circumvented by the use of heterologous cell-free systems such as the reticulocyte lysate (2), the ascites cell-free system (3), or the wheat germ system (4). Alternatively, lens polysomes or purified lens messenger RNA can be injected into living oocytes (5, 6) that represent a heterologous *in vivo* system.

Since the lens epithelial cells in the process of fiber differentiation lose nuclei and ribosomes (7), one may conclude protein synthesis occurs preferentially in the epithelium and the outer cortex. Actually the classical experiments by Reeder and Bell (8) have demonstrated that DNA synthesis is restricted to the epithelial layer, whereas RNA and protein synthesis still occur in the peripheral cortex.

5.2 LENS PROTEIN BIOSYNTHESIS *IN VIVO*

Reports on lens protein synthesis studies in the whole animal are scarce, presumably owing to technical difficulties. When rats were injected intraperitoneally with [^{35}S]-methionine only the outer cortex appeared to be labeled after 1 day. After a month this region was not labeled anymore, whereas a region of fibers inside the lens was radioactive. Only a small proportion of the total radioactivity could be found in the insoluble protein after 1 day (9). However, 7 weeks later about half of the activity appeared to be present in the insoluble fraction. This result indicates that at least the major part of newly synthesized crystallins exists in the soluble form and becomes insoluble by maturation or aging. Radioactivity could be detected in α-, β-, and γ-crystallin (9, 10).

Recent experiments in our laboratory revealed that part of the αA chains may interact with membrane constituents and become "insoluble" (11, 11a). Since these experiments were conducted *in vitro,* it is not apparent whether a small quantity of α-crystallin chains attaches immediately after synthesis to the plasma membrane and/or cytoskeleton *in vivo* as well. Incubation of isolated whole lenses revealed that protein biosynthesis takes place in the periphery of the organ(8). Roughly each of the three main crystallin groups is synthesized at a similar rate (12). Crystallin messenger RNA is stable in the lens fibers. This could be demonstrated by incubation in the presence and absence of actinomycin D (8, 13–16). In this connection it should be mentioned that a potent ribonuclease inhibitor is present in lens fibers, which might be at least in part responsible for the great stability of crystallin messengers (17–19).

McAvoy verified the distribution of the various crystallins in rat lens by the indirect immunofluorescence method (20).

A quantitative analysis of cell division and elongation in the lenses of 1-

day-old and 6- to 7-week-old rats showed that rat lens could be divided into a proliferation compartment in the epithelium of the anterior part of the lens and an elongation compartment in the posterior part. In the 6- and 7-week-old rat lens the proliferation could be divided into three subcompartments. It appeared that epithelial cells in the proliferation compartment contained only α-crystallin whereas all three crystallin classes were detected in the elongation compartment. α-Crystallin was detected first in those elongating cells that had stopped dividing. γ-Crystallin was detected later in the elongation process. McAvoy concluded that β- and γ-crystallin synthesis may depend on some factor(s) from the retina. Since detection of β-crystallin synthesis can be correlated with the cessation of cell division the mere fact of cessation of division may also be a crucial factor. There exists a correlation between ionic imbalances and changes in protein biosynthesis in the so-called Philly cataract in mice. In this hereditary cataract the observed reduced content of soluble crystallins is due both to differential reduction of synthesis and to degradation, in particular, of β- and γ-crystallin (20a).

5.3 PROTEIN BIOSYNTHESIS IN TISSUE CULTURE

Lens epithelial cells from various species can be kept in culture for many months (21–26). However, there is a striking difference in the patterns of the protein. Whereas the crystallins are formed in short-term culture the synthesis of these protein stops after a few weeks. Although elongation of the cells *in vitro* can be provoked, for example, by the addition of dexamethasone to the culture medium, no detectable synthesis of a specific crystallin accompanies the morphological change (26). This situation apparently differs from the situation in those lenses where γ-crystallin synthesis seems to be associated with the process of fiber elongation (27).

Experiments in my laboratory revealed the reason that no crystallins are synthesized upon prolonged culturing (28). The presence of the lens capsule appeared to be a prerequisite for the formation of α-crystallin (Figure 1).

When protein synthesis is studied in various parts of the epithelial monolayer the result of differential gene activation can be observed. It appears that in the central region more αB_2 is synthesized than αA_2 (Figure 2). This finding is quite astonishing since previously the αA_2 chain was considered to be quantitatively the most important α-crystallin subunit. Actually, experiments with fiber-free epithelium show that before the onset of fiber differentiation αB_2 is predominant. The ratio is then $B_2 : A_2 = 3 : 1$, changes to $2 : 3$ in the outer epithelium, but becomes almost $1 : 3$ in the fibers. Vermorken and Bloemendal, by preincubation with actinomycin D, ruled out that different messenger stabilities were responsible for the observed stoichiometry of αB_2 and αA_2 (28a).

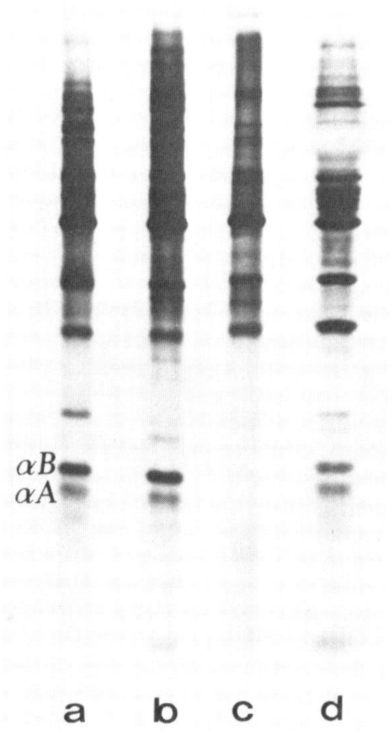

Figure 1 Sodium dodecyl sulfate–polyacrylamide gel electrophoresis of newly synthesized lens proteins. (*a*) Noncultured isolated central epithelium; (*b*) central epithelium cultured on the lens capsule; (*c*) central epithelium cultured on plastic (note that no α-crystallin synthesis occurred); (*d*) central epithelium cultured on capsules and preincubated with actinomycin D (5 μg/ml) for 16 hours. (The same pattern was obtained with noncultured cells.) Reprinted with permission from ref. 28.

Chick lens epithelium synthesizes a set of proteins that range from M_r 20,000 to more than 200,000. The major molecular weight species observed had values of about 20,000, 27,000, 45,000, 65,000, and 91,000. The principal component representing almost 35% of the total newly synthesized protein was the 45,000 M_r species (22). After 1 day of culturing this protein amounts to 50% and after 3 days in the fibers to 80%. One may assume that at least a major part of this 45,000 component is identical to actin.

Lens fibers appear in the form of so-called lentoid bodies during long-term clonal culture of lens epithelial cells from newly hatched chicks (29, 30). In this case accumulation of δ-crystallin occurs. De Pomerai et al. (31) observed that increasing levels of δ-crystallin in culture of 1-day-old epithelial cells do not approach those found in lens fibers from the same stage. However, the dividing epithelial cells of 1-day-old chick lens differentiate *in vivo* into the cortical fibers of the (young) adult lens. δ-Crystallin in these adult fibers is less prominent than in the embryonic fibers (32, 33). De Pomerai et al. (34) demonstrated that δ-crystallin synthesis during lentoid development is affected by the age and the gene type of the starting cells.

Lentoid bodies developed during long-term culture of 8-day embryonic neural retina contain crystallins. The proportions of these crystallins, however, do not resemble those of embryonic lens fibers. δ-Crystallin accumulation is much greater in transdifferentiating cultures of early embryonic neural retina, whereas α- and β-crystallins become proportionately more prominent in cultures of late embryonic neural retina. Traces of α- and β- but not δ-crystallin are detectable in 60-day cultures of 17-day

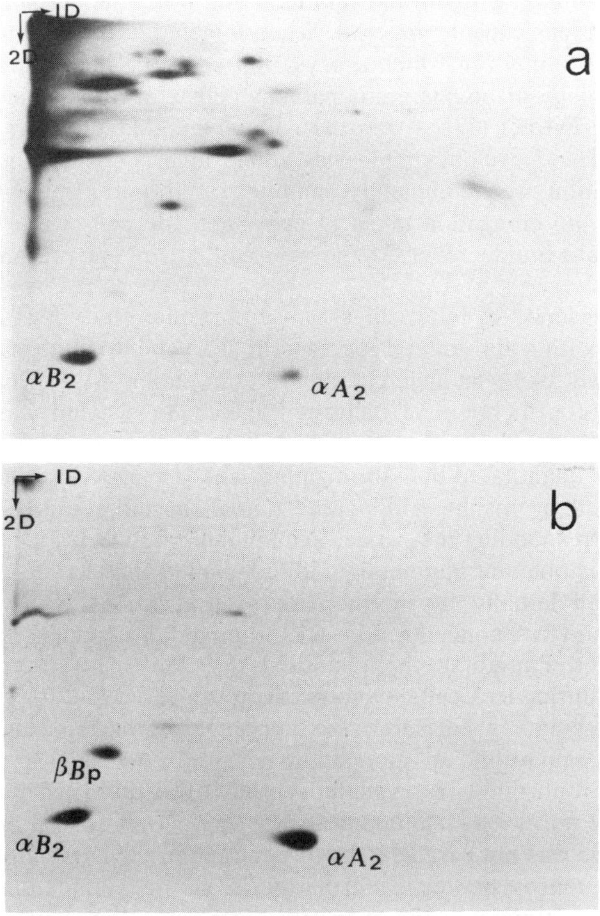

Figure 2 Autoradiography of two-dimensional polyacrylamide gel electrophoresis patterns of products synthesized *de novo* (*a*) in the central part of the lens epithelium (note that αB_2 is predominant whereas βB_p is absent) and (*b*) in the outer edge of the epithelium. Reprinted with permission from ref. 28.

embryonic neural retina (35). Thomson et al. provided evidence that the synthesis of crystallins in cultured embryonic chick neural retina and pigmented epithelium is under transcriptional control (35a). A possible relationship between crystallin mRNA expression in retina cells and their capacity to re-differentiate into lens cells was discussed by Clayton et al. (35b).

A striking observation has been reported by Courtois et al. (36). The bovine lens epithelial cells studied by these workers in long-term culture spontaneously acquired a number of characteristics of transformed cells such as unlimited growth and multilayering potential. *In vitro* these cells synthesized lens capsule material. When injected into nude mice, however, they formed tumors in which α-crystallin could be demonstrated. *In vitro* the morphology of bovine epithelial lens cells can be modified by addition of a retinal extract (37). When the cellular monolayers were treated with this extract and cycloheximide was added at a concentration of 5 μg/ml, cell elongation was completely inhibited. Without protein biosynthesis, apparently, no elongation takes place. When the cells were already elongated cycloheximide reverted the morphology in the presence of retinal extract.

In the presence of fetal calf serum embryonic chick lens elongation is associated with a differential increase in δ-crystallin synthesis, which can be accounted for by an accumulation of δ-crystallin mRNA during the fifth and twenty-fourth hours of culture. Labeled δ-crystallin was purified by gradient SDS–polyacrylamide gel electrophoresis whereas δ-crystallin mRNA was quantitated by hybridization with [H^3]-cDNA. When the serum was replaced by insulin it appeared that δ-crystallin synthesis increased and the corresponding messenger accumulated between 5 and 24 hours of culture. Milstone and Piatigorsky (38) concluded that δ-crystallin synthesis is stimulated initially by an increase in translational efficiency and an increase in mRNA concentration when the medium contains insulin.

Beebe and Piatigorsky (39) studied the control of δ-crystallin gene expression during lens cell development. It appeared that for stimulation of gene expression, as measured by δ-crystallin synthesis and δ-crystallin mRNA accumulation, no elongation of cultured cells was required. A decreased proportion of δ-crystallin synthesis was observed in the epithelial cells during normal development and during serum starvation. However, this decrease was not paralleled by a reduction in the number of δ-crystallin mRNA sequences per cell. Furthermore, serum stimulation of the cell division did not increase the proportion of δ-crystallin synthesis but could promote the accumulation of δ-crystallin messenger. Therefore the relative proportion of δ-crystallin synthesized during chick lens development is not only a function of the number of δ-crystallin mRNA sequences in the lens cells.

Shinohara and Piatigorsky (40) observed that the control of protein biosynthesis in cultured chick lenses is associated with changes in the ratio of intracellular Na^+ and K^+ ion concentrations. Alteration of δ-crystallin synthesis is due to an increase in the intracellular concentration of sodium and a decrease in the intracellular potassium concentration. That these ions affect the translational machinery both *in vivo* and *in vitro* has been demonstrated by other investigators (41, 42). Shinohara and Piatigorsky suggest that the regulatory step is at the level of initiation.

Lens cell cultures described so far in the literature were of the monolayer type. Recently we established a hamster lens cell line that grows upon transformation with simian virus (SV 40) in suspension culture (42a,b). In the two-dimensional electrophoresis pattern of the proteins from these cells crystallins are undetectable, whereas the cytosceletal protein subunits vimentin and actin are abundant. Cells from the transformed cultures provoke tumors in nude mice.

5.4 CRYSTALLIN SYNTHESIS IN CELL-FREE SYSTEMS

Since this section deals with *in vitro* translation of lens mRNA, the isolation of lens polyribosomes and the messengers derived therefrom must precede the biosynthetic data *per se*. As in other eukaryotes subcellular organization in lens is far more complex than in prokaryotes. One of the marked differences is the occurrence of a well-defined nucleus in which messenger RNA synthesis takes place in the nucleoplasm, whereas the formation of ribosomal RNA occurs in the nucleolus.

It is now well established that eukaryotic mRNA is synthesized as a precursor molecule, so-called heteronuclear RNA (HnRNA) (cf. ref. 43). HnRNA molecules and mRNA share a number of properties, for example, the poly(A) track at the 3′ end and high percentage of hybridization to DNA.

However, as far as the lens is concerned such a precursor–product relationship has not been demonstrated yet. We have recently described the purification of nuclei from calf lens. Because this method may be useful for the study of HnRNA, including crystallin pre-mRNA processing *in vitro,* the procedure is referred to here.

Viable lenticular nuclei can be isolated from the central part of the epithelial monolayer (44). Their polymerase activity is similar to that of rat liver nuclei. The use of 0.5% Nonidet P40 (nonionic detergent) to wash cytoplasmic contaminants such as plasma membrane fragments removes endoplasmic reticulum and fibrillar structures. Figure 3 shows the morphological features of the lens cell nuclei in the electron microscope.

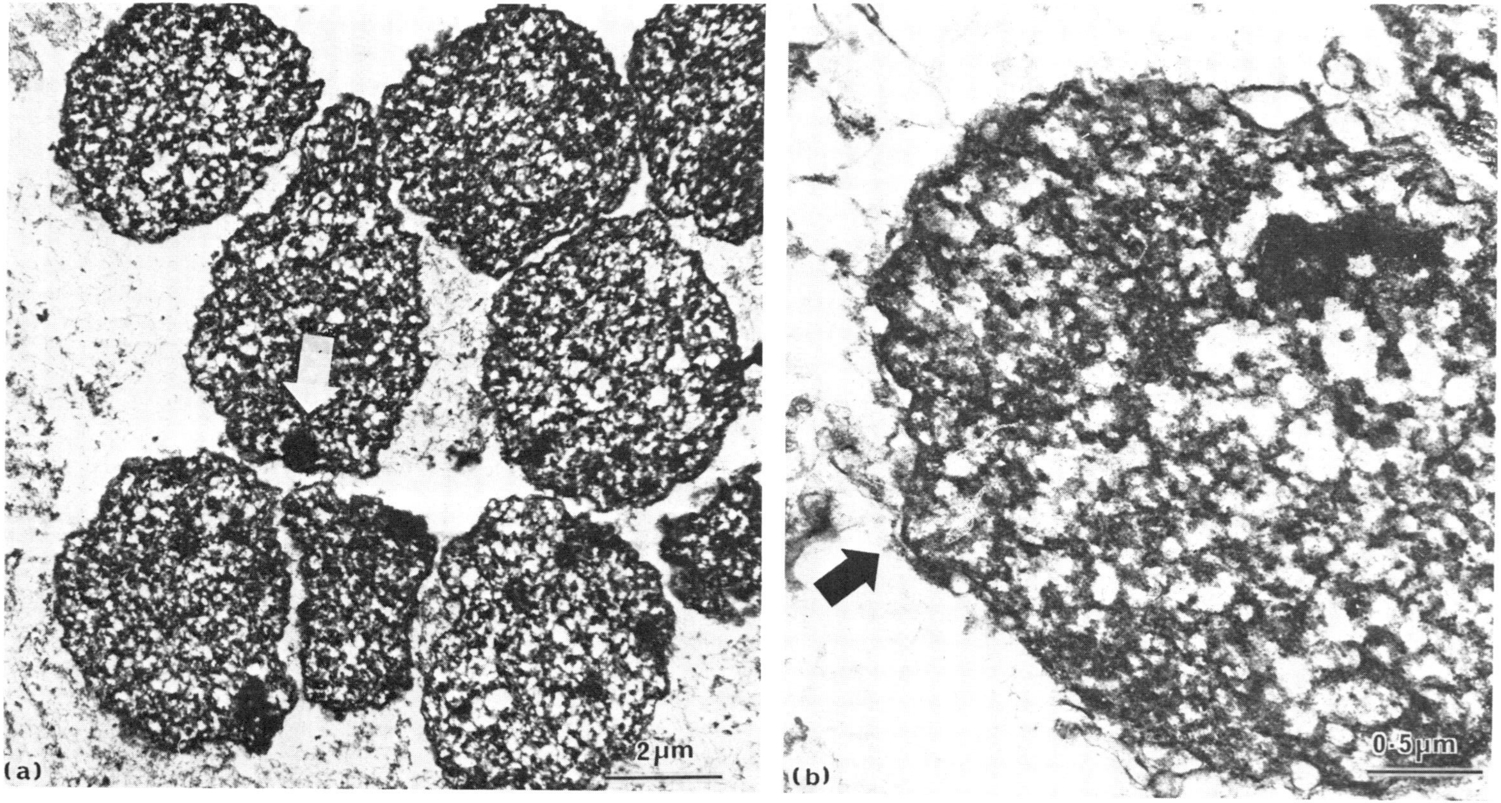

Figure 3 (*a*) Thin section stained with uranyl acetate–lead citrate of isolated lens nuclei washed with Nonidet P40. Well-preserved nuclei containing nucleoli (arrow) are visualized. (*b*) Higher magnification of nuclei not treated with Nonidet P40. The external nuclear envelope is preserved (arrow). Reprinted with permission from ref. 44.

5.5 CRYSTALLIN MESSENGER RNAs ARE "CAPPED"

The 5′-terminal cap structure $m^7G(5')ppp(5')N$ is required for most eukaryotic mRNAs (45). This structure may promote successful interaction between mRNA and some initiation factors. Translation of capped mRNAs can be almost completely inhibited by cap analogues such as $m^7G5'p$ (Figure 4) (46, 47). When this compound and lens messenger are simultaneously introduced into a nuclease-treated reticulocyte lysate, initiation of protein synthesis is almost completely inhibited (47, 48). However, variations in $m^7G5'p$-concentration needed to block the initiation step were observed for different lens messenger RNAs. This is in accordance with studies on globin synthesis. It appears that α-globin formation is more strongly inhibited by the cap analogues than β-globin synthesis (49). Initiation complexes were formed with both the 10 S and 14 S mRNA populations isolated from calf lens (compare next sections), after which the unprotected regions were subjected to RNA T_1 digestion. Single fragments of about 45 nucleotides were obtained from both mRNAs. These fragments retained the ability to form initiation complexes. A great similarity of the fragments was demonstrated, although marked differences in the (U)G region were observed.

The possible necessity of methylation for lens crystallin initiation was shown by addition of *S*-adenosyl-L-methionine, which enhanced the initiation complex formation from two to five times (50).

Lavers demonstrated the presence of an enzyme in chick lens that cleaves m^7GMP from $m^7GpppG(m)$. This enzyme has a broad specificity as to the nucleotide composition and sequence for N in the 5′-terminus of mRNA. Under the conditions used by Lavers, m^7GMP was not released from capped sequences longer than 100 nucleotides, for example, δ-crystallin mRNA (51).

In *Xenopus oocytes* protein biosynthesis of endogenous protein and translation of exogenous mRNA are not inhibited by microinjection of cap analogues (52).

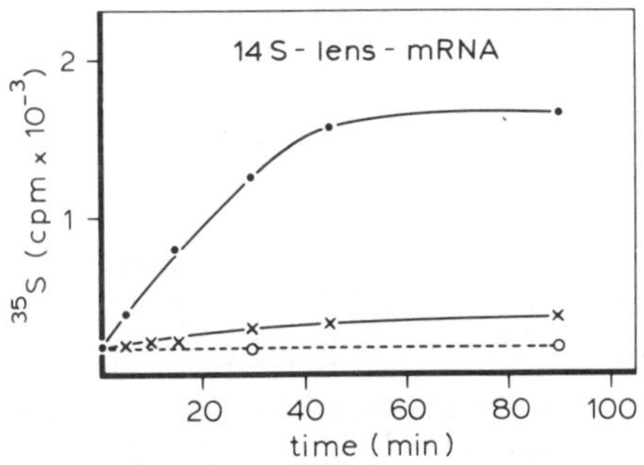

Figure 4 Time course of translation of lens 14 S mRNA. Purified mRNA was translated in the presence (X——X) or absence (•——•) of 0.8 m*M* $m^7G5'p$ in the mRNA-dependent reticulocyte lysate. A control incubation (no mRNA added) is also shown (○-----○). mRNA was present at a concentration of 8 μg/ml incubation mixture. Reprinted with permission from ref. 47.

5.6 LENS POLYRIBOSOMES

5.6.1 Preparation

5.6.1.1 Procedure 1

Lens polyribosomes can be translated as such in heterologous cell-free systems (for review, see ref. 53). Procedures for the isolation of calf lens polyribosomes have been described previously (54–56). The routine procedure used in the author's laboratory is as follows: After the animals are killed the eyes are removed as quickly as possible and transferred on ice from the slaughterhouse to the laboratory. The lenses are taken out. The lens capsules with the adhering epithelial cells are removed with the aid of forceps, and the outer cortex is obtained by punching out a layer 1 to 2 mm thick from the equator with the aid of a glass tube. The isolated tissue is homogenized in 1 volume (vol./wt.) of an ice-cold medium containing 5 m*M* $MgAc_2$, 25 m*M* KCl, 50 m*M* Tris-Cl, pH 7.6, and 0.35 *M* sucrose (medium A). A Teflon–glass homogenizer with a clearance of 0.15 to 0.2 mm is used; five strokes at 2000 rpm are applied. Nuclei, mitochondria, and cell debris are removed by centrifugation at 12,000 × *g* for 15 min in a rotor GSA of the Sorvall RC2-B preparative centrifuge. To the supernatant 0.1 volume of 10% (wt./vol.) sodium deoxycholate in 50 m*M* Tris-Cl buffer, pH 7.6 is added, and the solution is layered on top of a discontinous density gradient consisting of 10 ml 1.5 *M* sucrose and 10 ml 2 *M* sucrose, both dissolved in medium A without sucrose. Centrifugation is carried out in a rotor 30 of the Spinco L-50 ultracentrifuge at 78,000 × *g* for 17 hours at 2°C. For 15 ml of postmitochondrial supernatant 100 to 120 calf lenses are required. From 1000 calf lenses approximately 6 mg polyribosomal RNA is obtained. From 1-day-old rat lenses about twice this amount can be isolated. The profile of polyribosomes from rat lens fibers after density gradient centrifugation is shown in Figure 5.

5.6.1.2 Procedure 2

Alternatively, the polyribosomes can be isolated by a precipitation procedure with Mg^{2+} ions. This is a modification of the method reported by Palmiter (57). The lens tissue is homogenized for 30 sec in an equal volume of a buffer containing 50 m*M* Tris-Cl, pH 8 (at 4°C), 25 m*M* NaCl, and 5 m*M* $MgCl_2$. A Polytron homogenizer is used at a speed corresponding to 1.5 scale units. The homogenate is centrifuged in the GSA rotor of an RC2-B Sorvall at 10,000 rpm for 10 min. The supernatant is then supplemented with ¼ volume of 10% Triton X100 (wt./vol.) and heparin to a final concentration of 1 mg/ml. The solution is kept at 4°C for 10 min. Centrifugation is repeated and to the clear supernatant $^1/_{14}$ volume of 2 *M* $MgCl_2$ is

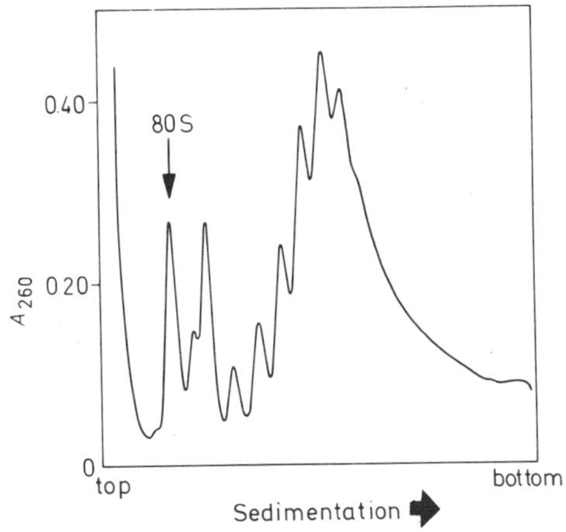

Figure 5 Profile of polyribosomes, isolated from 1-day-old rat lenses. Polyribosomes were dissolved in a buffer containing 0.05 *M* Tris pH 7.4, 5 m*M* $MgCl_2$, and 0.025 *M* KCl, and a sample was layered on an isokinetic 15 to 30% sucrose gradient in the same buffer. Centrifugation was carried out in the Spinco SW-41 rotor for 45 min at 200,000 × *g*. The absorption at 260 nm was recorded with a continuous-flow monitoring system using a Gilford spectrophotometer. Reprinted with permission from ref. 70a.

added, to a final concentration of 100 m*M* in Mg^{2+} ions. After standing for 1 hour the solution is centrifuged through a 1 *M* sucrose cushion in the buffer described above. Centrifugation at 15,000 rpm for 30 min can be carried out in an RC2-B Sorvall using a 5534 rotor. The pellets contain the lens ribosomes, which may be processed further for mRNA preparation.

5.7 LENS MESSENGER RNA

One of the first eukaryotic messenger RNA species to be isolated in a purified form was the 14 S mRNA from calf lens (58). Definitive evidence for the occurrence of mRNA in a given RNA population requires the translation of this material into a specific polypeptide. Lens messenger fractions can be isolated by zonal centrifugation through a sucrose gradient of lens polyribosomes dissociated by sodium dodecylsulfate (59). Further purification can be achieved by affinity chromatography through poly (dT) or oligo (dT) cellulose columns (60, 61) and semipreparative polyacrylamide gel electrophoresis (4). Translation of the separated fractions in a cell-free system yields a mixture of crystallin subunits among which is the αB_2 chain encoded by 10 S mRNA, whereas the 14 S mRNA carries the genetic information for the αA_2 chain (2, 62, 63).

It is striking that 14 S crystallin mRNA with an estimated molecular weight of about 500,000 (4) is required to direct the synthesis of a protein subunit of M_r 20,000. In 1972 (2) we envisaged two possible interpreparations:

1 Besides the poly(A) track at the 3′ end, the messenger contains untranslated regions of considerable length.
2 The 14 S lens mRNA might be bicistronic with either repeating sequences for αA_2 or a sequence of αA_2 and another unknown polypeptide.

Indeed, Chen and Spector (50, 64) claimed that the 14 S mRNA contains still another cistron than the αA_2 coding sequence. Their conclusion was derived from the following observations:

1 The mRNA retains its large size under denaturing conditions.
2 The mRNA directs the synthesis of a αB_2-sized polypeptide in addition to the αA_2 chain.
3 Molecular hybridization of the 14 S mRNA with 10 S mRNA-derived cDNA reveals that 2 mol of this DNA bind to 1 mol of 14 S mRNA.
4 The two-dimensional maps of RNAse A and T_1 digests of 10 S and 14 S lens mRNA indicated similarity. From all these data Chen and Spector suggest that the "additional" gene product is an α-crystallinlike polypeptide.

In contrast, in my laboratory purified calf lens 14 S mRNA, in agreement with earlier reports (4), was always translated into one gene product, namely, αA_2. One should also keep in mind that a bicistronic messenger implies the existence of two initiation sites on this mRNA. In view of a virtually complete inhibition of 14 S mRNA translation in the presence of cap analogues that exert their action presumably only at the 5′ terminus, this is not very likely (47).

Williamson et al. isolated crystallin mRNA from chick lens (65). As in calf lens tissue, a 15 S mRNA and a 9 S mRNA component could be identified. However, not only the smaller species, but also the 15 S mRNA appeared to be heterodisperse. This is due to the occurrence of δ-crystallin, which makes up about 80% of the protein. Zelenka and Piatigorsky (66) partially characterized the total cytoplasmic mRNA synthesized and accumulated during differentiation of embryonic chick fibers (15-day-old chick embryos). Actually, according to these authors no other functional mRNA species were present in detectable amounts. The molecular weight of the messenger appeared to be 680,000 as determined by its migration in polyacrylamide gels containing formamide (67). This value corresponds to about 2200 nucleotides and a sedimentation value of 17 S. Zelenka and Piatigorsky synthesized radioactive cDNA by AMV–RNA-dependent DNA polymerase. Hybridization analyses of the [H^3]-cDNA with excess δ-crystallin mRNA revealed that the messenger is composed primarily of a single species with a sequence complexity of about 2200–2500 nucleotides. From this result it can be concluded that the frequently observed heterogeneity of chick δ-crystallin is presumably due to some post-translational modification. By quantitation of δ-crystallin mRNA during lens induction Shinohara and Piatigorsky (68) arrived at the conclusion that δ-crystallin synthesis is regulated by the accumulation of the corresponding messenger in the embryo.

Cohen et al. (69) isolated mRNA from rat lens. Quite surprisingly the following observation was made. The mRNA that codes for the αA_2 subunit also directed the synthesis of an additional αA-polypeptide, designated as αA^{Ins} (70, 70a) with an approximate molecular weight of 24,000 (Figure 6). Sequence studies revealed that this was due to an insertion of 22 amino acid residues in the "normal" αA_2 chain between residues 63 and 64 (71). Not only rat crystallin contains this extra A chain but all members of the rodent family Muroidea appear to possess this peculiar polypeptide. It cannot be ruled out that this is the first detection of a gene product that arises due to a "failure" in the splicing mechanism, which is operative in messenger formation (72, 73).

So far the only purified crystallin messenger RNAs are the 14 S mRNA from calf lens coding for αA_2 and the chick δ-crystallin mRNA. Messengers for most of the β-crystallin chains and those coding for γ-crystallin are accumulated in the 10 S mRNA fraction. Careful examination of the patterns of β_H- and β_L-crystallin reveals that the major difference is the occurrence of three polypeptides that are not involved in β_L aggregate formation, namely, βB_{1a}, βB_{1b}, and βA (74, 75).

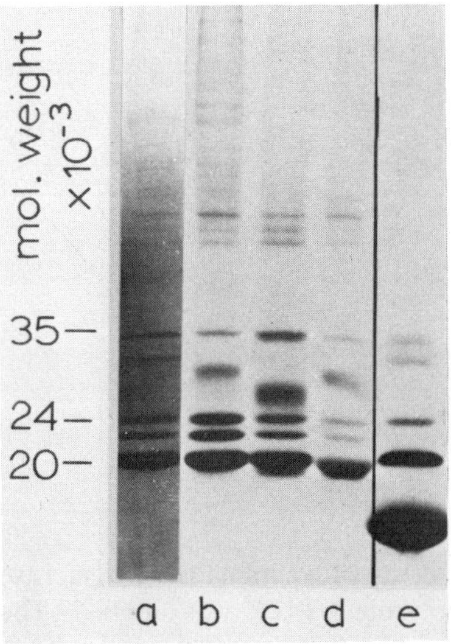

Figure 6 Dodecylsulfate–polyacrylamide gel electrophoresis of newly synthesized subunits of α-crystallin. The unlabeled bands were detected by staining (*a*), the labeled bands by autoradiography (*b–f*). (*a*) α-Crystallin isolated from fresh rat lenses; (*b*) α-crystallin labeled in epithelial cells of incubated rat lenses; (*c*) α-crystallin labeled in cortical cells of incubated rat lenses; (*d*) α-crystallin labeled in the rat lens cell-free system; (*e*) translation products of the reticulocyte cell-free system after addition of purified 14 S rat lens mRNA (predominantly αA_2 and αA^{Ins} are synthesized). The heavy labeled band at the bottom of the gel is globin. Reprinted with permission from ref. 70a.

βB_{1b} arises by some post-translational process from βB_{1a} (75). The following arguments can be given for this hypothesis:

1 No messenger RNA for βB_{1b} is detectable.
2 This polypeptide does not occur in very young calf lens.
3 In the cow lens nucleus βB_{1a} is virtually absent but βB_{1b} exists in relatively high concentration.

We were able to isolate the βB_{1a}-mRNA by a repeated centrifugation procedure. If a total lens polysomal RNA is separated on a sucrose gradient the various fractions can be translated in a reticulocyte lysate. It was observed that βB_{1a}, which corresponds to a molecular weight of 32,000, appears as a rather strong band (Figure 7). After recentrifugation an mRNA fraction with a sedimentation coefficient of 12.5 S could be obtained that directs almost exclusively the synthesis of βB_{1a} (Figure 8). For some

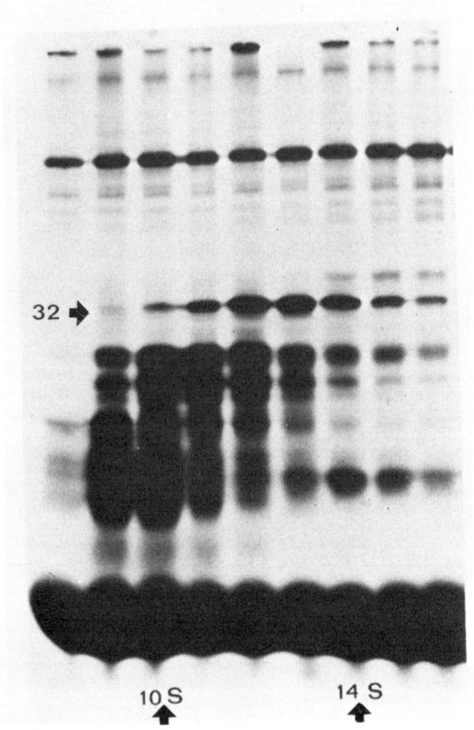

Figure 7 Autoradiograph of the translation products of fractions from a sucrose density gradient onto which total calf lens polysomal RNA was applied. The polypeptide with molecular weight 32,000 (βB_{1a}) and the 10 S and 14 S mRNA fractions are indicated by arrows. Reprinted with permission from ref. 75.

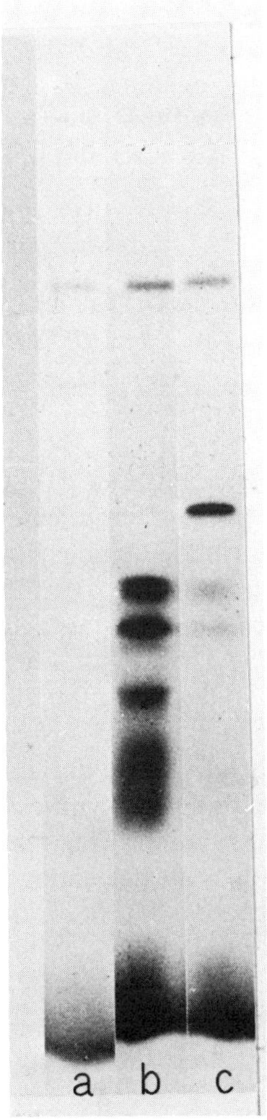

Figure 8 Autoradiograph of the translation products newly synthesized after addition of lens mRNA fraction to a reticulocyte cell-free system. (*a*) Blank incubation. Only globin is formed in addition to another endogenous protein of high molecular weight. (*b*) Translation products formed after addition of a 10 S mRNA fraction. (*c*) Translation products formed after addition of the purified βB_{1a} mRNA. Reprinted with permission from ref. 75.

unknown reason βB_{1a} and βB_{1b} are strongly stained with Coomassie brilliant blue but rather poorly with amido black in urea-containing polyacrylamide gels.

Thomson et al. (76) isolated messenger RNA from 1-day-old chick lens. The total mixture of poly(A) containing lens RNA was translated in a heterologous cell-free system and showed the same range of products that are also representative of the proteins synthesized in the intact lens. Poly(A)$^-$mRNA had a coding specificity almost similar to that of poly(A)$^+$mRNA. Translation products of mRNAs from 15-day embryo fibers and epithelial cells reflected the different protein compositions of both cell types.

This was also true for the translation patterns of mRNAs from lenses at different stages of development. From these observations Thomson et al. drew the important conclusion that control of translation of the various crystallins is regulated prior to their synthesis. Similar evidence can be derived from the results of Milstone et al. (77), who showed that fiber cell differentiation is paralleled by an increase in δ-crystallin synthesis and in δ-crystallin mRNA.

5.8 CRYSTALLIN BIOSYNTHESIS IN LIVING *XENOPUS LAEVIS* OOCYTES

The oocytes of the frog *Xenopus laevis* can be microinjected with controlled amounts of polysomes or isolated messenger RNA (5, 78, 81). In this system one can study how protein synthesis is regulated in an eukaryotic cell. If the total poly(A)-containing calf lens mRNA is microinjected into *Xenopus* oocytes, synthesis of all crystallin classes can be demonstrated (78). By quantitative immunoprecipitation it could be estimated that 20 to 30 polypeptide chains per hour and per mRNA molecule are synthesized. The kinetics of translation of the 14 S mRNA species shows that the messenger (Figure 9) is stable for at least 2 days (78). Presumably this stability is, as with globin mRNA (79), due to the poly(A) stretch which contains at least 50 adenyl residues (60, 80). When calf lens crystallin and rabbit globin mRNA are competing for factors that limit protein synthesis in the oocyte no mRNA is preferentially selected for translation. This may mean that the exogenous messenger is hardly competing with the endogenous messengers for accession of the translational apparatus (78).

The translation products of purified 14 S mRNA after microinjection are found exclusively in the form of crystallin aggregates. If 10 S lens mRNA is injected separately it seems that only αB subunits in the monomeric state are produced (81a).

5.9 THE LENS CELL-FREE SYSTEM

Important components of the lens protein synthesizing machinery, such as polyribosomes, were isolated and characterized years ago (54–56), so it seems rather amazing that a viable lens lysate was not introduced until the early 1970s (82–86). This homologous cell-free system can be prepared very easily from fresh eye lenses simply by removal of the capsule with the adhering epithelial monolayer. From the remaining lens body only the outer cortical region is used. This tissue is homogenized in 1 volume of distilled water followed by centrifugation of the homogenate at 15,000 × *g* for 10 min. The supernatant fraction is divided into small portions which should be stored in liquid nitrogen (84).

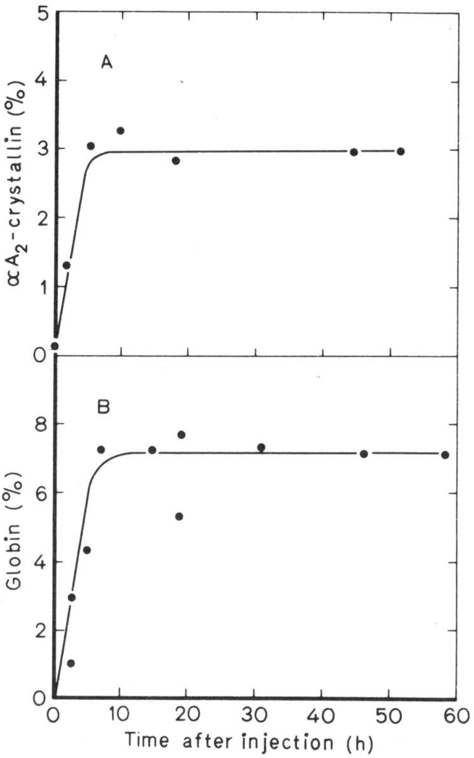

Figure 9 Kinetics of translation of mRNA in *Xenopus* oocytes. (*a*) 5 ng 14 S lens mRNA was injected into oocytes. Batches of 10 oocytes were labeled for 4-hour periods with [^{35}S]-methionine at different times after injection. The middle of the labeling period has been indicated in the figure. (*b*) Similarly, oocytes injected with 5 ng globin mRNA were labeled with [^{3}H]-histidine. The amount of αA_2-crystallin synthesized was measured by quantitative immunoprecipitation. Globin synthesis was assayed by determination of the radioactivity in the globin band of a polyacrylamide gel. The amount of specific polypeptides synthesized is expressed as a percentage of total acid-precipitable radioactivity. In control oocytes that had not been injected with globin mRNA a small amount (0.5%) of radioactivity migrated in the globin region of the gel. Reprinted with permission from ref. 6.

For incorporation studies the lysate must be supplied with the appropriate ions, energy, and amino acids. It could be demonstrated that the lens cell-free system is able to initiate, elongate, terminate, and eventually release and assemble completed crystallin polypeptide chains. The ionic conditions for cell-free translation are rather stringent. At 85 m*M* K^+ the optimal concentration of magnesium ions for initiation is 5 m*M*, and somewhat lower for elongation (Figure 10) (cf. 85).

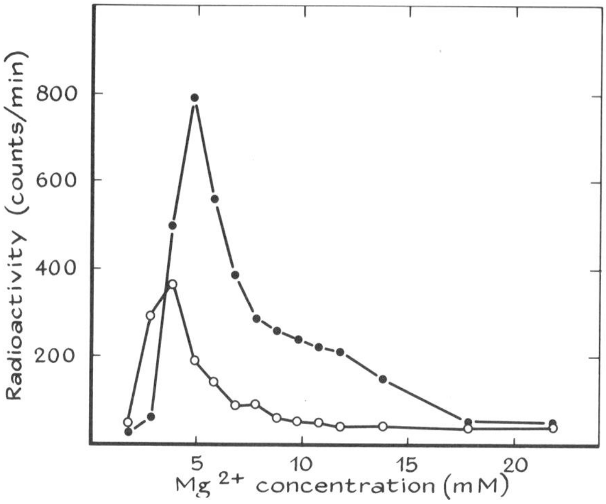

Figure 10 Optimal magnesium concentration for chain initiation (●——●) and elongation (○——○) in the lens cell-free system at a potassium concentration of 85 m*M*. Reprinted with permission from ref. 84.

The optimal potassium ion concentration for initiation is 60 m*M* but for elongation is 85 m*M* KCl. Both ionic conditions have been determined at a constant Mg^{2+} concentration of 5 m*M* (Figure 11).

After about 4 min of incubation a lens lysate supplemented with [^{35}S]-Met $tRNA_f^{Met}$ synthesizes completed α-crystallin polypeptides with an acetyl group on the radioactive *N*-terminal methionine residue. Approximately the same time is required to obtain completed chains in other systems.

Whereas α-, β_L-, and γ-crystallins are synthesized in the lens cell-free systems, β_H-crystallin is not formed (Figure 12) (84). This observation is paralleled by the finding that upon addition of lens polyribosomes to a reticulocyte lysate, α-, β_L-, and γ-crystallin can be detected after incubation but β_H-crystallin is absent (75). Also in organ culture β_H formation almost vanishes (84). Asselbergs et al. (81a) analyzed the newly synthesized multimeric crystallins after addition of lens messenger to the reticulocyte lysate or upon microinjection into *Xenopus* oocytes. (81) No carrier crystallins were added in these experiments. In this case no radioactivity at all could be detected in the β_H region. The reason for this lack in β_H synthesis is still not clear. Two possibilities may be envisaged for the *in vivo* situation:

1 The β_H aggregate arises gradually from β-polypeptide chains which are found first as aggregates of lower molecular weight or as monomers.

2 β_H biosynthesis takes place at a very slow rate as compared to the other crystallins.

The first possibility is sustained by the observation of van Kamp and Hoenders that β_H-crystallin increases during aging, whereas the amount of β_L diminishes (86a).

Our *in vitro* experiments provide some evidence for a more detailed explanation of β_H formation. As mentioned above, isolated β_H-crystallin contains two highly basic polypeptides, designated βB_{1a} and βB_{1b}, and an acidic subunit βA that do not occur in β_L. However, β_H- and β_L-crystallin have the major polypeptide chain βB_p in common. Quantitative and qualitative data from the cell-free and oocyte microinjection experiments showed that "native" β_H-crystallin as found after gel filtration of lens homogenates is never formed. But after dissociation and reassociation βB_p (from β_L-crystallin) is incorporated in a higher molecular weight aggregate.

Again two possibilities have to be considered in order to explain the transition of β_L constituents into β_H-crystallin upon reaggregation *in vitro* and to reconcile it with the *in vivo* situation:

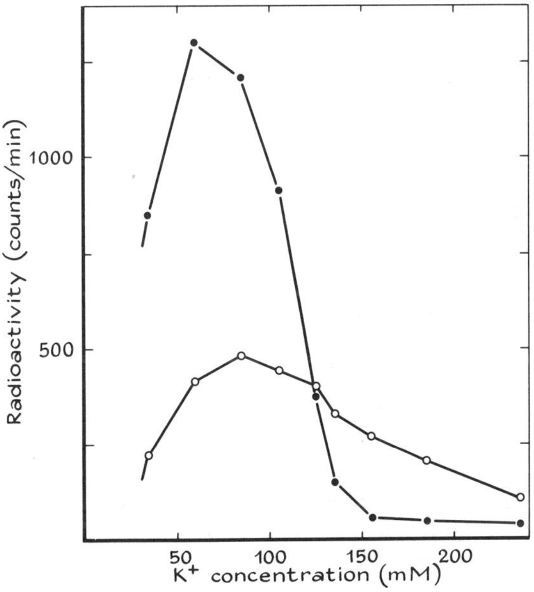

Figure 11 Optimal potassium concentration for chain initiation (●——●) and elongation (○——○) in the lens cell-free system at a Mg^{2+} concentration of 5 m*M*. Reprinted with permission from ref. 84.

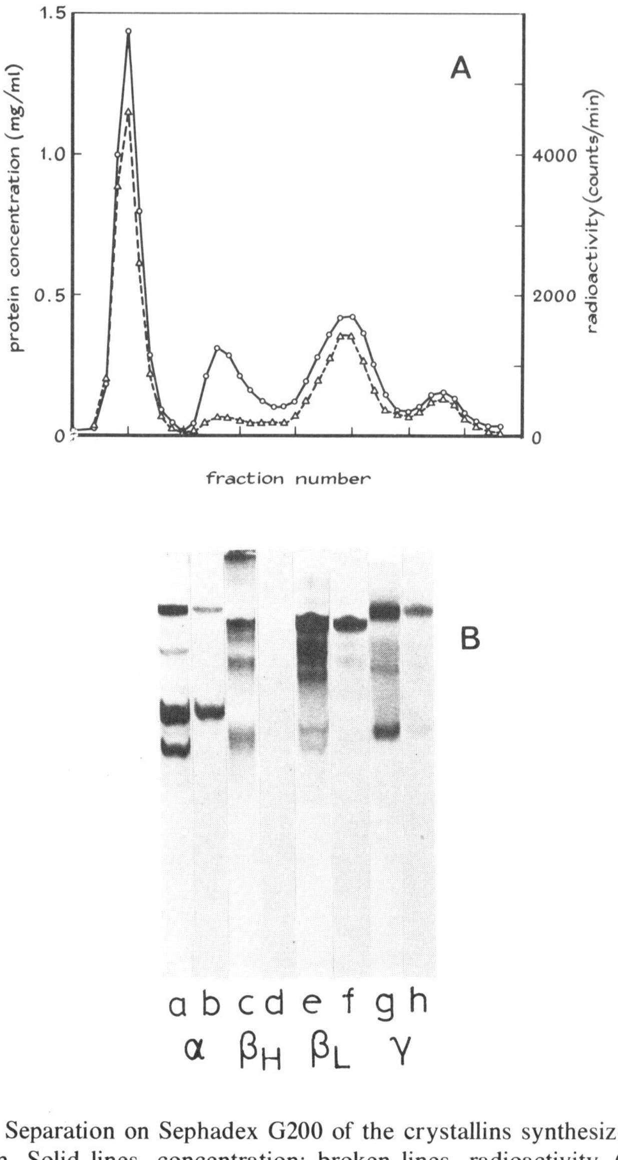

Figure 12 (*A*) Separation on Sephadex G200 of the crystallins synthesized in the cell-free system. Solid lines, concentration; broken lines, radioactivity. (*B*) Polyacrylamide gel electrophoresis and autoradiography of newly synthesized crystallin chains in the lens cell-free system. Notice that no β_H-crystallin subunits can be detected. (*a, c, e, g*) Stained gels; (*b, d, f, h*) autoradiographs of the longitudinally sliced gels. Reprinted with permission from ref. 84.

1 β_H-crystallin has a lower enthalpy than β_L-crystallin but a higher activation energy is required for its formation.

2 During maturation or aging modification of preexisting subunits occurs, which promotes the formation of β_H-crystallin.

The former possibility can be ruled out since in experiments with unlabeled β-crystallin no changes in the relative amounts of β_H- and β_L-crystallin are observed upon dissociation and reassociation irrespective of the initial ratio of both types of crystallins. The second possibility still leaves three interpretations open. All polypeptides composing β_L-crystallin have to be modified in order to fit the β_H-aggregate. If this assumption is true, the assumed modification should occur during reassociation. As a consequence the ratio of β_H- and β_L-crystallin in the unlabeled protein mixture should increase. This, however, is not the case. The remaining possibilities are as follows:

1 Only part of certain β-crystallin peptides has to be modified in order to enable incorporation of unchanged chains into β_H aggregates (exchange between newly synthesized radioactive and added unlabeled polypeptides).

2 Only one specific β_H-polypeptide is not synthesized *de novo* but has to be modified in a post-translational process.

We showed that the subunit βB_{1b} is a good candidate for this role. However, because β_H is still a rather poorly defined entity more structural work has to be done in order to fully understand β_H biogenesis.

5.10 *N*-TERMINAL ACETYLATION

The problem of *N*-terminal acetylation of α- and β-crystallin was solved with the aid of the lens cell-free system (87). The lens cell-free extract was labeled with [^{35}S]-Met-tRNA$_f^{Met}$ and the newly synthesized peptides of varying length assayed. It appeared that acetylation takes place while the peptides are still on the ribosome. The acetylation process starts when the chains are about 25 amino acid residues long and three or four amino acids protrude from the ribosome. Acetylation seems to be completed when the nascent polypeptides reach a length corresponding to about 50 amino acid residues. Acetyl–CoA is the acetyl group donor in this reaction. In the cell-free system it could also be shown that the initiating methionine is split off from β-crystallin at about the same moment that terminal *N*-α-acetylation starts (88).

The lens cell-free system was also used to establish whether or not acetyl Met-tRNA$_f^{Met}$ could be used in the initiation of lens crystallin biosynthesis.

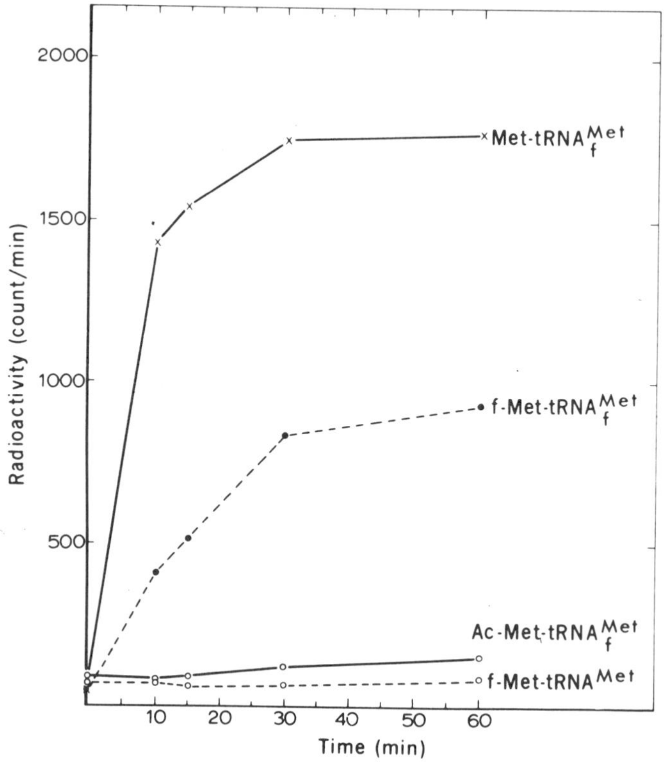

Figure 13 Kinetics of *in vitro* lens protein initiation with different [^{35}S]-Met-tRNA$_f^{Met}$ species. All incubation mixtures contained, in a final volume of 0.2 ml, the following components: 1 m*M* ATP, 0.5 m*M* GTP, 15 m*M* creatine phosphate, 0.01 mg creatine phosphokinase, 10 m*M* Tris-Cl, pH 7.4, 5 m*M* magnesium acetate, 85 m*M* KCl, 5 m*M* 2-mercaptoethanol, 0.1 m*M* of all 20 amino acids, and 360,000 cpm of one of the [^{35}S]-Met-tRNA$_f^{Met}$ species. The specific activity of [^{35}S]-methionine was 80 Ci/mmol. In order to prevent transfer of labeled methionine from tRNA$_f^{Met}$ to tRNA$_m^{Met}$ unlabeled methionine was present at a concentration of 1.2 m*M*. 0.12 ml of the lens cell-free extract was added. Incubation was at 30°C. At indicated times 0.02-ml aliquots were added to 0.5 ml 0.1 *M* KOH. After incubation at 37°C for 10 min trichloroacetic acid was added to 5%. The precipitate was collected on glass fiber filters and counted in a liquid scintillation counter at about 80% efficiency. tRNA specific for methionine was isolated from eye lens tissue, aminoacylated with [^{35}S]-methionine, and enzymatically formylated with transformylase from *E. coli*. Chemical formylation was carried out with *p*-nitrophenyl formate. Chemical acetylation was performed with *p*-nitrophenyl acetate (synthesized from *p*-nitrophenol and acetic acid in the presence of dicyclohexylcarbodiimide in dry ethyl acetate as solvent). In order to get 100% formylated or acetylated [^{35}S]-Met-tRNA$_f^{Met}$ unblocked species were hydrolyzed with Cu^{2+}. Removal of Cu^{2+} was performed by gel filtration on Sephadex G-25. X——X, [^{35}S]-Met-tRNA$_f^{Met}$; •----•, formyl-[^{35}S]-Met-tRNA$_f^{Met}$ (enzymatically formylated); ○-----○, formyl-[^{35}S]-Met-tRNA$_f^{Met}$ (chemically formylated); ○——○, acetyl-[^{35}S]-Met-tRNA$_f^{Met}$ (chemically acetylated). Reprinted with permission from ref. 94.

It turned out that this was not the case. Moreover, the free Met-$tRNA_f^{Met}$ appeared to be a better initiator than formyl-Met-$tRNA_f^{Met}$O (89). (Figure 13).

5.11 BIOSYNTHESIS OF NONCRYSTALLIN PROTEINS

The routine procedure now applied for the isolation of eukaryotic messenger is affinity chromatography on oligo (dT) cellulose (89a). As a rule, three fractions emerge from the column, which is equilibrated with 0.5 *M* NaCl, 0.01 *M* Tris-Cl, pH 7.5, 0.5% sodium dodecylsulfate, and 1 m*M* EDTA (Figure 14) (90). Nonabsorbed material is washed off with this

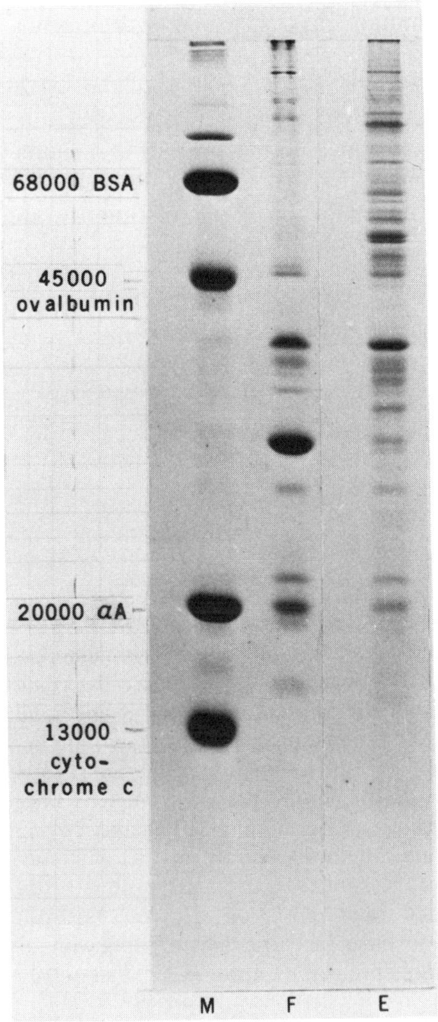

Figure 14 SDS gel electrophoretic pattern of lens plasma membrane proteins. F, isolated fiber plasma membranes; E, isolated epithelial plasma membranes; M, marker proteins αA, the A_2 chain of α-crystallin.

buffer, whereas the RNA retained is eluted with two buffers of reduced ionic strength. In the first one the NaCl concentration is reduced to 0.1 *M;* in the second no NaCl is added. As a rule the second fraction is neglected. Our studies revealed, however, that in particular messengers for the noncrystallin proteins are present in this fraction. Gozes et al. (91), who studied the translation *in vitro* of actin and tubulin messengers, also reported that in particular mRNAs coding for proteins of relatively high molecular weight are found in the second fraction. Whereas the crystallins have been studied intensively for almost a century (cf. refs. 92–94) the noncrystallin lens proteins have been rather neglected. Recently, however, the protein moiety of lens plasma membranes has roused the interest of various groups (95–102).

If prepared under proper conditions the lens fiber plasma membrane pattern is characterized by two predominant protein bands with approximate molecular weights of 34,000 and 26,000, respectively (Figure 14). On the other hand, lens epithelial plasma membranes seem to lack the MP26 component. During the process of elongation there is a slow but continuous accumulation of newly synthesized plasma membrane protein and lipids. For a better understanding of the biogenesis of fiber plasma membranes it is important to know whether the two major protein constituents are

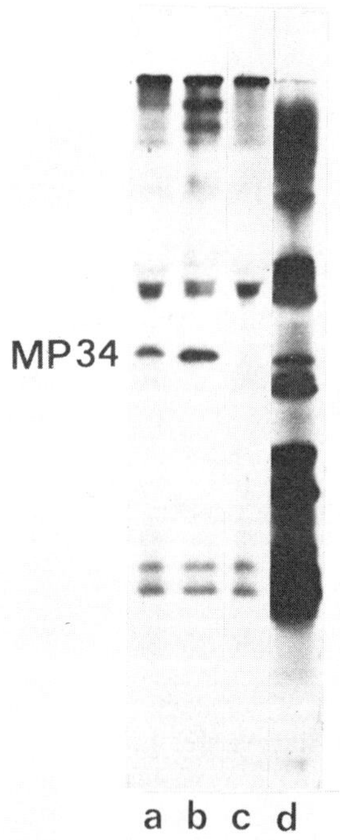

Figure 15 Autoradiograph of the SDS gel electrophoretogram of protein immunoprecipitated with total antiplasma membrane protein. (*a*) Antiserum directed against total plasma membrane protein from lens epithelium; (*b*) antiserum directed against total plasma membrane protein from lens fiber cells; (*c*) antiserum directed against bovine serum albumin (control); (*d*) the [^{35}S]-methionine-labeled translation products obtained after translation of "free" lens polyribosomes in a reticulocyte lysate without immunoprecipitation (for comparison).

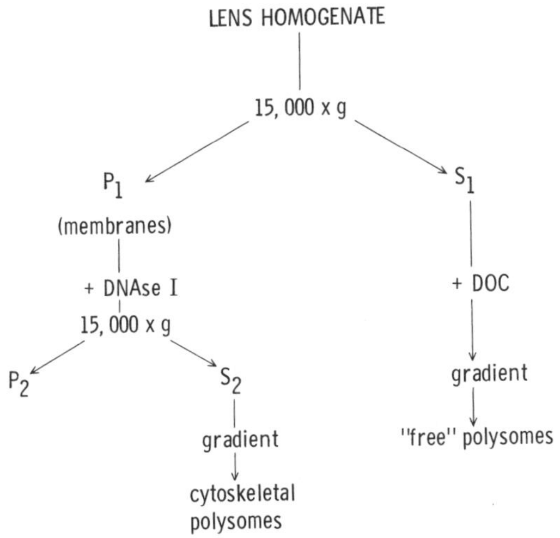

Figure 16 Schematic representation of the isolation of "free" and cytoskeleton-attached lens polyribosomes.

synthesized in the lens cortex. To verify this, lens cortical polyribosomes were translated in a reticulocyte lysate (103). After the incubation immunoprecipitation was carried out with an antiserum directed against the total plasma membrane protein.

Analysis on SDS–polyacrylamide gels revealed that predominantly MP34 was synthesized *de novo* (Figure 15). In addition, an unidentified component of molecular weight greater than 100,000 was detected after immunoprecipitation. Surprisingly the MP26 was absent.

The scheme represented in Figure 16 shows why the MP26 escaped detection upon translation of lens fiber polyribosomes in cell-free systems. The reason is that a population of polyribosomes is associated with the cytoskeleton–plasma membrane complex (104). This structure can be visualized in the electron microscope (Figure 17). Since the attached polysomes can be released by DNAse I treatment, it has been suggested that actin is involved in some way in the binding (105).

This assumption is based on two other findings. First, Lazarides and Lindberg showed that DNAse I depolymerizes actin (106). Second, Kibbelaar et al. demonstrated unequivocally by chemical characterization that G- and F-actin occur in lens fiber cells (107).

Ramaekers et al. (108) studied the translation of the released cytoskeletal lens polysomes in the reticulocyte lysate. From their results it appears that the messenger RNA encoding MP26 is localized on this class of polyribosomes (compare Figure 18).

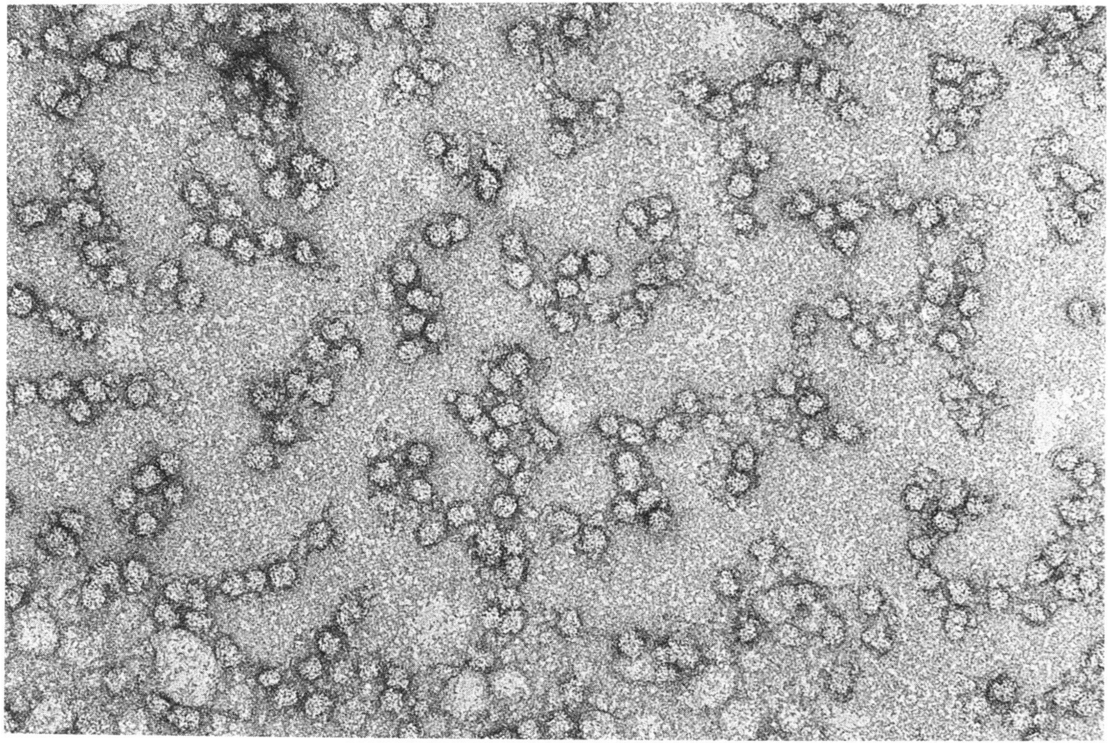

Figure 17 Negatively stained preparation of cytoskeleton-bound lens polyribosomes released by DNAse I treatment. The grid was washed once with 1% deoxycholate before staining (×200,000).

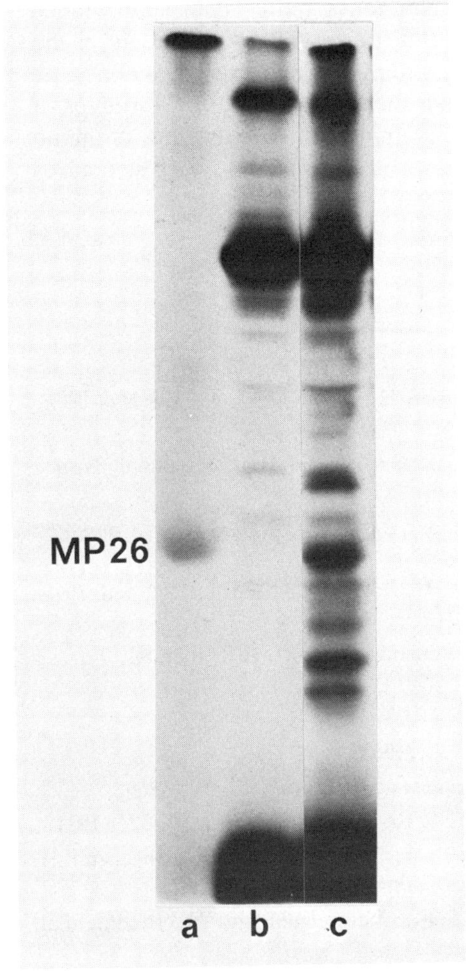

Figure 18 Autoradiograph of the gel electrophoretic pattern of products obtained after translation of cytoskeleton-bound lens polyribosomes in a reticulocyte cell-free system. (*a*) DNAse-released lens polyribosomes. The newly synthesized products were immunoprecipitated with anti-MP26 serum prior to gel electrophoresis; (*b*) no polysomes added (control); (*c*) for comparison the translation products of "free" polyribosomes are shown (no immunoprecipitation was applied prior to gel electrophoresis).

5.12 CONCLUDING REMARKS

The elucidation of the question how the biosynthesis of lens proteins is regulated requires knowledge of the structure of the corresponding genes. At this time nothing is known about the number and nature of the various crystallin messenger RNAs. Cloning of lens crystallin cDNA sequences will be extremely helpful in getting the desired answers.

Meanwhile the first papers on cloning of δ-crystallin cDNA in a bacterial plasmid are available (109, 110). A number of cDNA clones complementary to a variety of rat crystallin mRNAs have been obtained while this manuscript was in preparation (111). Preliminary sequence studies on the 14 S cDNA clone confirmed our previous assumption that lens αA crystallin

14 S mRNA is not bicistronic (112). No doubt more reports on cloned crystallin genes will appear in the near future, providing the student of the molecular biology of the lens with new tools to analyze crystallin gene organization and the control of crystallin mRNA translation. Moreover, it seems logical to expect that the time-consuming and tedious work of amino acid sequence determination of lenticular proteins whose primary structures are still unknown will become superfluous.

REFERENCES

1 W. M. Bonner and R. A. Laskey, *Eur. J. Biochem.* **46,** 83 (1974).

2 A. J. M. Berns, G. J. A. M. Strous, and H. Bloemendal, *Nature New Biol.,* **236,** 7 (1972).

3 M. B. Mathews, M. Osborn, A. J. M. Berns, and H. Bloemendal, *Nature New Biol.,* **236,** 5 (1972).

4 A. Berns, P. Janssen, and H. Bloemendal, *Biochem. Biophys. Res. Commun.,* **59,** 1157 (1974).

5 A. J. M. Berns, M. van Kraaikamp, H. Bloemendal, and C. D. Lane, *Proc. Natl. Acad. Sci. U.S.,* **69,** 1606 (1972).

6 F. A. M. Asselbergs, W. J. van Venrooij, and H. Bloemendal, *Eur. J. Biochem.,* **87,** 517 (1978).

7 T. Kuwabara, *Exp. Eye Res.,* **20,** 427 (1975).

8 R. Reeder and E. Bell, *Science,* **150,** 71 (1965).

9 R. W. Young and H. W. Fulhorst, *Invest. Ophthalmol.,* **5,** 288 (1966).

10 J. B. Coulter, A. A. Swanson, and T. J. Bond, *Ophthal. Res.,* **2,** 322 (1971).

11 A. J. M. Vermorken, M. A. Kibbelaar, J. M. H. C. Hilderink, and H. Bloemendal, *Biochem. Biophys. Res. Commun.,* **88,** 597 (1979).

11a F. C. S. Ramaekers, A. M. E. Selten-Versteegen and H. Bloemendal, *Biochim. Biophys. Acta,* **596,** 57 (1980).

12 S. G. Waley, in H. Davson, Ed., *The Eye,* Vol. 1, Academic Press, New York, 1969, p. 299.

13 J. Papaconstantinou, J. A. Stewart, and P. V. Koehn, *Biochim. Biophys. Acta,* **114,** 428 (1966).

14 J. A. Stewart and J. Papaconstantinou, *J. Mol. Biol.,* **29,** 357 (1967).

15 J. Papaconstantinou, *Science,* **156,** 338 (1967).

16 R. Reeder and E. Bell, *J. Mol. Biol.,* **23,** 577 (1967).

17 B. J. Ortwerth and R. J. Byrnes, *Exp. Eye Res.,* **12,** 120 (1971).

18 W. G. M. van den Broek, M. A. G. Koopmans, and H. Bloemendal, *Mol. Biol. Rep.,* **1,** 295 (1974).

19 H. Bloemendal, A. Zweers, M. Koopmans, and W. van den Broek, *Biochem. Biophys. Res. Commun.,* **11,** 416 (1977).

20 J. W. MacAvoy, *J. Embryol. Exp. Morphol.,* **44,** 149 (1978).

20a J. Piatigorsky, P. F. Kador, and J. H. Kiroshita, *Exp. Eye Res.*, **30,** 69 (1980).

21 C. V. Harding, J. R. Reddan, N. J. Unakar, and M. Bagchi, *Int. Rev. Cytol.*, **31,** 216 (1971).

22 J. Piatigorsky, H. deF. Webster, and S. P. Craig, *Dev. Biol.*, **27,** 176 (1972).

23 J. Piatigorsky, S. S. Rothschild, and L. M. Milstone, *Dev. Biol.*, **34,** 334 (1973).

24 R. C. Hughes, M. Laurent, M. O. Lonchampt, and Y. Courtois, *Eur. J. Biochem.*, **52,** 143 (1975).

25 W. J. van Venrooij, A. A. Groeneveld, H. Bloemendal, and E. L. Benedetti, *Exp. Eye Res.*, **18,** 517 (1974).

26 W. J. van Venrooij, A. A. Groeneveld, H. Bloemendal, and E. L. Benedetti, *Exp. Eye Res.*, **18,** 527 (1974).

27 E. E. Schubert, J. R. Trevithick, and M. J. Hollenberg, *Can. J. Ophthalmol.*, **5,** 353 (1970).

28 A. J. M. Vermorken, J. M. H. C. Hildrink, W. J. M. van de Ven, and H. Bloemendal, *J. Cell Biol.*, **76,** 175 (1978).

28a A. J. M., Vermorken and H. Bloemendal, *Nature,* **271,** 779 (1978).

29 T. S. Okada, E. Eguchi, and M. Takeichi, *Dev. Growth Diff.*, **13,** 323 (1971).

30 T. S. Okada, G. Eguchi, and M. Takeichi, *Dev. Biol.*, **34,** 321 (1973).

31 D. I. de Pomerai, D. J. Pritchard, and R. M. Clayton, *Dev. Biol.*, **60,** 416 (1977).

32 R. M. Clayton, in J. Davson and L. T. Graham, Jr., Eds., *The Eye,* Vol. 5, Academic Press, New York, 1974, p. 399.

33 I. M. Genis-Galvez, *Cell Diff.*, **4,** 49 (1975).

34 D. I. de Pomerai, R. M. Clayton, and D. J. Pritchard, *Exp. Eye Res.*, **27,** 365 (1978).

35 D. I. de Pomerai and R. M. Clayton, *J. Embryol. Exp. Morphol.*, **47,** 179 (1978).

35a I. Thomson, D. I. de Pomerai, J. F. Jackson, and R. M. Clayton, *Exp. Cell Res.*, **122,** 73 (1979).

35b R. Clayton, I. Thomson, and D. I. de Pomerai, *Nature,* **282,** 628 (1979).

36 Y. Courtois, L. Simonneau, J. Tassin, M. V. Laurent, and E. Malaise, *Differentiation,* **10,** 23 (1978).

37 C. Arruti and Y. Courtois, *Exp. Cell Res.*, **117,** 283 (1978).

38 L. M. Milstone and J. Piatigorsky, *Exp. Cell Res.*, **105,** 9 (1977).

39 D. C. Bebe and J. Piatigorsky, *Dev. Biol.*, **59,** 174 (1977).

40 T. Shinohara and J. Piatigorsky, *Nature,* **270,** 406 (1977).

41 M. Lubin, *Nature,* **213,** 451 (1967).

42 L. Carrasco and A. E. Smith, *Nature,* **264,** 807 (1976).

42a H. Bloemendal, G. A. Lenstra, H. Dodemont, F. C. S. Ramaekers, A. A. Groeneveld, E. L. Benedetti, and I. Dunia, *Exp. Eye Res.*, **31,** 513 (1980).

42b H. Bloemendal, F. C. S. Ramaekers, J. A. Lenstra, H. Dodemont, I. Dunia, and E. L. Benedetti In G. M. Williams et al., Eds., The Predictive Value of

Short-Term Tests in Carcinogenicity Evaluation, Elsevier, North Holland Biomedical Press, 1980, p. 199.

43 E. Derman and J. E. Darnell, *Cell,* **3,** 255 (1974).

44 A. J. M. Vermorken, J. M. H. C. Hilderink, H. Bloemendal, I. Dunia, and E. L. Benedetti, *Exp. Eye Res.,* **27,** 727 (1978)

45 A. J. Shatkin, *Cell,* **9,** 645 (1976).

46 B.Kemper, *Nature,* **262,** 321 (1976).

47 F. A. M. Asselbergs, W. H. M. Peters, W. J. van Venrooij, and H. Bloemendal, *Biochim. Biophys. Acta,* **520,** 577 (1978).

48 F. A. M. Asselbergs, W. H. M. Peters, W. J. van Venrooij, and H. Bloemendal, *Mol. Biol. Rep.,* **4,** 177 (1978).

49 H. Suzuki, *FEBS Lett.,* **79,** 11 (1977).

50 J. H. Chen and A. Spector, *Biochemistry,* **16,** 499 (1977).

51 G. C. Lavers, *Exp. Eye Res.,* **28,** 299 (1979).

52 F. A. M. Asselbergs, W. H. M. Peters, W. J. van Venrooij, and H. Bloemendal, *FEBS Lett.,* **94,** 195 (1978).

53 A. J. M. Berns and H. Bloemendal, in R. C. King, Ed., *Handbook of Genetics,* Vol. 5, Plenum Press, New York, 1976, p. 267.

54 H. Bloemendal, J. Schoenmakers, A. Zweers, R. Matze, and E. L. Benedetti, *Biochim. Biophys. Acta,* **123,** 217 (1966).

55 J. G. G. Schoenmakers, A. Zweers, and H. Bloemendal, *Biochim. Biophys. Acta,* **147,** 183 (1967).

56 E. L. Benedetti, A. Zweers, and H. Bloemendal, *Biochem. J.,* **108,** 765 (1968).

57 R. D. Palmiter, *Biochemistry,* **13,** 3606 (1974).

58 H. Bloemendal, A. Berns, G. Strous, M. Mathews, and C. D. Lane, *RNA Viruses/Ribosomes, Proceedings of the Eighth Meeting of the Federation of European Biochemical Societies, Amsterdam,* Vol. 27, North Holland/American Elsevier, Amsterdam, 1972, p. 237.

59 A. J. M. Berns, R. A. de Abreu, M. van Kraaikamp, E. L. Benedetti, and H. Bloemendal, *FEBS Lett.,* **18,** 159 (1971).

60 G. Piperno, U. Bertazzoni, A. J. M. Berns, and H. Bloemendal, *Nucleic Acid Res.,* **1,** 245 (1974).

61 J. H. Chen, G. C. Lavers, and A. Spector, G. Schutz, and P. Feigelson, *Exp. Eye Res.,* **18,** 189 (1974).

62 A. J. M. Berns, V. V. A. M. Schreurs, M. W. G. van Kraaikamp, and H. Bloemendal, *Eur. J. Biochem.,* **33,** 551 (1973).

63 H. Bloemendal, in: J. Paul, Ed., *Biochemistry of Cell Differentiation II,* Vol. 15, University Park Press, Baltimore, 1977, p. 195.

64 J. H. Chen and A. Spector, *Proc. Natl. Acad. Sci. U.S.,* **74,** 5448 (1977).

65 R. Williamson, R. Clayton, and D. E. S. Truman, *Biochem. Biophys. Res. Commun.,* **46,** 1936 (1972).

66 P. Zelenka and J. Piatigorsky, *Proc. Natl. Acad. Sci. U.S.,* **71,** 1896 (1974).

67 P. Zelenka and J. Piatigorsky, *Exp. Eye Res.,* **22,** 115 (1976).

68 T. Shinohara and J. Piatigorsky, *Proc. Natl. Acad. Sci., U.S.*, **73,** 2808 (1976).

69 L. H. Cohen, L. W. Westerhuis, D. P. Smits, and H. Bloemendal, *Eur. J. Biochem.*, **89,** 251 (1978).

70 H. Bloemendal, L. H. Cohen, and W. W. de Jong, *Interdiscipl. Topics Gerontol.*, **12,** 261 (1978)

70a L. H. Cohen, D. E. M. Smits, and H. Bloemendal, *Eur. J. Biochem.*, **67,** 563 (1976).

71 L. H. Cohen, L. W. Westerhuis, W. W. de Jong, and H. Bloemendal, *Eur. J. Biochem.*, **89,** 259 (1978).

72 A. J. Jeffreys and R. A. Flavell, *Cell,* **12,** 1097 (1977).

73 R. Breathnach, J. L. Nandel, and P. Chambon, *Nature,* **210,** 314 (1977).

74 P. Herbrink, H. van Westreenen, and H. Bloemendal, *Exp. Eye Res.*, **20,** 541 (1975).

75 A. J. M. Vermorken, P. Herbrink, and H. Bloemendal, *Eur. J. Biochem.*, **78,** 617 (1977).

76 I. Thomson, C. E. Wilkinson, J. F. Jackson, D. I. de Pomerai, R. M. Clayton, D. E. S. Truman, and R. Williamson, *Dev. Biol.*, **65,** 372 (1978).

77 L. M. Milstone, P. Zelenka, and J. Piatigorsky, *Dev. Biol.*, **48,** 197 (1976).

78 F. A. M. Asselbergs, W. J. van Venrooij, and H. Bloemendal, *Eur. J. Biochem.*, **94,** 249 (1979).

79 U. Nudel, H. Soreg, U. Z. Littauer, G. Marbaix, G. Huez, M. Leclercq, E. Hubert, and H. Chantrenne, *Eur. J. Biochem.*, **64,** 115 (1976).

80 G. C. Lavers, J. H. Chen and A. Spector, *J. Mol. Biol.*, **82,** 15 (1974).

81 F. A. M. Asselbergs, M. Koopmans, W. J. Van Venrooij, and H. Bloemendal, *Exp. Eye Res.*, **28,** 475 (1979).

81a F. A. M. Asselbergs, M. Koopmans, W. J. van Venrooij, and H. Bloemendal, *Eur. J. Biochem.*, **91,** 65 (1978).

82 G. Strous, J. van Westreenen, and H. Bloemendal, *FEBS Lett.*, **19,** 33 (1971).

83 G. Strous, A. J. M. Berns, J. van Westreenen, and H. Bloemendal, *Eur. J. Biochem.*, **30,** 48 (1972).

84 G. Strous, H. van Westreenen, J. van der Logt, and H. Bloemendal, *Biochim. Biophys. Acta,* **353,** 89 (1974).

85 H. Bloemendal and G. J. A. M. Strous, *Acta Biol. Med. Ger.*, **33,** 935 (1974).

86 G. J. A. M. Strous, J. van Westreenen, and H. Bloemendal, *Eur. J. Biochem.*, **38,** 79 (1973).

86a G. J. van Kamp and H. J. Hoenders, *Exp. Eye Res.*, **17,** 417 (1973).

87 G. J. A. M. Strous, A. J. M. Berns, and H. Bloemendal, *Biochem. Biophys. Res. Commun.*, **58,** 876 (1974).

88 H. Bloemendal and G. J. A. M. Strous, *Lipmann Symposium: Energy, Biosynthesis and Regulation in Molecular Biology,* Walter de Gruyter, Berlin, New York, 1974, p. 90.

89 G. J. A. M. Strous, Th. Bollen, and H. Bloemendal, *Mol. Biol. Rep.*, **1,** 471 (1974).

89a A. Aviv, and P. Leder, *Proc. Natl. Acad. Sci. U.S.*, **69,** 1408 (1972).

90 A. J. M. Vermorken, J. M. C. H. Hilderink, W. J. M. van de Ven, and H. Bloemendal, *Biochim. Biophys. Acta,* **454,** 447 (1976).

91 I. Gozes, J. Schmitt, and U. L. Littauer, *Proc. Natl. Acad. Sci. U.S.*, **72,** 701 (1975).

92 R. M. Clayton, *The Eye,* **5,** 400 (1974).

93 J. J. Harding and K. J. Dilley, *Exp. Eye Res.*, **22,** 1 (1976).

94 H. Bloemendal, *Science,* **197,** 127 (1977).

95 H. Bloemendal, A. Zweers, F. Vermorken, I. Dunia, and E. L. Benedetti, *Cell Diff.*, **1,** 91 (1972).

96 I. Dunia, C. S. Ghosh, E. L. Benedetti, A. Zweers, and H. Bloemendal, *FEBS Lett.*, **45,** 139 (1974).

97 E. L. Benedetti, I. Dunia, C. J. Bentzel, A. J. M. Vermorken, M. Kibbelaar, and H. Bloemendal, *Biochim. Biophys. Acta,* **457,** 353 (1976).

98 R. M. Broekhuyse and E. D. Kuhlmann, *Exp. Eye Res.*, **19,** 297 (1974).

99 A. Lasser and E. A. Balazs, *Exp. Eye Res.*, **13,** 292 (1972).

100 J. Alcalá, N. Lieska, and H. Maisel, *Exp. Eye Res.*, **21,** 581 (1975).

101 D. Roy, A. Spector, and P. Farnsworth, *Exp. Eye Res.*, **28,** 353 (1979).

102 J. Horwitz, N. P. Robertson, M. M. Wong, J. S. Zigler, and J. H. Kinoshita, *Exp. Eye Res.*, **28,** 359 (1979).

103 A. J. M. Vermorken, J. M. H. C. Hilderink, I. Dunia, E. L. Benedetti, and H. Bloemendal, *FEBS Lett.*, **83,** 301 (1977).

104 H. Bloemendal, E. L. Benedetti, F. C. S. Ramaekers, I. Dunia, M. A. Kibbelaar, and A. J. M. Vermorken, *Mol. Biol. Rep.*, **5,** 99 (1979).

105 H. Bloemendal, M. A. Kibbelaar, F. C. S. Ramaekers, A. M. E. Selten-Versteegen, I. Dunia, and E. L. Benedetti, in *Prot. Biol. Fluids Proc. 26th Colloq.*, Pergamon Press, Oxford-New York, Vol. 26, 1978, p. 499.

106 E. Lazarides and U. Lindberg, *Proc. Natl. Acad. Sci. U.S.*, **71,** 4742 (1974).

107 M. A. Kibbelaar, A. M. E. Selten-Versteegen, I. Dunia, E. L. Benedetti, and H. Bloemendal, *Eur. J. Biochem.*, **95,** 543 (1979).

108 F. C. S. Ramaekers, A. M. E. Selten-Versteegen, E. L. Benedetti, I. Dunia, and H. Bloemendal, *Proc. Natl. Acad. Sci. U.S.*, **77,** 725 (1980).

109 S. P. Bhat and J. Piatigorsky, *Proc. Natl. Acad. Sci. U.S.*, **76,** 3299 (1979).

110 S. P. Bhat, R. E. Jones, M. A. Sullivan, and J. Piatigorsky, *Nature,* **284,** 234 (1980).

111 H. J. Dodemont, P. M. Andreoli, R. J. M. Moormann, F. C. S. Ramaekers, J. G. G. Schoenmakers, and H. Bloemendal (1981) (in press).

112 R. J. M. Moormann, H. van de Velde, P. M. Andreoli, H. J. Dodemont, H. Bloemendal, and J. G. G. Schoenmakers (1981) (in press).

CHAPTER

6

Evolution of Lens and Crystallins

Wilfried W. de Jong

Department of Biochemistry, University of Nijmegen, Nijmegen, The Netherlands

6.1 INTRODUCTION

> Perle épithéliale transparante suspendue au centre de l'oeil pour donner des images nettes à la rétine, le cristallin est peut-être la pièce la plus étonnante de tout l'organisme.
>
> *Rochon-Duvigneaud, 1943*(1)

These sentences appropriately express the impression one gets when one fully appreciates the wonderful features of the crystalline lens of the vertebrates. One of the many intriguing questions about evolution is how such a delicate organ came into existence. Although this chapter can not shed much light on this particular problem, it tries to order the presently available data on vertebrate lens proteins and place them in an evolutionary perspective.

By studying and comparing the lenses of present-day vertebrates we can hope to obtain information about the evolutionary changes that have taken place in structure and composition of the lens. Despite its similar function in all vertebrate eyes, the lens varies considerably in shape and consistency. At the same time wide variations in protein content and composition occur. Some lenses, or parts of lenses, mainly contain low molecular weight basic proteins, and others are particularly rich in high molecular weight aggregates of lower isoelectric point. In this chapter the more recent advances in the knowledge of structure and properties of vertebrate lens proteins are surveyed from a comparative point of view. The main question to be answered by this kind of information is whether and how the observed differences in protein composition are related to various structural and functional properties of the lens, and to what extent the differences are just the result of adaptively neutral, nonfunctional, or so-called non-Darwinian, evolutionary changes.

An extensive review of the comparative aspects of lens proteins was provided by Clayton in 1974 (2). Harding and Dilley (3) have reviewed the information on mammalian lens proteins more specifically. This chapter does not deal with the proteins of invertebrate lenses. Even though the gross anatomy of the eyes of the Cephalopoda resembles that of the vertebrate eye, their ontogenetic development is entirely different and their lens proteins are totally unrelated to those of the vertebrates (2, 4).

6.2 EVOLUTIONARY ORIGIN OF EYE AND LENS

The eyes of all vertebrates are clearly homologous in their anatomy and ontogenetic development. This indicates that a similar type of eye must already have been present in the last common ancestor of all living vertebrates. These jawless fishlike ancestral vertebrates must have appeared during the Ordovician period, approximately 450 million years ago (Figure 1). It would be highly interesting to know the evolutionary events and forces which led to the development of such an intricate organ as the vertebrate eye, in which the lens forms an essential part.

Considering the absence of primitive or intermediate forms of the eye in present-day vertebrates, and lacking fossil remains of this soft organ, no satisfactory and falsifiable theory about the evolution of the eye is yet available. Indeed, Darwin (5) wrote, in relation to the evolutionary origin of the eye, that "to suppose that the eye with all its inimitable contrivances . . . could have been formed by natural selection, seems, I freely confess, absurd in the highest degree," although his reason told him he ought to conquer this feeling. The apparently sudden evolutionary development of the vertebrate eye, of which neither embryology, nor comparative anatomy or paleontology can provide a satisfactory reconstruction, has been compared to "the birth of Athena, full-grown and fully armed, from the brow of Zeus" (6). Nevertheless, several authors have elaborately speculated about the origin of the vertebrate eye. A wide range of opinions can be found in Walls (7), Polyak (8), and Duke-Elder (9).

In keeping with Haeckel's "first law of biogenesis," which proposes that ontogeny is to a certain extent a short repeat of phylogeny, several hypotheses have been inspired by the study of the embryological development of the eye. Most authors suppose that, as in ontogeny, the eyes–retinas developed in evolution as evaginations of the brain, sprouting from the central organ toward the surface in order to become accessible to the action of light. The earliest onsets of light-sensitive organs being internally located, associated with the primitive brain, required these ancestors of the vertebrates to be transparant, a condition as found in *Amphioxus* and ascidians.

Quite in contrast Polyak (8) provides strong arguments from embryology to show that the eyes, or the optic primordia, are ontogenetically and very probably also phylogenetically the first to appear and form a prerequisite and impulse for the subsequent development of the brain. Polyak stresses

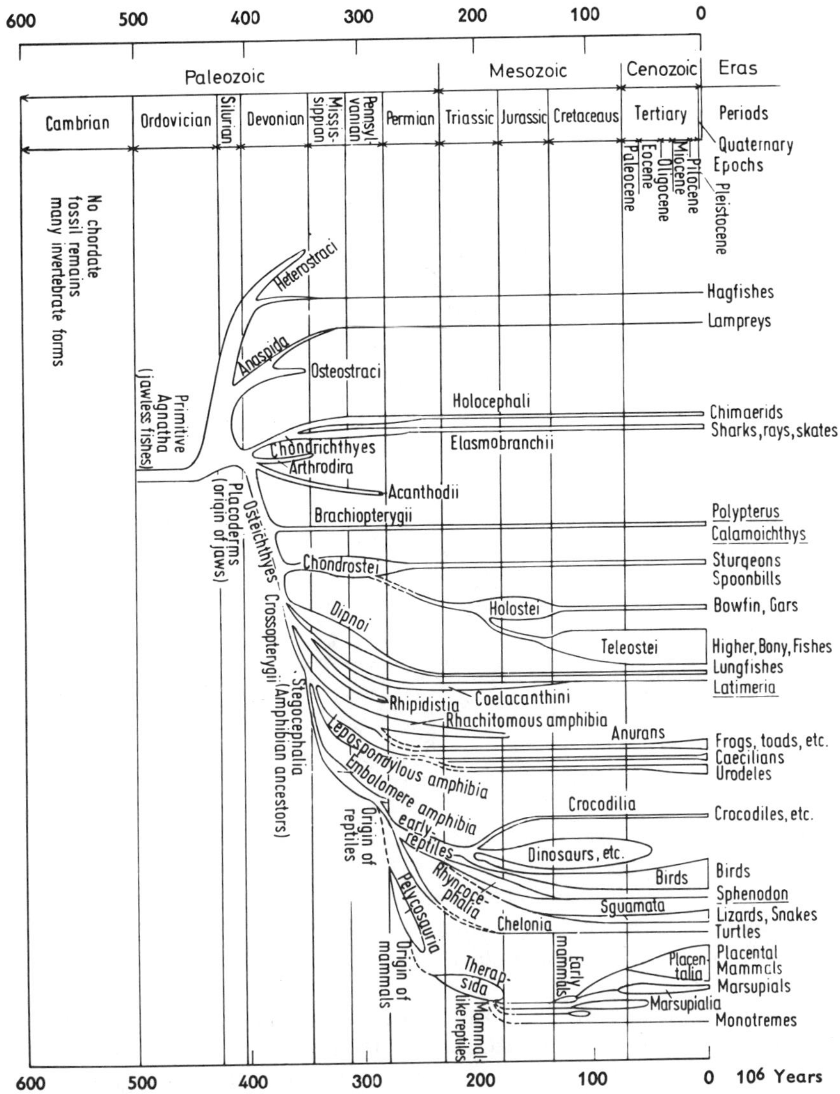

Figure 1 Schematic representation of current opinions on vertebrate phylogeny. Reprinted with permission from ref 10.

the early development and dominance of light-sensitive organs in the evolution of animal forms, preceding the origin of sophisticated central nervous systems. According to him embryology shows that the eyes originate from the very beginning on the surface of the body and preserve their original peripheral location. In accordance with this he has proposed a possible phylogenetic origin of the vertebrate eye as shown in Figure 2.

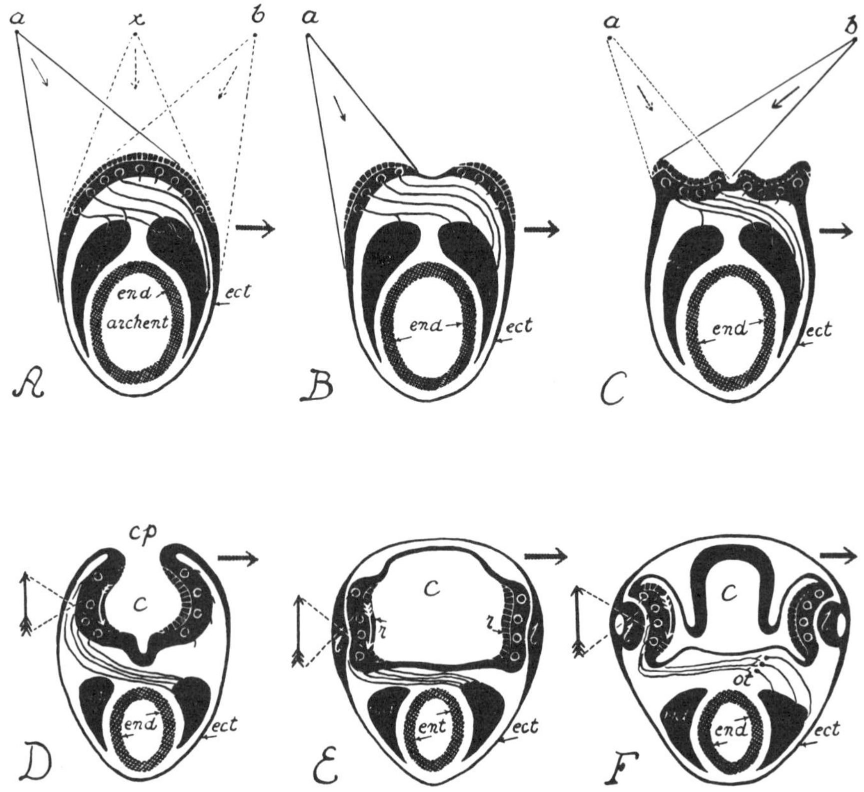

Figure 2 Proposed stages in the evolutionary development of the vertebrate eye, illustrated on cross-sections through the cephalic end of hypothetical vertebrate ancestors (8). A, open neural plate, with everted dorsal photoreceptor patch; B, neural groove with separation of photoepithelium in two lateral patches; C, beginning invagination of optic primordia into primitive optic cups; D, complete invagination of optoneural plate, with inversion of photoepithelium; E, closure of neural groove, appearance of lens placode; F, formation of essential parts of final vertebrate visual system. Abbreviations: a, x, b, left, central, and right viewed source of stimulating light, respectively; archent, archenteron or primitive gut; c, primitive cerebral vesicle; cp, cerebral pupil; ect, ectoderm; end, endoderm; l, lens primordium; md, mesoderm; r, retina. Reprinted, with permission, from S. Polyak, *The Vertebrate Visual System*, University of Chicago Press, 1957.

In fact the visual cells and visual pigments of invertebrates are very different from those of the vertebrates (10) and also provide no clue for the origin of the vertebrate eye. Only the very simple ocellus of larval tunicates contains photoreceptor cells, which are probably comparable to the vertebrate visual cells. The tunicates, or urochordates, are indeed considered, together with the cephalochordates (e.g., amphioxis), to be the nearest relatives of the vertebrates, and the three groups have been placed together in the phylum Chordata. Apart from its retinal cells the larval tunicate ocellus, consisting of about 20 cells, among which is a three-celled lens, and located within the wall of the posterior part of the cerebral vesicle, can hardly be compared to the vertebrate eye.

The photoreceptor cells of the pineal system of lower vertebrates closely resemble those of the lateral eyes. The pineal system develops embryologically also as an evagination of the diencephalon. In birds and mammals the photoreceptors are lacking, and the endocrine epiphysis is the only remaining component of the pineal complex. In the parietal eye, occurring in most lizards, the pineal system reaches its highest development as a photosensitive organ. Again, the lens of the parietal eye, forming its dorsal surface, is made up of tall columnar cells, and can not be compared to the lenses of the lateral eyes (11).

The evolutionary origin of the lens is even more difficult to envisage than that of the inverted retina of the vertebrates. One of the very few hypotheses relating to this problem is the proposal by Sharp, in 1885, that the lens derived from a lateral-line organ, as did the olfactory and otic organs. Initially functioning as an independent vesicular light-sensitive organ, it should secondarily have become a dioptric accessory after the origin of the neural retina. There is, however, no factual support for such an origin of the lens.

Embryologically the lens develops from the head ectoderm overlying the optic vesicle. It is therefore reasonable to assume that also phylogenetically the lens originated by local thickening of ectodermal cells. It is difficult, however, to envisage the intermediate stages of evolutionary development between a thickened ectodermal lens placode and a functionally adequate crystalline lens. The intermediate stages in embryological lens development, encompassing invagination of the placode and formation of a lens vesicle which becomes filled up by cellular elongation, can not be thought of as corresponding to functionally useful, or in Darwinian terms, adaptively favorable, steps in lens evolution.

Polyak's model (Figure 2) also provides no solution to this problem and apparently derives the lens vesicle in one sudden step from the lens placode. As Walls (7) stated, it still appears that "perhaps forever, the evolutionary origin of the vertebrate lens must remain a tantalizing mystery."

6.3 STRUCTURE AND ACCOMMODATION OF VERTEBRATE LENSES

6.3.1 Lens Morphology

A comparison of vertebrate lens proteins should evidently be made in relation to the differences in structure and properties among vertebrate lenses. It should be realized that despite the similar function and gross appearance of vertebrate lenses considerable differences exist (7, 12). The general plan for all vertebrate lenses is similar (compare Chapter 10). The most notable addition to this basic structure is the presence in birds and most reptiles of a ring-shaped thickening of epithelial cells around the equator, called the annular pad or ringwulst, which is an essential accessory for the accommodative mechanism in these animals.

The thickness and elasticity of the lens capsule varies depending on the level of development of the faculty to accommodate by changing the shape of the lens (13). In deformable lenses the thickness of the capsule varies in different regions around the lens.

The shape of the lens in different species varies from perfectly spherical to ovoid and flattened forms, where the anterior surface is usually less strongly curved than the posterior. Spherical lenses are found in aquatic and nocturnal animals. Vision under water requires a high refractive power of the lens to compensate for the absence of refraction at the water–cornea interface. Noctural land animals have nearly spherical lenses because the cornea is usually very large relative to the size of the globe. This demands a large lens to give a wide field of view to the eye and pick up every possible ray of light. A relatively large lens fills more of the eye, bringing its posterior surface nearer the retina, and consequently requires a higher refractive power, which is provided by a spherical shape.

The lens is perfectly spherical or only slightly flattened in cyclostomi and fishes. The lamprey lens is very solid. In fishes a solid nucleus is surrounded by a very soft, sometimes almost liquid, cortical part. Among the amphibians the anurans (salientians) have large, firm lenses, which are somewhat flattened. The ratio of equatorial to axial diameter (flatness index) is approximately 1.3. In agreement with their aquatic life the lenses of anuran tadpoles are spherical and close to the cornea. In urodeles (caudates) the overall size of the eye is reduced during phylogenesis, but the lens is relatively large and usually nearly spherical (14).

Among the reptiles the turtles have the softest and most pliable lenses of all vertebrates. The lenses have a weakly developed ringwulst and are flattest in land-living tortoises (flatness index 1.6) and virtually spherical in sea turtles. Crocodiles, as nocturnal animals, have more rounded lenses (flatness ratio 1.25) and a virtual disappearance of the ringwulst. Lizards have very soft lenses, with an extremely thin capsule and relatively thick ringwulst. The lens is flat in diurnal species and nearly spherical in nocturnal ones.

Snakes have eyes that are very different in structure from those of other reptiles, despite their supposed relationship to the lizard family Varanidae. It is assumed that the ancestral snakes lived underground, where the eye degenerated in size and lost most of its sauropsidian character. The eye of modern snakes had to be reconstructed from such a rudimentary organ and in doing so developed many new and original structures. The snake lens has no ringwulst but, in most species, a so-called anterior pad: a region of lens epithelium in which the cells are extremely tall instead of cuboidal. The snake lens is firmer than those of lizards and turtles; it is subspherical and yellow in color, and has sutures.

The relatively large eyes of birds are very uniform in structure throughout the class. The avian eye has no structures or properties that are not already seen in one reptilian order or another. The lens is usually very soft, with a flatness index of around 2.5 in diurnal birds, and 1.2 to 1.8 in crepuscular, nocturnal, and diving birds. A ringwulst is well developed, its thickness generally being related to the capacity for accommodation, although it is very thin in diving birds. The ringwulst is smallest in flightless birds.

Among the mammals the monotremes have relatively small lenses with perhaps a vestigial ringwulst. The spiny anteater (echidna) has, together with man, the flattest of all lenses (flatness index 2.75). Marsupials have relatively large and thick lenses (flatness index 1.05–1.30). Among the placentals man and the other diurnal primates have the flattest lenses. In seals and some toothed whales the lens is perfectly spherical, and very nearly so in several other aquatic or amphibious mammals and in murid rodents and other small-eyed nocturnal lower placentals.

6.3.2 Sutures

The spindle-shaped lens fiber cells are oriented between the anterior and posterior surfaces of the lens and form sutures where they meet at both poles. Different types of sutures can be distinguished and are related to the shape of the fiber cells (1). A central point-shaped "suture" can be found only when the fibers diminish strongly in width toward their ends, reaching one-third or one-fourth of the thickness observed at the center of the fiber. Such an "umbilic" suture is present in most birds, reptiles, salamanders, and certain teleosts.

A line-shaped suture, of variable length, is associated with fiber cells, which are less tapered, having half the thickness at the ends as compared to the equator. Usually the line is vertical at the anterior and horizontal at the posterior side. This turn of 90° results from the equal length of fibers in the same layer of a lens. This type of suture is found in many teleosts, sharks, and rays, in certain salamanders and small mammals (e.g., rabbit), and in frogs. More complex sutures (Y- or star-shaped, branching) are necessary where fiber cells diminish only slightly in thickness towards the poles. These are found only in larger mammals, being most complex in the human lens, where up to nine branches can be present.

6.3.3 Color, Consistency, and Refraction

Although most lenses are colorless, a yellow pigment is present in the lenses of several diurnal species from widely separated taxa: for example, man and other primates, tupaias, diurnal squirrels, snakes, gekkoes, and lampreys (7). Such yellow lenses probably serve the same purpose as the yellow corneas of some fishes and the yellow oil droplets in the retinal cones of certain diurnal frogs, turtles, lizards, and most birds. These intraocular color filters promote the visual acuity of diurnal animals by decreasing the effects of chromatic aberration, minimizing glare resulting from scattered light, and enhancing the contrast (7).

The consistency of the lens tissue varies from quite solid to very soft in different species but also in different layers of the same lens. This consistency largely corresponds with the protein contents of the lens. Increasing protein concentrations reduce the deformability of the lens and enlarge the refractive index.

The protein concentration, as measured by microspectrophotometry, in the nucleus of a 22-year-old human lens, is 4.7 times higher than at the equator and 2.7 times higher than at the poles (15). The highest protein content is found in fish lenses (trout: 50%), which must be as refractile as possible, because the cornea–water interface gives little refraction (16). Only 20% protein is found in the soft, deformable lens of a flying bird such as the pigeon. The decrease in refractive power by the low protein concentration is possibly compensated by the large amount of glycogen present in the lens nucleus of flying birds (17). Intermediate protein concentrations (30–40%) and intermediate consistencies are found in mammalian lenses. Differences in protein content are indeed reflected in the observed effective refractive index of the homogeneous lens in different species (Table 1).

In aerial vision the cornea becomes the most important refractive surface in the eyeball. The strong curvature of the cornea, together with the great difference in index of refraction between air and cornea, accounts for over 40 D of vergence of rays in the human eye. The anterior and posterior surfaces of the unaccommodated human lens contribute 5.0 and 8.3 D to the total refractive power of about 60 D of the eye. The remaining 6 D is provided by the refraction inside the lens. This is caused by the layers of

Table 1 Effective Refractive Index of the Homogeneous Lens in Different Species (18a)

Species	Refractive index
Monkey	1.463
Cat	1.554
Rabbit	1.6
Rat	1.683
Pigeon	1.408
Gekko	1.415
Iguana	1.4
Frog	1.65
Goldfish	1.69

increasing density toward the nucleus of the lens, which are also more and more strongly curved (18). There is indeed a gradient of refractive index from 1.386 near each surface to 1.406 in the center of the human lens (13). The higher refractive index and refractive index gradients in fish lenses contribute to the required greater refractive power. In mammals the refractive power ranges from 19 D in the unaccommodated human lens to 2 to 4 D in the almost spherical rat lens (18a).

6.3.4 Accommodation

Accommodation, that is, the capacity to adjust the optical system for near or distant vision, is not operative in all vertebrates. It is lacking in primitive fishes (chondrosteans, dipnoans, and coelacanths), in the reptile *Sphenodon,* and, most notably, in most mammals. Neither monotremes nor marsupials have any accommodation. Among the placentals a feeble accommodation has been demonstrated in squirrels and carnivores and an effective range is found only in the otter and in primates, especially man (9).

Small eyes, such as found in many rodents, insectivores, and bats, probably have no need for active accommodation. A small lens with a short focal length has a greater depth of focus than a large lens, whereas in the retina of a small eye the visual elements are relatively large. Therefore the image still falls within the layer of rods and cones when the object moves over a considerable distance.

Extension of the range of vision can be brought about by several mechanisms and structures outside the lens (7, 9). Well-known is the effect of a stenopaic or pinhole pupil, which is also attained by a slit pupil. Less frequent devices are the sloping ramp-retina, as found in horses and rays; the corrugated retina of the Megachiroptera; the use of a transparent nictitating membrane with high refractivity in some diving birds; the presence of unusually long receptor elements in the retina in many deep-sea fishes and in nocturnal gekkoes; and very rarely a duplicated retina in some deep-sea fishes. Change in curvature of the cornea and possibly retinal movement contribute to accommodation in avian eyes (19). In various amphibious vertebrates the lens provides a static solution for the change between aerial and aquatic vision. The optical asymmetry of a pyriform lens provides sharp vision under water along the long axis and above water along the short axis.

Dynamic accommodation involves either a movement of the lens as a whole or its deformation (7, 9, 12).

1 In lampreys the lens can be moved backward by corneal pressure, effected by a corneal muscle that flattens the cornea.

2 In teleost fishes the lens is moved backward and sideways, for distant vision, by a retractor lentis muscle.

3 A forward movement of the lens, for near vision, occurs in sharks, rays, and amphibians by the direct action of a special muscle, and in snakes by increase of the pressure in the vitreous cavity, resulting from contraction of the sphincter of the pupil.

Movement of the whole lens along the optical axis has also been reported to be probably the major accommodative mechanism in the cat eye (20).

Accommodation by deformation of the lens and consequent change in refractivity is found only in reptiles (except snakes), birds, and some mammals. Here the eye is at rest in distant vision.

1 A direct muscular pressure by the ciliary body deforms the lens in Sauropsida (except snakes). The ringwulst, surrounding the lens at the equator, transmits the pressure of the ciliary muscle evenly to the lens. Because the posterior side of the lens is steadied by the vitreous body, the squeezing results in a bulging forward through the iris of the central area of the anterior lens surface, forming a strongly curved lenticonus. In turtles most of the accommodation is effected by the formation in the very soft lens of such an anterior lenticonus by the direct action of the powerful sphincter pupillae. The annular pad and ciliary muscle are therefore relatively weakly developed in these animals. Crocodilians have only a rather weak accommodation, whereas lizards can accommodate very effectively. Birds have the most efficient accommodative mechanism among vertebrates, combining squeezing by the ciliary muscle with a powerful contraction of the pupillary sphincter for the formation of a lenticonus. The range of accommodation varies from 2 to 4 D in nocturnal birds, 8 to 12 D in chicken, pigeons, and most passerine birds, and reaches almost 50 D in aquatic birds. The accommodative power of the avian lens does not appear to decrease with age.

2 Those relatively few mammals that can accommodate by change of curvature of the lens do so by a unique mechanism. At rest the plastic lens retains its flattened shape by the molding effect of the elastic capsule stretched by the pull of the zonule and aided by the intraocular pressure which prevents the inward movement of the ciliary body. On accommodation the ciliary muscle contracts and overcomes the intraocular pressure, bringing the ciliary body toward the lens and releasing the tension on the zonule. The capsule then relaxes and allows the plastic lens to assume a more spherical shape. Bulging at the posterior side is restrained by the support of the vitreous, and the capsule, being thinnest at the poles, allows the greatest increase in curvature on the anterior surface in the form of a conuslike projection. Disadvantages of this system are the very limited deformation that can occur in large lenses and the impossibility of deformation when the lens looses its plasticity with age. The ciliary muscles of the lower mammals are vestigial or absent. Among rodents only an insignificant range of accommodation (1–1.5 D) is known among squirrels. In carnivores, excepting

the otter, accommodation is weak (1–3.5 D) and may even not be due to deformation of the lens (20). In cat, rabbit, and rat accommodating change of the lens is impossible because of the lack of compressibility together with the low capsular energy. Although they possess ciliary muscles, the ungulates are described as lacking accommodation, possibly because the lens is too stiff to alter its shape significantly (7). Primates as a group have the most effective range of accommodation among mammals, reaching 10 D in monkeys and up to 20 D in the human infant. This range, however, rapidly diminishes with age, decreasing to 10 D at 21 years, 2 D at 50 years, and disappearing around 75 years.

It is an interesting question, within the scope of this review, whether soft and moldable lenses (or parts of lenses) are characterized by particular protein properties, and whether the extremely high protein concentrations found in some lenses depend on certain protein structures.

6.4 STRUCTURE AND PROPERTIES OF VERTEBRATE CRYSTALLINS

6.4.1 General

This section deals with the water-soluble lens-specific proteins, the crystallins, which make up the bulk of the proteins in the lens fibers. The organ specificity of the crystallins is apparently not absolute; some crystallins have been demonstrated to occur outside the lens (see Section 6.5). Also the water solubility is not an absolute criterion for the crystallins: many crystallin polypeptide chains have been recognized in the urea-soluble and urea-insoluble protein fractions of the lens (cf. Chapter 1). The crystallins are generally referred to as the structural proteins of the lens, indicating that they are mainly thought to be responsible for the structural properties, consistency, and transparency of the lens. How the crystallins fulfill their functions is a question that is only beginning to be answered.

The classification of crystallins is becoming increasingly difficult with the advancing insight in the complexity of these lens proteins. Immunological cross-reactivity is no sure basis for classification of the lens proteins. Homologous proteins differing in more than about 40% of their sequence are no longer recognized as immunologically related (20a). Ideally the classification of lens proteins should be a phylogenetic one, that is, based on ancestral relationships as revealed by homologies in amino acid sequences. As will be seen this possibility is already starting to be approached. For the time being, and as far as is reasonable, the "classic" classification of the crystallins can still be used (see Chapter 1).

Species- and group-specific electrophoretic patterns of total lens proteins have been observed (21). Despite the apparent variability in molecular size

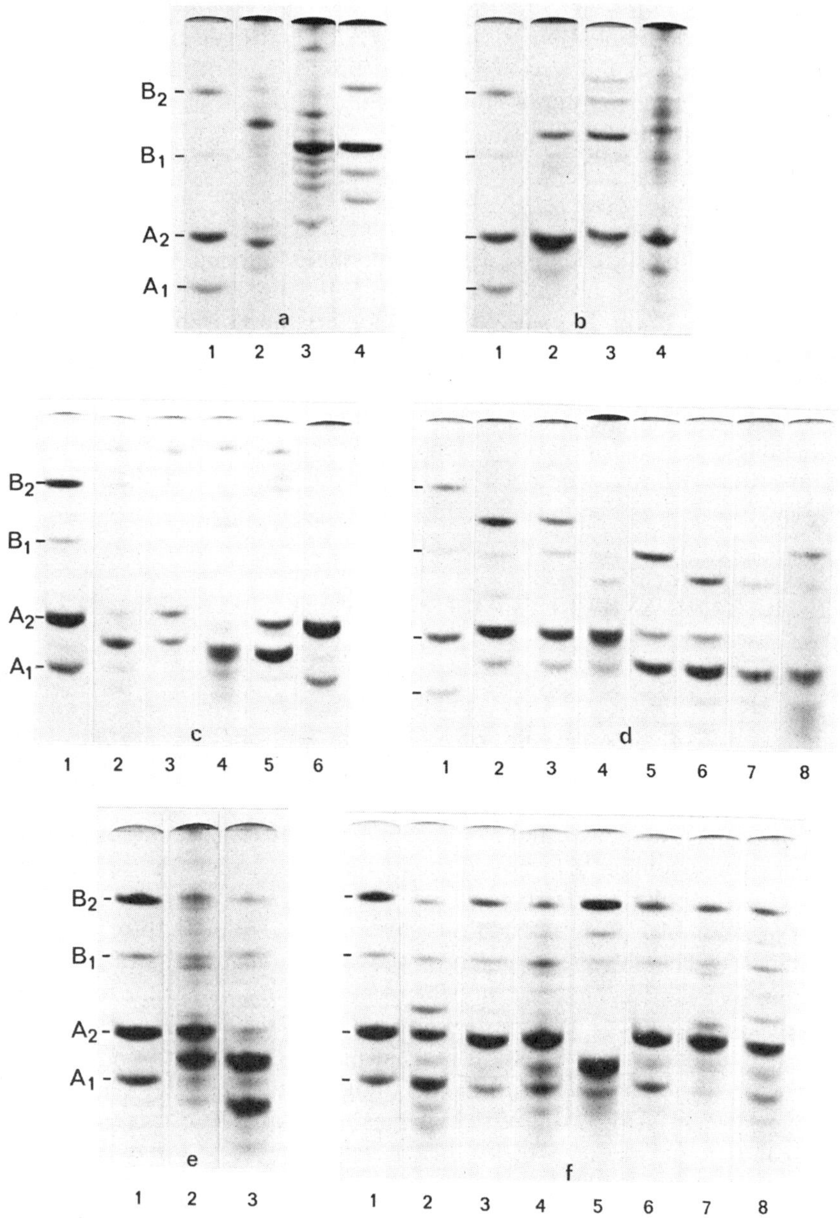

Figure 3 Polyacrylamide gel electrophoresis of α-crystallins in Tris–glycine buffer, pH 8.6, containing 6 *M* urea. (*a*) Fish and lamprey: 1, calf (reference); 2, dogfish; 3, cod (this material may not be α-crystallin); 4, lamprey. (*b*) Amphibians: 1, calf (reference); 2, *Rana temporaria*; 3, *Bufo bufo*; 4, *Xenopus laevis*. (*c*) Reptiles: 1, calf (reference); 2, *Tupinambis*; 3, *Cnemidophorus*; 4, *Gekko*; 5, *Basiliscus*; 6, *Python*. (*d*) Birds: 1, calf (reference); 2, chicken; 3, duck; 4, penguin; 5, crow; 6,

and charge, many lens antigens were found to be conserved to a considerable extent during vertebrate evolution. Immunochemical methods have been used to reconstruct the development of lens antigens during the phylogenetic history of vertebrate classes (22, 23).

6.4.2 α-Crystallin

The structure, biosynthesis, and post-translational modifications of bovine α-crystallin have been elucidated to a considerable extent (see Chapters 1, 5, and 7). High molecular weight aggregates with the characteristic low isoelectric point of α-crystallin and immunologically cross-reacting with bovine α-crystallin have been reported from all vertebrate classes (2). α-Crystallin is invariably found as an aggregate with a molecular weight of at least 400,000 in mammals (24, 25), birds (26, 27), frog (28), fish, and lamprey (29). In the lamprey some 18% of the soluble lens protein is α-crystallin (30). It has a simple subunit structure as judged from alkaline urea gel electrophoresis (Figure 3) and the authentic α-crystallinlike nature of the protein was confirmed by analysis of some tryptic peptides. It appears that the α-crystallin polypeptide chains are considerably longer than in other vertebrates: about 23,000 M_r.

The amount of α-crystallin has been reported to be very small in Chondrichthyes (2,31), although up to 11% α-crystallin was isolated from the lens of spiny dogfish (30) and smooth dogfish (32). In contrast to earlier reports (33), Mehta and Lerman (34) and Zigler and Sidbury (32) found no cross-reactivity between bovine and dogfish α-crystallin. Dogfish α-crystallins are separated in two or three subunits on electrophoresis (30, 34) or isoelectric focusing in the presence of urea (32) (Figure 3). The molecular weights of the subunits are similar to the bovine αA and αB chains, or slightly larger (30, 32), and it is suggested that there is more of the αB-like chain present than of the αA-like chain. Sedimentation values of lamprey and shark α-crystallins are similar to those of mammalians (29, 31). The amino acid composition of dogfish α-crystallin resembles that of other vertebrates (32, 34), and analysis of some tryptic peptides showed the homology with mammalian α-crystallin (30).

Most authors could isolate very little α-crystallin, if any, from Osteichthyes (30, 31, 35), but Zigler and Sidbury (36) isolated 41.6% α-crystallin from bluefish lenses, as identified by elution volume on Sephadex G-200, SDS gel pattern, and cross-reactivity with calf α-crystallin. Fish α-crystallin

---

pigeon; 7, cormorant; 8, buzzard. (*e*) Marsupials: 1, calf (reference); 2, opossum; 3, kangaroo. (*f*) Placental mammals: 1, calf; 2, horse; 3, dog; 4, whale; 5, elephant; 6, rabbit; 7, rat; 8, rhesus monkey. For complete species names see ref. 30.

shows only partial identity in immunodiffusion with mammalian and amphibian α-crystallin (22, 37).

In amphibians α-crystallin constitutes a considerable proportion of the soluble lens proteins (Table 2), although in larval stages the level may be very low (38). In *Xenopus laevis* (38) and the axolotl (39), α-crystallin occurs as a single component on isoelectric focusing, with an isoelectric point at pH 4.75. In some *Rana* and *Triturus* species, however, two or three differently charged α-crystallin populations can be distinguished by immunoelectrophoresis or column chromatography (40–43). As in mammalian α-crystallins these differences are probably due to changes in subunit stoichiometry of the native aggregates. The subunit patterns of α-crystallin from three anuran species are quite similar (Figure 3), consisting of a major αA chain, of approximately the same mobility as bovine αA, and a minor αB chain. The most acidic lens protein polypeptide, probably αA, of the urodele *Triturus cristatus* also migrates in the same position as the anuran αA chain (44), indicating little variation in charge of this chain among the amphibians. The primary structure of the αA chain of *Rana esculenta* is almost completely known (Figure 4) (45), although the experimental details have not yet been published. This chain shows 18% sequence difference with the bovine αA chain (46).

It appears that reptilian α-crystallin shares antigenic determinants with bird α-crystallin which are absent in other vertebrates (22). Lizard and alligator α-crystallins are immunologically more similar to each other than to turtle α-crystallin (22). Turtle α-crystallin has a lower sedimentation value than that of snake (47), the former probably corresponding in size to bird α-crystallin and the latter to mammalian α-crystallin (2). The subunit compositions of the α-crystallins from four lizards and one snake have been analyzed (Figure 3). It appeared that αB chains were absent in the investigated *Python* species, and only little αB chain of low electrophoretic mobility was present in lizards. The αA chains occurred either as a single band or in two electrophoretic forms. In the latter case the distances between the two bands correspond to one or two charge units, and thus make it likely that, as in the cow, one or two postsynthetic deamidation steps are responsible for the double αA bands.

α-Crystallin of birds have molecular weights that are lower than that of the cow (2) and show partial immunological identity with α-crystallins from other vertebrate classes (22). Native α-crystallin does occur in different aggregate sizes and charge groups both within the same species and between species. Again, most of these differences are due to quantitative shifts in the proportions of subunits that are electrophoretically identical in different species (48, 49). The subunit patterns of chick α-crystallin in alkaline urea gels resemble the general structure of mammalian α-crystallin: a major αA chain and a smaller amount of αB, both accompanied by a minor, supposedly deamidated, electrophoretically faster band (Figure 3) (30, 48, 50). The

Table 2 Proportions of Different Crystallins in the Water-Soluble Protein of Total Lenses from Adult Vertebrate Species[a]

Species	α	β_H	β	β_L	γ	LM[b]	δ	Method[c]	Ref.
Six mammalian species	30.2–53.6	13.9–29.8		9.2–22.5		11.2–33.8		A	(85)
Six mammalian species	22–35		42–71		7–25			B	(104)
Rabbit	27		63.5		8.5			B	(103)
Rat	20	15[d]		22[e]	41			C	(100)
Chicken	6.8		73.7		Trace		19.5	D	(147)
Chicken	36		39		Trace		25	B	(104)
Turtle	17.0	23.7		30.6		28.7		A	(36)
Leopard frog	25.4	14.1		13.5		32.5		A	(36)
Xenopus laevis	7.3		52.1		40.6			B	(38)
Axolotl	6.2		63.3		30.5			B	(39)
Bluefish	41.6	9.1		14.0		35.3		A	(36)
Goldfish	12		40		24			B	(104)

[a]It should be noted that identification of lens proteins as β- or γ-crystallin is often inadequate. Also the procedures to calculate the proportions of crystallins can frequently not be expected to yield accurate data. The values in this table are intended only to provide an impression of the great interspecies differences in crystallin composition.

[b]Low molecular weight proteins, probably largely consisting of γ- and monomeric β-crystallins.

[c]A, gel filtration (G200) followed by protein determination of the pooled fractions; B, densitometric scanning of stained isoelectric focusing or electrophoresis gels; C, separation by gel filtration (G200 super fine), quantitation not further indicated; D, surface determination of gel filtration (G200) elution pattern.

[d]Designated as fraction II by the authors (100).

[e]Actually separated in two β_L fractions of 12 and 10%, respectively.

```
                                  10                                        20
Calf  ac-Met-Asp-Ile-Ala-Ile-Gln-His-Pro-Trp-Phe-Lys-Arg-Thr-Leu-Gly-Pro-Phe-Tyr-Pro-Ser-Arg-Leu-Phe-
Chick ———————————————Thr——————————————————————————————Ala————————————Leu-Ile—————————————————————
Frog  ———————————————Thr——————————————————————————————Ala—————————————————————————Asn—————————————

                    30                                      40
-Asp-Gln-Phe-Phe-Gly-Glu-Gly-Leu-Phe-Glu-Tyr-Asp-Leu-Leu-Pro-Phe-Leu-Ser-Ser-Thr-Ile-Ser-Pro-Tyr-Tyr-
————————————————————————————————Leu————————————————————————Leu-Phe—————————————————————————————————
—————————Val————————————————Met———Asp—————————————Phe——————————————————————————————————————————————

      50                                      60                                      70
-Arg-Gln-Ser-Leu-Phe-Arg-Thr-Val-Leu-Asp-Ser-Gly-Ile-Ser-Glu-Val-Arg-Ser-Asp-Arg-Asp-Lys-Phe-Val-Ile-
—————————————————————————Ser—————————Glu—————————————————————————————————————————————————————
————————————Phe——————————Gly-Phe——————————————————————————————————————————————————Arg-

                    80                                      90
-Phe-Leu-Asp-Val-Lys-His-Phe-Ser-Pro-Glu-Asp-Leu-Thr-Val-Lys-Val-Gln-Glu-Asp-Phe-Val-Glu-Ile-His-Gly-
    ——————————————————————————————————————————————————Ser—————————Ile-Ile-Asp———————————————————————
            ——————————————————————————————————————————————————————Ile-Leu-Asp———————————————————————

     100                                     110                                     120
-Lys-His-Asn-Glu-Arg-Gln-Asp-Asp-His-Gly-Tyr-Ile-Ser-Arg-Glu-Phe-His-Arg-Arg-Tyr-Arg-Leu-Pro-Ser-Asn-
—————————Ser——————————————————————————————————————————————————————————————————————————————Ala———————
—————————Ser————————————————————————————————————————————————————————————————————————————————————————

                          130                                     140
-Val-Asp-Gln-Ser-Ala-Leu-Ser-Cys-Ser-Leu-Ser-Ala-Asp-Gly-Met-Leu-Thr-Phe-Ser-Gly-Pro-Lys-Ile-Pro-Ser-
——————————————————————Ile-Thr———————————————Ser————————————————————————————————————————Val—————————
-Leu-Asn-Glu————Ser-Ile————————————————————————————Ile————————————————————————————————Leu-Met——————

     150                                     160                                     170
-Gly-Val-Asp-Ala-Gly-His-Ser-Glu-Arg-Ala-Ile-Pro-Val-Ser-Arg-Glu-Glu-Lys-Pro-Ser-Ser-Ala-Pro-Ser-Ser-OH
-Asp-Met————Pro-Ser—————————————Pro———————————————————————————————————————Thr—————————————————————
-Ser-Leu————Ser-Ser————Gly——————Pro———————————————————————————————————————Thr—————————————————————
```

Figure 4 Amino acid sequences of the α-crystallin A chains of calf, chicken, and frog (45, 46). Positions 72–74 in the chicken αA chain and positions 71–76 in the frog αA chain have not been determined. The line indicates where chicken and frog αA chains are identical to the bovine αA chain.

intensity of the αA and αB bands varies between species of birds (30, 48) and also the mobility of the major αA and αB chains may differ.

The isoelectric points of chick α-crystallin aggregates range from pH 4.52 to 4.87 (51). The proportion of α-crystallin increases after hatching (51, 52). Immunologically the presence of α-crystallin has been shown in chick eye tissues outside the lens (53). In the chick α-crystallin makes up a variable proportion of the urea soluble protein (54).

The details of the subunit structure of chick α-crystallin is a matter of some disagreement. Beebe and Piatigorski (55) reported that α-crystallin from 15-day-old embryos and 10-week-old roosters contained the same four subunits, as revealed by alkaline urea gel electrophoresis. In SDS gel electrophoresis two bands were observed, approximately in the positions of calf αA and αB, more αA being present than αB. In contrast Thomson et al. (52) found on SDS gel electrophoresis of day-old chick α-crystallin a major band of apparent molecular weight 20,000 and a minor band at 19,000. On isoelectric focusing they found two predominant subunits, with isoelectric points similar to bovine αA_2 and αA_1. These authors suggested that very little αB chain is present in the day-old chick. Remarkable is the finding that a short pulse labeling resulted in incorporation of radioactivity in the presumed αA_1 and αB_1 chains of the chick, suggesting that these are primary gene products like αA_2 and αB_2 (55), unless a very rapid transformation of αA_2 and αB_2 into αA_1 and αB_1 takes place in the chick. In the calf no radioactivity is found in αA_1 and αB_1 after similar or even longer times of pulse labeling (56, 57). Thomson et al. (52) also conclude that both α-crystallin subunits are apparently primary gene products. The primary structure of chick αA chain shows 15% sequence difference with bovine αA (Figure 4). The mRNA for the chick αA chain has been isolated and was found to have a sedimentation value somewhat smaller than the 14 S mRNA for bovine αA, but much larger than the 10 S mRNA coding for the bovine αB chain (58).

Aggregate sizes of α-crystallins from different mammalian species show little variation. The $S_{20}W$ values of α-crystallins of 11 species ranged between 19.84 and 21.97 (59). Amino acid analysis and immunological data show only small differences between mammalian α-crystallins (2, 24, 34, 60). The quantitative complement fixation test gave identical values for bovine, sheep, and rabbit α-crystallin, using rabbit antisera specific for bovine α-crystallin (34). The human, monkey, and rat showed a somewhat lower degree of complement fixation, from 72 to 84%, with the antisera for bovine α-crystallin. In this system chick α-crystallin showed approximately 40% cross-reactivity with the bovine protein, and dogfish α-crystallin gave no detectable cross-reaction. Using a radioimmunoassay for α-crystallin Sandberg and Closs (61) could not detect differences in antigenic properties between bovine and sheep α-crystallin. Rabbit α-crystallin showed some antigenic differences with the sheep protein, and this difference was even

Table 3 Proportions of α- and γ-Crystallin in the Water-Soluble Protein from Human Lenses, Determined by Radioimmunoassay (62)

Source	α-Crystallin (%)	γ-Crystallin (%)
Stillborn, total lens	42	19
60 year old,		
Total lens	32	34
Cortex	54	9
Nucleus	20	48

greater between human and sheep. It was concluded that the α-crystallins of man and rabbit have different antigenic properties in common with sheep α-crystallin. Specific radioimmunoassays for human α- and γ-crystallins have also been used to determine the different proportions of these proteins with age and in different parts of the lens (Table 3) (62). Also in the mouse quantitation of α- and γ-crystallins has been obtained by radioimmunoassay (63). It showed that the level of α-crystallin remains at about 25% in the water-soluble fraction between 9 and 71 days after birth. The proportion of γ-crystallin dropped from 50 to 30% in 80 days after birth, owing to the relative decrease in γ-crystallin synthesis.

The subunit compositions of mammalian α-crystallins are very similar (24, 30, 34) (Figure 3) and vary mainly in the relative proportions of electrophoretically similar components. Platonov et al. (64) observed that, in incubated mouse lenses, the subunits of α-crystallin are first synthesized and form aggregates somewhat later.

Among 17 investigated species only the αA_2 chains of elephant, hyrax, and kangaroo differed in electrophoretic mobility at pH 8.6 from the bovine αA_2 chain (65, 66); no charge differences were found among the αB_2 chains. The amount of α-crystallin in mammalian species ranges from 11 to 54% of the total lens protein (30) (Table 2). Interspecies differences may not always be assessed very accurately, because the proportion of α-crystallin also increases with age (2), and different methods to determine the amount of α-crystallin can provide considerably different values. Also differences in electrophoretic mobilities of α-crystallins from different species, and from different ages, are usually explained by quantitative differences in subunit composition.

Considering the limited differences between mammalian α-crystallins, much of the knowledge available for bovine α-crystallin (Chapter 1) is therefore applicable to other species. Human α-crystallin has been studied by several authors. Newly synthesized human α-crystallin has a molecular weight between 4.0×10^5 and 7.5×10^5 and is composed of αA and αB chains which migrate in SDS gel electrophoresis with the same mobility as bovine αA and αB (67, 68), indicating molecular weights of 20,000 and

22,000, respectively. The ratio between αA and αB chains is approximately 3 : 1, reflecting a relatively greater proportion of αA chains as compared to bovine α-crystallin, where this ratio is 2 : 1 (68). Laser light scattering experiments and near-ultraviolet circular dichroism spectra of bovine and human low molecular weight α-crystallin (6–9 $\times$ 10^5 M_r) gave similar results, indicating similar secondary and tertiary structures in both proteins (68a).

Electrophoresis in the presence of urea reveals the presence of three polypeptides in fetal and neonatal human α-crystallin (Figure 5) (69, 70). A single αB chain is present in the position of bovine αB_2 and the major αA chain migrates approximately as bovine αA_2 does. The minor αA chain appears to differ only in a single charge unit from the major one, in contrast to bovine αA_2 and αA_1, which differ by two charge units. Both human αA chains become radioactively labeled after a 20-hour incubation of lenses in radioactive medium (68). Although this could indicate that both chains are

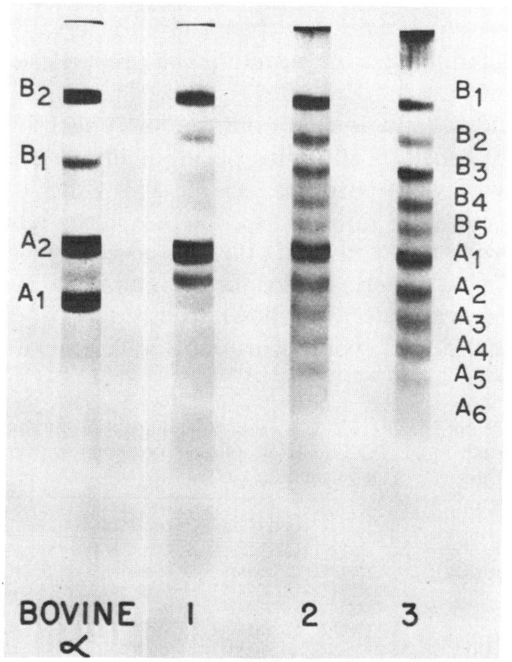

Figure 5 Alkaline urea–polyacrylamide gel electrophoresis of calf α-crystallin compared with human α-crystallin from (1) 6-week-old lens, (2) 65- to 70-year-old normal lens periphery, and (3) 65- to 70-year-old cataract lens periphery. Reprinted, with permission, from D. Roy and A. Spector, Human alpha-crystallin-III isolation and characterization of protein from normal infant lenses and old lens peripheries, *Invest. Opthalmol. Visual Sci.*, **15**, 394–399 (1976).

primary gene products, it is considered more likely that the minor αA chain arises by very rapid transformation of the major chain. α-Crystallin from adolescent and from cataractous or normal older human lenses shows a complex subunit pattern, owing to postsynthetic modifications of the original subunits (70–73). Up to 11 polypeptides can be distinguished (Figure 5), of which six are αA-like, on the basis of amino acid composition and peptide mapping, and five αB-like (71, 72). The nature of the differences between these chains is not yet clear. C-terminal degradations of αA chains are certainly involved (73–75).

After the elucidation of the amino acid sequences of the bovine αA_2 and αB_2 chains (46, 76) the αA chains of another 16 mammalian species (65, 66, 77, 78) and the αB chain of human α-crystallin (79) have been analyzed. As could be expected from earlier immunological data and amino acid compositions of mammalian α-crystallins, only relatively few amino acid replacements were observed between the αA chains of different species (Table 4). Human αB chain differs from bovine αB at only three positions: 40 Ala → Thr, 61 Ile → Phe, and 152 Ala → Val (79). All the substitutions can be attributed to single base substitutions in the αA and αB genes. Most amino acid substitutions are of a chemically conservative character, and can not be expected to alter the structure and properties of the α-crystallin molecule very much.

The amino acid substitutions are not randomly distributed over the αA chain (Figure 6). Most substitutions occur in the *C*-terminal part of the chain, especially between positions 145 and 157. Other parts of the chain, for example, positions 18 through 54, have not accepted a single substitution. It is reasonable to suppose that the conserved regions of the αA chain are somehow of vital importance for the structural and functional integrity of the α-crystallin aggregate. It is known from studies of other proteins (80) that surface areas are the evolutionary most variable parts. This may also

Table 4 Amino acid substitutions between the αA chains of different species (45). Only the variable positions in the αA chain are tabulated. The one-letter notation for amino acids is used; (.) not determined; (-) deletion.

	111111111111111111111111111111 11112233333455556777889990222222233334444455555555666 7 34367806123790256810246901312456789035826789012345682892
HORSE	IAAPFYSFLFELFLLTVDIKVFTVQEFNSVDQTALSVAMSIPSGMDAGHSEASGSS
RHINOCEROS	IATPFYSFLFELFLLSVDVKVFTVQEFNSVDQTALSLAMSIPSGMDAGHSEASSSS
PIG	IAAPFYSFLFELFLLTVDVKVFTVQEFNSVDQSALSLAMSVPSGVDAGHSEASSST
OX	IATPFYSFLFELFLLTVDIKVFTVQEFNSVDQSALSLAMSIPSGVDAGHSEASSSS
WHALE	IAAPFYSFLFELFLLTVDIKVFTVQEFNSVDQSALSLAMSVPSGMDAGHSEASSSS
DOG, CAT	IAAPFYSFLFELFLLTVDIKVFTVLEFNSVDQSALSLAMSVPSGVDAGHSEASSSS
HEDGEHOG	VTAPFYS.......LTVDIK..TVLEFS........LAMSVASGLDAGPSEASSSS
RAT, TUPAIA	VTAPFYSFLFELFLLTVDIKVFTVLEFNSVDQSALSLAMSVQSGLDAGHSEASSSS
RABBIT	VTTPFYSFLFELFLLTVDIKVFTVQEFNSVDQSALSLAMSVQSGLDAGHSEASSSS
HUMAN	VTTPFYSFLFELFLLTVDIKVFTVQDFNSVDQSALSLAMCIQTGLDA-HTEASTSS
MONKEY	VTTPFYSFLFELFLLTVDIKVFTVQDFNSVDQSALSLAMSIQTGLDA-HTEAASSS
ELEPHANT	VTAPFYSFLFELFLLTVDIQVLTVQDFNSVDQSALSLAMCIQSGMDASHSEASSSS
HYRAX	VTAPFYSFLFELFLLTVDIQLLTVLDFNSVDQSALSLAMCVQSGMDASHSEASSSS
KANGAROO	ITASLYSFLFELFLLTVEIKVFTVLDFSSVDQSAISLAMSIHSDMDASHSDSSTLS
OPOSSUM	ITASLYSFLFELFLLTVEIRVYTVLDYSSVDQSAISLAMSIHTNMESSHSDSSTLS
FROG	ITAPFYNVMFDFFLFGFDIR..TILDFSSLNESSISLAISLMSSLDSSHGEPSTSS
CHICKEN	ITAPLISFLLELLFLSVEIK..SIIDFSAVDQSAITLSMSVPSNMDPSHSEPSTSS

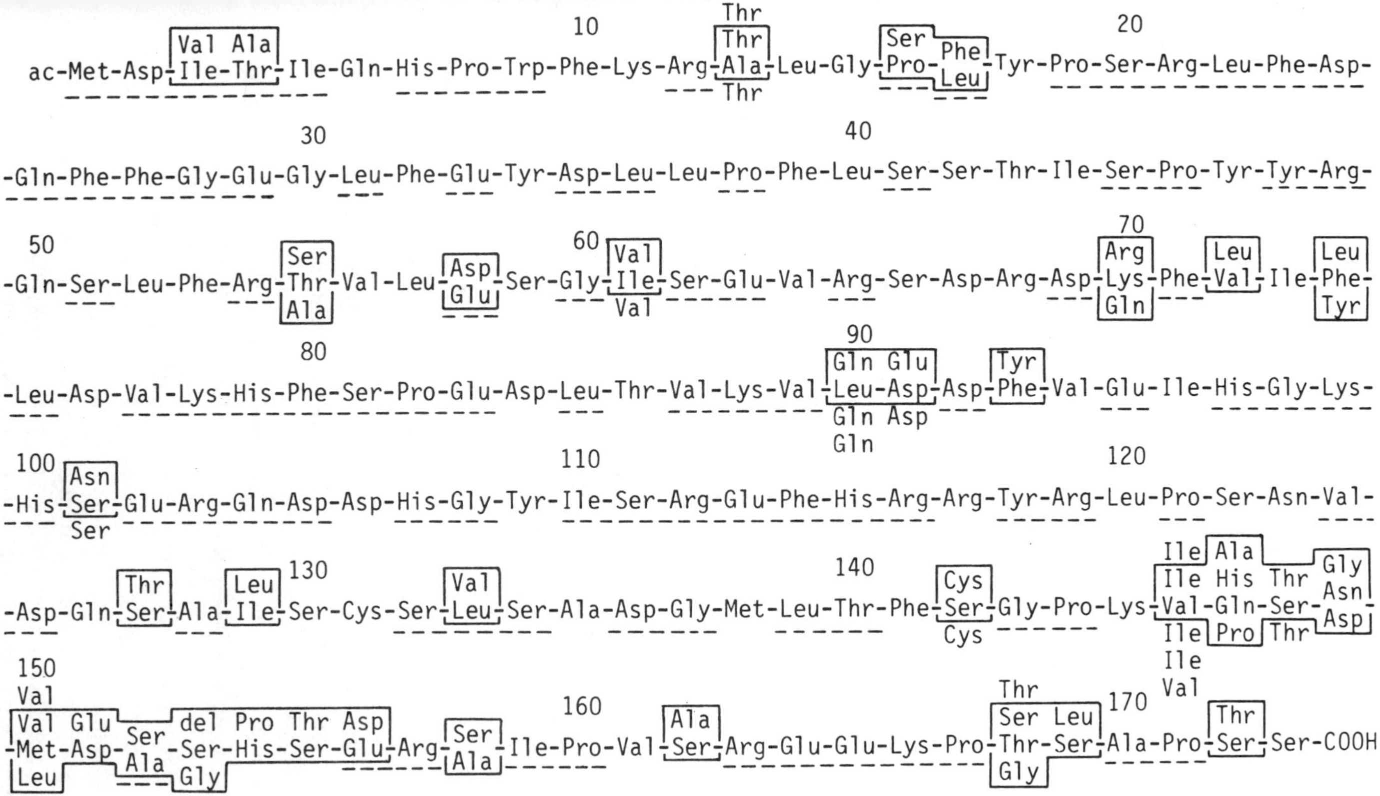

Figure 6 Composite sequence of the α-crystallin A chains of 17 mammalian species (45). Residues in line with the main sequence represent the most likely αA chain sequence at the time of placental–marsupial divergence. The variability actually observed at each mammalian position is represented by the residues enclosed in solid lines. The residues outside the solid lines represent probable convergent and back substitutions which have occurred during mammalian evolution. The dotted line below the sequence indicates those residues that are identical in the bovine αA and αB chains. It shows that substitutions in the mammalian αA chains mostly occur in regions where the bovine αA and αB chains are different.

apply to the variable regions of the αA chain. It has indeed been demonstrated by limited proteolysis of native α-crystallin that residues 156 to 163 and 168 to 169 must be at the surface of the aggregate (81).

6.4.3 β-Crystallin

The β-crystallins form a quite heterogeneous group of lens proteins which are characterized in the native state by an intermediate electrophoretic mobility (at alkaline pH) and intermediate isoelectric points (pH 5.7–7.0) (82), a high thiol content, and a blocked *N*-terminus. Whereas the higher molecular weight aggregates apparently contain little or no α-helix structure and some β-sheet structure (83), the monomeric β_S-crystallin possesses approximately 10% α-helix (84). The monomer molecular weights range from 19,000 to 31,000.

Lens proteins with the intermediate molecular weights and electrophoretic mobilities of β-crystallins have been reported from all classes of vertebrates, in some cases supported by immunological evidence (2). More detailed comparative data, especially at the level of subunit composition, is still scarce.

Most informative are the studies of Zigler and Sidbury (32, 36, 85, 86), who studied β-crystallins from all classes of vertebrates, apart from cyclostomi. The amounts of β-crystallins in various species are given in Table 2. For each species it is found that β_H and β_L are composed in large part of identical polypeptides, on the basis of SDS and alkaline urea gel electrophoresis. The molecular weights of the subunits were estimated to range from 24,000 to 29,500; in the β_H fraction some minor additional higher molecular weight bands were usually present.

On the basis of charge some four to six major subunits can mostly be distinguished. Isoelectric focusing patterns of native β-crystallins were complex and variable. The amino acid compositions of all β_H and β_L fractions were generally similar to the composition of mammalian β-crystallins, being highest in glutamic acid. Immunodiffusion with anti-calf β_L gave two precipitin bands with calf β_L. One of these bands was present with β-crystallins from fish, frog, turtle, and chick, indicating that antigenic determinants of β-crystallin are partially rigidly conserved. It appears that the polypeptides which are unique to β_H are the most variable of the β-subunits. β-Crystallin forms the major fraction in adult amphibian lenses (Table 2). On thin-layer isoelectric focusing native β-crystallins of axolotl and *Xenopus laevis* separate in approximately 11 components, with isoelectric points from pH 5.55 to 6.95 (38, 39) (Figure 7).

Chick β-crystallin has been studied in considerable detail. It can easily be isolated by gel filtration over AcA-34 (52) (Figure 8). The native protein occurs as heteropolymers of various size, which can be separated in about 10 fractions on isoelectric focusing (52, 87). The isoelectric points range

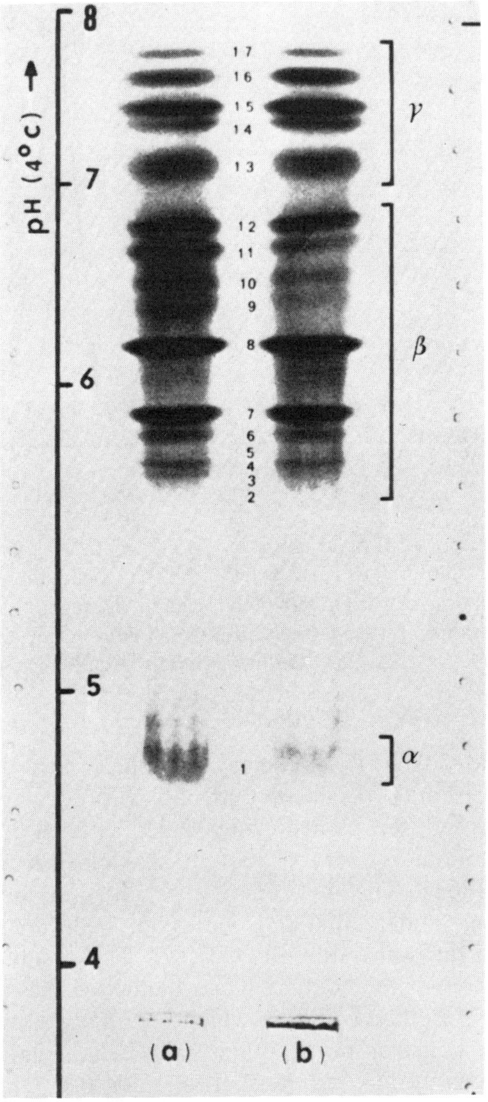

Figure 7 Isoelectric focusing patterns of soluble lens proteins from (*a*) cataractous and (*b*) normal axolotl lens. Reprinted with permission from S. K. Brahma and J. Bours, *Exp. Eye Res.*, **23,** 57 (1976). Copyright by Academic Press Inc. (London) Ltd.

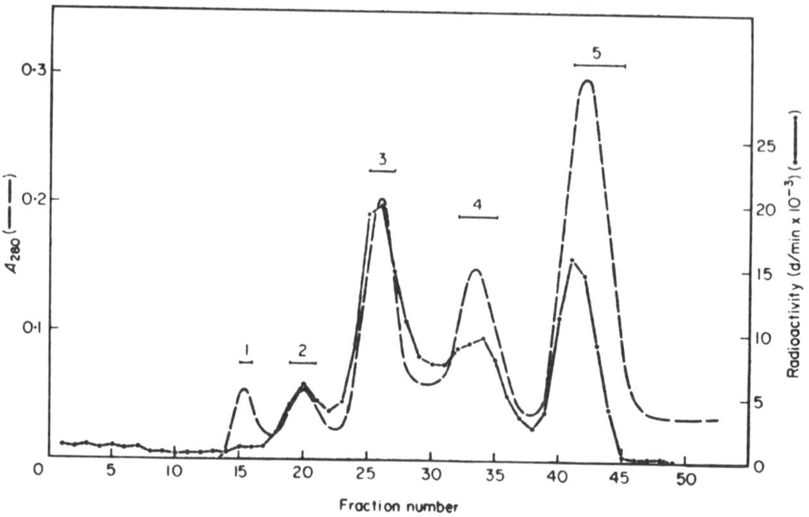

Figure 8 Gel filtration of day-old chick lens soluble proteins on Ultrogel AcA-34. Peak fractions were identified as (2) α-crystallin, (3) δ-crystallin, and (4) β-crystallin. Peaks 1 and 5 contain very high molecular weight aggregates and low molecular weight nonprotein material, respectively. Reprinted with permission from J. Thompson et al., *Exp. Eye Res.*, **26,** 351 (1978). Copyright Academic Press Inc. (London) Ltd.

between 5.69 and 7.03 (51). In embryonic chick lens epithelial cells β-crystallin is reported to occur in two native molecular weight classes: around 80,000 (composed of aggregates) and between 20,500 and 34,000 (consisting of monomers) (88). The two size classes are composed of different polypeptides. The different β-crystallin polymers isolated from embryonic, young, and adult chick lens fiber cells are immunologically similar and share the same subunits (51, 87). These subunits separate on SDS gel electrophoresis in about six major bands, with estimated molecular weights between 21,000 and 37,000, which are all apparently primary gene products (51, 89). Fourteen polypeptides have been isolated, by isoelectric focusing under denaturing conditions, from chick β-crystallin (90). Twelve of these are antigenically related (but not identical), having an overlapping distribution of antigenic determinants. Two of the polypeptides are not antigenically related to the others.

Immunofluorescence studies have shown that β-crystallin antigens first appear after 3½ days of incubation in epithelial cells of the lens vesicle, and at 4 days the complete lens reacts positively (91). More specifically, the cathodal β fractions, containing polypeptides of 25,000 and 27,000 molecular weight, are the first to appear, followed by the anodal β fraction, containing polypeptides of 37,000 M_r, which is detected in lens fibers at 5 days of incubation and in epithelial cells only at 16 days (92). The lens fibers made after hatching contain progressively less δ-crystallin and more β-crystallin (93), and in the adult chick β-crystallin is the major lens protein (Table 2). It does not appear that new β-crystallin components are synthesized in the post-hatch chick lens (51). β-Crystallin polypeptides of 25,000, 27,500, and 37,000 M_r also make up a considerable portion of the urea-soluble intracellular matrix of the fiber cells. The proportion increases from 6.0% in the 17-day embryo to 24.3% in the adult chick cortex (54). β-Crystallin is, like other crystallins, also observed in the chick cornea (94).

It should finally be mentioned that the γ-crystallin suggested to be present in birds (95, 96) (see Section 6.4.4) is considered to belong rather to the β-crystallins by McDevitt and Croft (97).

Early electrophoretic and immunological studies established that also in mammals the β-crystallins are the most complex lens proteins (21, 98). The native β-crystallin aggregates display indeed a wide range of molecular weights between approximately 4.10^4 and 2.10^5, and a complex pattern of bands on isoelectric focusing. There is considerable variation in native molecular weights and isoelectric points between mammalian β-crystallins The proportion between β_H and β_L varies greatly between species, but this is in part certainly due also to age differences; it is known that in calf β_H increases with age at the expense of β_L (99). In the rabbit a third molecular weight class of β-crystallin elutes on Sephadex G200 just in front of β_L (85). On Sephadex G200 superfine calf and rat β_L tends to separate into two fractions (100, 101). Also on Sepharose 6B rabbit β-crystallins elute in three size classes (102) and on Sephacryl S-200 even four β fractions can be separated (103) (Figure 9).

Usually some 10 components can be distinguished in isoelectric focusing or electrophoresis of native β-crystallins of rabbit (102), rat (100, 104) and other rodents (105, 106), cow (104), man (94), and dog (104). The isoelectric points of these native β fractions range between 5.7 and 7.0 (82). Considerable differences in the β-crystallin patterns can be seen (Figure 10). Most differences in the native β-crystallin patterns within species are due to changes in intensity of bands, but small differences in isoelectric points of β-crystallins have also been observed, as in dogs (104) and voles (106). On aging the isoelectric focusing patterns of the β-crystallins change considerably, mostly by differences in intensity of the various components (107–109).

On SDS gel electrophoresis β_L-crystallin appears to be composed of polypeptides with molecular weights between 24,000 and 29,000, whereas the same bands are present in β_H together with additional bands of 29,000 to 34,000 (85, 86, 110). The rabbit β-crystallin fraction eluting between β_H and β_L consists of aggregates of possibly four chains of molecular weight 33,000 (85, 103) and is antigenically related to β_H and β_L. A small amount of a 11,200 M_r band has been reported in rabbit β-crystallin (103). In human

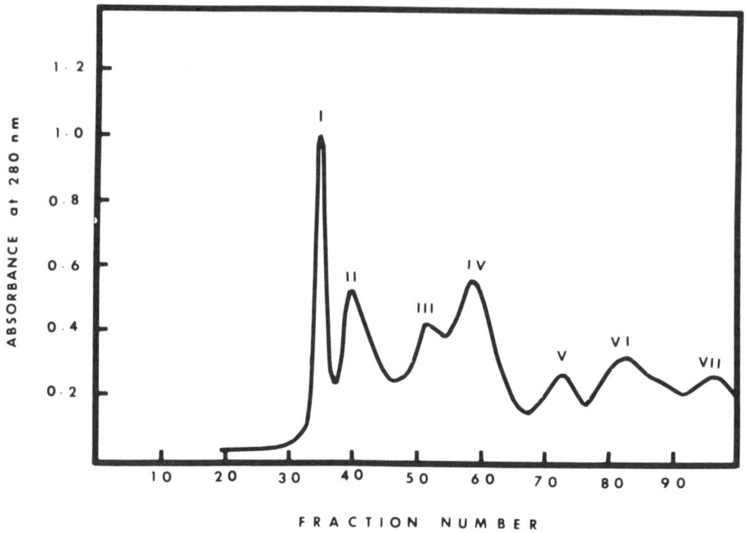

Figure 9 Gel filtration of rabbit lens cortex water-soluble proteins on Sephacryl S-200. Peak fractions were identified as α-crystallin (I), β-crystallins (II–V), and γ-crystallin (VI). Reprinted, with permission, from M. K. Mostafapour and V. N. Reddy, Studies on lens proteins. I. Subunit structure of beta crystallins of rabbit lens cortex, *Invest. Opthalmol. Visual Sci.*, **17,** 660–666 (1978).

β-crystallin a unique 60,000 M_r band is apparently present in considerable amount (86), although Coghlan and Augusteyn (67) mention only a major 29,500 and a minor 25,400 M_r band in human β-crystallin.

In the cow, marked age-related shifts occur in the relative abundance of polypeptides in β_H and β_L, probably mainly as a result of postsynthetic modifications (111). These changes in subunit stoichiometry can also explain the changes in native β-crystallin patterns with age (107–109). On the other hand, few differences that could be attributed to aging were found in human β-crystallin (67).

The mobilities of the β_L polypeptides are similar in calf, sheep, pig, rabbit, rat, rhesus monkey, and several other primates, but the proportions vary (85, 86). A polypeptide with the electrophoretic mobility of the principal calf β-crystallin chain, βB_p (112), is also the major polypeptide in the other mammals (85, 86, 110).

The amino acid compositions of β_H and β_L fractions of the investigated mammalian species are rather similar, the greatest degree of difference being found among the primates (85, 86). Immunodiffusion experiments with antiserum to calf β_L shows reactions of complete or partial identity with all other mammalian species. Calf, sheep, and pig β_H contains additional antigenic material that is not present in rat, rabbit, and primates.

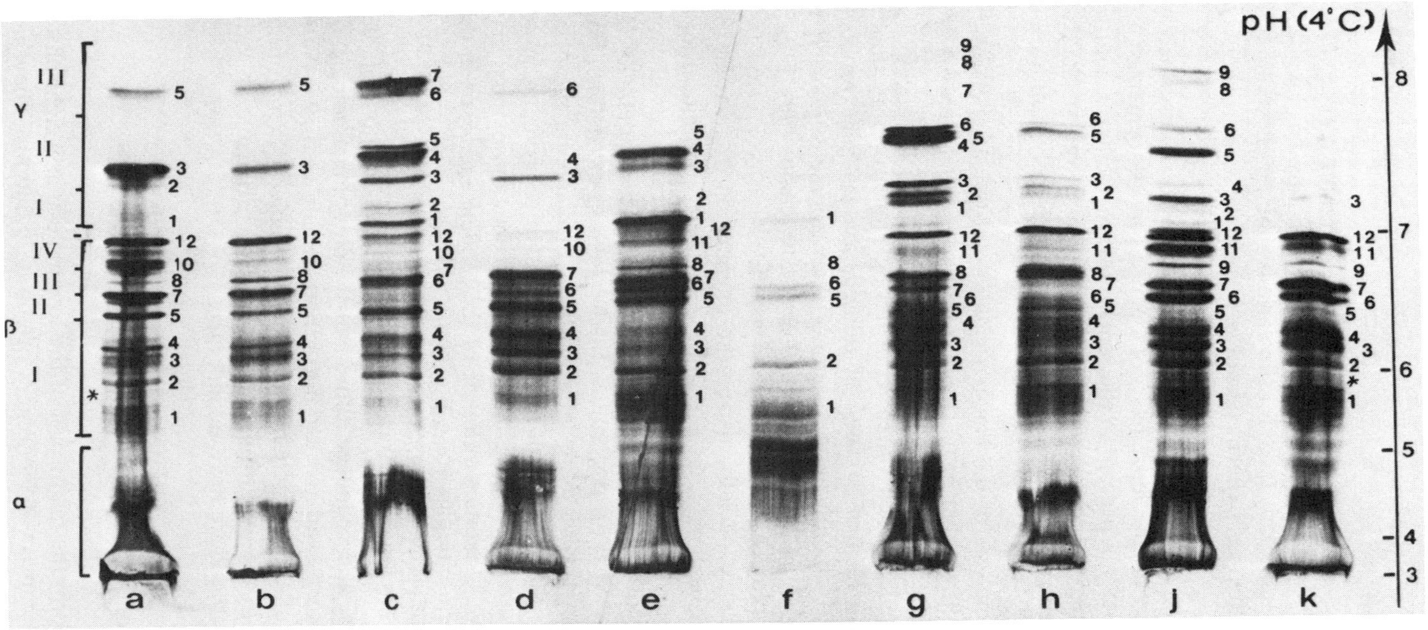

Figure 10 Thin-layer isoelectric focusing patterns of lens proteins from five mammalian species. (*a, b*) guinea pig, 0.5- and 2.5-year-old, respectively; (*c, d*) rat, 86- and 723-day-old, respectively; (*e, f*) human, 5.5- and 50-year-old, respectively; (*g, h*) bovine, 0.16- and 17.5-year-old, respectively; (*j, k*) dog, 0.77- and 9.8-year-old, respectively (107).

Immunologically the presence of a unique antigen in calf β_L could also be demonstrated (85).

The isoelectric points of the polypeptides of rabbit β-crystallins range from pH 5.9 to 7.0 (102). A simple oligomeric structure has been proposed by Mostafapour and Reddy (103) for the various rabbit β-crystallin fractions. Their published SDS gel patterns do not convincingly support such a model, however.

The occurrence of the monomeric β_S-crystallin, first described by Van Dam (113) in the bovine lens, has not been clearly documented in other species. A polypeptide present in the human low molecular weight fraction has been suggested to correspond to β_S (67, 114). In rabbit (102) and rat (100) low molecular weight lens proteins no such polypeptide has been distinguished.

The detailed analysis of the isolated subunits of bovine β-crystallin (see Chapter 1) has shown that in fact only a few primary gene products build up the β-crystallin molecules. These primary gene products may again be ancestrally related to each other, considering similarities in amino acid composition. The idea of a group of related primary gene products and their postsynthetic modification products is of course supported by the antigenic relationships between chick β-crystallin chains (90). In addition, however, immunologically unrelated polypeptides can be present in the β-crystallin fractions, especially also in the mammalian β_H fraction (85, 86). Such chains can be limited to certain species or taxonomic groups, and perhaps represent not genuine β-crystallin components, but just proteins which by virtue of molecular weight or charge are accidentally isolated together with the "real" β-crystallins. Such "real" β-crystallins, however, can only be defined when more information is available on the oligomeric structure and physiochemical properties of native β-crystallin molecules.

Primary structure data are available only for the major bovine β-crystallin chain βB_p. The most exciting outcome of the sequence analysis of this 204-residue chain is the obvious homology with the bovine γ-crystallin chains (115–117) (Figure 11). The sequence homology is, however, quite distant, approximately 25%, indicating an ancient ancestral relationship between the coding genes, and is indeed too small to allow immunological cross-reactivity between β- and γ-crystallins. Interestingly also, the *N*- and *C*-terminal halves of the βB_p chain show considerable sequence homology (Figure 11), and in the evolutionary history of the βB_p gene an intragenic duplication must therefore have taken place.

6.4.4 γ-Crystallin

γ-Crystallins form the major part of the low molecular weight proteins of the eye lens and consist of monomers of about 20,000 M_r. They have the highest isoelectric points among lens proteins, and have in contrast to α-

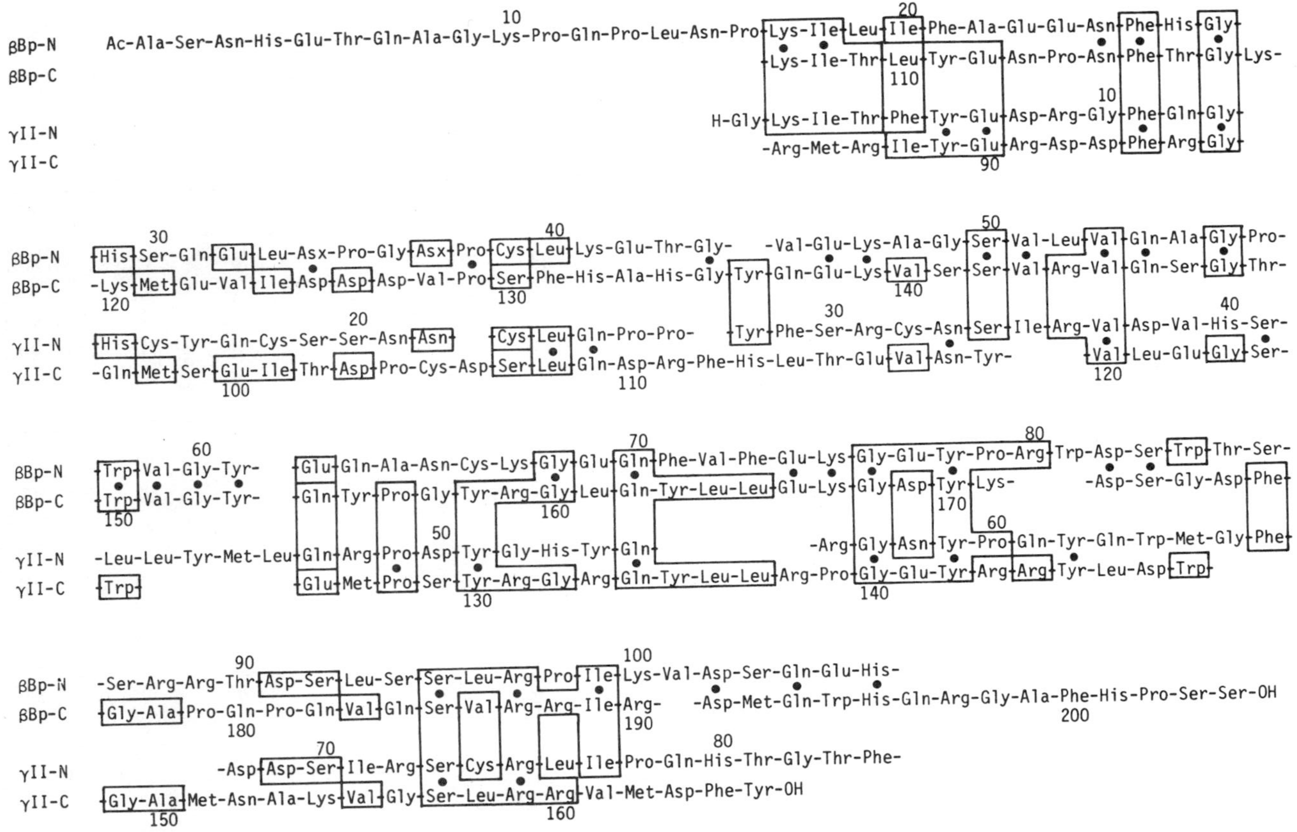

Figure 11 Sequence homology of bovine βB_p (116) and γ-crystallin fraction II (131). The *N*- and *C*-terminal halves of both chains are optimally aligned. Interchain identical residues in homologous positions of βB_p and γ_{II} half-chains are enclosed in boxes. Intrachain sequence homology between the *N*- and *C*-terminal parts of βB_p or γ_{II} is emphasized by dots.

Table 5 Amino Acid Composition of Various Crystallins (Residues/1000 Residues)

Amino Acid	α-Crystallin				γ-Crystallin					Pigeon "γ"	Calf βs	Chicken δ
	Calf	Human	Chicken	Dogfish	Calf	Human	Rat	Haddock Fr. IV	Dogfish			
Asp	86	96	90	103	116	112	116	120	132	88	89	71
Thr	34	49	38	34	22	18	30	18	22	53	28	76
Ser	103	95	125	71	42	51	83	92	80	65	61	96
Glu	105	103	117	129	131	129	106	113	97	122	126	130
Pro	81	55	81	107	58	60	37	32	53	47	37	23
Gly	60	89	49	58	93	84	90	73	91	106	81	56
Ala	44	38	30	36	32	51	19	0	10	58	57	80
Cys	5	13	n.d.	n.d.	n.d.	n.d.	n.d.	52	50	36	37	3
Val	58	58	46	56	48	42	46	18	38	66	49	80
Met	12	8	25	29	37	27	39	129	48	24	33	8
Ile	47	44	65	53	42	40	34	44	40	27	41	74
Leu	87	88	93	64	81	94	69	14	20	58	57	151
Tyr	31	28	21	39	76	80	100	88	127	63	65	8
Phe	76	80	57	66	48	33	58	57	69	55	57	22
Lys	48	46	49	48	24	22	16	19	20	50	49	72
His	39	35	35	30	37	40	36	24	15	20	41	12
Arg	73	76	77	76	111	117	119	111	53	62	73	39
Ref.	(68)	(68)	(34)	(34)	(24)	(70)	(100)	(123)	(32)	(97)	(28)	(140)

and β-crystallin a free N-terminal residue. γ-Crystallins have, like β-crystallins, a high cysteine content, but are distinct in the low level of alanine and lysine, and the higher content of arginine and tyrosine.

Circular dichroism measurements have shown the absence of α-helical conformation in γ-crystallin from calf (84, 118) and dogfish (119) but 22% β structure might be present in a purified calf γ_{II}-crystallin fraction (118). Four different γ-crystallin fractions from the calf have similar secondary structure parameters (84). The secondary and tertiary structure of γ-crystallin is efficiently disrupted by 1% SDS or 5 M guanidine HCl, but is hardly influenced by the presence of 8 M urea (84). Nevertheless the electrophoretic patterns of both bovine (120) and dogfish (32) γ-crystallins are remarkably different in the presence or absence of 7 M urea. It is suggested that this might be due to the removal of ions, and that ion binding may indeed be an important property of γ-crystallin (120).

γ-Crystallin is responsible for the phenomenon of "cold cataract" because it tends to precipitate reversibly in solution below 4°C. The indication "cryoprotein," however, is not justified because no change in conformation occurs between −20°C and +60°C (84). Calf γ-crystallin can readily be crystallized (121). The first results of X-ray diffraction studies at low resolution (5.5 Å) show that the calf γ_{II}-crystallin molecule is arranged in two globular domains, each approximately 25 Å in radius (122). The whole molecule is an ellipsoid with major axes of approximately 55 × 30 × 25 Å.

Although low molecular weight proteins are present in all investigated fish and lamprey lenses (29, 35), their identification as γ-crystallins has been made only in dogfish (32), cod, and haddock (114, 123). Dogfish low molecular weight protein shows a single band of 19,500 dalton in SDS gel electrophoresis, and five mostly basic bands on thin-layer isoelectric focusing. It does not cross-react with antisera against calf γ- or β-crystallin, but its amino acid composition is γ-crystallinlike (Table 5). Cod low molecular weight protein gives a single major band on cellulose acetate electrophoresis at pH 9.0 (114). Three main fractions could be isolated from haddock γ-crystallin (123), of which the major one was electrophoretically similar to calf γ_{IV}-crystallin, despite considerable differences in amino acid composition (Table 5). The calculated molecular weight of this fraction was 19,350 and the sequence of the N-terminal seven residues were identical to those of calf γ-crystallin chains. The C-terminal five residues, however, were quite different from the corresponding bovine ones.

γ-Crystallin is often the predominant lens protein in amphibians, especially in early stages of development (124). Levels of 40.6% in *Xenopus laevis* (38) and 30.5% in the axolotl (39) have been reported. Some five to seven γ-crystallin bands could be distinguished electrophoretically in 12 North American salamander species, which showed few interspecies differences in mobility, but great differences in relative proportions (125). About the same number of γ-crystallin bands are also present in axolotl and *X. laevis* (38, 39). Despite differences in isoelectric points between amphibian γ-crystallins, they have similar immunological properties (124). The isoelectric points of *X. laevis* and axolotl γ-crystallins range between 7.15 and 8.30 (38, 39). γ-Crystallin of *Rana pipiens* did not cross-react with antiserum against calf γ-crystallin (36).

The presence of γ-crystallins in reptilians is unclear. It was not reported to be present in *Anolis* (126), and no material with the sedimentation value of γ-crystallin was found in the rattlesnake (47). In turtles 55 and 29% of protein with the sedimentation value (47) or elution volume on Sephadex G200 (36) of γ-crystallin has been reported. However, no immunological or structural data are available, which allow the conclusion that reptilian low molecular weight proteins are homologous to the mammalian γ-crystallins.

In birds also the presence of genuine γ-crystallin is a subject of debate. Rabaey et al. (95) designated the low molecular weight, monomeric protein of birds as γ-crystallin. It is present in negligible to small amounts in different species, the chick having less than 1% in its total lens protein and reaching about 5% in the pigeon (95–97). It is electrophoretically homogeneous, migrating in the area where the β-crystallins generally migrate. The molecular weight is about 18,000 (95). It appears very late in the ontogeny of the chick lens and at 4 days in the embryonic pigeon lens (96). It is found not only in lens fiber cells but also in annular pad and epithelium. Specific antisera for this low molecular weight protein from pigeon give precipitin lines after immunoelectrophoresis of lens proteins of cow, tortoise, iguana, and a teleost fish, but not with *Bufo marinus* lens proteins (95). Remarkably the cross-reaction is much stronger with the low molecular weight proteins from bovine outer cortex than from the nucleus. McDevitt and Croft (97) have demonstrated that pigeon "γ-crystallin" lacks several of the characteristics of other vertebrate γ-crystallins. It has no low value for alanine and glycine (Table 5), it has no free *N*-terminal residue, and the tryptic peptide map is not characteristic of γ-crystallins, possibly being related to bovine β_S-crystallin.

In all investigated mammals multiple γ-crystallins are found (for summary of older data, see Table IV in ref. 2). It has best been characterized in calf (Chapter 1), rabbit (102, 127), rat (100, 128), and human (67, 70, 129). γ-Crystallin from these species can be resolved in up to eight components by electrophoresis, isoelectric focusing, or ion exchange chromatography. By isotachophoresis even up to 16 components could be distinguished in rat γ-crystallin (108). The isoelectric points of the different γ-crystallins generally range between pH 7 and 8.

Between mammalian species the differences in isoelectric focusing patterns are considerable (Figure 10), in both position and intensity of the bands. Even between fairly related species of rodents such differences occur (105). Within species only quantitative differences, in relative intensities of different γ-components, are observed (106). Such differences may be genetically determined, when comparing widely separated populations (106), but also occur with increasing age (107, 109, 114). Specific γ (and β) components may indeed disappear with age, and are absent in the cortex of older lenses (109). In human γ-crystallin few differences could be attributed to aging (67). In adult bovine lens cortex a γ-crystallin fraction is present that is not found in calf lens (130). It has a higher molecular weight than calf γ-crystallin: 24,000 as compared to 20,000. The ionic charge and immunochemical properties of cow cortex and calf γ-crystallin are different. Calf γ-crystallin is similar to the γ-crystallin found in the cow lens nucleus. Although the amino acid composition of cow lens γ is closer to that of β_S than to calf lens γ, it differs from β_S in having a free *N*-terminal glycine and a lower molecular weight and in charge and immunochemical properties.

In all species the total proportion of γ-crystallin in the water-soluble fraction drastically decreases with age (62, 63, 67, 108), and may even become absent in human cataractous soluble lens proteins (114).

The reported values for the molecular weights of γ-crystallin, as estimated by SDS gel electrophoresis, vary from 18,400 for rat (100), 19,100 for human (67), to 20,500 for rabbit (102). These are all reasonably close to the molecular weight, as derived from the amino acid sequence, of calf fraction γ_{II} 19,871 (131). It does not necessarily indicate that real differences in length are present; relatively small differences in amino acid sequence have been shown to influence the mobility of α-crystallin subunits in SDS gel electrophoresis appreciably (132).

Glycine is found as the *N*-terminal residue of the different γ-fractions of calf (121) and rat (128) although previously also alanine had been reported as *N*-terminus in rat γ-crystallin (133). The amino acid composition of γ-crystallins from calf (121), rabbit (127), rat (100, 128, 133), and human (70) are quite similar (Table 5). In agreement with this these crystallins are immunologically very similar, although by radioimmunoassay antigenic differences were found between sheep and rabbit γ-crystallin and even more between sheep and human γ-crystallin (61).

It can be concluded that at least four nonallelic genes are coding for the different calf γ-crystallins, which are differentially expressed at different stages of development (134). The sequence difference between the fractions γ_{II}, γ_{III}, γ_{IVa}, and γ_{IVb} apparently does not amount to more than about 15% (114).

6.4.5 δ-Crystallin

A lens protein limited to birds and reptiles was first identified by Rabaey in 1962 (135). He designated it as FISC ("first important soluble crystallin"), since it is the initial crystallin to appear during chick lens development, but it is now also commonly known as δ-crystallin. Using antisera to chick δ-crystallin it could be shown that similar, but not identical, lens proteins occur in several orders of birds, in the lizard *Lacerta* (135, 136), and in gekko, python, and caiman (136a). Considerable differences in electrophoretic mobility and relative proportions of δ-crystallin do exist in different species. The isoelectric points of the δ-crystallins range between pH 5 and 7. In general the reptilian δ-crystallins have the higher isoelectric points (136a).

From the results of Manski and Halbert (23) it seems probable that there is an immunological relation between turtle lens proteins and chick δ-crystallin. Demonstration of proteins with the spectroscopic peculiarities of δ-crystallin in the lenses of lizard and snake provided indirect evidence for the presence of this protein in these reptiles (137). This method failed to detect δ-crystallin in the lens of the snapping turtle, and indicated its presence as a minor component only in the lens nucleus of the caiman.

It appears that differences in electrophoretic mobilities of δ-crystallin from chick, quail, turkey, and pheasant, all belonging to the order Galliformes, are due to changes in the ratio of subunits of identical mobilities in gels containing urea (2). The subunits of δ-crystallin of duck and goose, belonging to the Anseriformes, are mostly of different electrophoretic mobilities as compared to the chick.

Most studies have been performed on δ-crystallin from the chick lens. It can easily be separated from α- and β-crystallins by a single passage of total lens extract through a column of AcA-34 (52) (Figure 8). Chick δ-crystallin can be resolved in seven immunologically identical components by thin-layer isoelectric focusing (138). The differences in isoelectric points, which range between pH 5.18 and 5.34, are attributed to slight differences in primary structure, possibly deamidations. The heterogeneity of native δ-crystallin increases with age. Post-translational modification of newly synthesized δ-crystallin in chick lens culture results in a decrease in the average isoelectric point of the native molecules (139).

Native δ-crystallin has a molecular weight near 200,000 in chicken (140) and several other avian species and the caiman (27, 136a). On acrylamide gel electrophoresis in the presence of sodium dodecylsulfate δ-crystallin usually shows a single band with a molecular weight variably estimated between 45,000 (52) and 50,000 (140). In SDS gels containing urea embryonic chick δ-crystallin can be resolved into two bands of molecular weights 48,000 and 50,000, occurring in a ratio of approximately 3 : 1 (141). Also in several other birds and in python and gekko, but not caiman, δ-crystallin can be resolved under these conditions into two components, of which the major one usually has the lower apparent molecular weight (136a). These two polypeptides from chicken as well as from gekko (136a) have very similar fingerprints, and pulse-chase experiments in chick lens cultures make a precursor–product relation between the two very unlikely (141).

Analysis by dodecylsulfate gel electrophoresis of peptides obtained by staphylococcal protease digestion of different avian and reptilian δ-crystallins revealed differences in primary structure (136a).

The chick δ-crystallin molecule is probably composed of four subunits which must be very similar, though not identical, in primary structure

(142). Electrophoresis or isoelectric focusing of δ-crystallin in the presence of urea demonstrates charge heterogeneity of the subunits (48, 52). It is unclear to what extent this heterogeneity is due to post-translational modifications or to the presence of different primary gene products. The presence of the 48,000 and 50,000 polypeptides in embryonic chick δ-crystallins, which moreover are differentially influenced by changes in culture conditions (143), suggests the presence of at least two closely related δ-crystallin loci.

The cloning of δ-crystallin cDNA in a bacterial plasmid (143a) has provided the opportunity to analyze the δ-crystallin gene structure. It could indeed be shown that at least two nonallelic but very similar δ-crystallin genes are present in chicken and turkey (143b). Interestingly, one of the δ-crystallins genes seems to be interrupted at least 14 times by intervening sequences, which is among the highest numbers yet observed in eukaryotic genes. Chick δ-crystallin cDNA has been used, by annealing with nuclear DNA, to detect the presence of sequences complementary to chick δ-crystallin mRNA in the genomes of different vertebrates (136a). No complementary sequences could be demonstrated in the genomes of fish, amphibians, and mammals, and reptilian DNAs annealed much less well than the avian DNAs.

As stated already δ-crystallin is the first crystallin to appear during embryonic development. Immunofluorescence (91) studies and radioimmunoassays (144) first show the presence of δ-crystallin in the basal part of chick lens placode cells after 50 to 54 hours of incubation. In the 3-day-old embryo the epithelium of the lens vesicle contains δ-crystallin and at 8 days the developing fiber cells show a positive reaction. The messenger-RNA for δ-crystallin can be detected already at 48 hours, which is some 8 to 9 hours after the initiation of lens induction by the optic vesicle (145). The 17S δ-crystallin mRNA has a molecular weight of about 680,000, accounting for approximately 2200 nucleotides. The sequence complexity of this mRNA has been analyzed by hybridization with cDNA and shown to be primilarily composed of a single species (146). This indicates that this mRNA preparation codes for a single polypeptide chain, or for two or more very similar gene products.

After hatching the synthesis of δ-crystallin in the chick lens rapidly decreases. Whereas δ-crystallin makes up from 60 to 80% of the soluble protein in embryonic chick lens fibers, the level decreases to around 20% in the adult chick (36, 51, 138, 147). Synthesis of δ-crystallin stops in the cortical fibers of older birds (52). The presence of δ-crystallin in adult bird lenses is almost entirely restricted to the nucleus (54, 91, 137). Already in the 5-week-old chick the indirect immunofluorescence method fails to detect δ-crystallin in the lens cortex (91), and only traces of δ-crystallin are detectable by electrophoretic analysis of soluble proteins from adult chick lens cortex (54, 148). In remarkable contrast δ-crystallin polypeptides of immunologically proven identity form 38% of the plasma membrane proteins from the outer chick lens cortex (148). The δ-crystallin subunits from embryonic and cortical adult lens membranes separate on SDS gel electrophoresis in two to three components of estimated molecular weight 43,000 to 48,000. Only the 43,000 δ-crystallin band is found in membranes from the adult nucleus.

The amino acid composition of chick δ-crystallin is given in Table 5. The amino acid compositions of reptilian δ-crystallins (148a) show the same characteristics, especially the remarkably high contents of leucine and isoleucine, which sets this protein clearly apart from the other crystallins (Figure 12). These high contents of leucine and isoleucine may contribute to the relative stability of δ-crystallin to 8 *M* urea and 5 *M* guanidine hydrochloride (140). Some change in conformation of δ-crystallin on treatment with 6 to 8 *M* urea must take place, however, because it results in an irreversible time-dependent loss of antigenicity (149).

Raman spectroscopy has shown that δ-crystallin, both *in situ* and after chromatographic isolation, is mainly in the α-helical conformation, in contrast to all other crystallins. Purified chick δ-crystallin was confirmed to

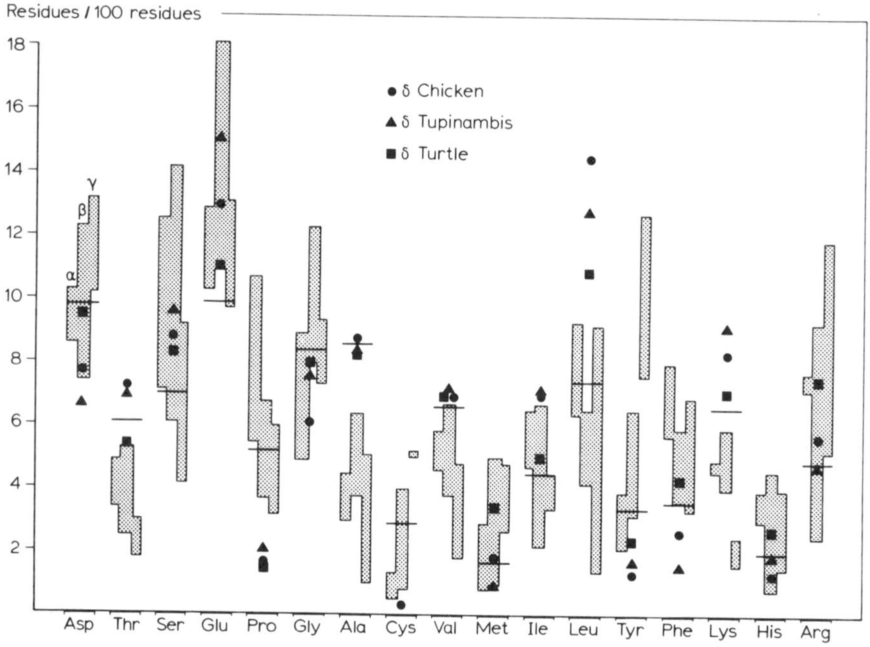

Figure 12 Comparison of amino acid compositions of δ-crystallin from chicken, teju (*Tupinambis teguixin*), and green turtle (*Chelonia mydas*) with values reported in the literature for α-, β-, and γ-crystallins from all classes of vertebrates (shaded blocks). Horizontal bars indicate the average occurrence of amino acids in proteins (148a).

contain 73% α-helix configuration by circular dichroism studies (137, 150). It is suggested that the α-helical conformation of δ-crystallin, together with the low protein concentration in the bird lens, is responsible for the soft consistency and easy deformability of this lens. The low content of tryptophan and cysteine, which both have been implicated in nuclear cataract formation in man, might be advantageous for maintaining the accommodative capacity of the aging bird lens.

The presence of δ-crystallin outside the ocular lens seems firmly established. Immunologically its presence is demonstrated in cornea, vitreous, aqueous, iris, and retina (53). The chick iris contains about 3% δ-crystallin which is identical to lens δ-crystallin both immunologically and by isoelectric focusing (138). Very low levels of crystallins and crystallin mRNAs have also been detected in 8-day-old chick embryonic neural retina and retinal pigmented epithelium (151, 152).

Very intriguing is the reported presence of δ-crystallin in the embryonic chick adenohypophysis (153). By indirect immunofluorescence, using antisera to chick α-, β-, and δ-crystallin, a positive reaction only for δ-crystallin was obtained in the primitive adenohypophysis at all stages between 60 and 156 hours of development. No δ-crystallin was detected in the adenohypophysis of the 5-day-old or adult chick. Immunoelectrophoresis showed a lower mobility for δ-crystallin from the adenohypophysis than for lens δ-crystallin.

6.4.6 Other Crystallins

Crystallins being defined as the water-soluble structural proteins of the lens, there remain a number of lens proteins that cannot be classified as α-, β-, γ-, or δ-crystallins.

The acidic bovine lens protein FM crystallin, previously called pre-α-crystallin (28, 113) is present only in small amounts in the bovine lens, and the sole reason for its being noticed as a separate component is its conspicious electrophoretic mobility (154).

An electrophoretically fast-moving component, designated as pre-α-crystallin, is also observed in some anuran and urodele amphibians (125, 155). This protein is not antigenically related to any bovine lens protein and no structural parameters have been determined. Indirect immunofluorescence studies have shown that this amphibian pre-α-crystallin appears ontogenetically immediately after the γ-crystallin and is likewise restricted to the fiber cells (156).

Another unclassified crystallin has been observed in *Rana pipiens* (36). It elutes on Sephadex G200 just behind β_L, making up 14.8% of the total water-soluble protein, and consists of 37.500 M_r monomers which separate in two bands on isoelectric focusing, between pH 7.5 and 8.5. This protein shows no cross-reaction with antisera against calf α-, β-, and γ-crystallin.

6.5 CRYSTALLINS IN LENS DEVELOPMENT, REGENERATION, AND AGING

Detailed information concerning these aspects can be found in earlier reviews (3, 49, 157, 158) and in Chapters 2, 4, 7, and 10. I want only to summarize here some general and recent findings that are pertinent to the purpose of placing the crystallins in an evolutionary perspective.

It is generally accepted that the development of the lens is induced by the optic vesicle, although structures other than the eye rudiment have also been shown to be capable of lens induction (159). Transplantation studies in amphibians have shown that most, but not all, regions of the general epidermis can be induced to form a lens, and that moreover the ability of the epidermis to respond to induction varies considerably in different species (160). There is a lag period between the moment of induction and the synthesis of the first crystallins, which can be considered as an objective criterion of lens differentiation. This lag period is longest in amphibians, whereas in chick embryos the synthesis of crystallin mRNAs already starts within a few hours after induction (161).

The capacity to regenerate a new lens after lentectomy is apparently present only in amphibians and some fishes. The new lens, induced by the neural retina, regenerates either from the corneal epithelium (in embryos and young larvae of the urodele *Hynobius* and in tadpoles of the anuran *Xenopus laevis*) or from iris epithelium at the pupillary margin of the dorsal iris (in adult newts of the family Salamandridae and in some fishes (162–164). In tissue culture, however, lenslike cells, forming lentoid bodies, can be obtained from retinal pigmented epithelium and neural retina of 8-day chick embryos (152, 165) and from pigmented cells of iris or retina and from neural retina of 12-week human fetuses (163, 166). The presence of crystallins in these lentoid bodies has been demonstrated by immunological methods.

The differentiation of lens epithelium into fiber cells is characterized both *in vivo* and *in vitro* by a sequential appearance of particular crystallins. In amphibians, chick, and mammals the synthesis of crystallins is first detected during ontogeny in the elongating cells of the invaginated lens placode in close proximity of the neural retina (38, 91, 124, 167–169). The synthesis of crystallins subsequently spreads over the whole developing lens.

Although α-crystallin appears early in embryonic development of amphibians, birds, and mammals, in lenses of 23-cm-long embryos of spiny dogfish no α-crystallin could be detected electrophoretically, the bulk of the protein being formed by γ-crystallins. Accordingly no α-crystallin is found in the hard, fibrous nucleus of adult spiny dogfish, whereas the soft cortex contains a considerable level of this protein (169a).

In the anurans γ-crystallin is one of the first lens proteins to appear (38, 124, 170, 171). In newts, however, α- and β-crystallins are reported as the first appearing lens proteins (170). The amphibian pre-α-crystallin appears directly after γ-crystallin (156) and is likewise exclusively localized in elongated fiber cells. With development the relative proportion of γ-crystallin decreases with a concomitant increase of β- and α-crystallin (38, 49, 172).

In the chick δ-crystallin is the first to appear, being detectable already at about 50 hours of incubation (91, 141, 144). The initiation of δ-mRNA accumulation already starts 8 to 9 hours after the initiation of lens induction (145). Although δ-crystallin can be detected by immunofluorescence in the lens epithelium in very early stages of development, it is absent from the epithelium at later stages and in fact is considered indicative of differentiation in chick lens (91, 157). Also in tissue culture the elongation of embryonic chick lens epithelial cells is associated with the synthesis of δ-crystallin (157). At 3 to 3½ days of incubation β- and α-crystallins can successively be detected (91, 169). α-Crystallin can first be demonstrated in the central lens fibers and β-crystallins in the epithelium (91).

The cathodal β-components are first to appear, followed after two days by the anodal β-component (92). Some β-crystallins, those that immunologically cross-react with β-crystallins from all other vertebrates, are present only at 7 to 8 days of incubation and mainly in the epithelium (173). In later embryonic development and post-hatching the synthesis of δ-crystallin decreases and α- and β-crystallin become the major products (51, 54, 91, 93). In the cortex of older chicks the synthesis of δ-crystallin has stopped (52). As a result the nucleus of an adult chick lens mainly contains δ-crystallin, whereas α- and β-crystallin replace the δ-crystallin in the cortex.

By indirect immunofluorescence the first lens antigens are detectable at 11 days in the mouse embryo (168). The first staining for γ-crystallin occurs at the twelfth day in the anterior cytoplasm of elongating primary lens fibers. In the 1-day-old rat lens the epithelial cells react positively for α-crystallin only by indirect immunofluorescence (174). β-Crystallin was first detected in elongating cells that had stopped dividing, and γ-crystallin appears still later during elongation. The synthesis of both β- and γ-crystallin thus appears to depend on the elongation process. Of the many known aspects of bovine lens cell development (Chapter 3) most relevant, perhaps, is the finding of changing ratios of subunit synthesis during ontogeny and differentiation. Examples are the far greater synthesis of αB chains, as compared to αA chains, in the central epithelium (175), and the very early appearance of the γ_{IV} chain during ontogeny (134).

Essential questions, of course, are which factors trigger the synthesis of particular crystallins and are these factors universal, following the same rules in different vertebrates. Lens induction and the synthesis of the first crystallins may well be brought about by inductor substances secreted by the optic vesicle cells. The developing retina loses its capacity for lens induction but stimulates the processes of fiber cells formation and crystallin synthesis (161). This is particularly clear from experiments in which the chick lens is reverted, resulting in the former epithelial cells starting to differentiate and produce δ-crystallin (176). Therefore, lens fiber formation not only is a function of age, but clearly also depends on a specific location within the eye. The influence of neighboring tissues is also obvious from the finding that embryonic chick lenses cultured with and without the vitreous attached synthesize δ-crystallins with different isoelectric points (143).

McAvoy (169) concludes that at 14 days of development the rat lens becomes separated in a proliferation compartment (the anterior epithelium) and an elongation compartment (without cell division), probably under the influence of factors transmitted by the neural retina. These factors too provide the signal for the synthesis of β- and γ-crystallins.

Summarizing, then, the spatial distribution of crystallins in normal vertebrate lenses is characterized by the presence of α-crystallin in both epithelial and fiber cells; β-crystallin seems more restricted to fiber cells in mammals but occurs also in the epithelium of amphibian (170) and chick lenses (177); γ-crystallin remains exclusive to fiber cells or presumptive fiber cells, even despite the occurrence of avian "γ-crystallin" in epithelial cells, because, as stated already, this protein is more likely to be a β-like crystallin (97).

Interesting in this respect is the observation that in giant tadpoles of *Xenopus laevis,* obtained by prevention of metamorphosis by propylthiouracil treatment, the lenses not only reach normal postmetamorphic size but also crystallin patterns which are indistinguishable from normal adult lenses (172). It is concluded that the crystallin pattern is correlated with lens diameter rather than with developmental body stage. This confirms Clayton's suggestion (49) that the adaptive value of ontogenetic changes in crystallin composition might be to maintain the appropriate refractile properties of the lens with increasing diameter. Such an explanation, on the other hand, is difficult to reconcile with the finding that regenerated *Xenopus* lenses may show either the normal full pattern of lens antigens or only a limited number, despite indistinguishable morphologies (178). Again, chicks of the Hy-1 strain, which are selected for high growth rate, have an anomalous lens morphology which is associated with differences in distribution and time of appearance of the crystallins (177). Also the lentoid bodies developed from neural retina of normal and Hy-1 chick embryos have different crystallin compositions, which both again are different from the crystallin pattern of embryonic lens fibers (151).

As already seen the crystallins are apparently not strictly organ-specific. The lens of the median or parapineal "eye" of the lizard *Anolis carolinensis* has been shown by indirect immunofluorescence to react positively for the presence of lateral eye lens crystallins (126). The apparent presence of δ-crystallin in the embryonic chick adenohypophysis (153) has already been mentioned. Extralenticular occurrence of crystallins in the eye has been reported by several authors in different species. Crystallins are reported from chick iris, cornea, vitreous, aqueous, and retina (53, 138, 151) and also from human embryonic vitreous (94). α-Crystallin determinants are probably present in bovine iris and choroid, but not in cornea and retina (178a). Crystallin mRNAs were detected at low levels in embryonic chick neural retina and pigmented epithelium (152). However, Sandberg and Closs (179) conclude that it is unlikely that crystallins are produced outside the lens in sheep eyes. The controversy regarding the existence of extralenticular crystallins has been reviewed in more detail by Clayton (180).

In all vertebrates investigated the crystallin composition not only changes during the ontogeny of the lens but, once full-grown, continues to change during aging. Whereas the ontogenetic changes are mainly due to differential gene expression, the aging effects are largely caused by postsynthetic modifications of the primary gene products. (see Chapter 7). Because the notoriously slow rate of turnover of the lens proteins is apparently approximately the same for α-, β-, and γ-crystallins (181), it may be expected that all three may undergo extensive modification. Thin-layer isoelectric focusing indeed shows considerable changes in crystallin patterns upon aging in different mammals (94, 107, 109, 182). There is not only a general decrease in content of γ-crystallin and an increase of β-crystallin, but also a profound change in the ratios of different β- and γ-components. Especially in man the extent of postsynthetic modifications is great and leads in the adult lens to an almost unresolvable complexity of crystallins (67, 69, 70, 72–75, 158, 183).

Despite the great similarity in primary structure of crystallins from different species, it seems that the rate and extent of postsynthetic modifications may vary greatly between species. This is obvious from inspection of the degree of deamidation of αA and βB chains in different species, as revealed by alkaline urea gel electrophoresis (30). Some species, like rat and elephant (despite an age of up to 40 years), have very little αA_1, whereas others, like horse and kangaroo, have half or more of their αA chain in the deamidated form. Likewise postsynthetic degradation leading to removal of the five *C*-terminal residues of the αA chain proceeds to greatly differing extents in different mammals, even taking into account age differences between the sampled individuals (184).

6.6 CRYSTALLINS IN TAXONOMIC AND PHYLOGENETIC STUDIES

Comparative studies of eye lens proteins, as of all other proteins, can be performed with different aims. From a biochemical point of view it is interesting to explore the variations that occur among lens proteins in different species, in an effort to relate these to structural and functional differences. How much protein variation is allowed and how can this variation be interpreted? At the same time such studies provide information about the processes of biochemical evolution of lens proteins. These aspects are dealt with in the next section.

The genetically or taxonomically oriented investigator can use comparative data of proteins to establish relationships between investigated species or higher taxa. Many authors use for this purpose immunological techniques, such as immunodiffusion and immunoelectrophoresis, but the most quantitative data are obtained by microcomplement fixation (185). All kinds of electrophoretic techniques are likewise frequently used, and sometimes chromatography as well. DNA hybridization is often used in evolutionary studies and may provide taxonomically valuable information (186). Most promising for the reliable reconstruction of evolutionary relationships is the comparative study of amino acid sequences of homologous proteins from different species (187, 188). It may become one of the most useful tools for reaching the goal of providing a systematic classification corresponding to phylogenetic relationships (189).

A general survey of biochemical and immunological applications in animal taxonomy can be found in Wright (190). Lens proteins have been used extensively for comparative immunological studies. Most spectacular are the results of Manski and Halbert (23), who were able to reconstruct a sequence of evolutionary changes of eye lens antigens within the whole vertebrate subphylum. The order of appearance of lens antigens in the different vertebrate classes was studied and could be used to infer the sequence of origin of different taxa. It thus was shown, for instance, that lenses of the shark contain antigens that are also found in all higher vertebrates, but not in the more ancient lampreys. Such antigens then must have originated after the branch leading to the sharks separated from the ancestral Agnatha. Manski and Malinowski (191) more recently estimated that approximately 42% of all antigenic determinants of calf α-crystallin are derived from primitive Agnatha, approximately 22% from subsequent aquatic vertebrates, approximately 21% from primitive land vertebrates, and only approximately 15% evolved with the mammals.

It is indeed the slow rate of evolution of eye lens proteins, in contrast to most serum proteins, that makes them particularly useful for studying interrelationships of higher taxa. This slow rate of change is also evident in the near absence of intraspecific variation (apart from age-dependent differences) among lens proteins (106). Only very few electrophoretic polymorphisms of eye lens proteins have ever been reported (192–194a). Differences in number and intensities of crystallin components on isoelectric focusing have been reported among different races of dogs (104). Maisel and Goodman (21) used two-dimensional paper-starch gel electrophoresis to analyze lens proteins from representatives of all vertebrate classes. They obtained complex patterns, which indeed showed group-specific characteristics, but did not endeavor to use these for taxonomic or phylogenetic conclusions.

Electrophoretic analyses of lens proteins have been performed in many fish species by different authors and are reviewed by O'Rourke (194). The results clearly show the considerable differences in native lens protein patterns, even between closely related species. In some cases the lens protein pattern turned out to be a valuable taxonomic character.

Lens proteins in amphibia and reptiles have hardly been studied from a comparative point of view (195). McDevitt and Collier (125) investigated the soluble lens proteins of 12 eastern North American salamander species by electrophoresis. They observed a considerable discriminative ability in the lens protein system and suggested its use as a "phylogenetic fingerprint" in urodele systematics.

Electrophoretic and immunoelectrophoretic comparisons of avian eye lens proteins have extensively been carried out by Gysels and Rabaey. Their studies included several hundred species, representing almost all avian orders and led to a considerable number of taxonomic proposals (27, 196, 197). Taxonomic decisions were often based on small differences in mobility and on the presence or absence of minor protein fractions. Sibley and Brush (198), using starch instead of agar gel electrophoresis, were unable, however, to find taxonomically informative patterns in the lens proteins of more than 400 species of birds. In an attempt to refute criticism on his work, Gysels (199) could show that one of the protein components, considered by him to be taxonomically meaningful, the so-called typical songbird component, has similar biochemical properties in three different avian orders. It thus appears to be a real homologous protein in the orders in which it occurs, and could therefore be considered as a "shared derived" (189) character in assessing the evolutionary relations among birds. The fact that this "typical songbird component" is resistant to urea, like chick δ-crystallin (140), possibly indicates a relation with δ-crystallin.

The comparison of lens protein has hardly been used in mammalian taxonomy and phylogeny (190). Day (105) has analyzed the lens proteins of 15 species of rodents by polyacrylamide gel electrophoresis. A classification of these species on the basis of their lens protein bands, using the methods of numerical taxonomy, was claimed to correspond reasonably well with established opinions. However, the rabbit, belonging to the order Lagomorpha, could not, on the basis of its eye lens protein pattern, be differentiated from the species belonging to the Rodentia. The finding that the α-crystallin A chain, which is relatively easy to isolate and analyze, evolves at a slow rate which makes it suitable for comparison at higher taxonomic levels (200), prompted an investigation of this chain among representatives of different mammalian orders. The phylogenetic relationships among the 20-odd mammalian orders are far from settled and protein sequence data might contribute to the elucidation of these relations. A comparison of the αA chains of 17 mammalian species, representing 11 orders, indeed confirmed some firmly established relationships and provided strong suggestions where relations were disputed (45) (Figure 13). Most interesting was the connection of the whale with the ungulates, and the joining of hyrax and elephant on the same branch, corresponding to the ideas of some taxonomists, but refuted by others. A more recent study (201) of the αA chain of sloths, anteaters, and pangolins clearly joined the first two species and placed the order to which they belong, the Edentata, as the oldest offshoot of the placental mammalian stem, The pangolins, of

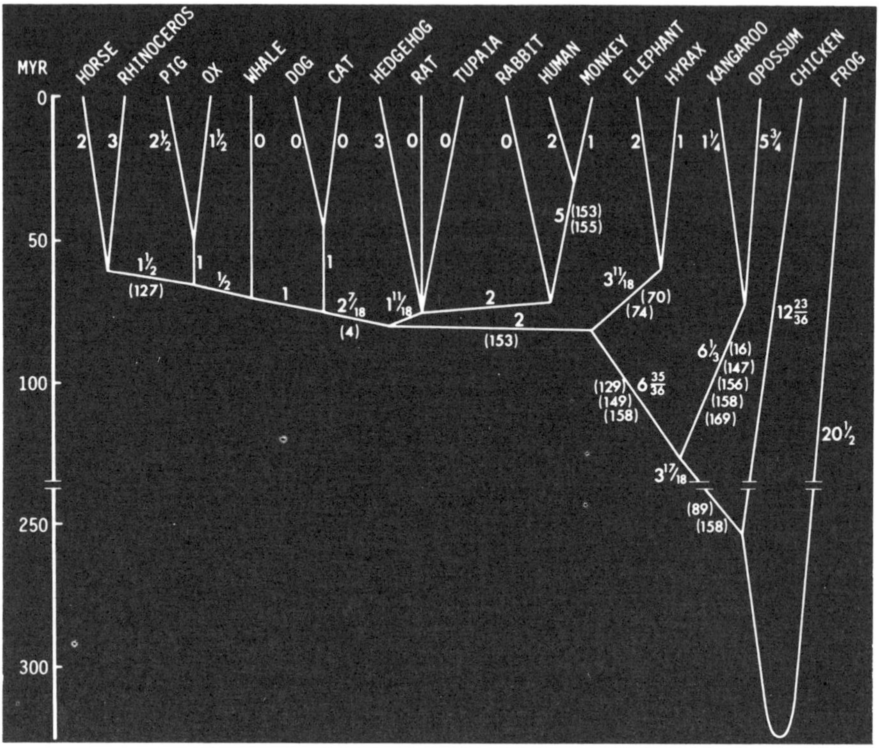

Figure 13 Phylogenetic tree constructed by a maximum parsimony procedure on the basis of the sequences of 19 α-crystallin A chains. Branch lengths are expressed in substitutions. Positions are shown (in brackets) of substitutions other than back, parallel, or trivial (giving no cladistic information). Estimated times of the branching points are taken from the paleontological literature (45).

which the evolutionary relations are entirely obscure, were found to resemble the carnivores most closely in their αA chain sequence.

6.7 EVOLUTIONARY INFERENCES

6.7.1 Primordial Lens and Crystallins

The last common ancestor of all present-day vertebrates, being a fishlike, aquatic animal, must have had a spherical lens of high refractive index. This lens was probably solid and unpliable, and accommodation, if present, was therefore possible only by displacement of the lens.

With the origin of the lens already existing genes came to expression, or were modified, to take care of the synthesis of specific lens proteins that were required in large quantities during fiber cell formation and that had to warrant the transparency and structural properties of the lens cells. Probably the crystallin genes developed from genes already actively engaged in the production of ectodermal proteins of a kind. However, comparison of the amino acid sequences of bovine α- and γ-crystallin chains with all other known sequences of animal proteins has not yet revealed any resemblances (202). In fact both γ-crystallin and the αA and αB chains of α-crystallin are placed in two protein superfamilies of their own. With the continuous expansion of known protein sequences we may hope one day to discover which family of genes provided the genetic material for the expression of the crystallins.

Considering the universal occurrence of α-, β-, and γ-crystallins the primitive lens must already have contained similar proteins. δ-Crystallin may have been present initially and subsequently been lost in all vertebrates except for the sauropsidans, or alternatively it may have originated in the common ancestor of present-day birds and reptiles, after the divergence of the mammalian ancestors (136a).

6.7.2 Gene Duplications in Crystallin Evolution

It is a well-accepted theory that gene duplication has played an important and universal role in protein evolution (203). The finding of multiple, immunologically related lens proteins in the same species led Zwaan and Ikeda in 1965 (204) to the proposal that the phylogeny of the crystallins also was based on the occurrence of repeated gene duplications. Originally single genes, for each category of crystallins, should have been duplicated repeatedly and subjected to continuous modification. This proposal has indeed been corroborated by many recent findings. The αA and αB chains of α-crystallin show 57% sequence homology and are therefore thought to have originated by duplication of a common ancestral gene (76).

Considering the rate of evolution of the αA chain [2.9 substitutions per 100 residues in 100 million years (45)] and the 43% sequence difference between αA and αB, it can be suggested that the divergence between the αA and αB genes might have taken place some 750 million years ago. One can object that studies on other proteins have shown wide fluctuations in the rate of change of a protein, depending on structural and functional requirements (45, 205). Especially shortly after a gene duplication the selective pressure on one of the copies might be relaxed and allow a faster rate of change. Nevertheless it is likely that both αA and αB chains were already present in the earliest vertebrate lens, and this is supported by the observation that even lamprey α-crystallin is probably composed of αA- and αB-like chains (30). It may be expected that the αA and αB genes are located on different chromosomes, as are the genes for hemoglobin α and β chains, which differ approximately 44% in sequence (206). Ancient genome or chromosome duplications, preceding the origin of the vertebrates, are probably responsible for such considerably divergent but still clearly homologous polypeptides (203).

Before the duplication of the ancestral α-crystallin gene an internal duplication may have occurred in this gene. Barker et al. (207) provided statistically significant evidence that positions 1 to 30 of the bovine αA chain are homologous to positions 31 to 60, although visual inspection hardly supports such a resemblance (Figure 14). A similar internal duplication would also be expected in the αB chain, but could not be demonstrated, possibly because of the occurrence of deletions of residues. Intragenic duplications are apparently rather common evolutionary events, as inferred from the many examples of proteins containing small or large regions of internal duplication (188).

There is as yet no evidence for further duplications of the αA and αB genes in mammals. In all investigated mammals the αA chains seem to be the product of a single αA gene. Apart from deamidations and degradations, no sequence heterogeneities have in general been found among the αA chains of a single species (45). The hyrax αA chain indeed has equal amounts of alanine and threonine in position 55, but this is most likely due to the polymorphic occurrence of two alleles of the αA gene in the investigated population (65). In the rat and closely related rodents the normal αA chain of 173 residues occurs together with a very small amount of an elongated À chain. This chain, αA^{Ins}, has 22 residues inserted between positions 63 and 64 of an otherwise normal αA chain (208). Although αA^{Ins} might be the product of a duplicated αA gene, in which somehow a sequence of 66 nucleotides of unknown origin has been inserted, it seems for the time being more plausible that αA^{Ins} is the result of defective splicing of a small proportion of the mRNA transcribed from a single αA gene.

The reported incorporation of radioactivity in the chick αA_1 and αB_1 chains after a short pulse labeling of lenses (55) would suggest that in the chick four genes are coding for different α-crystallin chains. The sequence determination of the chick αA chain, which was largely done on a fraction containing both αA_2 and αA_1, revealed, however, no sequence heterogeneity and therefore does not support the presence of two different αA genes in the chick (W. W. de Jong and E.C. Terwindt, unpublished data).

The multiple β-crystallin chains of the chick have been proposed to derive from a process of gene duplication (87). The relation between the different calf β-crystallin chains is less clear. The similarity in amino acid composition between two of the major primary gene products, βB_p and βB_2, in any case indicates that these are the products of duplicated genes (112). It has been shown that at least four nonallelic genes are coding for the different calf γ-crystallins (134). The results of sequence determinations have shown that the degree of sequence homology is higher than approximately 85% (114, 117), indicating a more recent history of gene duplication events than in the case of αA and αB. This is in agreement with the considerable variation in number of γ chains in different vertebrates (2). In

```
                                   10
Met-Asp-Ile-Ala-Ile-Gln-His-Pro-Trp-Phe-Lys-Arg-Thr-Leu-Gly-
-Leu-Phe-Glu-Tyr-Asp-Leu-Leu-Pro-Phe-Leu-Ser-Ser-Thr-Ile-Ser-
                                   40

                    20                                         30
-Pro-Phe-Tyr-Pro-Ser-Arg-Leu-Phe-Asp-Gln-Phe-Phe-Gly-Glu-Gly-
-Pro-Tyr-Tyr-Arg-Gln-Ser-Leu-Phe-Arg-Thr-Val-Leu-Asp-Ser-Gly-
                    50                                         60
```

Figure 14 Alignment of positions 1 to 30 and 31 to 60 of the bovine αA chain, which show a statistically significant sequence homology, indicating an ancient partial internal duplication (207). Dots are placed between residues which can be interchanged by a single base substitution.

some vertebrate lines more duplications may have taken place than in others, and duplicated genes may again have got lost in certain lines. Such a history of repeated, more or less recent, gene duplications, necessarily combined with eventual loss of expression of some of the genes, is a reminder of the relation between the multiple β-type hemoglobin chains in several vertebrates (209). As the genes for these β-like hemoglobin chains, which show up to 30% sequence difference (210), the genes for the γ-crystallins are probably still arranged in tandem on the same chromosome.

The clear but distant homology between calf βB_p and γ-crystallin (116) indicates again an ancient relationship by gene duplication, probably predating the origin of the vertebrates, considering the generally slow rate of evolution of the crystallins. The degree of homology between β- and γ-crystallins is certainly sufficient (202) to place them in the same superfamily of proteins. Before the gene duplication giving rise to the ancestral β- and γ-crystallin genes, an intragenic duplication must have caused almost a doubling of a still earlier gene. This is evident from the sequence homology between the *N*- and *C*-terminal halves of the βB_p chain, and also from the two domains of similar spatial configuration of which the γ_{II} chain is built up (122).

The matter of multiplicity of crystallin genes and their mutual relationships in the different vertebrates clearly still needs a lot of study. No doubt the possibilities of cloning and sequencing the crystallin genes will soon enable a better insight into the organization of the genes of not only α-crystallin but also the β-, γ-, and δ-crystallins.

6.7.3 Evolutionary Changes of Crystallins

The rate of evolutionary change of the αA chain places it among the slowly evolving proteins (211) (Table 6). The two investigated αB chains, from calf and human, give evidence for a similarly slow rate of change in this polypeptide (79). It is evident from the studies of Zigler and Sidbury

Table 6 Rates of Protein Evolution (211)

Protein	Unit evolutionary period[a]
Histone H4	400
Histone H2A	60
Glutamate dehydrogenase	55
Glucagon	43
Collagen α-1	36
Corticotropin	24
Lactate dehydrogenase H4	19
α-Crystallin A	17[b]
Cytochrome c	15
Insulin	14
Lactate dehydrogenase M4	13
Cytochrome b5	11
Histone H1	8
β-Lipotropin	8
Ferredoxin	6
Myoglobin	6
Trypsinogen	6
Prolactin	5
Parvalbumin	5
Carbonic anhydrase B	4
Hemoglobin α chain	3.7
Hemoglobin β chain	3.3
Albumin	3
Lysozyme	2.5
Ribonuclease	2.3
Lactalbumin	2.3
Insulin C-peptide	1.9
Immunoglobulin γ chains (C)	1.7
Casein (κ)	1.4
Fibrinopeptide B	1.1
Short neurotoxins (snake venom)	0.8
Immunoglobulin γ chains (V)	0.7

[a]Unit evolutionary period is the average time, in millions of years, required for a 1% difference in amino acid sequence to arise between two lineages. It is thus an inverse measure of rate of protein evolution.
[b]The value for α-crystallin A is based on ref. 45.

(32, 36, 85) that at least some of the β-crystallins, or certain regions of these chains, are also highly conserved during evolution. The rate of evolution might be even slower than that of α-crystallin, considering that β-crystallins, but not α-crystallins, of calf and dogfish show immunological cross-reactivity (32, 60). Among the several β-crystallin chains the β_H-specific polypeptides are apparently more variable (85, 110).

The evolutionary rate of the γ-crystallins is less well documented. Similarities in amino acid composition (Table 5) and very limited comparative sequence data (114, 123) support the idea that in general also the γ-crystallins evolve very slowly. A greater variation in the electrophoretic pattern of rodent γ-crystallins as compared to β-crystallins suggests that γ-crystallins evolve faster than β-crystallins (105). Lack of knowledge concerning the causes of such electrophoretic variation should, however, make one cautious in the interpretation of these data. The results of amino acid analyses and tryptic peptide mapping of avian and reptilian δ-crystallins indicate that this protein evolves considerably faster than α-crystallin (148b).

The generally slow rate of evolution of the crystallins is also in agreement with the observed absence or scarcity of electrophoretic polymorphisms in lens proteins of a variety of species (105, 106). Polymorphisms are only incidentally observed, and often involve minor components or quantitative differences, which cannot be interpreted in terms of the advanced molecular knowledge of the crystallins.

The reasons for the slow rate of evolution of the crystallins are difficult to envisage, especially since considerable variation exists in overall protein composition of different lenses. This variation occurs in the proportions of low and high molecular weight aggregates and in the charge properties of these aggregates. These variations are possible by the differential expression of the considerable number of crystallin genes that every species possesses, and by different degrees of post-translational modification. The occurrence of sets of duplicated genes thus enables the synthesis of chains of similar three-dimensional structure but differing in certain properties, like surface charge or tendency to aggregate.

The conservation of certain tertiary and quaternary structural properties may well be of great importance for the proper functioning of the crystallins. A regular spatial organization of different crystallins into large protein aggregates in the lens fiber cells has been made likely by birefringence and laser scattering studies (212, 213). Despite this ordering and the close packing of crystallin in the fiber cell, it appears that the magnitude of mean diffusivity of proteins in the normal intact lens is roughly the same as that of α-crystallin in solution (214).

Raman spectroscopy of intact bovine lenses has shown that lens proteins are organized in an antiparallel β-pleated sheet with the CONH bonds preferentially aligned in directions orthogonal to the lens optical axis. Indeed, no indication for the presence of either α-helix or random coil structure was found in the intact lens (215). Schachar and Solin (215) therefore suggest that β structure is necessary for transparancy of the lens. The predominance of α-helical structure in δ-crystallin, both *in situ* and in solution, has been thought responsible for the soft consistency of the bird lens and thus for the easy deformability required for rapid and efficient accommodation (137, 150).

The selective constraints imposed by the requirements of the spatial organization of lens proteins, including their close packing and resulting molecular interactions, are the main factors that can presently be thought to be related to their very slow rate of evolution.

6.8 CONCLUSIONS AND PROSPECTS

This chapter has shown the great variability of native lens proteins in properties such as charge and aggregate size and distribution. On the other hand, it has become clear that immunological determinants and amino acid sequences have been conserved to a considerable degree during the evolution of the vertebrate lens proteins. By virtue of duplications of a limited number of genes a range of related gene products, with slightly different physicochemical properties, exists in every species. Differential gene expression, combined with age- and species-dependent postsynthetic modifications of proteins, thus enable the production of a wide variety of lens proteins.

It is far from clear why such differences in lens protein patterns exist between different species. Nor is it understood how the variation in lens protein composition relates with the structure and functioning of the lens cell and the lens as a whole. Apparently the conservation of the primary structure, and hence the spatial configuration, of lens proteins is often so important that most amino acid replacements are selectively disadvantageous. However, it is as yet unknown which vital properties are affected by specific substitutions.

Unfortunately no conditions are known in man or domestic and laboratory animals in which genetically determined abnormalities in crystallin structure do occur, or where particular lens components are lacking, thus enabling the study of their function. If, as may be conceived, it is functional constraint that determines the slow rate of evolution of most lens proteins, one would predict a faster evolutionary change in genes for lens proteins in species that have been blind for long periods, such as moles and cave fishes. With the advent of molecular techniques for the direct isolation and analysis of particular genes, this question comes within the reach of experimental approach.

It is to be hoped that studies of the submicroscopic structure of the lens cell, of the arrangement and interaction of different crystallins, membrane components, and the cytoskeleton, will reveal the detailed molecular architecture of different types of lens in relation to their function. Comparative studies of lens proteins can certainly contribute to a deeper understanding of lens structure and function and of its evolutionary developments. The stage of more or less haphazard exploration of the variability of lens proteins among different species should, however, be over now. Future comparative work should be focused on clearly defined problems, inspired by a thorough and wide knowledge of lens morphology, physiology, and biochemistry.

REFERENCES

1 A. Rochon-Duvigneaud, *Les Yeux et la Vision des Vertébrés,* Masson & Cie., Paris, (1943).

2 R. M. Clayton, "Comparative Aspects of Lens Proteins," in H. Davson and L.T. Graham, Eds., *The Eye,* Vol. 5, Academic Press, New York, 1974, p. 399.

3 J. J. Harding and K. J. Dilley, Exp. Eye Res., *22,* 1 (1976).

4 S. K. Brahma, *J. Embryol. Exp. Morphol.,* **46,** 111 (1978).

5 C. R. Darwin, in F. Darwin, Ed., *The Autobiography of Charles Darwin and Selected Letters,* Murray, London, 1860.

6 A. Froriep, in O. Hertwig, Ed., *Handbuch der vergleichende und experimentelle Entwicklungslehre der Wirbeltiere,* Vol. 2, No. 2, G. Fischer, Jena, 1906, p. 139.

7 G. L. Walls, *The Vertebrate Eye and its Adaptive Radiation,* Hafner, New York-London, 1963 (reprinted from 1942 ed.).

8 S. Polyak, *The Vertebrate Visual System,* University of Chicago Press, 1957.

9 S. Duke-Elder, *System of Ophthalmology,* Vol. 1, *The Eye in Evolution,* H. Kimpton, London, 1958.

10 F. Crescitelli, in J. A. Dartnall, Ed., *Handbook of Sensory Physiology,* Vol VII, Part 1, Springer, Berlin-Heidelberg-New York, 1972, p. 245.

11 D. I. Hamasaki and D. J. Eder, in F. Crescitelli, Ed., *Handbook of Sensory Physiology,* Vol. VII, Part 5, Springer, Berlin-Heidelberg-New York, 1977, p. 497.

12 J. H. Prince, Comparative Anatomy of the Eye, Charles C Thomas, Springfield, Ill., 1956.

13 G. Westheimer, in M. G. F. Fuortes, Ed., *Handbook of Sensory Physiology,* Vol. VII, Part 2, Springer, Berlin-Heidelberg-New York, 1972, p. 449.

14 Grüsser-Cornehls and W. Hinstedt, in K. V. Fite, Ed., *The Amphibian Visual System,* Academic Press, New York-London, 1976, p. 203.

15 M. Bando, A. Nakajima, M. Nakagawa, and T. Hiraoka, *Exp. Eye Res.,* **22,** 389 (1976).

16 J. F. R. Kuck, in J. G. Bellows, Ed., *Cataract and Abnormalities of the Lens,* Grune & Stratton, New York, 1975, p. 69.

17 M. Rabaey, *Nature,* **198,** 206 (1963).

18 W. B. Lancaster, *Refraction and Motility,* Charles C Thomas, Springfield, Ill., 1952, pp. 97–108.

18a A. Hughes, in F. Crescitelli, Ed., *Handbook of Sensory Physiology,* Vol. VII, Part 5, Springer, Berlin-Heidelberg-New York, 1977, p. 613.

19 D. B. Meyer, in F. Crescitelli, Ed., *Handbook of Sensory Physiology,* Vol. VII, Part 5, Springer, Berlin-Heidelberg-New York, 1977, p. 549.

20 H. R. Sunderland and W. D. O'Neill, *Vision Res.,* **16,** 1151 (1976).

20a E. M. Prager and A. C. Wilson, *J. Biol. Chem.,* **246,** 5978 (1971).

21 H. Maisel and M. Goodman, *Am. J. Ophthalmol.,* **59,** 697 (1965).

22 H. Maisel, *Arch. Ophthal. N.Y.,* **72,** 829 (1964).

23 W. Manski and S. P. Halbert, *Invest. Ophthalmol.,* **4,** 539 (1965).

24 I. Björk, *Exp. Eye Res.,* **7,** 129–133 (1968).

25 W. E. Ferguson and V. L. Koenig, *Comp. Biochem. Physiol.,* **43B,** 151 (1972).

26 H. Hoenders, "Ueber die Proteinzusammensetzung der Augenlinse des Huhns," Thesis, University of Amsterdam, 1965.

27 M. Rabaey, A. Lagasse, and M. de Mets, *Acta Zool. Pathol. Antverpiensia,* **48,** 63 (1969).

28 A. F. van Dam, "Ooglenseiwitten, Isolering en Karakterisering," Thesis, University of Nijmegen, 1967.

29 W. F. Bon, G. Ruttenberg, A. Dohrn, and H. Batink, *Exp. Eye Res.,* **7,** 603 (1968).

30 W. W. de Jong, E. C. Terwindt, and G. Groenewoud, *Comp. Biochem. Physiol.,* **55B,** 49 (1976).

31 B. F. Cobb, L. Carter, and V. L. Koenig, *Comp. Biochem. Physiol.,* **24,** 817 (1968).

32 J. S. Zigler and J. B. Sidbury, *Ophthal. Res.,* **9,** 92 (1977).

33 W. Manski, S. P. Halbert, T. Auerbach-Pascal, and P. Javier, *Int. Arch. Allergy Appl. Immun.,* **31,** 38 (1967).

34 P. D. Mehta and S. Lerman, *Comp. Biochem. Physiol.,* **38A,** 637 (1971).

35 W. F. Bon, P. L. Swanborn, G. J. C. M. Ruttenberg, and A. Dohrn, *Pubbl. Staz. Zool. Napoli.,* **34,** 59 (1964).

36 J. S. Zigler and J. B. Sidbury, *Comp. Biochem. Physiol.,* **55B,** 19 (1976).

37 P. L. Swanborn, *Exp. Eye Res.,* **5,** 302 (1966).

38 S. K. Brahma and J. Bours, *Exp. Eye Res.,* **13,** 309 (1972).

39 S. K. Brahma and J. Bours, *Exp. Eye Res.,* **23,** 57 (1976).

40 J. R. Polanski and T. P. Bennett, *Biochem. Biophys. Res. Commun.,* **38,** 450 (1970).

41 R. Nöthiger, D. S. McDevitt, and T. Yamada, *Exp. Eye Res.,* **12,** 94 (1971).

42 T. Ogawa, *Embryologia,* **8,** 345 (1965).

43 D. S. McDevitt, I. Meza, and T. Yamada, *Dev. Biol.,* **19,** 581 (1969).

44 S. K. Brahma and W. J. van Doorenmaalen, *Exp. Eye Res.,* **8,** 168 (1969).

45 W. W. de Jong, J. T. Gleaves, and D. Boulter, *J. Mol. Evol.,* **10,** 123 (1977).

46 F. J. van der Ouderaa, W. W. de Jong, and H. Bloemendal, *Eur. J. Biochem.*, **39,** 207 (1973).

47 B. F. Cobb, L. Carter, and V. L. Koenig, *Comp. Biochem. Physiol.*, **26,** 519 (1968).

48 R. M. Clayton, *Exp. Eye Res.*, **8,** 326 (1969).

49 R. M. Clayton, in A. A. Moscona and A. Monroy, Eds., *Current Topics in Developmental Biology,* Vol. 5, Academic Press, New York, 1970, p. 115.

50 M. W. Rana and H. Maisel, *Exp. Eye Res.*, **8,** 216 (1969).

51 S. K. Brahma and H. van der Starre, *Exp. Cell Res.*, **97,** 175 (1976).

52 J. Thomson, C.E. Wilkinson, A. T. H. Burns, D. E. S. Truman, and R. M. Clayton, *Exp. Eye Res.*, **26,** 351 (1978).

53 J. Bours and W. J. van Doorenmaalen, *Exp. Eye Res.*, **13,** 236 (1972).

54 H. Maisel, J. Alcalá, N. Lieska, and N. Rafferty, *Ophthal. Res.*, **9,** 147 (1977).

55 D. C. Beebe and J. Piatigorsky, *Exp. Eye Res.*, **23,** 83 (1976).

56 J. Stauffer, C. Rothschild, T. Wandel, and A. Spector, *Invest. Ophthalmol.*, **13,** 135 (1974).

57 J. Delcour and H. Bouchet, *Exp. Eye Res.*, **26,** 191 (1978).

58 L. H. Cohen, D. P. E. M. Smits, and H. Bloemendal, *Eur. J. Biochem.*, **67,** 563 (1976).

59 B. F. Cobb, J. E. Price, and V. L. Koenig, *Comp. Biochem. Physiol.*, **28,** 841 (1969).

60 S. Lerman, *Can. J. Biochem.*, **47,** 1115 (1969).

61 H. O. Sandberg and O. Closs, *Exp. Eye Res.*, **27,** 61 (1978).

62 H. O. Sandberg and O. Closs, *Exp. Eye Res.*, **27,** 701 (1978).

63 P. Russell, D. Carper, and J. H. Kinoshita, *Exp. Eye Res.*, **27,** 673 (1978)

64 E. S. Platonov, L. P. Evseev, and B. V. Konyukhov, *Biokhimiya,* **40,** 923 (1975).

65 W. W. de Jong, E. C. Nuy-Terwindt, and M. Versteeg, *Biochim. Biophys. Acta,* **491,** 573 (1977).

66 W. W. de Jong and E. C. Terwindt, *Eur. J. Biochem.*, **67,** 503 (1976).

67 S. D. Coghlan and R. C. Augusteyn, *Exp. Eye Res.*, **25,** 603 (1977).

68 A. Spector, J. Stauffer, D. Roy, L. K. Li, and D. Adams, *Invest. Ophthalmol.*, **15,** 288 (1976).

68a J. Horwitz, *Exp. Eye Res.*, **23,** 471 (1976).

69 D. Roy and A. Spector, *Invest. Ophthalmol.*, **15,** 394 (1976).

70 K. J. Dilley and J. J. Harding, *Biochim. Biophys. Acta,* **386,** 391 (1975).

71 D. Roy and A. Spector, *Exp. Eye Res.*, **22,** 273 (1976).

72 D. Roy and A. Spector, *Biochemistry,* **15,** 1180 (1976).

73 J. A. Kramps, W. W. de Jong, J. Wollensak, and H. J. Hoenders, *Biochim. Biophys. Acta,* **533,** 487 (1978).

74 D. Roy and A. Spector, *Exp. Eye Res.*, **26,** 429 (1978).

75 D. Roy and A. Spector, *Exp. Eye Res.*, **26,** 445 (1978).

76 F. J. van der Ouderaa, W. W. de Jong, A. Hilderink, and H. Bloemendal, *Eur. J. Biochem.*, **49,** 157 (1974).

77 W. W. de Jong, F. J. van der Ouderaa, M. Versteeg, G. Groenewoud, J. M. van Amelsvoort, and H. Bloemendal, *Eur. J. Biochem.*, **53,** 237 (1975).

78 W. W. de Jong, E. C. Terwindt, and H. Bloemendal, *FEBS Lett.*, **58,** 310 (1975).

79 J. A. Kramps, B. M. de Man, and W. W. de Jong, *FEBS Lett.*, **74,** 82 (1977).

80 R. E. Dickerson, *J. Mol. Evol.*, **1,** 26 (1971).

81 R. J. Siezen and H. J. Hoenders, *Eur. J. Biochem.*, **96,** 431 (1979).

82 J. Bours, *J. Chromatogr.*, **60,** 225 (1971).

83 G. Armand, E. A. Balazs, and M. Testa, *Exp. Eye Res.*, **10,** 143 (1970).

84 J. Horwitz, I. Kabasawa, and J. H. Kinoshita, *Exp. Eye Res.*, **25,** 199 (1977).

85 J. S. Zigler and J. B. Sidbury, *Comp. Biochem. Physiol.*, **53B,** 349 (1976).

86 J. S. Zigler and J. B. Sidbury, *Invest. Ophthalmol.*, **15,** 673 (1976).

87 D. E. S. Truman and R. M. Clayton, *Exp. Eye Res.*, **18,** 485 (1974).

88 D. C. Beebe and J. Piatigorsky, *Exp. Eye Res.*, **22,** 237 (1976).

89 I. Thomson, C. E. Wilkinson, J. F. Jackson, D. I. de Pomerai, R. M. Clayton, D. E. S. Truman, and R. Williamson, *Dev. Biol.*, **65,** 372 (1978).

90 R. M. Clayton and D. E. S. Truman, *Exp. Eye Res.*, **18,** 495 (1974).

91 J. Zwaan and A. Ikeda, *Exp. Eye Res.*, **7,** 301 (1968).

92 P. R. Waggoner, N. Lieska, J. Alcalá, and H. Maisel, *Ophthal. Res.*, **8,** 292 (1976).

93 J. M. Génis-Gálvez, H. Maisel, and J. Castro, *Exp. Eye Res.*, **7,** 593 (1968).

94 J. Bours and O. Hockwin, *Klin. Mbl. Augenheilk.*, **170,** 51 (1977).

95 M. Rabaey, J. Rikkers, and M. de Mets, *Exp. Eye Res.*, **14,** 208 (1972).

96 S. K. Brahma and W. J. van Doorenmaalen, *Exp. Eye Res.*, **17,** 341 (1973).

97 D. S. McDevitt and L. R. Croft, *Exp. Eye Res.*, **25,** 473 (1977).

98 W. Manski, T. P. Auerbach, and S. P. Halbert, *Am. J. Ophthalmol.*, **50,** 985 (1960).

99 G. J. van Kamp and H. J. Hoenders, *Exp. Eye Res.*, **17,** 417 (1973).

100 P. R. Ocken, S. C. J. Fu, R. Hart, J. H. White, B. J. Wagner, and K. E. Lewis, *Exp. Eye Res.*, **24,** 355 (1977).

101 F. A. M. Asselbergs, M. Koopmans, W. J. van Venrooij, and H. Bloemendal, *Exp. Eye Res.*, **28,** 223 (1979).

102 K. N. Liem-The and H. J. Hoenders, *Exp. Eye Res.*, **18,** 143 (1974).

103 M. K. Mostafapour and V. N. Reddy, *Invest. Ophthalmol.*, **17,** 660 (1978).

104 J. Bours and O. Hockwin, *Berl. Muench. Tieraerztl. Wochenschr.*, **89,** 417 (1976).

105 T. H. Day, *Comp. Biochem. Physiol.*, **43B,** 1019 (1972).

106 T. H. Day and R. M. Clayton, *Biochem. Genet.*, **8,** 187 (1973).

107 J. Bours, in B. J. Radola and D. Graesslin, Eds., *Electrofocusing and Isotachophoresis,* de Gruyter, Berlin-New York, 1977, p. 303.

107a J. Bours, *Comp. Biochem. Physiol.*, **65B,** 215 (1979).

108 J. Bours, *Interdiscipl. Topics Gerontol.*, **12,** 196 (1978).

109 J. Bours, K. Doepfmer, and O. Hockwin, *Doc. Ophthalmol. Proc. Ser.*, **8,** 75 (1976).

110 F. C. S. Ramaekers, P. L. E. van Kan, and H. Bloemendal, *Ophthal. Res.*, **11,** 143 (1979).

111 J. S. Zigler, *Exp. Eye Res.*, **26,** 537 (1978).

112 P. Herbrink, H. van Westreenen, and H. Bloemendal, *Exp. Eye Res.*, **20,** 541 (1975).

113 A. F. van Dam, *Exp. Eye Res.*, **7,** 611 (1968).

114 L. R. Croft, *CIBA Found. Symp.*, **19** (new ser.), 207 (1973).

115 P. Herbrink, "The Polypeptide Composition of β-Crystallin," Thesis, University of Nijmegen, 1976.

116 H. P. C. Driessen, P. Herbrink, H. Bloemendal, and W. W. de Jong, *Exp. Eye Res.*, **31,** 243 (1980).

117 C. Slingsby and L. R. Croft, *Exp. Eye Res.*, **26,** 291 (1978).

118 L. R. Croft, *J. Chem. Soc. Chem. Commun.*, **1972,** 437.

119 H. A. Jones and S. Lerman, *Can. J. Biochem.*, **49,** 426 (1971).

120 L. K. Li, *Exp. Eye Res.*, **27,** 553 (1978).

121 I. Björk, *Exp. Eye Res.*, **3,** 254 (1964).

122 T. L. Blundell, P. F. Lindley, D. S. Moss, C. Slingsby, I. J. Tickle, and W. G. Turnell, *Acta Cryst.*, **B34,** 3653 (1978).

123 L. R. Croft, *Biochim. Biophys. Acta*, **295,** 174 (1973).

124 S. K. Brahma and D. S. McDevitt, *Exp. Eye Res.*, **19,** 379 (1974).

125 D. S. McDevitt and C. R. Collier, *Exp. Eye Res.*, **21,** 1 (1975).

126 D. S. McDevitt, *Science*, **175,** 763 (1972).

127 M. C. Hines and J. A. Olive, *Life Sci. (Part II)*, **9,** 1101 (1970).

128 B. J. Wagner and S. C. J. Fu, *Exp. Eye Res.*, **26,** 255 (1978).

129 A. Spector, J. Stauffer, and J. Sigelman, *CIBA Found. Symp.*, **19** (new ser.), 185 (1973).

130 I. Kabasawa, Y. Tsunematsu, G. W. Barber, and J. H. Kinoshita, *Exp. Eye Res.*, **24,** 437 (1977).

131 L. R. Croft, *Biochem. J.*, **128,** 961 (1972).

132 W. W. de Jong, A. Zweers, and L. H. Cohen, *Biochem. Biophys, Res. Commun.*, **82,** 532 (1978).

133 S. Zigman, J. Schultz, and T. Yulo, *Biochim. Biophys. Acta*, **221,** 576 (1970).

134 C. Slingsby and L. R. Croft, *Exp. Eye Res.*, **17,** 369 (1973).

135 M. Rabaey, *Exp. Eye Res.*, **1,** 310 (1962).

136 H. Gysels, *Experientia*, **20,** 145 (1964).

136a L. A. Williams and J. Piatigorsky, *Eur. J. Biochem.*, **100,** 349 (1979).

137 N. T. Yu, E. J. East, R. C. C. Chang, and J. F. R. Kuck, *Exp. Eye Res.*, **24,** 321 (1977).

138 J. Bours, in N. Catsimpoolas, Ed., *Isolectric Focusing*, Academic Press, New York, 1976, p. 209.

139 J. Piatigorsky, *Exp. Eye Res.,* **21,** 245 (1975).

140 J. Piatigorsky, P. Zelenka, and R. T. Simpson, *Exp. Eye Res.,* **18,** 435 (1974).

141 R. Reszelbach, T. Shinohara, and J. Piatigorsky, *Exp. Eye Res.,* **25,** 583 (1977).

142 J. Piatigorsky, *J. Biol. Chem.,* **251,** 4416 (1976).

143 J. Piatigorsky, *Exp. Eye Res.,* **27,** 227 (1978).

143a S. P. Bhat and J. Piatigorsky, *Proc. Natl. Acad. Sci. U.S.,* **76,** 3299 (1979).

143b S. P. Bhat, R. E. Jones, M. A. Sullivan, and J. Piatigorsky, *Nature,* **284,** 234 (1980).

144 T. Shinohara, *Dev. Growth Diff.,* **17,** 127 (1975).

145 T. Shinohara and J. Piatigorsky, *Proc. Natl. Acad. Sci. U.S.,* **73,** 2808 (1976).

146 P. Zelenka and J. Piatigorsky, *Exp. Eye Res.,* **22,** 115 (1976).

147 H. Watanabe and I. Kawakami, *Exp. Eye Res.,* **17,** 205 (1973).

148 J. Alcalá, H. Maisel, and N. Lieska, *Exp. Cell Res.,* **109,** 63 (1977).

148a W. W. de Jong, S. O. Stapel, and A. Zweers, *Comp. Biochem. Physiol.,* (in press).

149 H. Maisel, N. Lieska, and J. Alcalá, *Ophthal. Res.,* **7,** 416 (1975).

150 J. F. R. Kuck, E. J. East, and N. T. Yu, *Exp. Eye Res.,* **23,** 9 (1976).

151 D. I. de Pomerai, D. J. Pritchard, and R. M. Clayton, *Dev. Biol.,* **60,** 416 (1977).

152 J. F. Jackson, R. M. Clayton, R. Williamson, I. Thomson, D. E. S. Truman, and D. I. de Pomerai, *Dev. Biol.,* **65,** 383 (1978).

153 V. M. Barabanov, *Dokl. Biol. Sci. Sect. (Engl. Transl.,)* **234,** 181 (1977).

154 W. G. M. van den Broek, J. N. Leget, and H. Bloemendal, *Biochim. Biophys. Acta,* **310,** 278 (1973).

155 S. K. Brahma and W. J. van Doorenmaalen, *Experientia,* **32,** 930 (1976).

156 S. K. Brahma, *Exp. Eye Res.,* **25,** 311 (1977).

157 J. R. Reddan, in J. G. Bellows, Ed., *Cataract and Abnormalities of the Lens,* Grune & Stratton, New York, 1975, p. 29.

158 CIBA Found. Symp., *19* (new ser.) (1973).

159 M. J. Keating and C. Kennard, in K. V. Fite, Ed., *The Amphibian Visual System,* Academic Press, New York-London, 1976, p. 267.

160 L. B. Arey, in J. G. Bellows, Ed., *Cataract and Abnormalities of the Lens,* Grune & Stratton, New York, 1975, p. 42.

161 A. T. Mikha'ilov, *Ontogenez,* **9,** 211 (1978).

162 G. Reeve and A. E. Wild, *J. Embryol. Exp. Morphol.,* **48,** 205 (1978).

163 K. Yasuda, T. S. Okada, G. Eguchi, and M. Hayashi, *Exp. Eye Res.,* **26,** 591 (1978).

164 R. W. Reyer, in F. Crescitelli, Ed., *Handbook of Sensory Physiology,* Vol. VII, Part 5, Springer, Berlin-Heidelberg-New York, 1977, p. 309.

165 G. Eguchi and T. S. Okada, *Proc. Natl. Acad. Sci. U.S.,* **70,** 1495 (1973).

166 T. S. Okada, K. Yasuda, M. Hayashi, Y. Hamada, and G. Eguchi, *Dev. Biol.,* **60,** 305 (1977).

167 M. van de Kamp and J. Zwaan, *J. Exp. Zool.*, **186,** 23 (1973).

168 A. Ikeda, Y. Seki, and J. Sawano, *Kawasaki Med. J.*, **2,** 155 (1976).

169 J. W. McAvoy, *J. Embryol. Exp. Morphol.*, **45,** 271 (1978).

169a P. J. M. Leenen and W. W. de Jong, *Exp. Eye Res.*, (in press).

170 T. Yamada, *Am. Zool.*, **6,** 21 (1966).

171 J. R. Polansky and T. P. Bennett, *Dev. Biol.*, **33,** 380 (1973).

172 M. J. Doyle and N. Maclean, *J. Embryol. Exp. Morphol.*, **46,** 215 (1978).

173 A. T. Mikha'ilov and N. D. Mgvdeladze, *Ontogenez,* **8,** 297 (1977).

174 J. W. McAvoy, *J. Embryol. Exp. Morphol.*, **44,** 149 (1978).

175 A. J. M. Vermorken and H. Bloemendal, *Nature,* **271,** 779 (1978).

176 J. M. Génis-Galvez and J. M. Castro, *J. Exp. Zool.*, **177,** 313 (1971).

177 D. S. McDevitt and R. M. Clayton, *J. Cell. Biol.*, **67,** 273a (1975).

178 J. C. Campbell and D. E. S. Truman, *Exp. Eye Res.*, **25,** 101 (1977).

178a C. J. J. Brinkman, M. P. J. Oerlemans-van Zutphen, and R. M. Broekhuyse, *Exp. Eye Res.*, **29,** 609 (1979).

179 H. O. Sandberg and O. Closs, *Exp. Eye Res.*, **27,** 73 (1978).

180 R. M. Clayton, in B. I. Lord, C. S. Potten, and R. J. Cole, Eds., *Stem Cells and Tissue Homeostasis,* Cambridge University Press, London, 1978.

181 J. B. Coulter, A. A. Swanson, T. J. Bond, and J. Jeter, *Ophthal. Res.*, **4,** 367 (1972/73).

182 J. Bours and S. K. Brahma, *Exp. Eye Res.*, **16,** 131 (1973).

183 J. A. Jedziniak, H. Baram, and L. T. Chylack, *Exp. Eye Res.*, **26,** 377 (1978).

184 W. W. de Jong, F. S. M. van Kleef, and H. Bloemendal, *Eur. J. Biochem.*, **48,** 271 (1974).

185 A. B. Champion, E. M. Prager, D. Wachter, and A. C. Wilson, in C. A. Wright, Ed., *Biochemical and Immunological Taxonomy of Animals,* Academic Press, London-New York, 1974, p. 397.

186 R. Hinegardner, in F. J. Ayala, Ed., *Molecular Evolution,* Sinauer, Sunderland, Mass., 1976, p. 179.

187 M. Goodman and R. E. Tashian, Eds., *Molecular Anthropology,* Plenum, New York-London, 1976.

188 M. O. Dayhoff, Ed., *Atlas of Protein Sequence and Structure,* Vol. 5, National Biomedical Research Foundation, Silver Spring, Md., 1972.

189 W. Hennig, *Phylogenetic Systematics,* University of Illinois Press, Urbana, 1966.

190 C. A. Wright, Ed., *Biochemical and Immunological Taxonomy of Animals,* Academic Press, London-New York, 1974.

191 W. Manski and K. Malinowski, *Immunochemistry,* **15,** 781 (1978).

192 B. Blake, *Comp. Biochem. Physiol.*, **54B,** 441 (1976).

193 A. C. Smith, *Calif. Fish Game,* **51,** 163 (1965).

194 F. J. O'Rourke, in C. A. Wright, Ed., *Biochemical and Immunological Taxonomy of Animals,* Academic Press, London-New York, 1974, p. 243.

194a S. L. Beck and H. V. Bennet, *Exp. Eye Res.*, **29,** 323 (1979).

195 H. C. Dessauer, in C. A. Wright, Ed., *Biochemical and Immunological Taxonomy of Animals,* Academic Press, 1974, p. 177.

196 H. Gysels, *Natuurwet. Tijdschr.,* **46,** 43 (1964).

197 H. Gysels and M. Rabaey, *Ibis,* **106,** 536 (1964).

198 C. G. Sibley and A. H. Brush, *Auk,* **84,** 203 (1967).

199 H. Gysels, *Experientia,* **30,** 1253 (1974).

200 W. W. de Jong, J. M. van Amelsvoort, F. J. van der Ouderaa, and H. Bloemendal, *Nature New Biol.,* **246,** 233 (1973).

201 W. W. de Jong, A. Zweers, K. A. Joysey, J. T. Gleaves, and D. Boulter, in G.G. Montgomery, Ed., *The Evolution and Ecology of Sloths, Anteaters, and Armadillos,* Smithsonian Inst. Press, Washington, D.C. (in press).

202 M. O. Dayhoff, W. C. Barker, and L. T. Hunt, in M. O. Dayhoff, Ed., *Atlas of Protein Sequence and Structure,* Vol. 5, Suppl. 2, National Biomedical Research Foundation, Washington D.C., 1976, p. 9.

203 S. Ohno, *Evolution by Gene Duplication,* Springer, New York, 1970.

204 J. Zwaan and A. Ikeda, in J. W. Rohen, Ed., *The Structure of the Eye (II Symp.),* Schattauer-Verlag, Stuttgart, 1965, p. 149.

205 M. Goodman, G. W. Moore, and H. Matsuda, *Nature,* **253,** 603 (1975).

206 A. Deisseroth, A. Nienhuis, J. Lawrence, R. Giles, P. Turner, and F. H. Rudale, *Proc. Natl. Acad. Sci. U.S.,* **75,** 1456 (1978).

207 W. C. Barker, L. K. Ketcham, and M. O. Dayhoff, *J. Mol. Evol.,* **10,** 265 (1978).

208 L. H. Cohen, L. W. Westerhuis, W. W. de Jong, and H. Bloemendal, *Eur. J. Biochem.,* **89,** 259 (1978).

209 V. M. Ingram, *Nature,* **189,** 704 (1961).

210 R. A. Flavell, J. M. Kooter, E. de Boer, P. F. R. Little, and R. Williamson, *Cell,* **15,** 25 (1978).

211 A. C. Wilson, S. S. Carlson, and T. J. White, *Ann. Rev. Biochem.,* **46,** 573 (1977).

212 F. A. Bettelheim, *Exp. Eye Res.,* **21,** 231 (1975).

213 F. A. Bettelheim, *J. Colloid Interface Sce.,* **63,** 251 (1978).

214 T. Tanaka and G. B. Benedek, *Invest. Ophthalmol.,* **14,** 449 (1975).

215 R. A. Schachar and S. A. Solin, *Invest. Ophthalmol.,* **14,** 380 (1975).

CHAPTER
7

Aging of Lens Proteins

Herman J. Hoenders and Hans Bloemendal

Department of Biochemistry, University of Nijmegen, Nijmegen, The Netherlands

7.1 INTRODUCTION

For many years aging processes of eye lens proteins have aroused the interest of various investigators. Aging phenomena of the lens have been discussed in review articles by van Heyningen (1), Harding and Dilley (2), Hockwin et al. (3), Bloemendal (4, 5), Nordmann (6), and Dreyfus et al. (7). Whether we are confronted with age-related alterations, namely, secondary effects, rather than with primary mechanisms of aging, has still to be established. In other words, one can observe the changes that take place as a consequence of aging, but the primary impulses that induce this process have still to be found (4).

Owing to its unique growth pattern (cf. Chapters 2 and 10), the cellular, subcellular, and molecular history of the eye lens is preserved in the adult and old tissue. It has become clear that "aging" of the eye lens is reflected by pronounced changes on the level of its structural proteins, the crystallins. Therefore, besides aging studies of the immune system, liver, muscle and nerve cells, erythrocytes, fibroblasts, collagen, elastin, and the nematode *Turbatrix aceti,* investigation of lens proteins promises to shed light on the secrets of the primary causes, the mechanism, and the immediate effects of the aging process.

A multitude of theories on mechanisms of aging has been put forward. One can distinguish, however, two alternative general views. The first holds that aging is a consequence of a developmental program, which means that the time-related deterioration of biological systems is under direct genetic control. The second view postulates that aging is the result of the accumulation of random damage to cellular structures and to tissue desorganization (8). Although today experimental evidence is available both in support of the genetic, "programmed" and the stochastic, "error accumulation" theory of aging, conclusive evidence in favor or against either type of view is still lacking.

The possible genetic origin of aging seems to be logical but so far no adequate experiments have been designed to prove or to reject this theory.

Examples of questions that should be answered are (9): Is there a locus on the genome that programs the senescence and death of an organism? Can the same genes that regulate development also lead to senescence when development is completed? If aging and longevity are inherited traits, how did they evolve?

Aging may commence at the transcriptional level, where it can be manifested as a mutation, DNA deletion, or cross-linking of DNA. It may also initiate at the translational level where it can be manifested by altered RNA polymerase, tRNA, or tRNA synthetase, for example. Furthermore, it may arise at the post-translational level, where it can be manifested by stochastic alteration of certain vital, slowly turning-over macromolecules (10).

The well-known "error catastrophe" hypothesis of Orgel (11) proposes that age-related changes in proteins occur during translation of mRNA. A low frequency of errors during protein synthesis would induce a rapidly increasing number of errors in proteins synthesized thereafter. The error theory can claim few nondisputable facts to support it, since most observations can be better explained by post-translational changes. It deserves credit, however, as being experimentally testable (12).

Post-translational modifications can account for most abnormalities found in proteins of aged cells. At present post-translational modifications do not constitute a theory of aging and they may be only a transition step in the normal catabolism of proteins (12). The possible mechanisms are still hypothetical. Only one reaction is well defined, namely, the deamidation of asparagine and glutamine residues. According to Robinson et al. (13–15) controlled deamidation of proteins may serve as a general biological timer or clock. The longevity of a protein would be determined by loss of amide groups via conformational change and increased susceptibility to proteolysis. Proteins with many unstable amide groups would have short half-lives and *vice versa*.

Harding (16) attempted to relate changes found in lens proteins in aging and in cataract to both the translational error catastrophe hypothesis and the deamidation hypothesis on the basis of experimental findings. His conclusion was that the changes described in the lens appear to be post-translational and not transcriptional or translational.

It is now generally recognized that not only the biosynthesis but also the intracellular breakdown of a protein is a fundamental process (17). The rate of degradation seems to be related to certain properties of the protein such as the compactness of folding, the subunit structure, the hydrophobicity, the charge, and the size.

The crystallins are good examples to show that protein degradation is highly specific (18). But not only selective shortening of polypeptide chains has been observed. A number of chemically and physically defined, partly interrelated processes have been described in lens proteins upon aging. The best-known age-related processes are as follows:

- Superaggregation.
- Insolubilization.
- Decrease of water content.
- Deamidation of asparagine and glutamine residues.
- Selective degradation from the *C*-terminus.
- Racemization of aspartyl residues.
- Formation of disulfide bonds.
- Formation of nondisulfide cross-links.

Moreover, radical reactions as intermediate steps in photooxidation and lipid peroxidation are put forward as possible causes of aging of lens proteins. Also, the formation of lipofuscin (aging pigment) has been correlated with aging of the lens. Finally, one may wonder whether chalones, antimitotic chemical messengers, may be involved in the aging process (19).

In general only the literature after 1975 has been considered in this chapter because preceding reports on aging of lens proteins have been surveyed by Harding and Dilley (2). Aging *in vitro* (lens culture) and aging of histones and other chromosomal proteins are not discussed in this chapter. For aging of fiber membrane polypeptides, the reader is referred to Chapter 1.

7.2 CHANGES IN DISTRIBUTION AND COMPOSITION OF WATER-SOLUBLE PROTEINS DURING AGING

7.2.1 Bovine and Dog Lens

In the past the distribution of the crystallins was particularly well studied by application of isoelectric focing. Hockwin and his colleagues especially made great efforts to determine carefully the crystallin composition as a function of age, using bovine lenses up to an age of 25 years.

By means of isoelectric focusing it was possible to discern 12 β- and four γ-crystallins (3, 20–22). The absolute amounts of several (groups of) crystallins in 0- to 25-year-old bovine lenses are plotted in Figure 1. A drawback of the method used is the precipitation of α-crystallin, which causes the appearance of a smear on the gel after staining. Furthermore, the high molecular weight crystallins do not penetrate into the gel and cannot be quantified.

Separation of the lens into four parts, equator, anterior cortex, posterior cortex, and nucleus, revealed a more detailed picture with respect to the age-related concentration changes of the various crystallins and their subgroups (Figure 2). The authors conclude that the continuous change of the crystallin composition during aging is caused by three different processes:

1 Aggregation of α- and probably β-crystallins to higher molecular weight complexes.
2 Loss of the ability of the equator and cortex to synthesize certain (especially γ-) crystallins.
3 Postsynthetic loss or conversion of single crystallin components.

As far as the study of aggregation is concerned, it is clear that isoelectric focusing is not the adequate method.

By gel chromatography on Sepharose 6-B it was possible to discern between the high and low molecular weight α- and β-crystallins (23). The distribution of the crystallins and the water-insoluble fraction in four parts of 2- to 10-year-old bovine lenses are summarized in Figure 3. Because of the different methods of separation and the different ways of presentation it is virtually impossible to compare the results obtained by isoelectric focusing and gel chromatography. Moreover, no conclusion can be drawn concerning the distribution of γ-crystallins since ribonucleic acids of comparable molecular weight such as tRNA interfere in gel chromatography. Dog (beagle) lenses of approximately 1 to 10 years were investigated by isoelectric focusing (20, 24). In general, isoelectric focusing and distribution patterns similar to those of bovine lenses were obtained.

7.2.2 Rabbit Lens

In a study on the development of X-ray cataract in New Zealand rabbits (25), unirradiated lenses of 3-, 4-, and 5-month-old rabbits showed some significant changes in the composition of the water-soluble fraction (Table 1). Using additional data, the authors were able to show that the absolute amounts in the total water-soluble fraction of 4- and 5-month-old lenses showed no change for α-crystallin (50.5/49.7 mg) and γ-crystallin (29.3/29.9 mg) and an increase for HM crystallin (7.66/9.72 mg) and β-crystallin (72.0/80.1 mg).

7.2.3 Rat Lens

Age-related changes in the distribution of the crystallins were studied by isoelectric focusing. Bours et al. (26) compared 98-, 158-, 259-, 325-, and 781-day-old Wistar rat lenses. Out of eight γ-crystallin components five

---

Figure 1 Crystallin concentration (in mg) in bovine lenses as a function of age. (*a*) Plot of concentration of α-, β-, and γ-crystallins as total; Σ, total of the soluble crystallins. (*b*) Plot of the concentration of β-crystallin components: β_{1-9} with high molecular weight, β_{10}, β_{11}, β_{12} with low molecular weight. (*c*) Plot of concentration of γ-crystallin components γ_1, γ_2, γ_3, and γ_4, and total of γ-crystallin δ_T. Reprinted with permission from ref. 21.

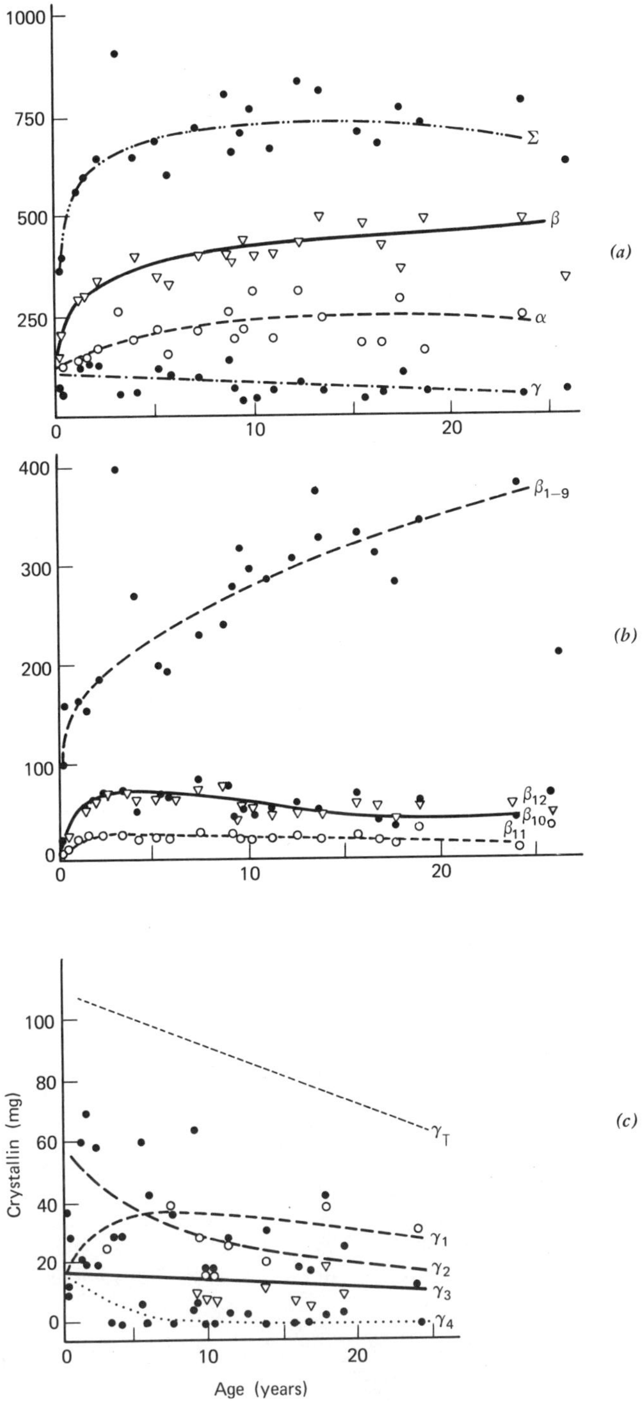

1000
750
500
250
Σ
β
α
γ
0
10
20
(a)
400
300
200
100
β1–9
β12
β10
β11
(b)
Crystallin (mg)
100
80
60
40
20
γT
γ1
γ2
γ3
γ4
(c)
Age (years)

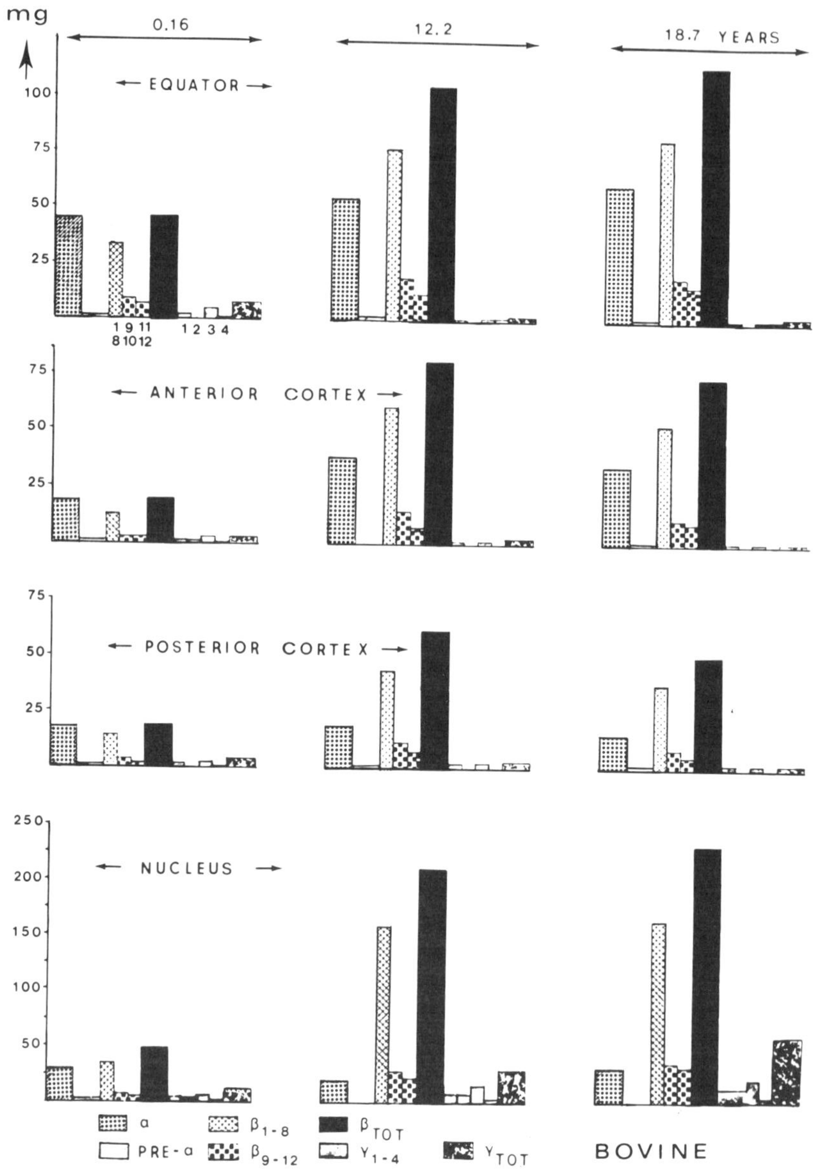

Figure 2 Absolute amount of crystallin (in mg) in different parts of the bovine lens in dependence on age. Values are obtained by evaluation of densitometer data after isoelectric focusing, with respect to the dry weight of the soluble crystallins. Reprinted with permission from ref. 24.

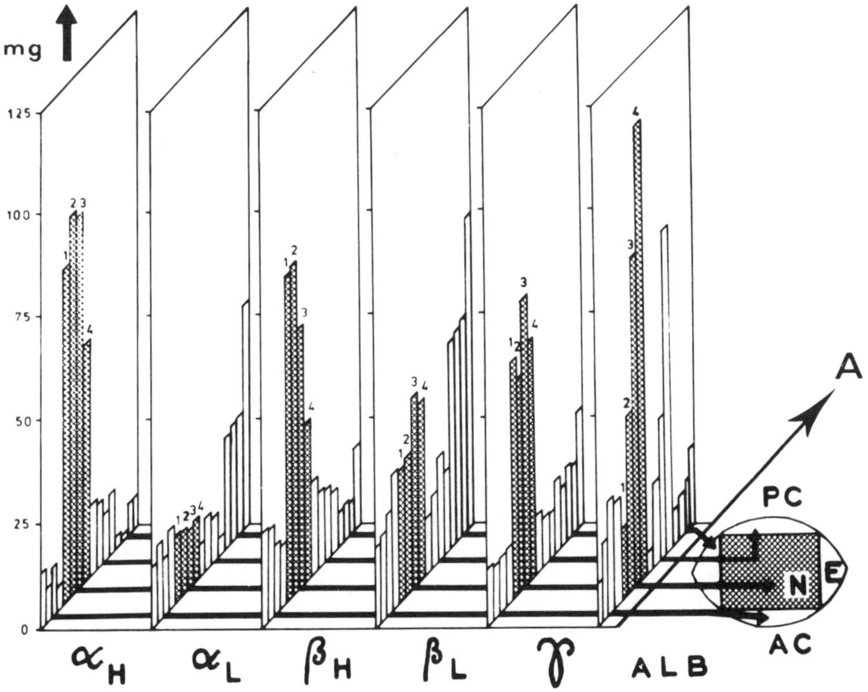

Figure 3 Topographic distribution of the water-soluble crystallins and albuminoid (ALB) in milligram amounts in one bovine lens of indicated age. AC, Anterior cortex; N, nucleus; PC, posterior cortex; E, equator; 1, 2.26(2.17)-year-old cow; 2, 3.83-year-old cow; 3, 5.18(5.60)-year-old cow; 4, 10.45-year-old cow; A, the direction of increasing age. Reprinted with permission from ref. 23.

Table 1 Relative Proportion (%) of Crystallins in Cortex and Nucleus of Rabbit Lenses

Lens part	Crystallin	Lens age (months)		
		3	4	5
Cortex (outer 40%)	HM	0.90	1.26	0.60
	α	18.9	18.1	17.4
	β	59.7	61.9	63.4
	γ	20.5	18.7	18.6
Nucleus (inner 20%)	HM	7.8	11.7	20.4
	α	16.2	15.8	11.5
	β	37.8	41.3	42.4
	γ	38.2	32.1	25.7

Taken from ref. 39.

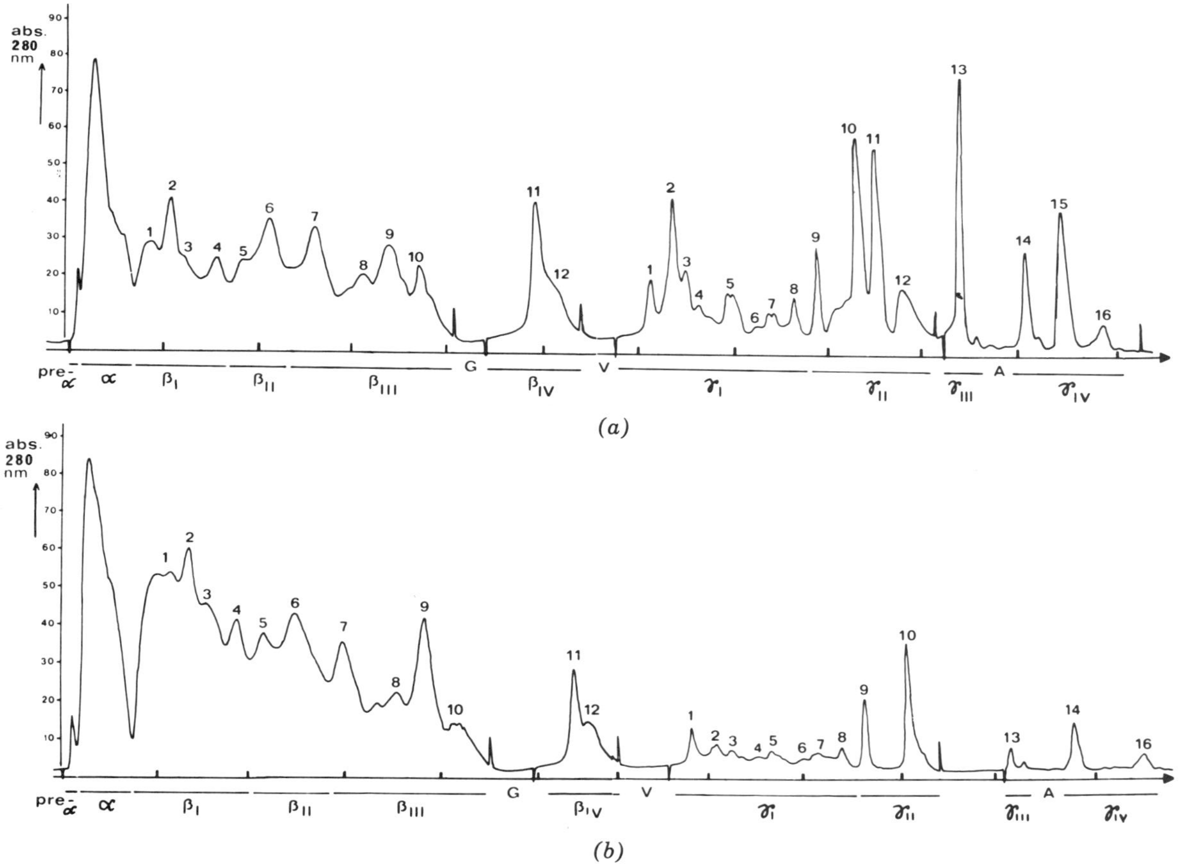

Figure 4 Patterns of capillary isotachophoretic analysis of (*a*) 86-day-old rat lens and (*b*) 723-day-old rat lens. Both samples, amount of protein 60 μg. β_{I-IV} and γ_{I-IV} denote the groups of β- and γ-crystallin components. The Arabic numerals 1 to 12 and 1 to 16 indicate the single crystallin components separated. Pre-α, pre-α-crystallin; α, α-crystallins; β, β-crystallins; γ, γ-crystallins; G, Glycine; V, Valine; A, β-alanine. Reprinted with permission from ref. 27.

disappeared with aging, and a decrease in the β-crystallin components of higher isoelectric points (β_{10}–β_{12}) was observed.

In another investigation Bours (27) applied isoelectric focusing and isotachophoresis. The latter method revealed a much higher resolution of the β- and γ-crystallin components than the first one. Moreover, the drawbacks of isoelectric focusing, namely, precipitation of α-crystallin and the inaccessibility of the polyacrylamide gel for high molecular weight crystallins, were not observed in capillary isotachophoresis. At least 30 components were detected and analysis times ranged from 12 to 30 min. Thus isotachophoresis may be considered as a promising technique for the analysis of lens proteins. An example of the patterns obtained with 86- and 723-day-old lenses is given in Figure 4. A similar study of 30- to 793-day-old mouse lenses has recently been reported by Bours and Delmotte (27a).

7.2.4 Human Lens

Changes in the distribution of α-, β-, and γ-crystallin as a function of age have been given by Lerman (28) (Table 2).

Coghlan and Augusteyn (29) studied changes in the distribution of proteins in 28 normal human lenses of ages from 0 to 90 years by means of fractionation on DEAE cellulose. They found that the rate of extraction varied with age (Figure 5) and they adapted their extraction procedure to this phenomenon. Often ignored differences in absorption coefficient at 280 nm were avoided by the use of the Lowry method and amino acid analysis for protein concentration determinations. The fractionation on DEAE cellulose could be scaled down to 0.5 mg and the protein eluted with NaOH (increasing from 10% at birth to 25% at 90 years) was taken into account. It was found that the levels of the crystallins decreased linearly with age whereas the protein eluted with NaOH increased (Figure 6). The amino acid composition of the NaOH fraction most closely resembled that of the

Table 2 Age-Related Changes in the Concentration of Human α-, β-, and γ-Crystallins

	Concentration (%)[a]		
Age (years)	α-crystallin (%)	β-crystallin (%)	γ-crystallin (%)
10 to 19	37	38	25
60 to 69	55	33.6	11.4
80 to 89	57.6	33.7	8.7

Taken from ref. 28.
[a]Expressed as percentage of total soluble proteins.

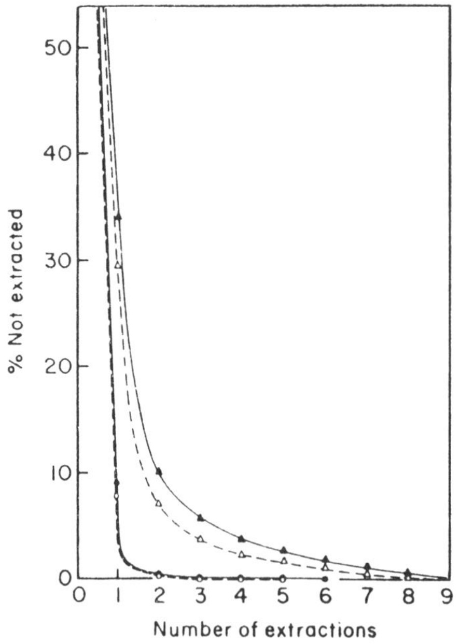

Figure 5 Extraction of soluble proteins and glutathione reductase from 0- and 78-year-old lenses as a function of the number of extractions. ▲—▲, 78-year-old protein; ●——●, 0-year-old protein; △——△, 78-year-old glutathione reductase; ○——○, 0-year-old glutathione reductase. Results are expressed as percentage of the total protein or glutathione reductase activity that could be solubilized with 10 extractions. Reprinted with permission from ref. 29. Copyright by Academic Press Inc. (London) Ltd.

α-crystallins, but certain differences in amino acid levels (Glu, Gly, Tyr) suggested that this fraction might be a mixture of all the crystallins.

Roy and Spector (30) examined old human lenses (60 to 70 years) by gel chromatography on Bio-Gel A-1.5m and found no low molecular weight α-crystallins in the inner 30 to 40% (according to these authors, the nucleus), whereas the periphery contained normal levels of this protein. Since in the young human lens low molecular weight α-crystallin represents approximately 20% of the total soluble protein, the question arises: what has been its fate upon aging? In addition to the obvious suggestion that it has shifted to high molecular weight aggregates and to the water-insoluble fraction, the authors discuss the possibility that the αA chains are subject to preferential modification or destruction and thus they might be the source for the reported 43,000 and 11,000 dalton polypeptides.

The absence of low molecular weight α-crystallins in the nucleus of old human lenses (30) was not found by other authors (31–33). Extracts of cortex and nucleus from normal human lenses, ranging in age from 50 to 80 years, were fractionated on a Bio-Gel A-5m column. These extracts (as well as those from cataractous lenses) showed essentially the same ratio of high and low molecular weight α-crystallin. Fractionation of cortical and nuclear water-soluble proteins on Sephadex G200 revealed five fractions (α-, β_1-, β_2-, β_3-, and γ-crystallin) and a final fraction containing mainly nonprotein material. Comparison of the cortical and nuclear extracts shows considerable differences in elution patterns (Figure 7). Essentially the same distribution patterns have been obtained by Ringens et al. (34).

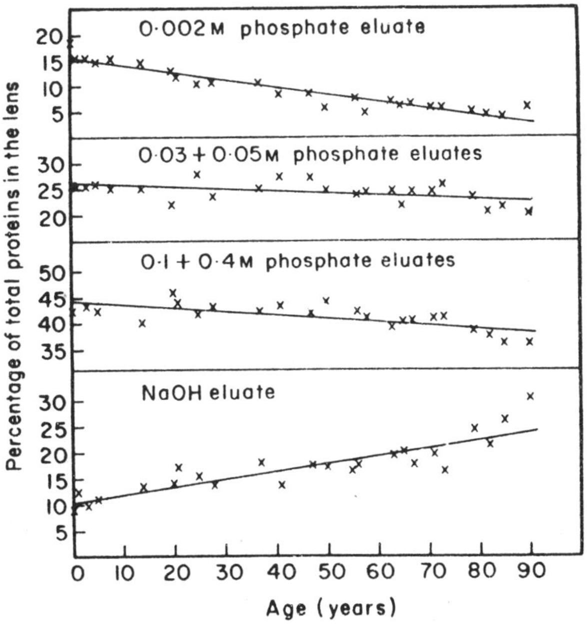

Figure 6 Distribution of the soluble crystallins as a function of age. Proteins were separated on DEAE cellulose. Results are expressed as percentages of the total proteins in the lens and are based on amounts of protein determined with the Lowry method. Regression lines were determined by the method of least squares and linear correlation coefficients (r) were calculated. The slopes calculated for the lines were 0.002 M eluate, -0.139 ($r = -0.95$); (0.03 + 0.05) M eluates, -0.040 ($r = -0.57$); (0.1 + 0.4) M eluates, -0.068 ($r = -0.74$); NaOH eluate, 0.151 ($r = 0.87$). Reprinted with permission from ref. 29. Copyright by Academic Press Inc. (London) Ltd.

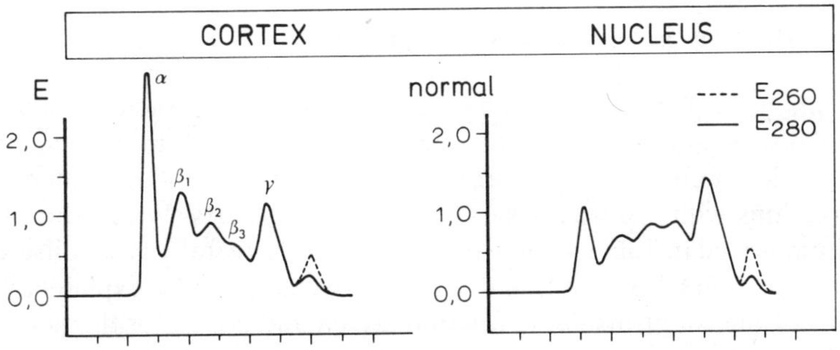

Figure 7 Chromatography on Sephadex G200 of cortical and nuclear water-soluble crystallins, isolated from normal and cataractous lenses. Taken from ref. 32.

In a study on brunescent cataracts, Zigler et al. (35) determined the amino acid composition of the soluble protein fraction of individual normal human lenses, ranging in age from 42 to 62 years. The mean values ± SD for the amino acid composition of four normal lenses are reported; obviously, no age-related changes were observed.

The low molecular weight crystallin distribution in 6- and 53-year-old as well as in other young and old human lenses was studied by Kabasawa et al. (36) using Sephadex G75 column chromatography. With aging, high molecular weight γ-crystallin (γ_H; 24,000 dalton by SDS gel electrophoresis) and β_S-crystallin (28,000 dalton) showed increased proportions, the low molecular weight γ-crystallin (γ_L; 20,000 dalton) a decreased proportion.

The most extensive study of the size distribution of the native water-soluble crystallins in 23- to 83-year-old human lenses (in comparison with cataractous ones) has been performed by Jedziniak et al. (37). By application of Sephacryl S200 recycling chromatography, six protein fractions with molecular weights >200,000, 140,000, 83,000, 40,000, 17,000 and 11,000 dalton were obtained. As the lens ages, the >200,000 and 140,000 dalton fractions increase in concentration, the 17,000 and 11,000 dalton fractions decrease by roughly the same order of magnitude, and the 83,000 and 40,000 dalton proteins remain at a constant concentration (Figure 8).

7.2.5 Comparison of Six Species

The crystallin composition of young and old lenses from guinea pig, rat, mouse, man, ox, and dog has been compared by Bours (37a) utilizing isoelectric focusing. Besides the known species specificity for mammals, some age-dependent characteristics could be demonstrated.

7.3 CHANGES IN DISTRIBUTION AND COMPOSITION OF WATER-INSOLUBLE FRACTIONS DURING AGING

7.3.1 Cattle, Dog, Rabbit, Rat, and Dogfish Lens

The amounts of the water-soluble and water-insoluble fractions of various parts of bovine and dog lenses (cf. Sections 2.1 and 2.2) in dependence on age were determined by Bours et al. (24) (Tables 3 and 4). An increase of both fractions with age was observed in most lens parts. With regard to the data summarized in Table 4 one wonders how the generally great differences between the sum (WS + WI) and DW (dry weight) can be explained. High amounts of the water-insoluble fraction are characteristic for the nucleus of the older bovine and dog lenses.

The amounts of the water-insoluble and the urea-insoluble fractions in

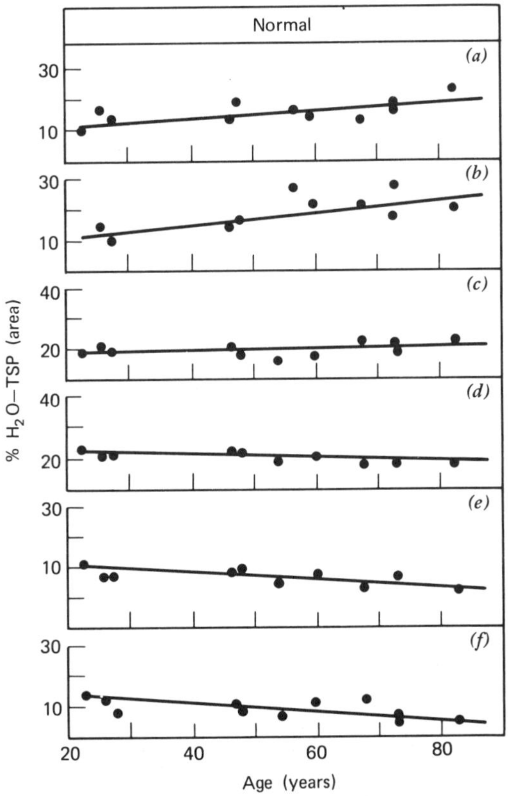

Figure 8 Concentration variation of protein components of H_2O–TSP (total soluble protein) from human lenses with age. The percent H_2O–TSP, that is, area of each of the six proteins, to the total percent H_2O–TSP, that is, sum of areas from human lenses, is shown in descending order of size on the vertical axis: (*a*) >200,000 mol. wt.; (*b*) 140,000; (*c*) 83,000; (*d*) 40,000; (*e*) 17,000; (*f*) 11,000. Taken from ref. 37. Copyright by Academic Press Inc. (London) Ltd.

lenses of 3-, 4-, and 5-month-old rabbits were determined by Liem-Thé et al. (38, 39) (Table 5). A drastic increase of the urea-soluble fraction can be inferred from these data. As determined by isoelectric focusing in the presence of urea and SDS–polyacrylamide gel electrophoresis, the composition of this urea-soluble fraction was 70% α-crystallin and 30% β-crystallin. In this connection it should be mentioned that Satoh and Bando (40) claim that the urea-soluble fraction of 24-month-old rabbit lenses is formed by a partial combination of β- and γ-crystallins. This was based on the results of amino acid analyses as well as isoelectric focusing.

Hockwin (41) initiated an investigation on Wistar rat lenses with regard to the content of water-insoluble proteins within the framework of a

Table 3 Determination of the Amount of Protein Extract in the Different Parts of the Bovine Lens in Dependence on Age.

Age (years)	Lens portion	FW (g)	WS (mg)	WI (mg)	WS/WI (ratio)
0.16	EQ	—	102	2	46
12.2	EQ	—	163	9	18
18.7	EQ	—	179	14	13
0.16	AC	—	37	3	14
12.2	AC	—	122	12	10
18.7	AC	—	110	21	5
0.16	PC	—	39	1	39
12.2	PC	—	86	19	4
18.7	PC	—	67	14	5
0.16	N	—	91	3	31
12.2	N	—	260	35	7
18.7	N	—	298	55	5
0.16	WL	1.125	269	9	30
12.2	WL	2.686	631	75	8
18.7	WL	2.819	654	105	6

Taken, by permission, from ref. 24.

Legend: FW Wet weight lens.
WS Water-soluble proteins: crystallins.
WI Water-insoluble proteins: albuminoid.
— Not determined.
EQ Lens equator.
AC Anterior cortex.
PC Posterior cortex.
N Lens nucleus.
WL Whole lens.

longevity project on eye changes. Results concerning the ratio water-soluble–water-insoluble protein distribution are not yet available, but it seems noteworthy that after an age of 400 days sex differences in lens weight were detected. Therefore, one wonders whether sex-linked differences in the distribution of water-soluble and water-insoluble fraction ought to be considered. The same phenomenon has been found for human lenses ("men have heavier lenses than women of the same age") by Harding et al. (42).

Age-related alterations in the distribution of water-soluble, high molecular

Table 4 Determination of the Fresh Weight and the Amount of Protein Extract in the Various Parts of the Dog Lens in Dependence on Age.

Age (years)	Lens portion	FWP (mg)	DW (mg)	WS (mg)	WI (mg)	WS/WI (ratio)
1.34	EQ	211	63	43	10	4
7.8	EQ	261	78	64	8	8
9.8	EQ	278	90	71	10	7
1.34	AC	84	32	15	2	9
7.8	AC	101	38	26	8	3
9.8	AC	73	28	19	6	3
1.34	PC	94	46	17	9	2
7.8	PC	124	50	29	11	3
9.8	PC	147	62	36	19	2
1.34	N	156	62	34	3	12
7.8	N	186	95	43	37	1.2
9.8	N	188	96	40	47	0.8
1.34	WL	595	203	109	24	4
7.8	WL	696	261	162	64	2.5
9.8	WL	716	276	166	82	2.0

Taken, by permission, from ref. 24.

Legend: FWP Wetweight lens portion.
DW Dry weight lens portion.
WS Water-soluble proteins: crystallins.
WI Water-insoluble proteins: albuminoid.
EQ Lens equator.
AC Anterior cortex.
PC Posterior cortex.
N Nucleus.
WL Whole lens.

Table 5 Gravimetric Data of Rabbit Lenses

Age (months)	Wet weight (mg)	Dry weight (mg)	Water-insoluble (mg)	Urea-insoluble (mg)
3	325	119	3.9	1.5
4	408	162	2.7	1.3
5	500	182	12.6	2.1

Taken from refs. 38 and 39.

weight and water-insoluble proteins of the lens of smooth dogfish, *mustelus canis*, were measured by Zigman and Yulo (42a). The ages of the animals ranged approximately from 2 to 50 years, during which time the lenses grew from 100 to 1500 mg wet weight. In this species the water-insoluble protein accumulates to the extremely high level of greater than 50% of the total protein at maturity.

7.3.2 Human Lens

The distribution of the proteins of individual normal human lenses between the water-soluble, water-insoluble, and high molecular weight protein fraction as a function of age (10 to 80 years) has been determined by Zigman et al. (43) (Figure 9). It can be seen that the level of soluble protein declines with a plateau between 30 and 60 years of age, and that the high molecular weight fraction increases up to age 50, followed by a decrease over age 60, which is accompanied by a rapid rise of the level of the insoluble protein. A negative relationship between the percentage of high molecular weight protein and the water-insoluble protein level for aging normal (and cataractous) lenses was observed. The subunit composition of the sulfonated water-soluble, high molecular weight and water-insoluble protein fractions of an 18- and a 65-year-old lens were compared by polyacrylamide gel electrophoresis in 8 *M* urea. Ten distinct bands and several others, not easily distinguishable, were present in all three fractions. The bands of the older lens, however, smeared and increased in the more electronegative region of the gel. The authors attribute this diminished sharpness and greater spread to an increase in heterogeneity of specific subunits with age.

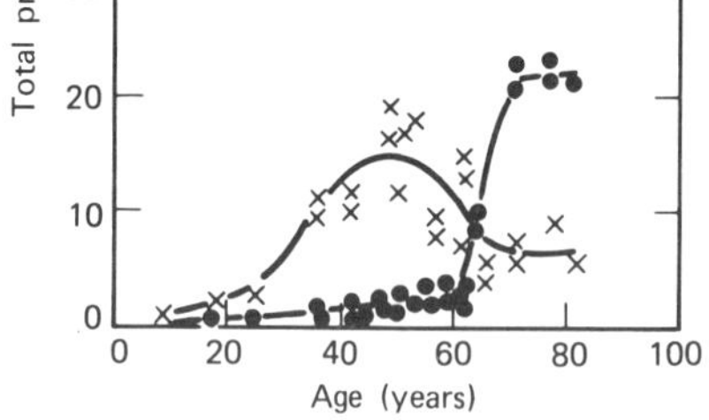

Figure 9 Distribution of the proteins of individual normal human lenses among the soluble (○), insoluble (●), and high molecular weight (X) protein fraction as a function of age. Reprinted with permission from ref. 43. Copyright by Academic Press Inc. (London) Ltd.

The amino acid composition of the water-insoluble and water-soluble protein fractions of four individual normal lenses, 42- to 62-year-old, was determined by Zigler et al. (35). The mean values obtained for both fractions showed no significant differences. Low concentrations (0.2–0.4 mol %) of kynurenine were observed in the acid hydrolysates.

The quantities of cortical and nuclear water-soluble, urea-soluble, and urea-insoluble fractions of 50- to 80-year-old, pooled normal (and nuclear cataractous) lenses, expressed as percent of wet weight, have been given by Kramps et al. (31–33). In order to rule out artifactual results, anaerobic isolation procedures were applied. The protein content in the nucleus (inner 30% of wet weight) as well as in the cortex was approximately 35%. Comparing cortex with nucleus, the water-soluble and the urea-insoluble fraction are slightly decreased, whereas the urea-soluble fraction is increased in the latter (approximately 10% versus 15% of wet weight). The urea-soluble fractions were analyzed by isoelectric focusing in the presence of 6 *M* urea and revealed essentially the same polypeptide patterns, containing at least 16, partly trailing bands. The presence of very acidic polypeptides (isoelectric point about 5), previously observed by Liem-Thé (38), is striking. Chains with the mobilities of the α-crystallin subunits were present in considerable amounts. The molecular weights of the polypeptide chains of the urea-soluble and urea-insoluble fractions were obtained by SDS gel electrophoresis (32, 33). Major bands in the 18,000 to 20,000 molecular weight region and a minor band of 30,000 dalton were seen with the urea-soluble fractions. The cortical and nuclear urea-insoluble fractions, on the other hand, showed two polypeptides with molecular weights of 27,000 and 31,000, respectively. It should be kept in mind that strongly loaded SDS gels revealed 10 bands for the urea-soluble fractions, comprising the molecular weight range from $<$10,000 to 52,000. Finally, it should be mentioned that no difference was found between cortex and nucleus with relation to their low content of covalently linked protein ($>$200,000 dalton) when gel chromatography was performed under denaturing conditions of the delipidated, reduced, and aminoethylated lens parts (44).

The distribution of four lens protein fractions, separated on the basis of their solubility in different media, was determined by Coghlan and Augusteyn (29) for 28 lenses of ages between 0 and 90 years. A linear decrease of the water-soluble and a linear increase of the urea-soluble fraction were found. The solubilized (with 8 *M* urea plus 0.1 *M* mercaptoethanol) urea-insoluble material (SUI) and the insoluble residue (UI) remained constant throughout life. The percentages of total protein are given in Table 6. It seems noteworthy that the urea-insoluble material was isolated in a similar way (extraction in the presence of 0.1% 2-mercaptoethanol and 8 *M* urea) by Maisel (45). This author subjected the pellet to electron microscopy and showed that largely cell membranes were seen, whereas occasionally granular material was noted within membrane-bound vesicles.

Table 6 Changes with Age in the Proportion of Proteins of Normal Human Lenses

	Proportion (%)	
Fraction	0 years	90 years
WS	96	89
US	2	10
SUI	1.3	1.3
UI	0.2	0.2

Taken from ref. 29.

Much higher values for the proportion of water-insoluble protein in the old lens than those shown in Table 6 have been given by Lerman and Borkman (28, 46). For lenses up to the age group 40 to 49 years approximately 3% of the total protein was found to be insoluble, but then a strong increase is observed, ending with 40% insoluble protein for the group 80 to 89 years. Unfortunately no experimental details have been given in these reports.

Water-insoluble, covalently (presumably disulfide-) linked high molecular weight (CHMW) proteins were isolated by Takemoto and Azari (47) from totally guanidine-solubilized lens proteins after alkylating the sulfhydryl groups under anaerobic conditions. The authors employed Bio-Gel A-50m chromatography in the presence of 0.1% SDS and sucrose gradient centrifugation in the presence of 8 *M* urea. Within the age span of 14 to 87 years, normal human lenses showed no age dependence with regard to their CHMW content, which was found to be about 3% of total protein. The molecular weight of the CHMW fraction from normal lenses seems to be of the order 2×10^6 dalton.

The expected increase of the amount of insoluble protein in the normal human nucleus upon aging has been determined by Anderson and Spector (48). Using 16 lenses in the age range 18 to 63 years, they found that the proportion of water-insoluble protein (WI) increases from 20 to about 55% (Table 7). The color classification used was that of Chylack (49), in which 1 means colorless and 4 yellow. Considerably higher values for the proportion of water-insoluble protein were found in nuclei of various types of cataract (60–93%) (compare Chapter 8). Taking into account that in this study the water-insoluble pellet contains the high molecular weight protein, the ratio WI/total protein for the older lenses (55%) is in good agreement with the value that can be calculated from data of Kramps et al. (33). In a subsequent paper Anderson et al. (49a) reported that in normal human lenses between 18 and 80 years of age the proportions of water-insoluble proteins increase from 20 to 44% of the total cortical protein and from 20 to 55% of the total nuclear protein.

To elucidate the origin of the water-insoluble protein fraction, Roy and Spector (50) investigated its constituent polypeptide chains. Using lenses older than 60 years, polypeptides with the following approximate molecular weights on the basis of SDS gel electrophoresis were found:

$>$100,000   43,000   27,000   22,000   11,000 dalton

Marked charge heterogeneity was observed in all samples investigated. Although no changes as a function of age were studied, it seems important to note that the authors, on the basis of similarity in chemical and physical properties and increasing insolubility, propose the following relationship: high molecular weight protein → urea-soluble but water-insoluble protein → urea-insoluble protein. The absence of any free *N*-terminal residue in the water-insoluble fraction suggests that γ-crystallin is not present or that its *N*-terminal amino acid has become blocked. In a subsequent paper the same authors (51) were able to show that the above presumptive 11,000 molecular weight component of the water-insoluble fraction has an apparent molecular weight of 9600 and is very heterogeneous in charge (DE-52 chromatography in 8 *M* urea). Analysis of tryptic peptides and other methods revealed that the generation of the 9600-dalton polypeptide is at least partly a result of post-translational *C*-terminal shortening of the α-crystallin A chain. The same conclusion is drawn by van Haard et al. (51a) for a possibly identical, heterogeneous polypeptide chain fraction, isolated from normal or nuclear-cataractous lenses and showing a mean molecular weight of about 10,000 and an isoelectric point range of 4.6 to 4.7.

Age-dependent changes in the abundance of polypeptides in the water-soluble and water-insoluble fractions of human lenses have been studied by Garner and Spector (51b). They found that dramatic changes occur in the

Table 7 Water-Insoluble Proteins in Normal Human Lens Nucleus

Age (years)	Number of (pooled) lenses	Wet weight (mg)	Color class	WI protein (% of total)
18	1	164	1	20
33, 35	2	217	2	37
35, 39	2	209	2	37
60	1	224	3	56
63 ± 1	8	231	4	58
62	2	255	4	47

Taken from ref. 48.

first few years of life. From approximately 10 to 40 years of age gradual changes take place. After this period the relative concentrations of the different subunits are fairly constant, except for a continued marked decrease in the 20,000 and increase in the 10,000 dalton water-insoluble components. Amino acid incorporation studies confirmed the conclusion that the 43,000 dalton polypeptide is directly synthesized and the 10,000 dalton component arises by post-translational degradation.

7.4 WATER CONTENT OF AGING LENSES

The water content of the lens is of great importance for its function. This parameter can be related to the ability of the lens to accommodate and to its refractive power. Water represents the medium for all chemical reactions in the lens and participates in most of them. It is essential for physicochemical phenomena such as osmotic pressure, viscosity, and pH value. With regard to proteins, it is clear that their conformation may be influenced by the degree of hydration. One might suppose that the formation of high molecular weight protein aggregates and their insolubilization depends on loss of water. Moreover, there may exist a relationship between deamidation and degradation processes in the lens proteins and their concentration. It is known that the water content differs in the various parts of the lens of most animals, suggesting that age and water content of the lens are interrelated. It is characteristic for human lenses, however, that they possess the same water proportion in cortex and nucleus, independent of their age.

Van Heyningen (52) reported that she could not find any marked changes in the water content of the human lens after the age of 30. Moreover, it was shown by Fisher and Pettet (53) that in the age range from birth to 80 years the total water content of the human lens both for nucleus and cortex did

Table 8 Water Content of Whole Bovine Lenses during Aging

Age (years)	Water content (%)
0–1	68.6
1–2	65.4
3–4	64.4
5–10	63.7
10–16	58.4 ± 0.79, $n = 8$
16–22	57.1 ± 0.22, $n = 4$

Taken, by permission, from ref. 55.

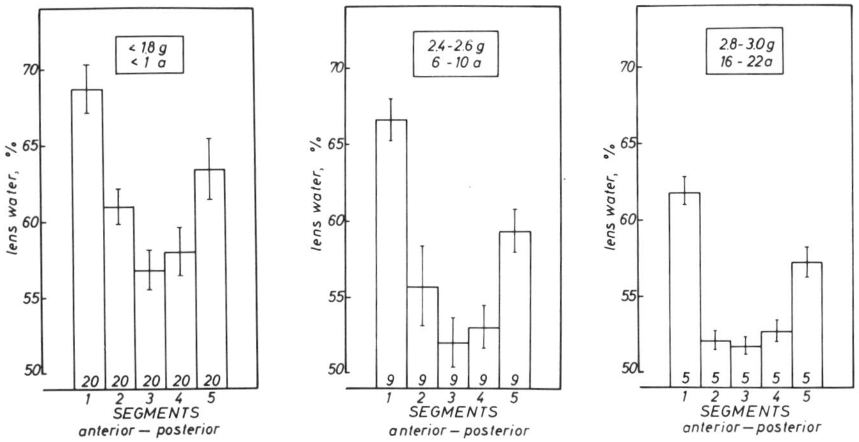

Figure 10 Water content in five segments of bovine lenses of different age. 1, Anterior; 5, posterior. Reprinted with permission from ref. 55.

not significantly vary. The authors concluded that "sclerosis" of the aging lens due to loss of water does not occur.

Klethi (54), studying the relative water contents of whole human lenses (19 to 78 years), found values of 67 to 70% but no correlation with age could be observed. Kramps et al. (33) found 65% water in cortices and nuclei; even in dark brown cataractous nuclei, which are very hard in consistency, no significant dehydration was observed.

The water content in bovine lenses was studied by Rink (55). The data obtained for whole bovine lenses during aging by Ely (56) and Rink are given in Table 8. Rink also determined the water content in five segments of a central core of 80% of the equatorial lens diameter, obtained with a cork borer. The results for three age groups are shown in Figure 10. The water content diminishes with age in all five segments. In this context it may be recalled that van Kleef (57) observed no continuous increase of the protein content in concentric layers of the nucleus of calf lenses. The protein content was highest in the outer nucleus and significantly lower in the inner nucleus.

The water content in cortex and nucleus of bovine, guinea pig, rabbit, and rat lenses as a function of age was determined by Hockwin et al. (58). The values obtained for bovine lenses correspond to those of Rink (55). For guinea pig lenses, ranging in weight from 115 to 140 mg, no age dependence of the water content was found; a strong difference between cortex and nucleus was observed, however (approximately 70 and 50%, respectively). In the case of rabbit lenses, the cortex of 5- to 6-week-old animals showed a water content of nearly 75% decreasing to about 69% in rabbits of 6 to 8 months, whereas the nucleus showed a decrease from initially 61 to 50%. Fisher and Pettet reported earlier (53) the water contents of 2- to 4-month-old and 1- to 3-year-old rabbit lenses; changes with age in the cortex were not significant, whereas the nucleus showed a very significant loss of water as the lens aged. The rat lenses examined by Hockwin et al. (58) showed remarkable differences between cortex and nucleus. In the cortex of 4-week-old rat lenses a water content of 60% was found, remaining constant up to the age of 250 days; the nucleus of lenses of 26 days had a water content of only 47% and decreased to 40% by 250 days.

The age-dependence of the water content of beagle dog lenses (210-day-to 14-year-old) has been studied by Hockwin et al. (58a). The lenses of the female animals turned out to be of less weight than those of males of the same age. In both males and females the water content in the lens equator is higher than in the nucleus and decreases in both regions with aging.

For the dry weight of dogfish lenses in the age range from 2 to 50 years a value of 50% of the wet weight has been reported (42a).

Thus there is sometimes an age-dependent decrease of the water content (bovine, rabbit, rat nucleus, beagle dog), sometimes not (human, guinea pig, rat cortex). At the moment it is not yet possible to draw definite conclusions concerning age-dependent changes of the water content of the lens because the age ranges investigated were rather limited. If a relationship between age-dependent changes of lens proteins and water content exists at all, it might turn out to be rather complicated.

7.5 PROPERTIES OF AGED CRYSTALLINS AND THEIR SUBUNITS

7.5.1 Light Absorption and Scattering in Intact Human Lenses

Maintenance of transparency of the lens as such forms a major interest of investigators in this field. It may therefore be considered as remarkable progress in lens research that light absorption and scattering properties of intact lenses are now being investigated and compared with those of *in vitro* extracts and fractions.

Zigman et al. (43) demonstrated that in the wavelength range of 440 to 700 nm the optical density of intact, 30- to 80-year-old, normal human lenses increases with age. The absorption at 440 nm (where yellow pigment absorbs strongly) and at 700 nm (minimal absorption by the pigment) was studied in greater detail. The absorption at 440 nm was corrected for scattering, representing then a measure of the color of the lens. This "color" increased only slightly in normal lenses until the fifth decade (absorbance 0.11 → 0.15), at which time it increased much more rapidly up to the seventh decade (absorbance 0.15 → 0.42). The isolated water-insoluble protein fraction contained the greatest amount of yellow pigment, the high molecular weight material had less color, and the soluble proteins had little.

Applying the advanced method of quasi-elastic laser light-scattering spectroscopy in combination with an autocorrelator, Jedziniak et al. (59) demonstrated that high molecular weight aggregates are really present in intact normal human lenses. Although certain assumptions had to be made in their calculations, these authors could convincingly show that the concentration of high molecular weight protein increases monotonically with age from 5 months up to 43 years. The values for its concentration, as obtained after isolation, were in good agreement with those of the scattering experiments. In the infant lens no high molecular weight protein is present and at 50 years it represents about 2% of the total soluble proteins.

7.5.2 High Molecular Weight Protein

The proteins discussed in this section are the types 1 and 2 as defined in the review of Harding and Dilley (2). They are separated from α-crystallin and from the unseparated soluble proteins of the lens, respectively, and they have molecular weights greater than 10^6 dalton. In a subsequent paper Harding (60) designated both types as HM1 and HM2 and this nomenclature is sometimes used in this chapter (compare Chapter 8).

It is well established that high molecular weight protein species are formed with aging in lenses of different animals. Examination of the role of these proteins is strongly hampered because they do not present an unambiguous defined fraction of the lens homogenate. They seem to form a broad population of particle weights somewhere between 10^6 and 10^9 dalton. The border between high molecular weight and water-insoluble fraction is fluent and the definition of these fractions is operational and more or less arbitrary. Special attention is therefore paid to this problem in the next paragraphs.

7.5.2.1 Human Lens

After removal of insoluble proteins from human lens homogenates by centrifugation for 25 min at 12,000 $\times$ g at 10°C, Zigman et al. (43) sedimented the high molecular weight (HM2) fraction by centrifugation for 55 min at 100,000 $\times$ g and 10°C. Fractionation of the lens extract on Agarose A-5m, after the 12,000 $\times$ g spin, revealed two peaks with molecular weights greater than 2×10^6 dalton. As judged by the yields, which differed less than 10%, approximately the same material was isolated by both methods. The high molecular weight protein showed less color and fluorescence than the water-insoluble fraction, but more than the low molecular weight ($<10^6$ dalton) crystallins. Some data with regard to the subunit composition of the high molecular weight fraction are discussed in Section 7.3.2 in connection with that of the water-soluble and water-insoluble fractions. Applying the differential centrifugation procedure just mentioned, Zigman et al. (61) reported on the nontryptophan fluorescence and the polypeptide chain composition of the high molecular weight fraction from normal human lenses, 40-, 60-, and 70 to 80-year-old. In this order relative fluorescence values of 245, 375, and 385 were measured (360 nm excitation; 440 nm emission); separated nuclei and cortices from three lenses with an average age of 54 years revealed values of 289 and 449, respectively, for the high molecular weight fraction. On SDS–polyacrylamide gel electrophoresis polypeptides with apparent molecular weights of

50,000 45,000 31,000 27,000 21,000 and 11,000

were observed for the high molecular weight (as for the water-soluble and water-insoluble) fraction; only the 11,000- and 21,000- to 31,000-dalton chains showed stronger fluorescence for the nuclear than for the cortical high molecular weight fraction.

Differential centrifugation in order to isolate high molecular weight protein from human lenses (cataractous, 65 to 70 years of age) was also described by Roy and Spector (62). Three centrifugation steps were applied, each for 15 min at 4°C, under the following conditions:

$$9900 \times g \qquad 26{,}400 \times g \qquad 59{,}400 \times g$$

The composition of the pellets and supernatants obtained was studied by gel chromatography on Bio-Gel A-15m and A-50m. The supernatant of the 59,400 $\times$ g centriguation was found to be almost completely free of any high molecular weight protein, whereas the corresponding pellet contained only this species. On the basis of their observations the authors defined the water-insoluble fraction as the precipitate obtained by centrifugation of a solution containing 0.7 to 1.0 g (wet weight) of lens per 10 ml of buffer at an average centrifugal force of 9,900 $\times$ g for 15 min. The high molecular weight fraction is then that part of the supernatant which is spun down at an average force of 59,400 $\times$ g for 15 min.

The next, more systematic investigation of centrifugation conditions was undertaken by Harding (60). He isolated the high molecular weight fraction (HM2) from clear human lenses by gel chromatography on Bio-Gel A-5m after centrifugation at two different forces: 9900 $\times$ g and 26,000 $\times$ g. The amount of HM2 obtained by direct weighing was 4.1% of the total protein at 9900 $\times$ g, but only 1.0% after the faster centriguation. Groups of lenses, varying from 59- to 82-year old, were compared but no clear-cut age dependence could be detected.

Surveying the data mentioned in the preceding paragraphs of this section, it seems advisable for future research that a standardized centrifugation scheme, for instance, the one according to Roy and Spector (62), be applied

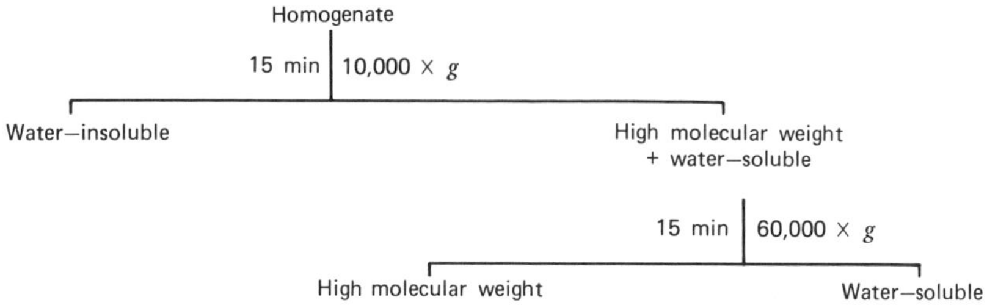

Figure 11 Proposal for standardization of the isolation of lens protein fractions.

in order to more clearly define three main fractions and to enable comparison of the results of different investigators (Figure 11). The temperature should be below 10°C. The more subtle question of possible artifactual interchain and protein–protein disulfide formation during isolation should be kept in mind and considered for each case. As shown by Harding (60), one should not overlook the unreliability of determinations of the high molecular weight protein content on the basis of the absorbance of 280 nm and Folin-Ciocalteu measurements. Moreover, the rather strange finding of Dilley (63), that only 50% of HM2 is protein, should also be taken into consideration.

The polypeptide chain composition of the high molecular weight and water-insoluble fraction of human lenses led Roy and Spector (50, 62, 64), on the basis of similarity in chemical and physical properties and decreasing solubility, to the proposal that the following precursor relationship may exist:

$$\text{high molecular weight} \rightarrow \text{urea-soluble} \rightarrow \text{urea-insoluble protein}$$

Polypeptides with molecular weights

$$>100{,}000 \quad 43{,}000 \quad 27{,}000 \quad 22{,}000 \quad 20{,}000 \text{ and } 9600$$

were found. The high molecular weight polypeptide (>100,000 dalton), representing 40 to 50% of the high molecular weight and water-insoluble fraction, could be partly dissociated to units of 43,000 dalton by electrophoresis in a strongly deaggregating medium, that is, 0.1% SDS + 6 *M* urea. Isolated polypeptide fractions (>100,000; 43,000; 27,000–20,000; and 11,000 dalton) revealed, surprisingly, few differences in amino acid composition. It is remarkable in this work that the 27,000-dalton polypeptide, which is supposed to be a β-crystallin chain, shows a molecular weight of 20,000 in SDS–urea electrophoresis.

Great similarity between urea-soluble and high molecular weight fractions with regard to their polypeptide chain composition was also found by other authors (34, 38, 65) for pooled, 60- to 80-year-old cortices and nuclei. Both lens parts showed only quantitative differences on polyacrylamide gel electrophoresis in the presence of SDS and DTE (dithioerythritol). Nine largely corresponding chains with molecular weights in the range from 11,000 to 50,000 in addition to a considerable amount of material at the top of the gels were found in the high molecular weight and urea-soluble fraction. An additional $<10{,}000$-dalton polypeptide was present in minor amounts in the urea-soluble protein fraction. An 11,000-dalton component was also observed in the γ- but not in the α- and β-crystallin fractions of the same lenses. From these and immunochemical results the authors concluded that it seems reasonable to assume that high molecular weight protein in older human lenses contains α-, β-, and γ-crystallins (or at least part of their constituent chains) and might be an intermediate in the insolubilization process.

7.5.2.2 Bovine Lens

Since the publication of the review of Harding and Dilley (2) little attention has been paid to the properties and subunit structure of the high molecular weight proteins of bovine lenses and the reason(s) why and how low molecular weight α-crystallin molecules aggregate upon aging. At that time it was known that α-crystallin appears to be the only constituent of bovine high molecular weight aggregates. In the cortex low molecular weight α-crystallin predominates, whereas toward the older lens nucleus a gradual transformation to two higher molecular weight populations is observed. Furthermore, it had been established that the polypeptide composition changes progressively from the outer cortex to the inner nucleus as the proportions of the *de novo* synthesized A_2 and B_2 chains decrease whereas those of *C*-terminal shortened chains increase. The hypothesis that degradation of the polypeptide chains induced aggregation to high molecular weight α-crystallin was abandoned when it was found that *in vitro* degradation was not followed by increased aggregation.

The possible role of shortened polypeptide chains in the aggregation process was, however, reinvestigated by Siezen et al. (66). Nuclear α-crystallin from calf lenses was separated into the three known molecular weight populations on Ultrogel AcA-22. Electron microscopy revealed that the two higher molecular weight classes, which are formed on aging of the lens, are chainlike oligomers and polymers of spherical monomeric α-crystallin molecules of 15 nm average diameter and 1.1×10^6 average molecular weight, which form the lowest molecular weight class. Combined sedimentation and electron microscopic data made the alleged nonequilibrium state within the populations questionable. From the subunit composition it was inferred that the total amount of *C*-terminal shortened chains increases progressively from 33 to 53% from monomeric to polymeric α-crystallin. Reassociation of different types of isolated shortened αA-chains in various mixtures indicated a tendency for formation of higher molecular weight reaggregates with increasing relative amount of degraded chains. The involvement of degraded chains in polymerization of α-crystallin has, therefore, been postulated again, although it is stressed that it is not likely to be the only factor causing aggregation.

7.5.2.3 Mouse Lens

High molecular weight aggregates were found to increase with age (17 to 90 days) in the normal mouse lens (66a). The high molecular weight fraction showed an increase from 0.4 to 1.6% of total lens protein, whereas the water-insoluble protein increased from 5 to 19%. By immunodiffusion α- and β- but no γ-crystallin could be detected in the high molecular weight material.

7.5.3 α-, β-, and γ-Crystallin

In this section "α-crystallin" corresponds to the low molecular weight α-crystallin population of the molecular weight range 6 to 9 $\times$ 10^5 dalton. Fluorescence and immunochemical properties are dealt with in Sections 7.5.4 and 7.5.6, respectively. More detailed data concerning postsynthetically changed polypeptide chains are described in Section 7.6.

7.5.3.1 Human Lens

It seems worth mentioning that the values for the absorption at 280 nm of 1% solutions ($A_{280}^{1>}$) of α- and γ-crystallin from young and old lenses turned out to differ (46). A comparison of the age groups 1 to 10 years and 70 to 80 years showed $A_{280}^{1>}$ values of 10.2 and 15.6 for α-crystallin and 15.7 and 17.3 for γ-crystallin, respectively.

An interesting new technique applied by Lerman et al. (67, 68) is photoacoustic spectroscopy (PAS), which allows the recording of absorption spectra of amorphous, light scattering, and opaque samples. The significant advantage of PAS over transmission techniques is that in PAS only light absorbed by the sample contributes to an acoustic signal; transmission of light is not necessary and scattered light effects are negligible. Whole lenses were placed in the photoacoustic cell and the capsular and cortical material were scraped to the side in order to expose the nucleus. Normal lenses revealed a significant absorption band between 320 and 360 nm, increasing in magnitude with age.

Recycling chromatography on Sephacryl S200 revealed that the molecular sizes of the human β- and γ-crystallin fractions do not vary with age (37, 69). The charge distribution within each fraction, however, undergoes changes with age and cataract. A shift toward electronegative charge is observed.

On polyacrylamide gel electrophoresis in 8 *M* urea of the sulfonated proteins the whole of α-, β-, and γ-crystallin subunits, but without those of the high molecular weight fraction, was shown to be extremely complex. Comparison of an 18- and a 65-year-old lens showed the bands of the older one to smear and to increase in the levels of more electronegative species (43).

Polypeptide chain analysis was performed by Kramps et al. (31, 33) by isoelectric focusing in 6 *M* urea of the separated crystallins α, β_1, β_2, β_3, and γ (cf. Figure 7) characterized by sedimentation coefficients of 18.9, 8.3, 5.1, 3.6, and 2.2 S, respectively, from cortex and nucleus (inner 30%) of 50- to 80-year-old human lenses. As shown in Figure 12, mainly quantitative differences between the cortical and nuclear preparations can be noticed. The polypeptides of the human β-crystallins have isoelectric points in the same range as those of bovine β-crystallins, namely, between pH 5.5 and 7.5. The γ-crystallin fraction revealed chains with isoelectric points ranging

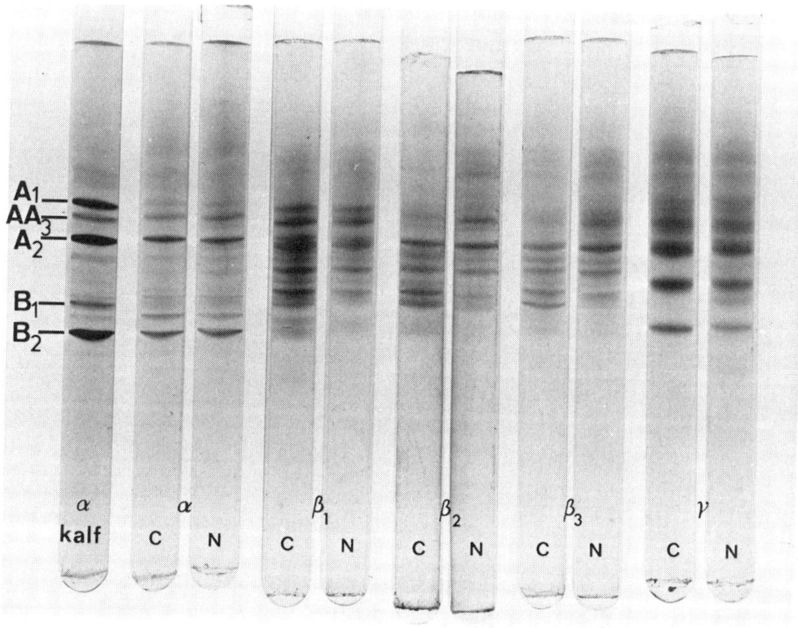

Figure 12 Isoelectric focusing in the presence of 6 *M* urea of cortical and nuclear water-soluble protein fractions. Ampholine in a pH range of 3.5 to 10 was used (top to bottom). The acrylamide concentration was 3.3%. α-Crystallin isolated from calf lenses is shown for comparison. C, cortex; N, nucleus. Taken from ref. 33.

Table 9 Molecular Weights of Polypeptide Chains of Water-Soluble Human Lens Crystallins Obtained by Dodecylsulfate Gel Electrophoresis[a]

α	β_1	β_2	β_3	γ
	28,000	28,000	28,000	
22,000	(23,000)	(23,000)	(23,000)	
20,000	18,000	18,000	18,000	18,000
				(two bands)
				(10,000)

Taken from ref. 33.
[a]Parentheses indicate minor chains.

from pH 5.0 to 7.0. It is remarkable that all crystallins from these lenses contained minor amounts of chains with very low isoelectric points (pH less than 5.0). The molecular weights of the polypeptide chains of the same crystallin fractions, obtained by SDS gel electrophoresis, are given in Table 9. It should be noted, however, that stronger loaded gels revealed more components (34):

Three for α-crystallin, ranging from 18,000 to 22,000 dalton.
Six for the β-crystallins, ranging from 19,000 to 51,000 dalton.
Six for γ-crystallin, ranging from 11,000 to 29,000 dalton.

These investigations showed that no essential differences exist in the properties, including subunit structure, of the α-, β-, and γ-crystallins from cortex and nucleus of the age group used.

It was shown by Roy and Spector (70, 71) that α-crystallin from a normal 6-week-old human lens is composed of only three major polypeptides, whereas 11 were found in α-crystallin from 65- to 70-year-old periphery (outer 50–60%). On the basis of amino acid analyses and tryptic peptide mapping six bands appeared to represent αA, the other five, αB chains. The α-crystallin from the 6-week-old lenses showed on SDS gels only two bands corresponding to molecular weights of 22,000 and 20,000; the older material contained an additional 17,000-dalton component. Comparison of these data with an earlier paper from Spector's group (72) suggests that the three major α-crystallin polypeptides from the young lens are identical with those of newly synthesized α-crystallin. Whether the latter three polypeptides were independently synthesized or arose by rapid transformation of one of the αA chains could not be determined yet.

The amino acid composition of α-crystallin was claimed to vary with age (29, 73). For instance, large differences were found in the levels of glutamic acid, glycine, alanine, tyrosine, leucine, and phenylalanine when whole lenses of 45- and 80-year-old humans were compared to a 1-year-old lens

(29). The α-crystallin from the old lenses had amino acid compositions similar to those of the β-crystallins. The meaning of this striking result is still obscure. In this study it was also found that 20 to 30% of the total protein from old lenses is so firmly bound to DEAE cellulose that it has to be eluted with NaOH; this NaOH fraction closely resembled α-crystallin in amino acid composition.

The near- and far-ultraviolet circular dichroism wavelength profiles of α-crystallin from lenses of various ages were shown to be identical by Horwitz (73), which implies identical conformation.

Analyses of the low molecular weight proteins from young and old human lenses by Kabasawa et al. (36, 74) revealed a remarkable age-related change of γ-crystallins. A higher molecular weight γ-crystallin (γ_H, 24,000 dalton, not identical with β_S!) turned out to be more abundant in older lenses than a 20,000-dalton γ-crystallin (γ_L); moreover, γ_H was found in larger amounts in the cortex of an older lens than in its nucleus.

7.5.3.2 Bovine Lens

The polypeptide composition of bovine α-crystallin strongly depends on the age of the aggregate owing to deamidation and specific degrad ation processes (cf. Section 7.6). Data concerning this ever-changing composition, as obtained since 1973 using fetal, calf, and 6- to 8-year-old cattle lenses, were surveyed by Hoenders (75).

Age-related changes in the polypeptide composition of β-crystallin were studied by Zigler (76). The properties of the constituent polypeptide chains were characterized by polyacrylamide gel electrophoresis in the presence of SDS and 6 *M* urea and by isoelectric focusing in 6 *M* urea. Applying the first method, three to four polypeptides were found in β_L-crystallin and seven to eight in β_H-crystallin. Both β_L- and β_H-crystallin showed marked, age-related shifts in the relative abundance of their component polypeptide chains. Zigler's data suggest that post-translational modifications may occur in bovine β-crystallin.

Continued investigation of low molecular weight crystallins from young calf and older cattle lenses by Kabasawa et al. (77) supported the earlier reported presence of a distinct cattle lens γ-crystallin of 24,000 dalton. In the cortical fibers of the cattle lens this protein is present, whereas calf γ-crystallin of 20,000 dalton is almost absent. On the other hand, the 24,000-dalton γ-crystallin is present in only minor amounts in the calf whole lens extract. The 24,000-dalton γ-crystallin in the newer fibers of the cattle cortex was found to differ in charge and immunochemical properties from the 20,000-dalton γ-fraction characteristic for the young lens.

The established data of the age-dependent alterations of the α-crystallin subunits is reflected by a pronounced size and charge microheterogeneity of this multisubunit protein. The nature and degree of this heterogeneity and the physicochemical properties of the heterogeneous population formed have been thoroughly studied by Siezen et al. (78, 79). Since only calf cortex α-crystallin was used, it is beyond the scope of this chapter to further discuss the results. Possible relationships between the age-dependent degradation of polypeptide chains and the aggregation state of α-crystallin in the nucleus of calf lenses were also studied by Siezen et al. (66) and were mentioned in Section 7.5.2.2.

7.5.4 Fluorescence

Although fluorescence of the lens was first described more than 100 years ago, it was not until the end of the 1960s that research in this field became more intensive. In a long series of papers (28, 46, 67, 68, 80–85b) by Lerman et al, the field of photochemistry and lens aging has been covered up to the present. In this section only fluorescence properties insofar as they are relevant to the aging lens proteins are considered.

Lens fluorescence may be "intrinsic" or "extrinsic." Most proteins are endowed with an intrinsic ultraviolet fluorescence because they contain aromatic amino acid residues (phenylalanine, tyrosine, and tryptophan). Phenylalanine has a low fluorescence quantum yield and its contribution is virtually negligible. But also the fluorescence of tyrosine is overshadowed by that of tryptophan, even if only one residue of it is present in a protein containing 10 or 20 tyrosine residues. The tryptophan fluorescence maximum of proteins varies from 330 to 352 nm, depending on the protein. Whole lenses of all ages (human and rat) show a tryptophan fluorescence maximum at 332 nm if excited at 280 to 300 nm (28, 46, 84). This fluorescence seems not to be age-dependent.

The extrinsic fluorescence of the normal human lens is clearly age-dependent and seems to be caused by at least two fluorophors. From the work of Lerman et al. (28, 46, 67, 68, 80, 82, 84) it can be inferred that these fluorophors are characterized by excitation/emission wavelength maxima of, respectively,

360 to 370/420 to 444 nm
and
400 to 435/500 to 520 nm

A new fluorophor with maximal excitation at 514.5 nm, showing a broad emission peak with a maximum of 556.4 nm, was recently discovered by Kuck and Yu (86), using a Raman spectrometer with laser as a fluorometer. This fluorophor was absent in lenses 7 years of age and younger, detectable at 23 years, and clearly present in older lenses. Still another, red fluorescence (excitation/emission at 647.1 and 672 nm, respectively) turned out to be exhibited by clear human lenses after the seventh decade of life; in this study intact lenses with ages between 1 and 95 years have been investigated with a laser Raman spectrometer (86a). Regarding the chemical nature of these fluorophors, so far only presumptions could be put forward.

The yellow to brown coloration of the human lens appears to be due to fluorophors which are associated with protein (87, 88). The process of increasing coloration of the lens nucleus with age seems to be induced by prolonged exposure of the lens to ultraviolet radiation above 295 nm, where the cornea no longer absorbs. Moreover, it has been suggested that the fluorophors may serve as cross-links between crystallins (43), thus possibly participating in aggregation and insolubilization with aging (84, 88, 89).

The increase of fluorescence, as well as the increase in insoluble protein levels and a decline of sulfhydryl group concentration, may be considered as a specific aging parameter (28, 46, 68, 84). In normal human lenses within the first year of life no nontryptophan fluorescence was present but thereafter it became manifest (82, 85).

The age-related production of protein-associated, nontryptophan fluorescence in the human lens turned out to be most pronounced in the water-insoluble and the high molecular weight fraction. Within these fractions it was mainly associated with a 43,000 to 45,000 dalton and a 9,600 to 11,000-dalton polypeptide fraction (51, 61, 87, 90). The nature and structure of the fluorescent groups in these polypeptides is so far unknown, but a number of characteristics of these polypeptides and their proteolytic fragments have already been described (51, 87, 89–91).

The blue fluorescence of bovine α-crystallin seems to be due to at least two fluorophors, characterized by excitation peaks at about 350 and 370 nm and fluorescence peaks at about 410 and 435 nm; Fujimori (92) found that, upon aging, in the nuclear region of bovine lenses low molecular weight α-crystallin shifts to noncovalently linked, higher molecular weight aggregates which are also blue fluorescent.

Rat and mouse lenses also show a direct relationship between the presence of a fluorophor (440–444 nm emission) and age (68, 81). It is striking that in avian lenses, which do not become presbyopic and lack a water-insoluble protein fraction, no fluorescence could be demonstrated (81).

7.5.5 Complex Formation

The crystallins show a tendency to form complexes. This property seems to increase with age: crystallin fractions obtained from young lenses are generally more homogeneous than those from old lenses.

Extracts from 5-, 11-, and 60-year-old clear human lenses were fractionated by Sandberg and Closs (93), applying ion exchange and gel chromatography. Analysis of the fractions obtained was performed by immunoelectrophoretic methods. The α- and β-crystallin fraction of young lenses showed traces of γ-crystallins. The α-crystallin fraction of old lenses contained significant amounts of β- and γ-crystallins, the β-crystallin fraction significant amounts of γ-crystallin.

Manski et al. (94) studied the >300,000 dalton crystallin fraction obtained by ultrafiltration of extracts from young and mature cattle lenses. They demonstrated complex formation of α-crystallin with β- and γ-crystallin by immunoelectrophoretic analysis. The >300,000 dalton fraction from young lens tissue was separated by gel chromatography into free α-crystallin followed by an α-β-crystallin complex, that from mature lenses into two complexes of quantitatively different α-β-γ-crystallin composition. Specific dissociation of the complexes was obtained by the use of anti-α- and anti-β-crystallin immunoadsorbents; the results point to a potentially central role of β-crystallins in complex formation. The age-dependence of the stability of the crystallin complexes was subsequently studied by exposure to acid or alkaline pH and to high salt concentrations (94a). An increase of the stability with progressive aging was observed, but in a constant electric field the complexes dissociated into free α-, β-, and γ-crystallin. In this connexion it should be kept in mind that the phenomenon of complex formation can, at least partially, be interpreted in terms of the known heterogeneity of high molecular weight protein (2).

7.5.6 Immunochemical Characteristics

7.5.6.1 Bovine Lens

Certain changes in the immunoelectric behavior of soluble crystallins from bovine lenses with age were reported by Bours and Hockwin (21). The results obtained by isoelectric focusing, as given in the same paper, were much more detailed and more relevant to age-dependent changes of crystallins than those obtained by immunoelectrophoresis (cf. Section 7.2.1). It is striking that the bovine nucleus revealed a rather simple immunoelectrophoretic pattern (one α-, one β- and one γ-crystallin precipitation line), whereas the equator and cortex showed the additional presence of pre-α-crystallin (FM crystallin) and two to three β-crystallin precipitation lines.

The crystallin composition of bovine lens epithelium related to aging has been studied immunoelectrophoretically by Bours et al. (94b). Applying polyvalent and specific antisera, two α-, one pre-α-, three β-, and one γ-crystallin were observed. With aging, one of the two α-crystallins disappeared. The epithelial γ-crystallin differed structurally from the γ-crystallins in the equator and the nucleus.

Studying β_H- and β_L-crystallin from bovine lenses of widely varying age, Zigler (76) demonstrated marked age-related shifts in polypeptide chain composition by means of electrophoretic methods. His data suggest that post-translational modifications, comparable with those found for α-crystallin, are responsible for the age-related changes. Immunodiffusion studies, using antisera specific for β_H and β_L-crystallin preparations, however, did not reflect these age-related changes.

7.5.6.2 Human Lens

Crossed immunoelectrophoresis and fused rocket electrophoresis were applied by Sandberg and Closs (93) to identify the various antigens in chromatographic fractions obtained from homogenates of individual young (5, 11 years) and old (60 years) lenses. A mixture of two antisera prepared against α- and γ-crystallin was used for these experiments. The authors were able to define 16 antigenic components, three with α-, four with β- and nine with γ-crystallin mobility, but they make the reservation that some of these antigens may not be crystallins but are instead various cell constituents (in order to reduce the number of noncrystallin antigens, the lenses had been decapsulated). Much more complicated crossed immunoelectrophoretic patterns were obtained for the old lenses in comparison with those for the young ones. The data obtained are interpreted as complex formation between crystallins which increases with age (cf. Section 7.5.5).

Investigating only the age range of 60 to 70 and 70 to 80 years, Ringens et al. (34) concluded from their immunodiffusion results that no essential changes can be detected in the properties and composition of the soluble proteins.

7.5.7 Size and Molecular Weight of α-Crystallin

Except for the polypeptide composition, data concerning the architecture of low molecular weight α-crystallin and especially its changes with age are scarce (2). From the data available one inclines to the idea that the original, newly synthesized, and assembled α-crystallin is subject to a growth process on aging until it has reached a certain size, after which superaggregation starts.

It has been shown by Bloemendal et al. (95), van Kamp et al. (96), and Li (97) that the molecular weight of the older nuclear α-crystallin is greater than that of the cortical α-crystallin. By means of electron microscopy, Siezen et al. (66) observed an increase in average diameter of calf α-crystallin from 13.9 to 15.0 nm going from cortex to nucleus, agreeing with an increase in average sedimentation constants from 18.8 to 23.3S. These data correspond to a molecular weight increase of about 25 to 35%, thus leading to an average molecular weight of 1.1×10^6 for nuclear α-crystallin. The globular units in the higher molecular weight, "oligomeric" α-crystallin population from calf lens nucleus showed the same average diameter, and it seems that on aging "monomeric," low molecular weight α-crystallin increases in size concentrically until a critical diameter of about 15 nm is reached, after which it begins to polymerize.

A number of new data regarding aspects of the architecture of α-crystallin from calf cortex have been published since 1977 (78, 79, 98–102c) but it is beyond the scope of this chapter to discuss them.

7.6 CHANGES ON POLYPEPTIDE LEVEL

7.6.1 Deamidation and Degradation

In the last 10 years it has become clear that *de novo* synthesized A_2 and B_2 polypeptide chains of bovine α-crystallin are subject to two types of postsynthetic modifications: deamidation and specific *C*-terminal degradation (chain shortening) (Figure 13). The deamidation products A_1 and B_1 appear only in the later stages of fetal life and reach their highest level in the old bovine cortex, where another putative deamidation product also becomes prevalent (103). The degradation of α-crystallin chains is apparently a postsynthetic process which starts early in embryonic life. The number of degraded chains is, however, clearly largest in the old bovine nucleus (103). It seems justified, therefore, to consider these degradations as an aging process. Although several features make it unlikely that the degradations observed are the result of enzymatic processes, the latter assumption cannot be ruled out completely. The present state of knowledge of the bovine α-crystallin polypeptides is summarized in Table 10.

Changes of the α-crystallin subunits were also observed in 3- to 5-month-old rabbit lenses applying isoelectric focusing in the presence of urea (39, 75). So far no data concerning the primary structure of these polypeptides

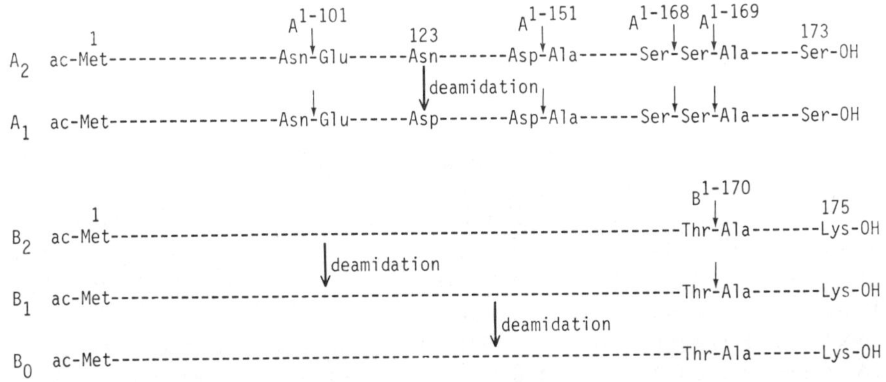

Figure 13 Polypeptide chains of bovine α-crystallin and their postsynthetic modifications. Taken from ref. 18.

Table 10 Polypeptide Chains of α-Crystallin from Calf and Bovine Lenses

Name	Character	Occurrence	Degradation site
A_2	Primary gene product	Total calf + bovine lens	
A_2^{1-169}	Degraded A_2	Prenatal lens	↓ 170 -Ser-Ser-Ala-Pro-
A_2^{1-168}	Degraded A_2	Lens nucleus	↓ 170 -Pro-Ser-Ser-Ala-
A_1	Deamidated A_2	Calf + bovine lens cortex	
A_1^{1-168}	Degraded A_1	Lens nucleus	↓ 170 -Pro-Ser-Ser-Ala-
A_1^{1-151}	Degraded A_1	Lens nucleus	150 ↓ -Val-Asp-Ala-Gly-
A^{1-101}	Degraded A	Lens nucleus	100 ↓ -His-Asn-Glu-Arg-
A_2^{1-151}	Degraded A_2	Lens nucleus	150 ↓ -Val-Asp-Ala-Gly-
B_2	Primary gene product	Total calf + bovine lens	
B_1	Deamidated B_2	Total calf + bovine lens	
B_1^{1-170}	Degraded B_1	*In vitro*	170 ↓ -Val-Thr-Ala-Ala-
B_0	Deamidated B_1	Bovine lens cortex	
B_2^{1-170}	Degraded B_2	Lens nucleus	170 ↓ -Val-Thr-Ala-Ala-

Taken from ref. 103.

are available and therefore it cannot be judged whether the changes observed are postsynthetic in nature.

The polypeptides of α-crystallin from 60- to 80-year-old human eye lenses were studied to determine whether they have been subject to similar postsynthetic alterations as known from bovine α-crystallin (104, 105). The human and bovine αA_2 chains differ in 10 out of 173, the αB_2 chains in three out of 175 positions (106, 107). Other major polypeptides turned out to be deamidated A_2 of normal length (172 residues), shortened A_2 chain ($A_2{}^{1-151}$), shortened B_2 chain ($B_2{}^{1-170}$), and deamidated B_2 of normal length

(175 residues). Additionally, a group of polypeptides with very acidic isoelectric points, showing a heterogeneous molecular weight distribution below 18,000 dalton and a high degree of similarity with the A chains as judged by tryptic peptide mapping, was found in low proportions. The authors concluded that, in spite of several differences in subunit composition, comparable deamidation and degradation processes take place in human and bovine α-crystallin as a function of age.

Determination of the amide groups of human α-crystallins showed that almost one-quarter of the amide groups of the fetal protein is lost before birth and that a further small loss may occur after birth, but no significant loss between the ages of 34 and 79 years (16, 108). In an interesting attempt to relate changes found in lens proteins in aging to theories of aging, Harding (16) stated that the changes described in the lens certainly appear to be post-translational and that hypotheses which ascribe aging changes to translational or transcriptional events, are not relevant (compare Chapter 8).

A remarkable finding regarding the age-related formation of high molecular weight β-crystallin (B_H) was reported by Vermorken et al. (109) and Asselbergs et al. (109a). These authors hypothesized that the formation of the β_H aggregate requires the presence of a specific polypeptide (βB_{1b}) which presumably originates from post-translational modification of a precursor (βB_{1a}) of which the messenger RNA could be isolated. The results shown suggest that βB_{1a} is subject to degradation.

The polypeptide composition of β_H- and β_L-crystallin from bovine lenses of widely varying age was studied electrophoretically by Zigler (76). Marked age-related shifts in relative polypeptide abundance, both among the polypeptides common to β_H and β_L and those unique to β_H, were observed. Although the occurrence of postsynthetic modifications could not be demonstrated by these experiments, Zigler believes that they may play a role.

7.6.2 Oxidation of Cysteine

According to Lerman (28) a decline in sulfhydryl groups and a concomitant increase in the level of disulfide bonds in human and rat lenses would be definitely age-related. The experimental evidence provided is limited, however.

The state of sulfhydryl groups in the proteins from the nucleus of 16 normal human lenses (18 to 63 years) was investigated by Anderson and Spector (48). The insoluble proteins, which increase in proportion from 20 to 55% with age (cf. Section 7.3.2), showed a total level of sulfhydryl + disulfide groups ranging from 100 to 170 μmol/g protein and no trend of this level with aging. The soluble proteins, regardless of age, had a much higher level (200 to 290 μmol/g protein). Definite changes in the distribution of protein SH and SS groups were found to accompany aging. In the 60-year-old lenses disulfides in the insoluble or soluble protein fraction averaged higher than in the younger ones and amounted to 15 to 30% of the SH + SS level. The total amount of the nuclear SH + SS groups turned out to remain reasonably constant. It could not be deduced from these experiments whether the SH groups of the soluble proteins are oxidized prior to or after they become insoluble.

In a later paper (48a) the above data have been extended to the cortical region. In normal human lenses between 33 and 80 years of age, protein and total SH+SS levels of cortex and nucleus remain fairly constant. Protein disulfides, not detectable in the normal 18-year-old cortex, accumulate by 63 years of age to a level of about 9% of the total cortical SH+SS and 17% of the total nuclear SH+SS groups. Nonprotein disulfides are present at low levels in the normal aging lens.

Van Haard et al. (109b) confirmed that no significant correlation between age and SH content of normal human nuclei and cortices exists in the age range from 40 to 90 years. Moreover, these authors demonstrated a significant relationship between the distribution of nonprotein SH compounds among the lens parts and age.

The notorious point of possible artifactual oxidation of sulfhydryl groups during sample preparation was avoided by East et al. (110). In this impressive pioneering study the authors applied laser Raman spectroscopy to intact fresh rat and mouse lenses in a nutrient medium. In this technique the Raman spectrum of a selected small zone (50 μm diameter $\times$ 0.6 mm length) of the lens could be obtained. It could be shown that in the aging rat and mouse lens nucleus a quantitative conversion of SH to SS groups takesplace which can only be a postsynthetical process. In the aging cortex a similar decrease in SH concentration with much less formation of SS was correlated with changes in tyrosine/phenylalanine ratio, indicating marked alterations in the ratios at which the various crystallins are synthesized. Another interesting point, the accessibility of SH groups as a function of age, was also studied by Raman spectroscopy after exchange of SH for SD in heavy water. Accessibility of SH groups was more rapid and nearly complete in the young rat lens as compared to the older one.

Quantitative measurements of the SH Raman intensity along the visual axis of intact rat and mouse lenses of different ages have been reported by the same group in a later paper (110a). In general, the SH profiles exhibited two maxima in the cortex (one anterior, one posterior) and a central minimum, except the youngest, 26-day-old rat lens, which showed a bell-shaped curve. A plot of SH level versus age for certain distances from the center along the visual axis indicated a steady decrease in SH levels. According to the authors, the Raman technique may provide a means whereby *in vivo* spectroscopic measurements can be made. For the present, the method can be used to prepare profiles of other protein "subgroups," particularly disulfide, tryptophan, and tyrosine.

7.6.3 Racemization of Aspartic Acid

An interesting new aspect of lens protein aging, namely, racemization of aspartyl residues, was recently discovered by Masters et al. (111, 112). The D/L enantiomeric ratios for alanine, glutamic acid, proline, phenylalanine, and leucine were also determined, but no racemization within the limits of detection by the technique used (gas chromatography of diastereomeric dipeptides) was observed. D-Aspartyl residues were shown to accumulate in normal human lens nuclear proteins at a rate of 0.14% per year. In their second paper (112) the authors proposed that D-aspartic acid should be distributed in a predictable manner if racemization is to any degree responsible for the structural changes in lens proteins during aging. It should occur predominantly in the nucleus and D/L Asp ratios ought to be higher in the insoluble than in the soluble fraction. Indeed, high D/L Asp ratios (about 0.18) were observed in high molecular weight aggregates and in the water-insoluble fraction from whole lenses and their cortical and nuclear regions. Purified α-, β-, and γ-crystallins showed low D/L ratios (0.04–0.08). Within the water-insoluble fraction high D/L ratios occurred in degraded and cross-linked material (about 0.21). The authors conclude that racemization may trigger aggregation of lens proteins, which are then converted by other reactions (degradation, covalent cross-linking) to water-insoluble material seen in aging lenses. As a possible mechanism they propose that conformation changes resulting from racemization of a few aspartyl residues would expose the hydrophobic amino acids that compose a very high proportion of the total amino acids in α-crystallin, the major component in the high molecular weight fraction; this would lead to aggregation through hydrophobic interactions.

Racemization of aspartic acid in human lens proteins was also studied by Garner and Spector (113). They found a D/L Asp ratio of 0.036 for the water-soluble fraction of whole lenses, remaining quite constant with age. For the total water-insoluble fraction (including the high molecular weight protein fraction) an increase of the D/L Asp ratio with 0.17% per year was determined. As far as these experiments are comparable with those of Masters et al. (112), similar results were obtained in both studies. Remarkable differences between the D/L Asp ratios of polypeptide fractions, isolated from the water-soluble and the water-insoluble fractions of cataractous lenses, throw doubt upon the conclusion of Masters et al. (112) that racemization may be a requirement for cross-linking or degradation of polypeptide chains (cf. Chapter 8).

7.6.4 Covalent Nondisulfide Cross-Links

It was assumed by Pirie (114) and clearly demonstrated by Buckingham (115, 116) that cataractous lenses with dark-brown nuclei contain nondisulfide covalently cross-linked protein of relatively high molecular weight. The proportion of this material is low in old normal lenses but increases progressively with coloration in nuclear cataracts (44). The brown color is largely associated with the cross-linked protein fraction and a correlation seems to exist. Nontryptophan fluorescence, which increases during development of nuclear cataract, is also associated with this cross-linked protein (117–119). In spite of many efforts it has not yet clearly been shown what the chemical nature of the cross-link(s) is. Although most of the recent investigations aiming at its elucidation were performed with cataractous lenses, some new findings may be relevant to the aging of proteins in normal lenses.

Products with β-carboline structure (Figure 14) were isolated from cataractous lenses by Dillon et al. (120). Proteins obtained from 45- and 77-year-old normal human lenses contained structures Ia and Ib of Figure 14. Essentially none of these components was detected in the protein of young human and calf lenses. Structure III might in the opinion of the authors be an attractive cross-linking agent between polypeptide chains. It is probable that tryptophan is the precursor for the formation of the β-carboline structure.

Moreover, 3,3'-bityrosine and anthranilic acid (Figure 15) were isolated from cataractous lenses (121, 122) after proteolytic digestion of the insoluble yellow protein fraction (123) by means of ion exchange, gel, and high-pressure liquid chromatography. Anthranilic acid was absent from the digest of normal lenses and was suggested to be an oxidation product of tryptophan during nuclear cataract formation. It was also suggested that anthranilic acid is derived from a fluorescent precursor present in the protein and that this precursor is transformed only after proteolytic digestion (122). No bityrosine was detected in any of the proteins from decapsulated human nuclear cataracts (123a); a small amount of bityrosine was, however, found in the capsule of a normal human lens.

Whether β-carbolines and an anthranilic acid precursor virtually serve as

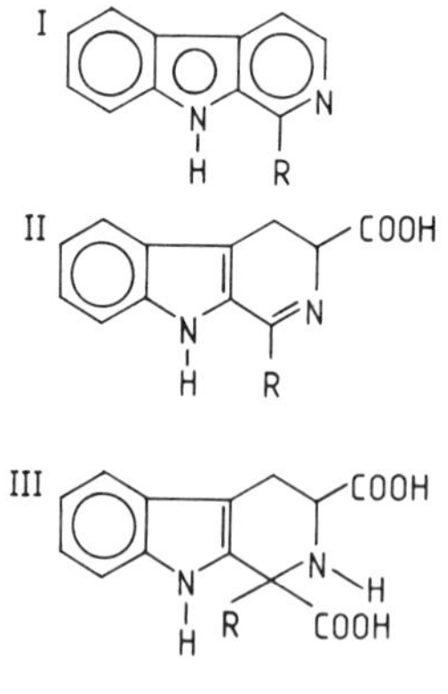

Figure 14 β-carboline structures. I, Unsaturated β-carbolines; II, 3,4-dihydro-β-carboline-3-carboxylic acid; III, 1,2,3,4-tetrahydro-β-carboline-1,3-dicarboxylic acid. (a) R = H or (b) CH_3. Reprinted with permission from ref. 120.

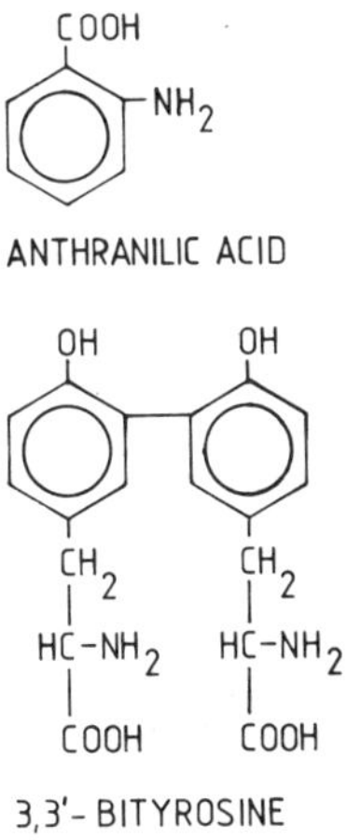

Figure 15 Structural formulas of 3,3′-bityrosine and anthranilic acid.

a cross-links is still an open question. It should also be noted that van Haard et al. (91) investigated the presence of collagenlike cross-links, isopeptide bonds, and bityrosine bonds in nuclear cataracts but that their results were negative.

7.6.5 Photooxidation of Tryptophan

It is likely that lens proteins, as do other proteins, undergo certain changes by ultraviolet irradiation. Therefore, part of age-related changes in lens proteins may be attributed to sunlight or artificial light sources. Electromagnetic vibrations in the wavelength range of 300 to 400 nm are suspected of causing damage to the lens. Radiant energy emerging from shorter wavelengths is absorbed by the cornea and the aqueous humor. For mammals approximately one-third of the 300 to 400-nm radiation energy is absorbed by the cornea and some 2% by the aqueous humor (124). The main target for this near-UV radiation seem to be tryptophan residues and the primary photooxidation product is *N*'-formylkynurenine (125). It is hypothesized that the photooxidation of tryptophan results in the formation of yellow, nontryptophan fluorescent compounds and covalent cross-links.

It is possible that brown nuclear cataract is a consequence of photooxidation of lens proteins induced by sunlight. The reader is referred to Chapter 8 where the literature concerning photooxidation processes in the lens is dealt with in more detail. As to aging of normal lenses some more recent publications are very briefly described below.

The changes taking place in nuclear cataract formation, such as increase of water-insoluble protein, intensification of the yellow color of the lens, and increase of the nontryptophan fluorescence also occur but to a much lesser degree with aging of the normal human lens (87, 88, 126–128).

Using normal human lenses of all ages as well as rat and mouse lenses, Lerman and Borkman (28, 46, 82–84, 129, 130) investigated the mechanism of UV-induced changes and concluded that at least one fluorophor is derived from tryptophan by a free radical induced photooxidation reaction. Further data regarding the role of tryptophan in age-related changes of lens proteins were provided by Zigman (124, 131). It was stated by Griess et al. that calf γ-crystallins are particularly sensitive to photomodification and that this process may be associated with age-related conformational changes (132).

Product analysis after exposition of *N'*-formylkynurenine, the primary photooxidation product of tryptophan (125), to sunlight at neutral pH led to the identification of *N*-formylanthranilic acid and 4-hydroxyquinoline (133). These compounds were, however, not detected after photooxidation of the dipeptides Gly-Try or Try-Gly, although the indole ring was split. According to the authors it seems unlikely that 4-hydroxyquinoline could be formed within the protein chain whereas *N*-formylanthranilic acid, if formed from *N'*-formylkynurenine within a protein, would be split out.

7.7 CONCLUDING REMARKS

The mechanisms underlying the process of aging are still far from being understood. As an inevitable result one cannot yet discriminate between causes and consequences of senescence. If one pursues the fate of lens proteins during the life-span of an animal, it seems very likely that post-translational events such as aggregation, insolubilization, racemization, and selective polypeptide chain shortening are involved in aging, but it is still too early to integrate these phenomena into a well-founded interpretation. However, in our opinion the crystallins are suitable models for aging studies owing to the unique features of the eye lens. There is presumably no other tissue in which one can, in principle, distinguish so simply between post-translational modificaton and *de novo* biosynthesis as in the lens.

It has to be taken for granted that still much more solid experimental evidence concerning alterations of lens proteins in dependence of age is required. For instance, the following questions must be answered:

- Do proteins disappear from the lens?
- To what extent do crystallins participate in the high molecular weight and the water-insoluble fractions?
- Does a balance exist between catabolized and newly synthesized proteins?
- Are certain crystallins synthesized only at certain stages of the life-span?

It is of utmost importance that for a well-defined characterization of the high molecular weight and the water-insoluble fractions standardized procedures be adopted by the individual students of the lenticular proteins. The origin of many polypeptides, occurring in older lens parts, has still to be elucidated. There is also a need to establish the exact role of nonprotein substances in the insolubilization process of crystallins. Another question that still awaits an answer is whether an alteration in water content is due to increasing hydrophobicity accompanied with changes in conformation of crystallins, or *vice versa*. Photooxidation of aromatic amino acid residues in lens proteins deserves attention as a possible principle responsible for radical formation and subsequent cross-linking processes.

Modern physical methods, for instance, quasi-electric laser light scattering (59) and laser Raman spectroscopy (86, 110), applied to intact lenses represent most useful extensions of current analytical tools in protein biochemistry, since interference by artifact formation is avoided.

Fundamental problems, such as

- What is the real nature and significance of covalent cross-links?
- Which factors induce superaggregation and insolubilization?

have to be solved in order to obtain a better insight into the pathways of "physiological aging" of lens proteins. Only then might it become clear whether senile cataract can be considered as precocious aging and what stresses over and above the normal effects of aging (1) are responsible for it.

REFERENCES

1 R. van Heyningen, *Sci. Am.*, **70,**70 (1975).

2 J. J. Harding and K. J. Dilley, *Exp. Eye Res.*, **22,** 1 (1976).

3 O. Hockwin, H. R. Koch, Ch. Ohrloff, and J. Bours, *Klin. Monatsbl. Augenheilkd.*, **169,** 165 (1976).

4 H. Bloemendal, in J. Paul, Ed., *Biochemistry of Cell Differentiation II,* Vol. 15, University Park Press, Baltimore, 1977, p. 195.

5 H. Bloemendal, *Science,* **197,** 127 (1977).

6 J. Nordmann, *Adv. Ophthalmol.*, **34,** 1 (1977).

7 J. C. Dreyfus, J. Banroques, L. Poenaru, H. Skala, and M. Vibert, *Interdiscipl. Topics Gerontol.*, **12,** 147 (1978).

8 B. L. Strehler and C. H. Barrows, in O. A. Schjeide and J. de Vellis, Eds., *Cell Differentiation,* van Nostrand Reinhold, New York-London, 1970, p. 266.

9 E. L. Schneider, Ed., *The Genetics of Aging,* Plenum, New York-London, 1978, p. 2.

10 T. Makinodan, in E. Yunis, Ed., *Immunology and Aging,* Plenum, New York-London, 1978, pp. 1–7.

11 L. E. Orgel, *Proc. Natl. Acad. Sci. U.S.*, **49,** 517 (1963).

12 J. C. Dreyfus, H. Rubinson, F. Shapira, A. Weber, J. Marie, and A. Kahn, *Gerontology,* **23,** 211 (1977).

13 A. B. Robinson, J. H. Mc Kerrow, and P. Cary, *Proc. Natl. Acad. Sci. U.S.,* **66,** 753 (1970).

14 A. B. Robinson and C. J. Rudd, *Curr. Topics Cell Reg.,* **8,** 247 (1974).

15 A. B. Robinson, *Mech. Ag. Dev.,* **9,** 225 (1979).

16 J. J. Harding, *Inserm,* **60,** 215 (1976).

17 A. L. Goldberg and A. C. St. John, *Ann. Rev. Biochem.,* **45,** 747 (1976).

18 F. S. M. van Kleef, W. W. de Jong, and H. J. Hoenders, *Nature,* **258,** 264 (1975).

19 M. J. Voaden, *Exp. Eye Res.,* **7,** 313 (1968).

20 J. Bours and O. Hockwin, *Berl. Muench. Tierärztl. Wochenschr.,* **89,** 417 (1976).

21 J. Bours and O. Hockwin, *Klin. Monatsbl. Augenheilkd.,* **170,** 51 (1977).

22 J. Bours, A. Weber, and O. Hockwin, *Inserm,* **60,** 205 (1976).

23 J. Bours, A. Wieck, and O. Hockwin, *Interdiscipl. Topics Gerontol.,* **12,** 205 (1978).

24 J. Bours, K. Doepfmer, and O. Hockwin, *Doc. Ophthalmol. Proc. Ser.,* **8,** 75 (1976).

25 K. N. Liem Thé, A. L. H. Stols, and H. J. Hoenders, *Exp. Eye Res.,* **20,** 307 (1975).

26 J. Bours, L. Gruber, O. Hockwin, and J. Harris, *Interdiscipl. Topics Gerontol.,* **12,** 127 (1978).

27 J. Bours, *Interdiscipl. Topics Gerontol.,* **12,** 196 (1978).

27a J. Bours and P. Delmotte, *Science Tools,* **26,** 58 (1979).

28 S. Lerman, *Doc. Ophthalmol. Proc. Ser.,* **8,** 241 (1976).

29 S. D. Coghlan and R. C. Augusteyn, *Exp. Eye Res.,* **25,** 603 (1977).

30 D. Roy and A. Spector, *Proc. Natl. Acad. Sci. U.S.,* **73,** 3484 (1976).

31 H. A. Kramps and H. J. Hoenders, *Inserm,* **60,** 237 (1976).

32 J. A. Kramps, Thesis, University of Nijmegen, (1977).

33 H. A. Kramps, H. J. Hoenders, and J. Wollensak, *Biochim. Biophys. Acta,* **434,** 32 (1976).

34 P. J. Ringens, K. N. Liem-Thé, H. J. Hoenders, and J. Wollensak, *Interdiscipl. Topics Gerontol.,* **13,** 193 (1978).

35 J. S. Zigler Jr., J. B. Sidbury, Jr., B. S. Yamanashi, and M. Wolbarsht, *Ophthal. Res.,* **8,** 379 (1977).

36 I. Kabasawa, *Abstr. 3rd Int. Congr. Eye Res., Osaka, May 1978,* XI-11.

37 J. A. Jedziniak, H. Baram, and L. T. Chylack, Jr., *Exp. Eye Res.,* **26,** 377 (1978).

37a J. Bours, *Comp. Biochem. Physiol.,* **65B,** 215 (1980).

38 K. N. Liem Thé, Thesis, University of Nijmegen, (1975).

39 K. N. Liem-Thé, P. J. Ringens, J. C. W. M. Holtslag, and H. J. Hoenders, *Interdiscipl. Topics Gerontol.,* **12,** 105 (1978).

40 K. Satoh and M. Bando, *Doc. Ophthalmol. Proc. Ser.,* **8,** 113 (1976).

41 O. Hockwin, *Interdiscipl. Topics Gerontol.,* **13,** 88 (1978).

42 J. J. Harding, K. C. Rixon, and F. H. C. Marriott, *Exp. Eye Res.,* **25,** 651 (1977).

42a S. Zigman and T. Yulo, *Comp. Biochem. Physiol.,* 63B, 379 (1979).

43 S. Zigman, J. Groff, T. Yulo, and G. Griess, *Exp. Eye Res.,* **23,** 555 (1976).

44 J. A. Kramps, H. J. Hoenders, and J. Wollensak, *Exp. Eye Res.,* **27,** 731 (1978).

45 H. Maisel, *Exp. Eye Res.,* **24,** 417 (1977).

46 S. Lerman and R. F. Borkman, *Interdiscipl. Topics Gerontol.,* **13,** 154 (1978).

47 L. J. Takemoto and P. Azari, *Exp. Eye Res.,* **24,** 63 (1977).

48 E. I. Anderson and A. Spector, *Exp. Eye Res.,* **26,** 407 (1978).

49 L. T. Chylack, Jr., *Arch. Ophthalmol.,* **96,** 88 (1978).

49a E. I. Anderson, D. D. Wright, and A. Spector, *Exp. Eye Res.,* **29,** 233 (1979).

50 D. Roy and A. Spector, *Exp. Eye Res.,* **26,** 429 (1978).

51 D. Roy and A. Spector, *Exp. Eye Res.,* **26,** 445 (1978).

51a P. M. M. van Haard, G. J. M. Martens, J. J. L. Mestrom, H. J. Hoenders, and J. Wollensak, *Ophthal. Res.,* **11,** 433 (1979).

51b W. H. Garner and A. Spector, *Docum. Ophthalmol. Proc. Series 18,* 91 (1978).

52 R. van Heyningen, *Exp. Eye Res.,* **13,** 155 (1972).

53 R. F. Fisher and B. E. Pettet, *J. Physiol.,* **234,** 443 (1973).

54 J. Klethi, *Doc. Ophthalmol. Proc. Ser.,* **8,** 213 (1976).

55 H. Rink, *Interdiscipl. Topics Gerontol.,* **12,** 271 (1978).

56 L. O. Ely, *Am. J. Ophthalmol.,* **32,** 215 (1949).

57 F. S. M. van Kleef, Thesis, University of Nijmegen, (1975).

58 O. Hockwin, F. Rast, H. Rink, J. Munninghoff, and H. Twenhoven, *Interdiscipl. Topics Gerontol.,* **13,** 102 (1978).

58a O. Hockwin, F. Rast-Czyborra, W. Schnitzlein, and H.-J. Müller, *Ophthal. Res.,* **11,** 136 (1979).

59 J. A. Jedziniak, D. F. Nicoli, H. Baram, and G. B. Benedek, *Invest. Ophthalmol.,* **17,** 51 (1978).

60 J. J. Harding, *Ophthal. Res.,* **9,** 123 (1977).

61 S. Zigman, J. Groff, and T. Yulo, *Photochem. Photobiol.,* **26,** 505 (1977).

62 D. Roy and A. Spector, *Exp. Eye Res.,* **22,** 273 (1976).

63 K. J. Dilley, *Exp. Eye Res.,* **20,** 73 (1975).

64 D. Roy and A. Spector, *Abstr. 3rd Int. Congr. Eye Res., Osaka, May 1978,* XI-2.

65 P. J. Ringens and H. J. Hoenders, *Abstr. 3rd Int. Congr. Eye Res., Osaka, May 1978,* XI-3.

66 R. J. Siezen, J. G. Bindels, and H. J. Hoenders, *Exp. Eye Res.,* **28,** 551 (1979).

66a P. Russell, S. G. Smith, D. A. Carper, and J. H. Kinoshita, *Exp. Eye Res.*, **29,** 245 (1979).

67 S. Lerman, B. S. Yamanashi, R. A. Palmer, J. C. Roark, and R. Borkman, *Ophthal. Res.*, **10,** 168 (1978).

68 S. Lerman, R. F. Borkman, and B. S. Yamanashi, *Abstr. 3rd Int. Congr. Eye Res., Osaka, May 1978,* XI-8.

69 J. A. Jedziniak, *Abstr. 3rd Int. Congr. Eye Res., Osaka, May 1978,* XI-5.

70 D. Roy and A. Spector, *Biochemistry,* **15,** 1180 (1976).

71 D. Roy and A. Spector, *Invest. Ophthalmol.,* **15,** 394 (1976).

72 A. Spector, J. Stauffer, D. Roy, L. K. Li, and D. Adams, *Invest. Ophthalmol.,* **15,** 288 (1976).

73 J. Horwitz, *Exp. Eye Res.,* **23,** 471 (1976).

74 I. Kabasawa, G. W. Barber, and J. H. Kinoshita, *Jap. J. Ophthalmol.,* **21,** 87 (1977).

75 H. J. Hoenders, *Doc. Ophthalmol. Proc. Ser.,* **8,** 105 (1976).

76 J. S. Zigler Jr., *Exp. Eye Res.,* **26,** 537 (1978).

77 I. Kabasawa, Y. Tsunematsu, G. W. Barber, and J. H. Kinoshita, *Exp. Eye Res.,* **24,** 437 (1977).

78 R. J. Siezen, J. G. Bindels, and H. J. Hoenders, *Eur. J. Biochem.,* **91,** 387 (1978).

79 R. J. Siezen and H. Berger, *Eur. J. Biochem.,* **91,** 397 (1978).

80 S. Lerman, J. F. Kuck Jr., R. F. Borkman, and E. Saker, *Ann. Ophthalmol.,* **8,** 558 (1976).

81 S. Lerman, J. F. Kuck, Jr., R. F. Borkman, and E. Saker, *Ophthal. Res.,* **8,** 213 (1976).

82 S. Lerman and R. F. Borkman, *Ophthal. Res.,* **8,** 335 (1976).

83 R. F. Borkman, A. Dalrymple, and S. Lerman, *Photochem. Photobiol.,* **26,** 129 (1977).

84 S. Lerman and R. Borkman, *Acta Ophthalmol.,* **56,** 139 (1978).

85 R. F. Borkman and S. Lerman, *Exp. Eye Res.,* **26,** 705 (1978).

85a S. Lerman and R. F. Borkman, *Metab. Pediat. Ophthalmol.,* **3,** 27 (1979).

85b S. Lerman, in:. F. Regnault, O. Hockwin and Y. Courtois, Eds., *Ageing of the Lens,* Elsevier/North Holland Biomedical Press, Amsterdam-New York-Oxford, 1980, p. 263.

86 J. F. R. Kuck Jr. and N. T. Yu, *Exp. Eye Res.,* **27,** 737 (1978).

86a N.-T. Yu, J. F. R. Kuck, Jr., and C. C. Askren, *Invest. Ophthalmol.,* **18,** 1278 (1979).

87 A. Spector, D. Roy, and J. Stauffer, *Exp. Eye Res.,* **21,** 9 (1975).

88 M. Bando, A. Nakajima, and K. Satoh, *Exp. Eye Res.,* **20,** 489 (1975).

89 M. Bando and A. Nakajima, *Abstr. 3rd Int. Congr. Eye Res., Osaka, May 1978,* XI-7.

90 J. Dillon, S. Garcia-Castineiras, M. Garner, and A. Spector, *Abstr. 3rd Int. Congr. Eye Res., Osaka, May 1978,* XI-6.

91 P. M. M. van Haard, J. A. Kramps, H. J. Hoenders, and J. Wollensak, *Interdiscipl. Topics Gerontol.*, **13,** 212 (1978).

92 E. Fujimori, *Biochim. Biophys. Acta,* **534,** 82 (1978).

93 H. O. Sandberg and O. Closs, *Exp. Eye Res.*, **27,** 457 (1978).

94 W. Manski, K. Malinowski, and G. Bonitsis, *Exp. Eye Res.*, **29,** 625 (1979).

94a K. Malinowski and W. Manski, *Exp. Eye Res.*, **30,** 537 (1980).

94b J. Bours, H.-D. Zauzig, and H. Rink, *Ophthal. Res.*, **11,** 351 (1979).

95 H. Bloemendal, T. Berns, A. Zweers, H. Hoenders, and L. Benedetti, *Eur. J. Biochem.*, **24,** 401 (1972).

96 G. J. van Kamp, L. H. M. Schats, and H. J. Hoenders, *Biochim. Biophys. Acta,* **295,** 166 (1973).

97 L. K. Li, *Exp. Eye Res.*, **18,** 383 (1974).

98 L. K. Li and A. Spector, *Exp. Eye Res.*, **26,** 419 (1978).

99 R. J. Siezen and H. J. Hoenders, *FEBS Lett.*, **80,** 75 (1977).

100 R. J. Siezen and H. J. Hoenders, *Interdiscipl. Topics Gerontol.*, **12,** 221 (1978).

101 J. G. Bindels, R. J. Siezen, and H. J. Hoenders, *Interdiscipl. Topics Gerontol.*, **12,** 253 (1978).

102 R. J. Siezen, F. G. M. Coenders, and H. J. Hoenders, *Biochim. Biophys. Acta,* **537,** 456 (1978).

102a R. J. Siezen and H. J. Hoenders, *Eur. J. Biochem.*, **96,** 431 (1979).

102b R. J. Siezen, J. G. Bindels, and H. J. Hoenders, *Eur. J. Biochem.*, **107,** 243 (1980).

102c J. G. Bindels, R. J. Siezen, and H. J. Hoenders, *Ophthal. Res.*, **11,** 441 (1979).

103 F. S. M. van Kleef, W. Willems Thijssen, and H. J. Hoenders, *Eur. J. Biochem.*, **66,** 477 (1976).

104 J. A. Kramps, W. W. de Jong, H. J. Hoenders, and J. Wollensak, *Interdiscipl. Topics Gerontol.*, **13,** 225 (1978).

105 J. A. Kramps, W. W. de Jong, J. Wollensak, and H. J. Hoenders, *Biochim. Biophys. Acta,* **533,** 487 (1978).

106 W. W. de Jong, E. C. Terwindt, and H. Bloemendal, *FEBS Lett.*, **58,** 310 (1975).

107 J. A. Kramps, B. M. de Man, and W. W. de Jong, *FEBS Lett.*, **74,** 82 (1977).

108 K. J. Dilley and J. J. Harding, *Biochim. Biophys. Acta,* **386,** 391 (1975).

109 F. J. M. Vermorken, P. Herbrink, and H. Bloemendal, *Eur. J. Biochem.*, **78,** 617 (1977).

109a F. A. M. Asselbergs, M. Koopmans, W. J. van Venrooij, and H. Bloemendal, *Exp. Eye Res.*, **28,** 475 (1979).

109b P. M. M. van Haard, B. M. de Man, H. J. Hoenders, and J. Wollensak, *Ophthal. Res.*, **12,** 118 (1980).

110 E. J. East, R. C. C. Chang, N. T. Yu, and J. F. R. Kuck, *J. Biol. Chem.*, **253,** 1436 (1978).

110a C. C. Askren, N.-T. Yu, and J. F. R. Kuck, Jr., *Exp. Eye Res.,* **29,** 647 (1979).

111 P. M. Masters, J. L. Bada, and J. S. Zigler, Jr., *Nature,* **268,** 71 (1977).

112 P. M. Masters, J. L. Bada, and J. S. Zigler, Jr., *Proc. Natl. Acad. Sci. U.S.,* **75,** 1204 (1978).

113 W. H. Garner and A. Spector, *Proc. Natl. Acad. Sci. U.S.,* **75,** 3618 (1978).

114 A. Pirie, *Invest. Ophthalmol.,* **7,** 634 (1968).

115 R. H. Buckingham, *Biochem. J.,* **124,** 54P (1971).

116 R. H. Buckingham, *Exp. Eye Res.,* **14,** 123 (1972).

117 R. C. Augusteyn, *Jap. J. Ophthalmol.,* **18,** 127 (1974).

118 R. C. Augusteyn, *Ophthal. Res.,* **7,** 217 (1975).

119 K. J. Dilley and A. Pirie, *Exp. Eye Res.,* **19,** 59 (1974).

120 J. Dillon, A. Spector, and K. Nakanishi, *Nature,* **259,** 422 (1976).

121 R. J. W. Truscott, K. Faull, and R. C. Augusteyn, *Ophthal. Res.,* **9,** 263 (1977).

122 S. Garcia-Castineiras, J. Dillon, and A. Spector, *Exp. Eye Res.,* **26,** 461 (1978).

123 R. J. W. Truscott and R. C. Augusteyn, *Exp. Eye Res.,* **24,** 159 (1977).

123a M. K. McNamara and R. C. Augusteyn, *Exp. Eye Res.,* **30,** 319 (1980).

124 S. Zigman, J. Groff, and T. Yulo, *Photochem. Photobiol.,* **26,** 437 (1977).

125 A. Pirie, *Biochem. J.,* **125,** 203 (1971).

126 H. Mach, *Klin. Mbl. Augenheilk.,* **143,** 689 (1963).

127 R. Clark, S. Zigman, and S. Lerman, *Exp. Eye Res.,* **8,** 172 (1969).

128 K. Satoh, *Exp. Eye Res.,* **14,** 53 (1972).

129 R. F. Borkman, *Photochem. Photobiol.,* **26,** 163 (1977).

130 R. F. Borkman and S. Lerman, *Exp. Eye Res.,* **25,** 303 (1977).

131 S. Zigman, *Doc. Ophthalmol. Proc. Ser.,* **8,** 267 (1976).

132 G. A. Griess, S. Zigman, and T. Yulo, *Mol. Cell Biochem.,* **12,** 9,(1976).

133 A. Pirie and K. J. Dilley, *Photochem. Photobiol.,* **19,** 115 (1974).

CHAPTER

8

Changes in Lens Proteins in Cataract

John J. Harding

Nuffield Laboratory of Opththalmology, University of Oxford, Walton Street, Oxford, Great Britain,

8.1 INTRODUCTION

Many manifestations of cataract are exaggerations of changes found in the aging lens (1), so this chapter has some common ground with the chapter on aging of lens proteins (Chapter 7). Membrane proteins and enzymes are discussed elsewhere in this book, leaving changes in the intracellular structural proteins of the lens during cataract formation to be dealt with here. Results from experimental cataract and human cataract are given side by side, with emphasis on human cataract because it has been studied in more detail recently.

The lens substance consists largely of protein and probably has a higher protein content than any other tissue, rising to almost 70% at the center of a rat lens but to only 35% at the center of a human lens (2). Irregularities in protein distribution resulting in sharp changes of refractive index could be the basis for the opacities in some cataracts (3, 4). The human lens grows throughout life without shedding cells. Cells from the epithelium divide and elongate at the equator, starting to synthesize β- and γ-crystallins only as elongation occurs (5–8), and the cells build up on each other to form a younger cortex outside the older nucleus of the lens (Figure 1). Further information on the crystallins of normal lens is found elsewhere in this book and in the reviews of Harding and Dilley (9) and Bloemendal (10). In some places I have omitted early references that can be found in these reviews.

Synthesis of proteins cannot be detected in the central region of calf (11) and human lens (12) and there is no evidence for degradation, so protein may remain there throughout adult life, perhaps for more than 60 years in the human lens. In spite of this longevity few changes have been detected in the proteins of clear lenses although they are exposed to the full range of visible light in an avascular tissue. This is in sharp contrast to the instability of the proteins after the lens has been homogenized in aqueous buffers. After homogenization there is some protein in solution, which may

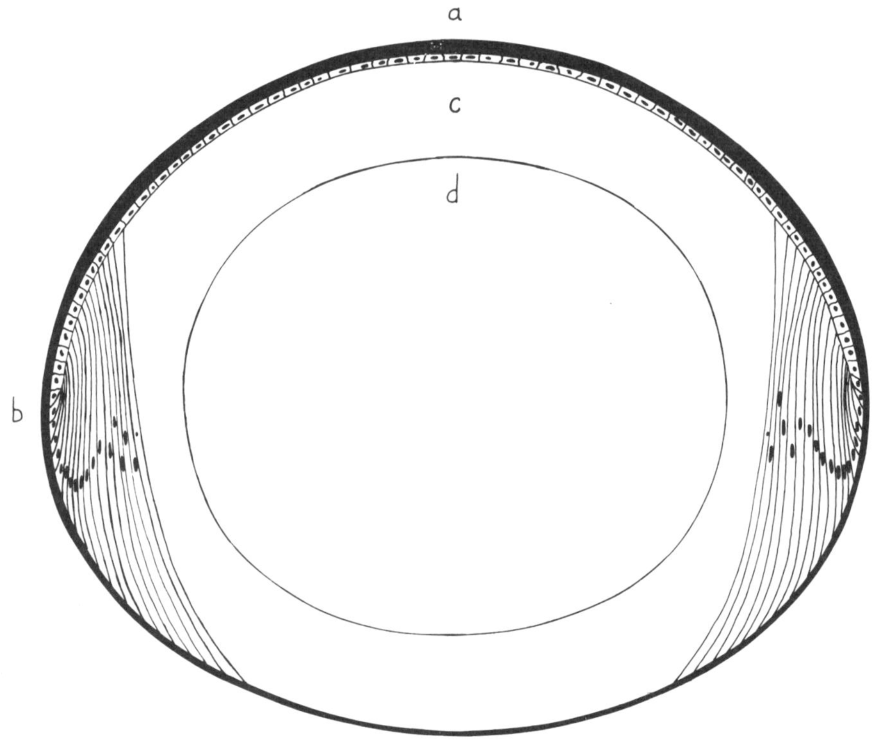

Figure 1 Schematic cross-section of adult rat lens showing *(a)* central epithelium and capsule, *(b)* equator showing differentiation of epithelial cells to fiber cells, *(c)* cortex, and *(d)* nucleus (9). Figure provided by J. O. Habgood.

be separated by a variety of chromatographic and electrophoretic methods into three major fractions α-, β-, and γ-crystallins (9) (see also Chapter 1) and there is some protein out of solution. The water-soluble proteins of normal and cataractous human lens appear to be antigenically and chemically similar (13, 14).

The decreasing amount of γ-crystallin and the increasing amounts of aggregated soluble and insoluble fractions in cataract are discussed later, but first we must address ourselves to the problem of classification of cataractous lenses.

8.2 CLASSIFICATION OF CATARACTS

Experimental cataracts in animals are readily classified by the nature of the toxic agent, the dose, and period of administration, but classification of human cataracts is a difficult and controversial problem.

Human cataract lenses vary greatly in appearance but for many purposes they must be classified, that is, assigned to classes. Lenses are usually classified according to the properties of the individual lens itself although sex, age, and any condition (e.g., diabetes) or treatment (e.g., steroids, phospholine iodide) that might be related to cataract should be noted. There are classifications based on the appearance of the lens when examined *in vivo* with a slit lamp, defining site and shape of the opacities; others are based on the appearance of the lens after removal. Other authors have used other properties, including sodium ion content (15).

For the present I wish to discuss two methods of classification by apperance of the lens *in vitro:* that of Pirie and van Heyningen (16, 17), which with or without modification has been used by many authors; and the new method of Chylack adopted by the Cooperative Cataract Research Group in the United States (18). It seems to me that the two systems are not rivals but serve different purposes. The Pirie–van Heyningen system is designed to aid those pooling lenses or pooling results from different lenses; it does not give a precise description of each lens, although in Oxford added descriptive information is recorded based on observation of the lenses on arrival in the laboratory and on the hospital notes. The Chylack system permits a strict definition of each lens but does not help pooling—in the strictest sense it is not yet a classification system.

The Pirie–van Heyningen system originally classified all lenses into four groups according to the visibility and color of the lens nucleus (16). This was subsequently modified to remove the small number of totally opaque and Morgagnian lenses into separate groups (17). Dr. Pirie related the original method to the clinical classification of Mach (19, 20). The four main groups were numbered I to IV. Group I lenses are a pale uniform yellow; group II lenses have a pale cortex and a visible nucleus, group III lenses have a pale cortex with a hazel brown nucleus; group IV lenses have a deep brown nucleus. In general, group I lenses are those the clinicians call cortical, plus various subcapsular cataracts. Groups II, III, and IV are nuclear cataracts with or without cortical involvement. The suggestion of a progression from type II through III to IV was apparent when the system was introduced—the amount of insoluble protein isolated from the lens increased steadily from normal (and group I) through groups II, III, and IV. The fraction of the insoluble protein that was insoluble in concentrated urea rose in the same way (16). The amount of lens material insoluble in guanidinium chloride with or without dithiothreitol also increased with group number (16). Subsequently progressive changes in many lens components and protein fractions were related to the increase in nuclear color (see Table 1).

Thus there is now a large body of evidence suggesting that this simple classification system is related to many of the changes that occur in cataract and helps to define the sequence and progression of these changes. The method is quick and inexpensive. Undoubtedly it will be superseded but when this happens one hopes that the new system will be supported by equally good evidence of the sequence of events in cataractogenesis.

Table 1 Progression of Changes in Lenses Classified by the Pirie–van Heyningen System

Component	Change	Groups	Refs.
Nuclear color	Increase	I to IV	16
Water-insoluble protein	Increase	I to IV	16, 21, 22
Urea-insoluble protein	Increase	I to IV	16, 22, 23
Guanidine-insoluble protein	Increase	II to IV	16
Guanidine-dithiothreitol insoluble protein	Increase	II to IV	16
Wet weight	Decreases	I to III	24
Dry weight	Decreases	II to III	24
Sodium	Increases	I to IV	24, 25
γ-Crystallin (% of soluble protein)	Decreases	I to III	24
Inositol	Decreases	I to III	24
Glutathione	Decreases	I to III	24, 26, 27
Disulfide-bonded protein of high molecular weight (Fraction HM3–see later)	Increases	I to III	28 and unpublished results
A high molecular weight fraction cross-linked by covalent bonds that are not disulfides (HM4–see later)	Increases	I to IV	29, 30
Protein fluorescence	Increases	I to IV	16, 31
Protein of high molecular weight in urea–mercaptoethanol	Increases	I to IV	31
Protein thiol	Decreases	I to IV	27
Protein-bound glutathione	Increases	I to II	26, 27
Protein-insoluble in urea mercaptoethanol	Increases	I to IV	23
Methionine sulfoxide in lens proteins	Increases	I to IV	32
Calcium	Increases	I to III	25
Glutathione reductase	Decreases	I to III	33
Aspartyl racemization	Increases	III to IV	34

The system described by Chylack relies on photographic reproduction and does not yet permit allocation of lenses to a small number of classes; it is complex, requires a great deal of expensive equipment, has high running costs, and would occupy a large part of the time of a valuable technician in a Western hospital; it would not be suitable in an eye camp in Pakistan with more than 300 lenses being removed each week. Of course any pooling of lenses, though useful and necessary for many studies, will obscure some subtle differences between cataracts (34a).

8.3 DECREASE IN SMALL PROTEINS IN CATARACT

One of the earliest changes observed in human lens proteins in cataract was the decrease in the proportion of protein of lowest molecular weight and electrophoretic mobility, usually assumed to be γ-crystallin (see refs. 9, 22, 35). A similar change was found in lenses of rabbits fed on dimethyl sulfoxide (36,37), in lenses of rats injected with *p*-chlorophenylalanine (38), in cataracts from axolotl (39), in X-ray cataracts in rabbits (40, 41), in triparanol cataract in rats (42), in hereditary cataract in mouse (43), and in galactose cataracts in rats (44). In the last-named paper subfractions of γ-crystallins were also compared.

Recently a greatly decreased crystallin synthesis has been found in the lens of mice with inherited cataracts (Nakano mice) in conjunction with an altered polypeptide composition (45).

The authors showed that protein leaked from cultured Nakano mouse cataract lenses into the medium. In tryptophan deficiency cataract it is the content of the high molecular weight β-crystallin that decreases (46).

It is difficult to interpret a decrease in γ-crystallin when it is usually expressed as a percentage of the soluble protein and the proportions of the soluble and insoluble fractions are themselves changing. However, less γ-crystallin can be isolated from the soluble fraction of most cataract lenses and this could be the result of any of at least four processes: (1) it may have leaked out through the capsule (47, 48, 48a); (2) it may have been converted to insoluble protein (49–51); (3) it may have been cross-linked to a higher molecular weight (52); or (4) it may not have been synthesized in the first place (9). The amounts of insoluble protein and some large soluble aggregates isolated from cataract lenses exceed those from clear lenses, as discussed in detail below, and these aggregates consist of mixtures of lens proteins. Very recently Sandberg and Closs (53), in a careful study of human lens proteins using radioimmunoassay techniques, showed that after conventional fractionation of proteins from 60-year-old human lens by ion exchange chromatography, the fractions usually considered to be β-crystallin consisted largely of γ-crystallin, and even the α-crystallin contained 10% γ-crystallin. These findings give some support to the suggestion that γ-crystallin is cross-linked to a fraction of high molecular weight (52).

Sandberg and Closs (53) did not find a decrease in γ-crystallin antigen in the soluble fraction of cataract lens.

8.4 INSOLUBLE PROTEIN IN CATARACT

The water-insoluble fraction of lenses is partly the lens membrane, but mostly consists of the same proteins as are found in the soluble fraction. In bovine lens the insoluble protein is derived largely from α-crystallin but that from human, rat, rabbit, and dogfish lens is derived from a mixture of crystallins (see ref. 9).

The division of the protein between the insoluble and soluble fraction depends on the species, the age of the animal, and whether the lens is cataractous, but also on the method of preparation.

An author using one method has usually isolated more water-insoluble protein from older animals than from younger animals (13, 19, 54–65), more from nucleus than cortex (23,58), and more from most types of cataractous lenses than from normal lenses (13, 16, 21–23, 37, 40, 41, 50, 63, 66–69; however, see ref. 70) but the amounts isolated from apparently similar lenses by different authors vary widely, as shown for normal human lenses in Figure 2. Note that some curves rise very little before the age of 50 whereas others rise steeply during infancy; that two authors have isolated more than 20 times as much insoluble protein from young adult lenses as another author; that the general shapes of the curves differ enormously; and that these results are for clear lenses without the problems of classification of cataracts. When one author obtains much more insoluble protein than another from similar lenses it is unlikely that the material they both call water-insoluble protein is identical. The conditions of centrifugation and pH for extraction give no clue as to the basis of these discrepancies. When Krause (71) isolated far less insoluble protein from bovine lens than Mörner (72) had found, he gave three possible explanations: Mörner's cattle were older than Krause's; they were European, not American; and Mörner had allowed his lens extract to stand for several days. Age differences are accounted for in Figure 2 and there is no evidence of a racial difference; Krause's third suggestion perhaps gives us a better lead. The conditions of isolation determine the amount of insoluble protein isolated and hence determine its nature.

Most authors isolate relatively little insoluble protein from bovine lens and it is mostly derived from α-crystallin, although other crystallins are present; that from rat, dogfish, rabbit, and normal and cataractous human lens is derived from a mixture of lens crystallins (see ref. 9).

In many studies the insoluble protein has been fractionated by extraction with concentrated urea solutions into urea-soluble and urea-insoluble fractions. The small amount of urea-insoluble protein isolated from bovine lens

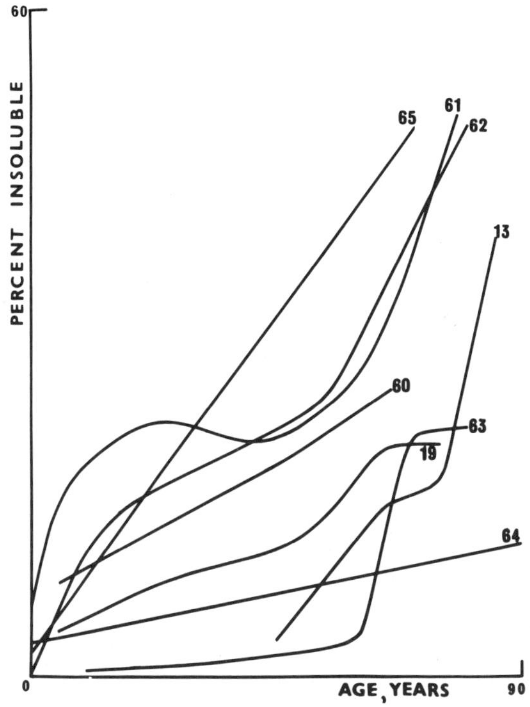

Figure 2 Proportion of the total protein of clear human lenses isolated as the insoluble fraction. The lines are labeled with the numbers of the references from which the original data were taken.

(73–75) consists almost entirely of the lens plasma membranes (74, 76, 77). Much more urea-insoluble protein is isolated from rat, normal human lens, and most cataractous human lenses. This protein fraction includes lens membranes but it is mainly held insoluble by disulfide bonds (16, 23, 52, 78, 79). The normal human lens has little if any protein–protein disulfide (52) and so these disulfide cross-links must have been introduced during isolation. In fact, intact rat lens, normal human lens, and most human cataractous lenses dissolve almost completely in 6 *M* guanidinium chloride, a powerful dissociating agent (16, 52, 79). Much less urea-insoluble protein is isolated using anaerobic methods to extract rat, normal human, and cataractous human lenses (52, 75, 79) and the lowest curve of Figure 2 at age 10 to 60 years was obtained by the only authors who attempted to prevent artifactual oxidation by the use of dithiothreitol and extraction in a glove bag filled with nitrogen. Other authors isolate less insoluble protein when a reducing agent is present during homogenization (62).

I conclude that most of the urea-insoluble protein isolated by most methods from rat and human lenses is an artifact caused by oxidation of reactive thiol groups of the crystallins during isolation, and the variability of this process is the cause of the variability in amounts of total insoluble protein isolated from human lens (Figure 2). The high proportion of disulfide bonds in both soluble and insoluble proteins isolated in air (80, 81) compared with the small amounts of protein–protein disulfides determined in lens after protection of thiol groups (52) demonstrates the extent of disulfide formation during isolation.

We are left with the consistent observation that the amount of insoluble protein increases in most forms of cataract, and indeed it is the urea-insoluble fraction that increases in human nuclear cataract (16, 23).

Although the insoluble fraction found after homogenization has been studied intensively we have no reason to believe that all the "soluble protein" was in solution *in vivo* in lenses that have up to 70% protein in the center.

Much of the urea-insoluble protein isolated from rat and human lens by conventional procedures is an artifact caused by oxidation of exposed protein thiol groups to disulfides during extraction in air (52, 79). The oxidation was prevented by isolation in the absence of air. Recently the oxidation of protein thiols in homogenates of human cortical cataracts was demonstrated directly (82). The tiny amount of urea-insoluble protein isolated from calf lens is not an artifact and consists largely of cell membranes (74, 82); it is probable that the small amounts of insoluble protein isolated by Zigman et al. (63) from younger human lenses after extraction in the presence of a reducing agent (Figure 2) and the small amounts of urea-insoluble protein isolated from human cataracts by Kramps et al. (22), who homogenized under nitrogen, and by Truscott and Augusteyn (23) are not artifactual, but most authors have isolated much greater amounts after conventional aerobic procedures. Perhaps as Kuck (70) remarked, cataractous lens protein insolubility is a ghost that will never be laid to rest.

The conformational changes that are the basis of the increased susceptibility of certain lenses, notably nuclear cataract lenses, to oxidation are discussed below.

The increased insoluble protein has been found in nuclear cataracts but not in cortical cataracts (16). The darker nuclear cataracts yield a fraction that is held together by covalent bonds that are not disulfides, that are mostly in the nucleus of the lens, and that appear to be associated with the nuclear color and fluorescence (16, 23, 29, 62, 83, 84), although it has been suggested that the color and fluorescence have different origins (85).

These cross-links have not been identified in spite of the efforts of various workers. Dilley and Pirie (83) found only small amounts of kynurenine, which was probably derived from *N*-formylkynurenine, in hydrolysates of a protease-resistant core from the guanidine–dithiothreitol-insoluble fraction of dark brown cataract lenses. Free kynurenine and the *O*-β-D-glucoside of 3-hydroxy-L-kynurenine had already been identified in human lens and were shown to be formed enzymically from tryptophan (86). The amounts of the free compounds did not increase in cataract and there was no evidence that 3-hydroxykynurenine became attached to the brown material from nuclear cataracts, although the glycosides can act as photosensitizers *in vitro,* promoting the yellowing of lens proteins by sunlight.

The presence of small amounts of kynurenine attached to lens proteins, especially in dark brown lens (brunescent, groups III and IV by the Pirie–van Heyningen classification), has been confirmed (87, 88) and Zigler et al. found about one residue of kynurenine per 200 amino acids, which would be sufficent to act as a cross-link either as kynurenine or as *N*-formylkynurenine.

Methods similar to those of Dilley and Pirie (83) were used by Truscott et al. (88) to isolate anthranilic acid from the water-insoluble protein of brown cataract lenses but not from normal human lenses. The identity of anthranilic acid was convincingly demonstrated by mass spectrometry. After proteolysis the anthranilic acid was free and not part of the resistant core and corresponded to about 4% of the total tryptophan, that is, only about 0.13 residues per 200 amino acid residues (the size of a typical lens polypeptide chain) using the tryptophan content from Dilley and Pirie (83). This is insufficient to be the sole cross-linking agent. The anthranilic acid results from the breakdown of other fluorescent moieties (89).

Fluorescent components have been isolated from human cataracts (62) and identified as β-carbolines (90). The amounts isolated correspond to about 3 mol per thousand polypeptide chains, taking a chain weight of 20,000 dalton, so these compounds and their *in vivo* precursors are probably not responsible for the cross-linking of protein in dark brown lenses.

Pirie isolated fluorescent substances that give a typical β-carboline fluorescence spectrum from an acid hydrolysate of bovine γ-crystallin (91).

More recently Garcia-Castineiras et al. (89) isolated unspecified amounts of another fluorescent component from human cataracts. They identified this component as bityrosine but van Haard et al. (92) looked for bityrosine in lens proteins and failed to find it. Perhaps only one or two of the large number of lenses necessary for this type of study contained the bityrosine found. Collagenlike cross-links derived from lysine and isopeptide bonds could not be found in lens proteins (92).

Obviously the proteins in dark brown cataracts have been altered beyond recall but what set them on the downhill path? The changes to the proteins indicate that conformational changes have occurred and this process is discussed in a later section, after which I discuss the possible causes of the conformational change. First I review our present knowledge of high molecular weight aggregates that may be obtained in solution.

8.5 HIGH MOLECULAR WEIGHT SOLUBLE AGGREGATES

High molecular weight fractions have been isolated from the lens by four different approaches. In this chapter I refer to these four products as HM1, HM2, HM3, and HM4. An outline of their preparation is shown in Figure 3. The first, HM1, is that part of α-crystallin that can be isolated as a high molecular weight fraction after gel chromatography on agarose (93), whereas HM2 is that part of the complete lens supernatant that is isolated as a large aggregate after gel chromatography on agarose (94) or centrifugation (95). The other two aggregates are isolated by gel chromatography in denaturing solvents. Fraction HM3 is held together by disulfide bonds (28), whereas fraction HM4 isolated from dark cataracts is held together by covalent bonds that are not disulfides (29).

Fractions HM1 and HM2 from bovine lenses seem to be identical (9) but the human and rabbit fraction HM2 consists of other crystallins aggregated with α-crystallin (9, 35, 41, 63, 96). In the past few years there have been many studies of the fraction HM2 whereas study of fraction HM1 declined.

8.5.1 High Molecular Weight Fraction of Lens Soluble Proteins (HM2)

The literature on this fraction (HM2) was reviewed in 1976 (9) and the authors voiced their worries about the widely variable amounts isolated by different authors from apparently similar lenses. That worrying aspect is still with us, and is illustrated by the figures for the amounts of HM2 isolated by different authors from clear human lenses of a similar age (Table 2). Values from 1.2 or 1.9% to 29 or 34% for HM2 isolated by apparently similar techniques require explanation before the changes in cataract can be considered. The first attempt to elucidate these discrepancies was the determination of the amounts of HM2 isolated in the presence and absence of dithiothreitol, which reduces disulfide bonds (69). Less HM2 was isolated from nuclear cataracts when dithiothreitol was present.

Subsequently fraction HM2 was isolated by centrifugation at 100,000 × *g* (63), and it was suggested that the conditions of centrifugation were critical (96). Further investigation showed that more HM2 was in the supernatant after centrifugation at 9,900 × *g* than after centrifugation at 26,000 × *g* (100); however, considerable amounts of high molecular weight protein have been found in supernatants after centrifugation at 26,000 × *g* or greater from human lens (69, 98, 101, 102) and from bovine lens (93). Variation of the temperature for the isolation of fraction HM2 by gel chromatography appeared to have no effect on yield (100).

Another problem results from the measurement of the absorbance of the protein at 280 nm, the method most commonly used. Fraction HM2 does not have a typical protein absorbance peak at 280nm (Figure 4) and so its measurement in this way cannot be accurate (100). Furthermore, this

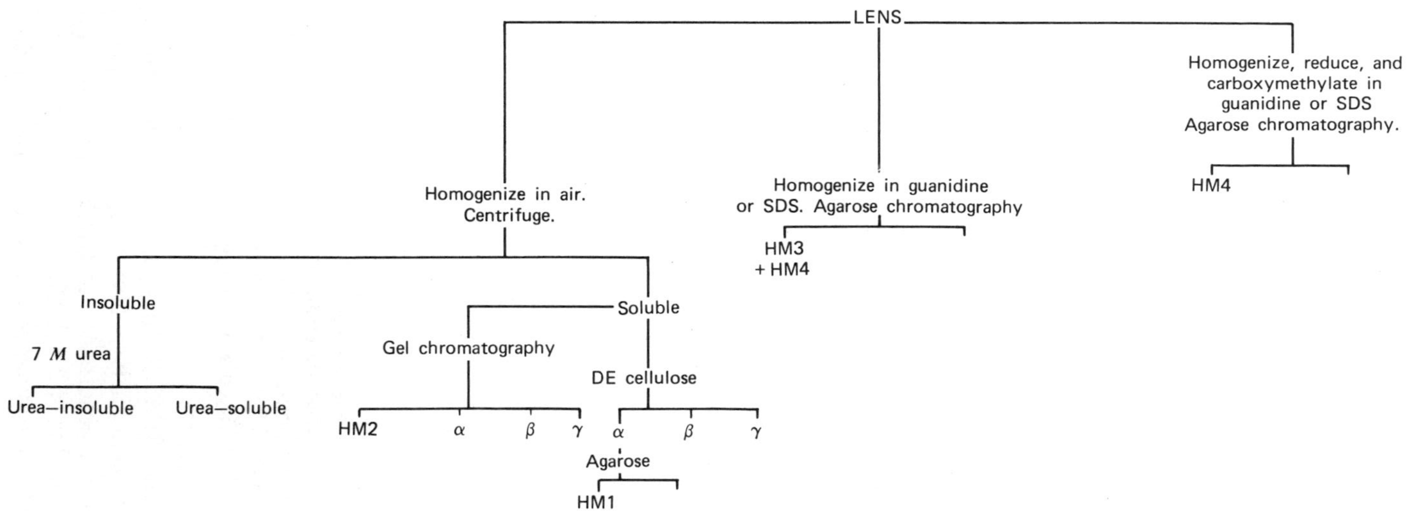

Figure 3 Fractionation of lens proteins to obtain the high molecule weight fraction.

Table 2 Proportion of the Soluble Protein Isolated as Fraction HM2 from Clear Human Lens Aged 55 to 70

Age (years)	HM2 (%)	Ref.
64	1.9	
65	5.8	97
64	6.1	
68	7.7	
57	29	98
61	3.0(2.8)[a]	
64	5.4(5.9)	
65	3.3(3.2)	69
66	6.9(2.8)	
69	6.9(1.7)	
61	9.4	99
59	8.6	
66	34	62
56	7	
56	9	
61	7	
61	13	
61	15	63
65	3	
65	5	
70	5	
70	7	
59–65	2.2	100[b]
59–63	1.2	
70	2.5	
62	1.6	
63	4.7	
68	3.3	

[a]Figures in parentheses were for fractions prepared in the presence of dithiothreitol.
[b]Centrifuged at 9900 × *g*.

spectrum indicates that HM2 is not entirely protein, and indeed the low recovery from amino acid analysis (69) would indicate that only 50% of it is protein. In mouse the absorbance of HM2 at 280 nm overestimates the protein content by a factor of 2.3 (102a).

One must conclude that much of the material isolated by some authors as high molecular weight protein (HM2) is artifactual, entirely dependent on experimental conditions, and that part of it is not protein. Nevertheless some authors isolated more HM2 from cataract lenses than from normal lenses of the same age both in human cataract (69, 97, 102) and in X-ray cataract in rabbits (40, 41). However, other authors do not find this difference (62, 63, 103). Bushell and Duncan (103) suggest that the great increases in HM2 reported in cataract in the past are the result of the methods chosen for presentation of the data and recommend that results should be expressed as weight of HM2 per lens, not as percent of the soluble fraction. These discrepancies undermine any hypothesis of cataractogenesis related to fraction HM2. Indeed when up to 50% of the protein of *clear* lenses has been isolated as an insoluble fraction (Figure 2) and up to 39% of the remaining soluble fraction has been found as fraction HM2 (98), it is difficult to discern a direct role for either HM2 or insoluble protein in cataract.

8.5.2 Disulfide-Cross-linked Protein of High Molecular Weight (HM3)

Normal human and most cataract lenses appear to dissolve almost completely in 6 *M* guanidinium chloride, a powerful dissociating medium, and these solutions can be fractionated by gel chromatography in the same solvent to give two major peaks, a high molecular weight component (HM3) and low molecular weight polypeptides (28).

About 2% of the total protein of normal human lens and 7% of that from nuclear cataracts (group II) is found as this high molecular weight fraction.

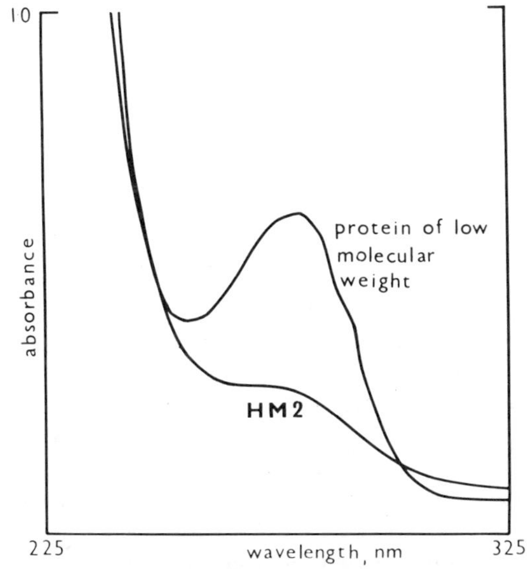

Figure 4 Absorption spectra of proteins of high molecular weight (fraction HM2) and low molecular weight from normal human lens (100).

The latter figure is brought down to 2% if the lens proteins are reduced and carboxymethylated before chromatography, indicating that 5% of the protein in these lenses was held together by disulfide bonds. These cataractous lenses contain enough protein–protein disulfide bonds (52) to give this amount of high molecular weight aggregate, and fraction HM3, unlike HM2, is found only in nuclear cataract lenses. Fraction HM3 was isolated only from the nucleus (Figure 5) and since the proteins of cortex and nucleus are generally similar and both are completely unfolded in guanidine, it is unlikely that fraction HM3 is an artifact. The amino acid composition of fraction HM3 indicated that it consisted of a mixture of crystallins.

Similar results are obtained if the thiol groups are first blocked under anaerobic conditions to prevent artifactual disulfide cross-linking and then fractionated on agarose in sodium dodecylsulfate, another dissociating medium (104). The presence of disulfide bonding was demonstrated by sodium dodecylsulfate gel electrophoresis with and without mercaptoethanol. The disulfide bonded aggregate was again found only in nuclear cataracts (Pirie's groups II and III), constituting an average of 14% of the

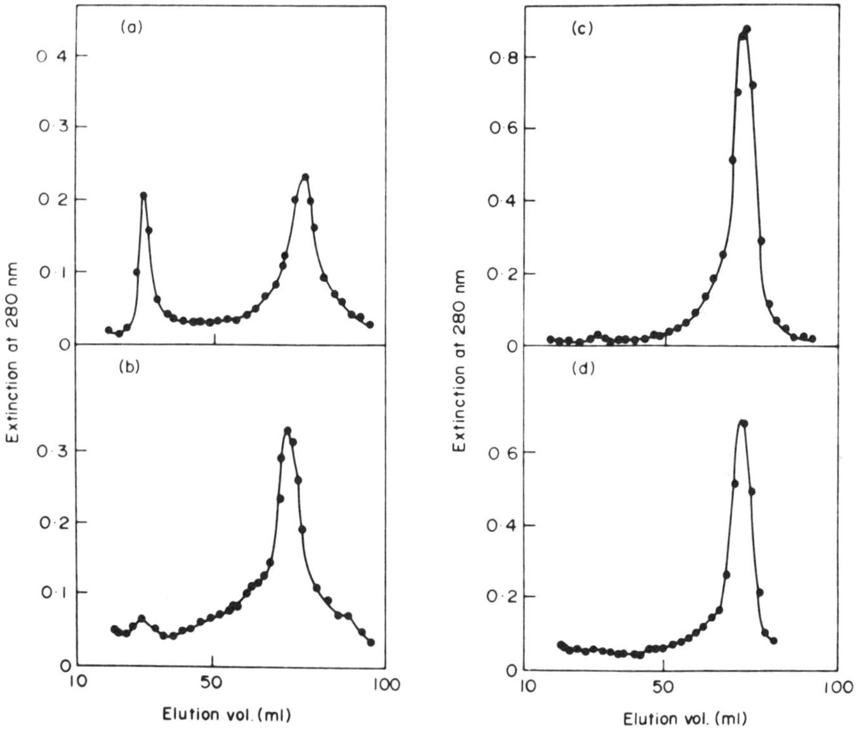

Figure 5 Gel filtration on agarose in 6 *M* guanidinium chloride of *(a)* unmodified nucleus; *(b)* reduced and carboxymethylated nucleus; *(c)* unmodified cortex; *(d)* reduced and carboxymethylated cortex, of one group II lens (28).

total protein. It was also isolated by sucrose gradient centrifugation in 8 *M* urea (104), which showed a wide molecular weight distribution, mostly between 2 and 33 $\times$ 10^6 dalton. Takemoto and Azari (104), took great care to prevent artifacts and confirmed the view that the material isolated is held together by disulfide bonds in the lens itself, where it could be associated with a larger mass of protein by noncovalent bonds. Similar methods have been used to identify disulfide cross-linked protein of high molecular weight in mature hereditary cataract in the rat (105). These authors also produced disulfide-cross-linked protein by exposure of normal rat lens to pure oxygen, demonstrating again the susceptibility of lens proteins to oxygen.

Recently it was shown that the disulfide-bonded protein of high molecular weight (HM3) is found in the water-insoluble protein fraction (106), and examination by sodium dodecylsulfate gel electrophoresis confirmed that fraction HM3 is formed from a mixture of lens proteins. These authors found very little disulfide in the guanidine-solubilized water-insoluble fraction of normal human lens but considerable amounts in the same fraction of cataractous lens, confirming that the disulfide bonds of fraction HM3 are not introduced by treatment with guanidine.

8.5.3 Protein of High Molecular Weight Cross-linked by Covalent Bonds Other than Disulfides (Fraction HM4)

The disulfide-cross-linked fraction HM3 is found in most nuclear cataracts (group II by the Pirie–van Heyningen classification), although more is found in darker lenses (J. J. Harding, unpublished results). In these darker lenses (groups III and IV) some protein is cross-linked into large aggregates by covalent bonds that are not disulfides. Buckingham (29) found these proteins by gel chromatography of reduced lens proteins in sodium dodecylsulfate, by gel chromatography of reduced and carboxymethylated lens proteins in 6 *M* guanidinium chloride, and by sodium dodecylsulfate gel electrophoresis of reduced proteins, and their presence in dark brown lenses has been confirmed by Takemoto and Azari (104).

Fraction HM4 is found exclusively in the lens nucleus (Figure 6; refs. 29, 30), and after conventional fractionation of proteins into water-soluble and insoluble fractions it is found in the water-insoluble fraction (23). These findings are compatible with the finding that most of the insoluble protein of human cataract lenses is found in the nucleus.

8.6 THE RELATIONSHIP OF PROTEIN AGGREGATION TO CATARACT

Various authors have suggested that insoluble protein and all the high molecular weight aggregates may impede the transmission of light to the retina, and may be the direct cause of the visual impairment in cataract. All

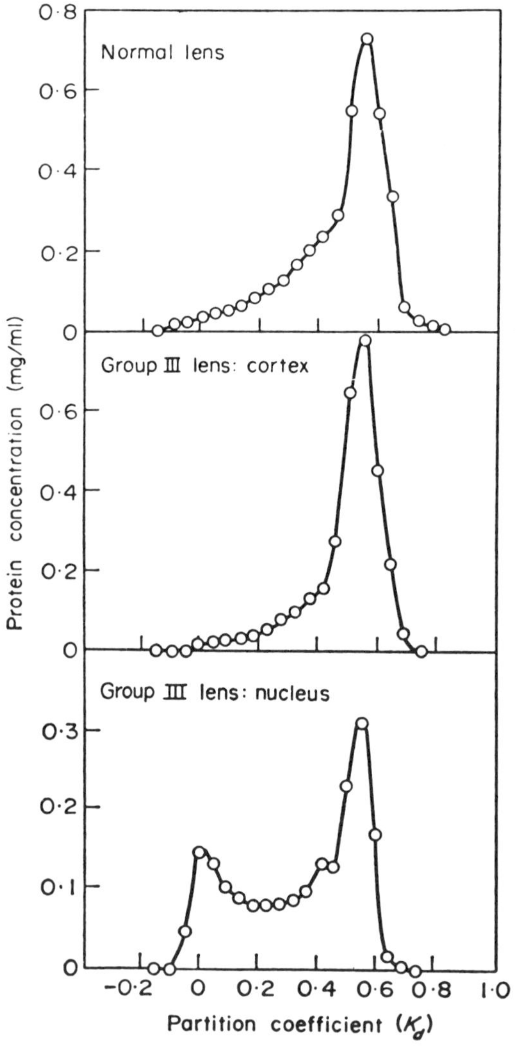

Figure 6 Gel chromatography on agarose of complexes with sodium dodecylsulfate of reduced proteins from normal lenses and the nucleus and cortex of group III cataractous lenses. Blue dextran (K_d = 0) and oxidized β-mercaptoethanol (K_d = 1) were used to determine the K_d of the eluted protein (29).

these protein fractions appear to be derived from mixtures of lens crystallins.

It was predicted that large aggregated proteins in the lens would cause scattering of light and impair vision (107, 108) and Benedek, after assuming values for the refractive indexes of the insoluble protein and of the lens substance, calculated that if 20% of the lens protein was insoluble with a molecular weight greater than 5×10^7 dalton the lens would be turbid. Subsequent authors have suggested that their own aggregated protein fraction, whether water-soluble or water-insoluble, guanidine-soluble, urea–dithiothreitol-soluble, yellow, white, or brown protein, was the fraction that really mattered. Our ignorance of the physical relationship between proteins in the lens, our ignorance as to whether the proteins are in solution in the lens (109) irrespective of their solubility in aqueous buffers, and the assumptions behind Benedek's bold and stimulating prediction were forgotten. Because of the discrepancies in amounts isolated and the large amounts isolated from clear lenses I have already discarded insoluble protein and fraction HM2 as isolated by many authors as candidates for the light scattering bodies in cataract lenses. Fraction HM1 can be discounted for similar reasons (9). This leaves fractions HM3 and HM4, both of which appear to have a real existence in the lens as cross-linked proteins. Fraction HM4 is colored as well as cross-linked and so could both absorb and scatter light. These two fractions would probably be attached noncovalently to a much larger body of protein in the lens and could easily represent aggregates large enough to fit Benedek's prediction. However, we have no knowledge of the refractive index of the aggregates relative to the remaining protein, and so cannot say if they are sufficiently different to cause the scattering of light in cataract. Certainly both fractions are present in increasing amounts in darkening human cataract lenses and neither is present to a significant extent in clear lenses, which supports their candidacy. As it happens neither is present in cortical cataracts (group I of the Pirie–van Heyningen classification), and it is more likely that water clefts and similar disruptions cause these cataracts with no significant change in protein structure. Changes of this nature have been observed in cortical cataracts by electron microscopy but they were not seen in nuclear cataract; nuclear cataracts contained aggregates, presumably protein, 50 to 100 nm in diameter (110).

Physical techniques applied to whole lenses should help to elucidate the nature of the light scattering bodies in cataract. The light backscatter from normal human lenses increases with age (111). Tanaka and Benedek (112) measured protein diffusivity in intact bovine and human lenses and found no evidence for aggregates larger than α-crystallin in two intact clear human lenses, whereas they did find larger aggregates in the cataract lens, indicating that the light scattering entities are to be found only in cataract lenses. The two clear lenses gave results within 10% but later experiments (113) gave diffusivity results that differed by about 34% at the same age, and from these results the authors concluded that large aggregated proteins are present in clear lenses and increase with age to about 1.4% at age 43 years. The calculation of the proportion of aggregated protein is very sensitive to differences in the measured diffusivity; for example, if the value for a clear lens of a 43-year-old is taken from the earlier paper (112), one obtains a figure of 0.7% high molecular weight protein. Furthermore, the earlier diffusivity value for a cataract lens gives a negative number for the proportion of high molecular protein. Clearly there are problems with the interpretation of these results.

Returning to the changes in nuclear cataract, what lies behind the increase in aggregated fractions HM3 and HM4 in the lens, and the increase in insoluble protein isolated from cataractous lens? They appear to be the result of conformational changes in the proteins, and these are discussed in the next section.

8.7 CONFORMATIONAL CHANGES IN HUMAN LENS PROTEINS IN NUCLEAR CATARACT

The proteins of nuclear cataract lenses appear to be more susceptible to oxidation than those from normal lenses and it was subsequently shown (Figure 7) that the reactivity of the thiol groups of nuclear cataract lens is greater than in the normal lens (114). In the same paper it was shown that proteins of nuclear cataracts are more susceptible to proteolysis by trypsin. Both these methods have been widely used as probes of protein conformation, and the results suggest that a large proportion of the lens protein has unfolded during cataract formation to a more open structure with newly

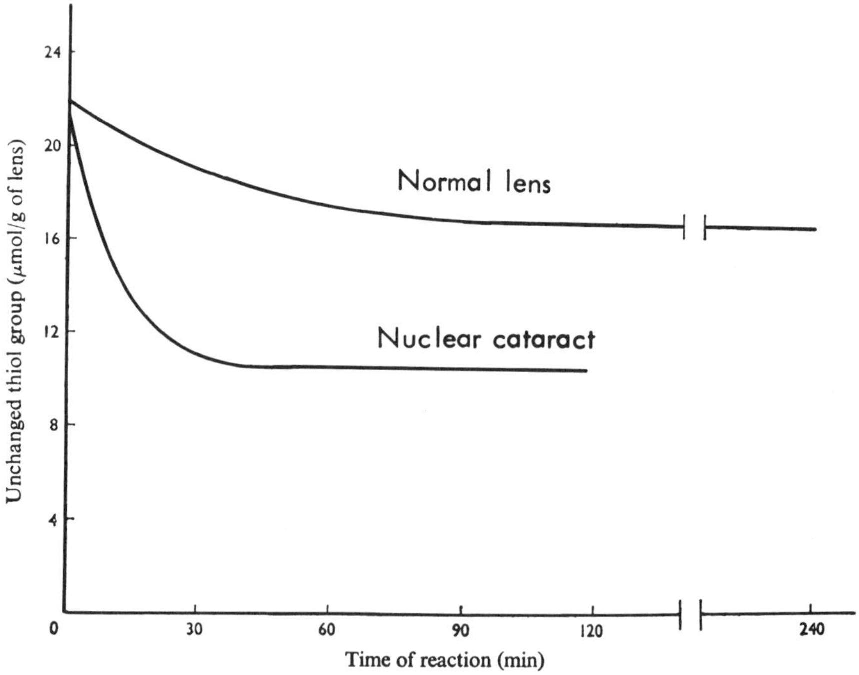

Figure 7 Rate of carboxymethylation of lens protein thiol group (114).

exposed and reactive thiol groups and with newly exposed peptide bonds susceptible to trypsin. Labeling the reactive thiol groups showed their presence on α-, β-, and γ-crystallins, which suggests that the conformational change is a general phenomenon and may affect other lens proteins including enzymes. It appears that the change of conformation also permits the binding of extra calcium (115).

The protein thiols of bovine lens are less reactive than those of clear human and rat lenses, suggesting a difference in conformation (116) that would be consistent with the greater yield of urea-insoluble protein from human lens.

Many authors now discuss the changes they find in cataract in terms of conformational change to lens proteins (27, 32, 34, 41, 83, 105, 117–118b).

Physical methods should have provided further information on the conformation change but applications of such methods are few. Studies of the phosphorescence emission of normal and cataractous lens show a difference of emission lifetime in brunescent cataract, indicating either a new phosphorescent species or an altered environment for tryptophan within the proteins (119). An altered environment for tryptophan is consistent with the protein unfolding to expose buried tryptophan to a hydrophilic environment. The decrease of protein-bound water in cataract is also consistent with conformational changes (119a).

The unfolding of lens proteins in nuclear cataract with exposure of reactive thiol groups leads to other changes drawn schematically in Figure 8. Part *a* represents the tightly folded proteins of normal human lens with the thiol groups tucked away inside the molecule. On formation of nuclear cataract the proteins unfold to more open structures with reactive thiol groups (Figure 8*b*). *In vivo* these newly exposed thiol groups appear to react in one of two ways: they react with the tripeptide thiol glutathione to form mixed disulfides (Figure 8*c*) which have been detected chemically in these lenses (26, 27, 120, 121) and they also react together to form protein–protein disulfides (Figure 8*d*), which are also found in nuclear cataracts (52). The disulfide bonding joins together polypeptide chains of lens proteins that may be isolated as a disulfide-cross-linked high molecular weight fraction (fraction HM3) under denaturing conditions (28). Further changes in darker lenses result in a protein fraction held together by covalent bonds other than disulfides which may be isolated as fraction HM4 (29).

On the right of the figure (8 *f*) I have depicted the oxidation of reactive thiol groups that occurs after aerobic extraction of lenses into aqueous solutions. This is shown as urea-insoluble protein but an identical process would give a high molecular weight but "soluble" fraction such as fraction HM2.

Because the mixed disulfides, the disulfide-bonded protein of high molecular weight (Figure 8), and the urea-insoluble protein are isolated in greater amounts from the lens nucleus (27, 28, 83, 122), it is probable that most of

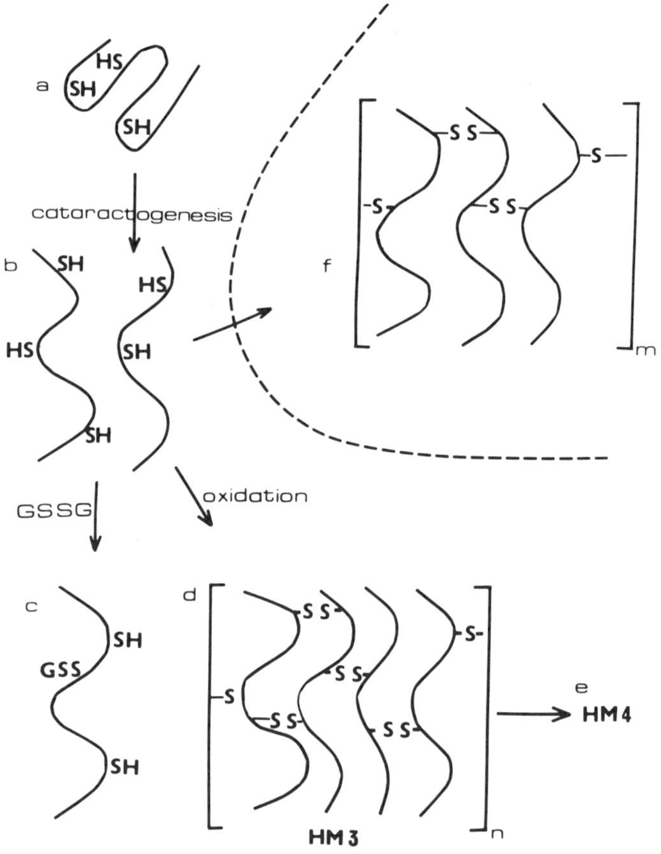

Figure 8 Changes in lens proteins in the development of human nuclear cataract. *(a)* Native protein; *(b)* unfolded protein; *(c)* mixed disulfide; *(d)* disulfide-bonded protein of high molecular weight fraction HM3; *(e)* fraction HM4 (9); *(f)* urea-insoluble protein and fraction HM2.

the unfolded protein is in the central region of the lens. This suggests that the unfolded protein is old protein that was synthesized 50 or more years earlier, and this in turn indicates that the unfolding is not caused by defective synthesis of proteins but is the result of a postsynthetic change that may occur extremely slowly in the center of the human lens.

8.7.1 The Cause of Unfolding

The three-dimensional structures of many proteins are known; they are usually compact structures with most of the hydrophilic and charged groups on the outside able to interact with water, ions, and other proteins. Most of the hydrophobic amino acid side chains are on the inside. This arrangement can be disrupted *in vitro* by heat, by extremes of pH, by detergents such as sodium dodecylsulfate, by high concentrations of guanidinium chloride or urea, and in other ways. It is unlikely that any of these are relevant to human cataract, with the possible exception of the involvement of elevated temperatures in microwave or infrared (glassblowers') cataract. One would also expect that the structure could be disturbed by the introduction of charged groups into the hydrophobic core of the molecule, or perhaps even by the addition of hydrophobic molecules to the hydrophilic exterior. Nine factors proposed by different authors as the cause of the conformational changes are discussed below. All involve direct changes to the proteins rather than a change in their environment, and most are chemical modifications to the protein that would involve alterations in charge or polarity.

8.7.1.1 Deamidation

Because asparagine and glutamine are among the least stable uncharged amino acids, Dilley and Harding (123) investigated the possibility that deamidation of these amino acids to release carboxylic acid groups occurs in aging and cataract. Postsynthetic loss of amide groups had already been reported in bovine α-crystallin (124, 125), cytochrome C (126), and aldolase (127), and indeed one theory of aging was based on deamidation of proteins (128, 129). It was found that some amide groups are lost from α-crystallin during fetal life and more during infancy, but that there was no loss between the ages of 34 and 80 years (123). The mixed proteins from the center of the lens, the oldest part of the lens, showed no loss of amide groups during adult life and no loss in a nuclear cataract. The gross chemical method used could not rule out the loss of a few amide groups from lens proteins that might be important, but there is no positive evidence to implicate deamidation in the unfolding process.

8.7.1.2 Oxidation of Methionine

Another structural change that could cause the conformational changes is the oxidation of hydrophobic methionine to the polar sulfoxide or sulfone. Truscott and Augusteyn (32) found a decreasing amount of methionine in hydrolysates of proteins from lens nuclei of cataracts of increasing color (30% low in group IV lenses). They then used three independent methods to determine methionine sulfoxide in lens proteins, and found traces in group II lenses and much greater amounts in groups III and IV (Pirie–van Heyningen classification), corresponding approximately to the deficit in methionine. Previous authors had not observed a loss of methionine by amino acid analysis possibly because they used HCl hydrolysis, which can convert the sulfoxide to methionine, especially if cysteine is present, rather than toluene-4-sulfonic acid used by Truscott and Augusteyn. However, Harding (28) used toluene-4-sulfonic acid hydrolysis of fraction HM3 from group II and III lenses and found normal levels of methionine; Dilley and Pirie (83) using the same method found normal methionine values in whole brown nucleus (group IV), conflicting with the results of Truscott and Augusteyn (32). In spite of the care taken by the latter authors the presence of methionine sulfoxide needs confirmation. Because it was found in only trace amounts in most nuclear cataracts (group II) it probably does not cause the conformational changes already present in such lenses.

8.7.1.3 Racemization of Aspartic Acid

Racemization of aspartic acid in lens proteins, meaning the progressive conversion of the L-aspartyl residues originally present to D-aspartyl residues, was first reported by Masters et al. (34). They had earlier used the extent of racemization of aspartyl residues, which requires cleavage of one of the four bonds of a carbon atom, to date fossil bones (130) and had shown that D-aspartyl residues accumulate with age in enamel (131) and dentin (132).

The proportion of D-aspartyl residues in lens nucleus increased steadily with age by about 0.1% per year, and the findings for cataracts of groups I and II were similar (34), but much more D-aspartic acid was found in the nuclei of brunescent cataract (group IV of Pirie's classification). Recent results of Masters et al.(117) showed that the increased racemization of aspartyl residues in the lens nucleus did not occur in lens periphery, and was associated with protein that ended up in the water-insoluble fraction and in fraction HM2. Fractionation of the water-insoluble protein showed the highest proportions of D-aspartic acid in a fraction corresponding to HM4, that is, an aggregate held together by covalent bonds other than disulfides, and in the small amounts of polypeptide of 11,000 dalton. In both papers Masters et al. discussed the chicken and egg problem: did the conformational change cause the racemization, or did the racemization of aspartyl residues cause the conformational changes to proteins in human nuclear cataracts? Perhaps both occur, giving a positive feedback system so that racemization starts the conformational change, which alters the environment of further aspartyl residues making them more susceptible to racemization, and so on. However, the occurrence of excess racemization only in the darker lenses whereas the conformational changes were found in most nuclear cataracts (114) would support the view that racemization is the result of the conformational change rather than the cause of it. Masters et al. (117) tentatively conclude that racemization causes the conformational changes found in nuclear cataract; they had suggested that racemization would cause conformational changes before their work on lens. Very recently Garner and Spector (65) confirmed some of these results but they find less extensive racemization in insoluble protein from cataract lenses possibly due to differences in isolation procedures. They find the 10,000-dalton polypeptide (presumably corresponding to 11,000 dalton by Masters et al.) in the soluble fraction with unremarkable D-aspartyl content and so conclude that racemization is not necessary for its production by postsynthetic degradation of the A chain of α-crystallin.

8.7.1.4 Glycosylation

In diabetes the high level of blood glucose results in the nonenzymatic addition of glucose to the amino terminal valine of hemoglobin A (133). The initial reaction produces an aldimine bond but this may undergo Amadori rearrangement to a more stable form:

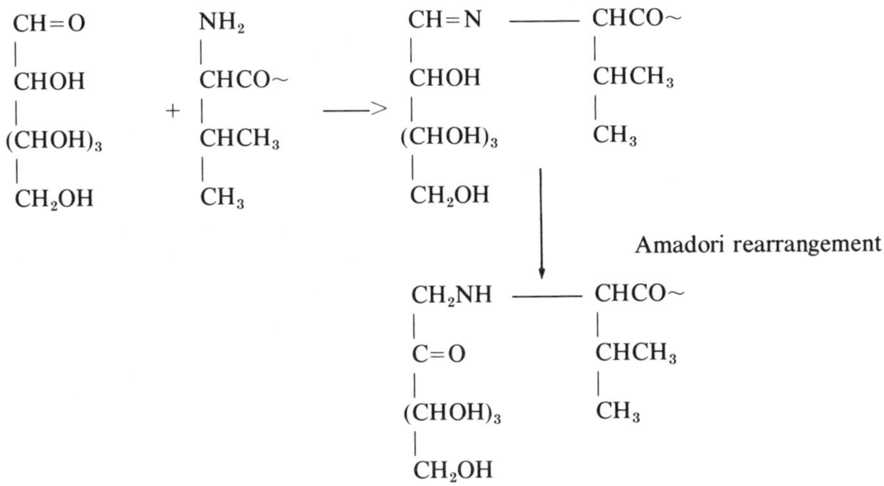

These changes are similar to those that occur in collagen cross-linking, where aldehydes produced by enzymic oxidation of the ϵ-amino groups of lysine and hydroxylysine react with other ϵ-amino groups to form aldimines that, if derived from hydroxylysine, can undergo Amadori rearrangement to a stable form (134). Cerami investigated the possible role of glycosylation in the formation of diabetic cataract and showed first that the development of opalescence in solutions of lens proteins was greatly accelerated by 50 m*M* [^{14}C]-glucose or 5 m*M* [^{14}C]-glucose-6-phosphate, and second that the label was incorporated into all the major groups of lens proteins (118). The solutions containing glucose-6-phosphate became opaque more rapidly in the presence of oxygen even though glycosylation had occurred to a comparable extent in the presence of nitrogen. Reducing agents not only prevented the development of turbidity but could reverse the process. The authors concluded that glycosylation of lens proteins with glucose or glucose-6-phosphate increased the susceptibility of their thiol groups to oxidation—a change characteristic of the conformational change to lens proteins in cataract (114). Incubation with glucose-6-phosphate also produced disulfide-bonded protein of high molecular weight. Evidence for glycosylation of lens proteins was found in rat lenses and in bovine lens proteins incubated in media containing glucose, and in lenses from diabetic rats (118, 135). Glycosylated lysine was identified in acid hydrolysates of borohydride-reduced crystallins from diabetic cataracts and galactose cataracts in rats (118, 135a) and has been reported to increase in human cataracts (135a) at least in diabetic cataract (135b).

8.7.1.5 Addition of Glutathione to Form Mixed Disulfides

When protein-bound glutathione was first found to increase in nuclear cataracts it was thought that its formation was caused by the decreasing glutathione content of cataractous lenses (26). Later it was thought that both the decreasing glutathione content and the unfolding of lens proteins promoted the formation of protein-bound glutathione (9, 28); subsequently it has been suggested that the addition of glutathione to lens proteins promotes the unfolding (136). Presumably formation of protein–protein disulfide bonds (52) could also be effective. Binding of glutathione to lens protein is found in most nuclear cataracts (group II of Pirie–van Heyningen) (26), and thus seems to occur in the earlier stages of cataract with the conformational changes, rather than later with the final browning reactions.

8.7.1.6 Degradation of Lens Polypeptide Chains

A variety of truncated chains of α-crystallin are found in extracts of clear bovine lens (see refs. 9 and 10) and at least one proteinase is present in bovine and human lens (137–143). Short polypeptide chains (less than 20,000 dalton) have been found in human lens protein fractions, and cleavage of peptide bonds could cause conformational changes and their sequelae, but does the proportion of truncated chains increase in cataract?

Truncated chains were found in high proportion in α-crystallin isolated from the nucleus of normal calf and adult bovine lens (144, 145) and a picture of stepwise degradation of bovine α-crystallin from the *C*-terminal end has evolved (10; see also Chapters 1 and 7). Truncated α-crystallin chains have also been identified in the nucleus of normal rabbit lens by gel electrophoresis (146) but there is no evidence for an increase in truncated chains in X-ray cataract (41, 147).

For human lens Kramps et al. (22) showed that for five soluble protein fractions and two water-soluble fractions there was no change in electrophoretic pattern on sodium dodecylsulfate gels, and they concluded that degeneration of α-crystallin played no part in human cataract. Roy and Spector (101, 102) reported that essentially all the polypeptides present in α-crystallin from human lens periphery were unchanged in cataract, and concluded that there was no evidence that modified polypeptides of α-crystallin are related to the development of cataract. The view that a change in polypeptide pattern and in particular an increase of degraded chains is not a feature of cataract has recently received further support (35, 84, 106).

Van Heyningen (148) has proposed a different way in which proteolysis may be involved in cataract; she suggested that when a lens becomes leaky at an early stage of cataract the lens proteinase could be released from inactive complexes with degraded peptides and α-crystallin, with loss of the peptides from the lens and activation of the proteinase. The high calcium content of many cataract lenses would also activate the enzyme, and the proteolysis would account for the low dry weight of most cataract lenses (2, 24).

8.7.1.7 Sunlight, Free Radicals and Cataract

The possible role of sunlight in cataract formation has been discussed for many years based on the observations that cataract is common in sunny parts of the world and on the knowledge that irradiation by ultraviolet light has been shown to cause lens opacities in animals.

Some early findings on the common and early occurrence of cataract in areas where the proportion of ultraviolet light in sunlight is high were discussed by Duke-Elder (149).

Fuchs (150) studied the geographic distribution of senile cataract by finding the proportion of cataract cases in 367 eye hospitals around the world. He concluded that cataract is very common and develops early in parts of India and Pakistan but not throughout these countries. Because the conditions of drought and strong ultraviolet light are present in many parts of the world where the incidence of cataract is much lower, Fuchs suggested that factors in the food or soil were responsible. Another survey based on records for different regions of the United States calculated cataract prevalence relative to other eye diseases (151). They found an increasing ratio of cataract to other eye diseases in regions with more sunshine. Another survey reported that a higher proportion of dark brown lenses was removed at a free clinic in the Philippines than in a private hospital in Rochester, New York (151a).

It is obviously better to determine the prevalance of cataract by painstakingly surveying a large random sample of different populations, and in this way the high prevalence in parts of India was confirmed (152, 153). By examination of 19,500 people they found 4.2% of the population of the dry belt of East Punjab and Haryana with cataract or aphakia, 13.4% of the population aged over 30 years; on the plains a staggering 31% of those aged more than 60 years had cataract or aphakia. The prevalence in those living at an altitude of about 3350 meters in the Himalayas was less than for those populating the hot plains, although cloudy weather is rare in the higher region and there are few trees. Curiously the prevalence of pterygium, a lesion apparently caused by sunlight (154), is low in the plains population (152). Chatterjee(153) mentions several possible causes of the high prevalence in this part of India including climatic effects, racial differences, the high mineral intake caused by drinking up to 5 l of the local water a day, and the heavy consumption of yogurt. Chatterjee pointed out that the influence of genetic factors could be examined by studying Punjabis and their children living in temperate climates. The prevalence of cataract also appears to be high in rural Africans in Rhodesia (155).

The rate of cataract extraction appears to be higher in Israel than in Oxford (156, 157), but in addition there is a marked difference between Israelis of European origin and those from the Middle East and North Africa.

Lenses removed at eye camps in Pakistan are generally darker brown than lenses removed in the Oxford Eye Hospital (155).

Bachem (158) exposed the right eyes of 78 rabbits and 24 guinea pigs to ultraviolet light of different wavelengths and observed the following changes: keratitis, corneal opacity, palpebral and bulbar conjunctivitis, iritis, turbidity of the aqueous humor, and reversible blurring of the exposed lens capsule and front of the lens. Irreversible lenticular opacities (cataract) occurred in the exposed frontal layers of the lens after repeated, excessive irradiation with wavelengths 297 to 365 nm. These reactions occurred in many guinea pigs and one rabbit. An ultraviolet laser (325 nm) produced anterior subcapsular and cortical cataracts, but not nuclear cataracts, in rabbits (159). Laser radiation (337 nm and 351 + 364 nm) caused lens opacities in monkeys, which the authors suggested were caused by a thermal damage mechanism (160).

When solutions of proteins which contain tryptophan, including proteins from bovine and human lens, are exposed to sunlight the tryptophan residues are oxidized to N' - formylkynurenine (91), giving them an altered fluorescence not shown by proteins lacking tryptophan (161); Pirie suggested that this change may be related to the changes in the lens proteins in brown cataracts. The photooxidation process *in vitro* causes cross-linking of the lens proteins with production of protein held together by covalent bonds other than disulfides, that is, fraction HM4 (162).

Sunlight and near-ultraviolet light caused a yellowing of human lens, and of various proteins, in the presence of added tryptophan (163, 164), but purified lens crystallins did not turn yellow after exposure to near-ultraviolet light in the absence of tryptophan under the same conditions (164). So two hypotheses for the role of sunlight in causing nuclear pigmentation in cataract were developed: one that tryptophanyl residues in the proteins were photooxidized leading to yellowing, increased fluorescence, and cross-linking; the other that small molecules were photooxidized and then became bound to the proteins to cause brunescent cataract. Zigman pointed out that the presence of reducing agents in lens must be considered because ascorbic acid is a potent inhibitor of the photooxidation of tryptophan (165, 166).

Irradiation of carboxymethyllysozyme with simulated sunlight caused loss of tryptophan, yellowing, change of fluorescence, and formation of covalent cross-links, whereas none of these changes occurred on irradiation of carboxymethylribonuclease, a protein that lacks tryptophan (167).

It is a peculiar coincidence that *N*-formylkynurenine not only is the primary photooxidation product of tryptophan but also is produced enzymatically from tryptophan in the human lens and is subsequently converted to kynurenine, hydroxykynurenine, and a glucoside of hydroxykynurenine (168, 169). These compounds could cause at least some of the photooxidative changes, and additionally could act as photosensitizers; indeed it was shown that lens proteins exposed to sunlight in the presence of these low molecular weight fluorescent compounds at physiological concentrations became brown much more rapidly than proteins exposed to the sunlight without the fluorescent compounds (170). Yellowing was accompanied by destruction of tryptophan and several other amino acids and by an increase in resistance to proteolysis. There was no obvious difference in yellowing between the soluble protein from normal and cataractous lens, or from cortex and nucleus. This similarity of yellowing of cortical and nuclear proteins was confirmed by Dilley and Pirie (83). The concentrations of the fluorescent compounds are less in senile cataract lenses (171).

Irradiation with near-ultraviolet light was shown to increase the electronegativity of calf γ-crystallin and Zigman et al. (166) suggested that the change in charge could change the conformation of the protein. Irradiation also increased the amount of insoluble protein isolated from the lenses (172).

Weiter and Finch (173) concluded from an electron spin resonance study that both normal and cataractous human lens nucleus have a paramagnetic species derived from tryptophan; and that ultraviolet irradiation of lens produced unidentified free radicals. Lerman and Borkman (174) found differences in electron spin resonance signal between normal lenses and cortical cataracts after ultraviolet irradiation; nuclear cataracts gave a signal similar to normal lenses of the same age.

In 1974 Dilley and Pirie (83) critically reviewed the evidence that sunlight causes brown cataracts comparing and contrasting experimental browning with cataracts, pointing out that there seems to be no less tryptophan, and only traces of kynurenine or *N*′-formylkynurenine in brown cataract lens, and that yellowing progresses equally well in solutions of cortical proteins, whereas *in vivo* the lens nucleus becomes yellow, then brown much more rapidly than the cortex.

Examination of the human lens nucleus had yielded no hard evidence of photooxidized proteins (175), and Dilley (176) pointed to two major arguments against the hypothesis that sunlight causes brown nuclear cataract: first, the destruction of tryptophan found after *in vitro* photooxidation of the lens proteins (162) is not found in brown cataracts (83), and second, in brown nuclear cataract only the lens nucleus is pigmented, although most of the supposedly damaging light would be absorbed at the front of the lens if it were transmitted by the cornea.

It has not yet proved possible to produce experimentally a brown nucleus in a pale cortex by irradiation of lenses (177). There have been attempts to answer these two arguments. Weiter and Finch (178) suggested that the tryptophan in the lens proteins *in vivo* might absorb the radiation but then transfer the energy to other components such as lipids, which could produce a damaging free radical; Lerman and Borkman (174) suggested that so few tryptophan residues are destroyed that their loss was not noticed by amino acid analysis. These suggestions would indicate that the protein changes found *in vitro* and *in vivo* are produced by different mechanisms. Attempts to answer the second point depend on the presence of more free radical scavenger, notably glutathione, in the cortex than nucleus (174), but differences in glutathione content in human lens are not that great even compared with differences between animals; furthermore, if the damaging light was absorbed in the anterior cortex it would not produce free radicals in the nucleus that would have to be scavenged there. Elaboration of the sunlight hypothesis with postulated inhibition of cell division and destruction of enzymes by photoproducts (179) leaves the arguments against the sunlight hypothesis unanswered. There are no reports of cataracts produced experimentally by sunlight or simulated sunlight.

Pterygium and climatic droplet keratopathy (Bietti's corneal degeneration) are ocular lesions apparently caused by prolonged exposure to solar radiation. They occur chiefly in those who work outdoors in countries where the face is exposed to a relatively high intensity of solar ultraviolet radiation (154,180–184). Freedman (181) and Rodger (182, 183) draw attention to the particular dangers of ultraviolet reflection from sand or snow. Pterygium is virtually unknown among the native English population (154) and is not associated with cataract (C. Peckar, personal communication); its incidence is lower in spectacle wearers. There is no evidence that climatic droplet keratopathy is associated with cataract; indeed the proportion of blindness attributable to cataract is unremarkable in the Dahlak Islands where sunlight damaged corneas are found in 46% of males (182), and pterygium is unimportant in East Punjab and Haryana where cataract afflicts 13% of the population aged over 30 years (152).

Cataracts have been produced in guinea pigs and mice exposed to near-ultraviolet light after administration of a photosensitizer, methoxsalen, but the cataracts, as one would expect, were initially confined to the anterior cortex and other ocular tissues were severely damaged (185, 186). After 10 months, when the majority of the mice had died, a few survivors had mature cataract always associated with devascularized iris. Other photosensitizers such as chlorpromazine act similarly, with the principal change occurring in the anterior cortex of the lens (187).

In summary the balance of evidence at present suggests that brown nuclear cataract is not caused by sunlight, and certainly not by sunlight alone.

There remains the possibility that free radicals other than those formed by photolysis could cause senile cataract, but although this is an attractive idea there is as yet no evidence to support it. Free radicals are presumably important in cataracts caused by high-energy radiation and those caused by photosensitizing drugs plus sunlight. Diquat, a weed killer that causes cataract in rats, forms free radicals that react with air to form superoxide radicals (188). The lens contains superoxide dismutase (189–191), which converts the superoxide radical to the less dangerous hydrogen peroxide, which in turn can be reduced by the glutathione peroxidase reaction (192) or by catalase (193). However, the superoxide was also destroyed by concentrations of ascorbic acid and glutathione (189) far below those found in normal and cataractous human lens (26, 66). Nevertheless it is possible that free radicals play a role in human cataract and the lowered level of glutathione, a free radical scavenger, in most forms of cataract would be consistent with the radical involvement, although in cataract caused by diquat free radicals the level of glutathione remains high (194). Malondialdehyde, a product of lipid peroxidation by a free radical mechanism, colors and cross-links lens proteins (92), but peroxidation products of lipids could not be found in brown cataract lenses (195).

8.7.1.8 Carbamylation, Uremia and Diarrhea

Because the sunlight hypothesis does not provide a convincing explanation for the high prevalence of cataract in some tropical countries it seemed useful to consider alternative explanations that might also relate to the conformational changes. Several authors have suggested that malnutrition causes some of this excess cataract (196, 197), although cataract has not been associated with the recognized malnutrition diseases. However, cataract is found in tryptophan-deficient animals (198), lens changes occur in vitamin A deficiency (199), and a low protein diet increases the incidence of galactose cataract in rats (200, 201).

In many tropical countries there is another common condition besides malnutrition that causes millions of deaths annually—diarrhea. Several side effects of diarrhea have been associated with cataract: malnutrition itself, acidemia (202), osmotic imbalance (202, 203), and uremia. It has been

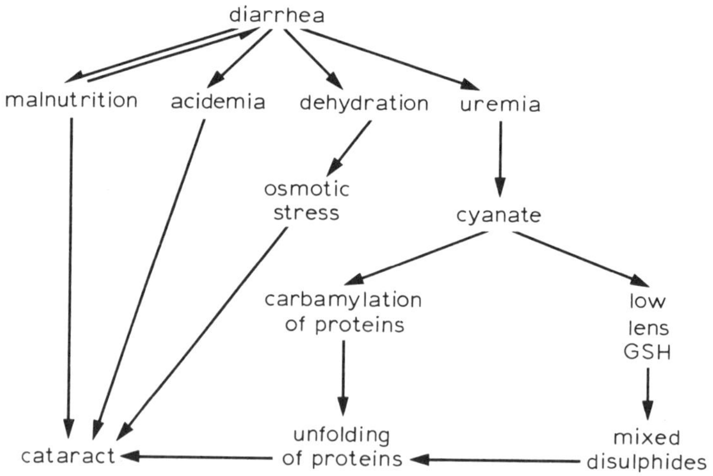

Figure 9 Speculative scheme indicating possible relationships between diarrhea and cataract.

suggested that in these ways diarrhea may be responsible for the high prevalence of cataract in some tropical countries (118b).

The possible relationship between diarrhea and cataract is shown in Figure 9.

The concentration of urea in aqueous humor parallels that in blood (204, 205), and uremic patients tend to develop cataracts especially after dialysis (206, 207), probably due to the acute osmotic change, but lens opacities have been seen before dialysis, possibly caused by uremia itself (208). Increased urea levels in the lens could have several effects: osmotic effects in both directions as aqueous urea concentration rose and fell, a direct effect in perturbing proteins, and perhaps more importantly, the formation of cyanate, which can carbamylate proteins (209). Cyanate formation and carbamylation of lens proteins were found in urea solutions used for electrophoresis (210, 211), and direct binding of label from [^{14}C]-KCNO or [^{14}C]-urea to lens proteins has also been demonstrated (118b).

At equilibrium 250 m*M* urea at 37°C, pH 7.4, contains 2m*M* cyanate (212) and so in severe diarrhea with anuria or oliguria one might expect up to 1 m*M* cyanate in the blood. Rabbit lenses incubated in 10 m*M* cyanate overnight lost glutathione, and became swollen and opaque, and as little as 1 m*M* cyanate produced a ring of opacity (213). Oral sodium cyanate caused cataracts in man and in dogs, lowering the glutathione level in the dog lens (214, 215), showing that cyanate formed in the blood could damage the lens, possibly by lowering the glutathione concentration and by carbamylating protein thiol and amino groups. Carbamylation of the lens proteins would change the charge profile of the proteins and could thus cause the conformational changes found in nuclear cataracts. Recently we found evidence for carbamylation of lens proteins from cataracts removed in Pakistan (216). Cyanate could also inactivate enzymes such as glucose-6-phosphate dehydrogenase (217, 209) and carbamylate glutathione, thus promoting mixed disulfide formation.

It seems possible that these various manifestations of diarrhea could cause cataract individually or perhaps acting synergistically, but only good epidemiological surveys of the prevalence of cataract in populations with different incidences of diarrhea will show whether or not repeated episodes of severe diarrhea are a major cause of cataract in some countries.

8.7.1.9 Summary: The Cause of the Conformational Changes

I have discussed a number of possible causes of the conformational changes in proteins in nuclear cataracts but there is no unequivocal proof for any of them: glycosylation is an interesting possibility and has been related to the unfolding of rat lens proteins, but has yet to be studied in human lens; racemization looked exciting but appears to be the result of a conformational change rather than the cause of it; binding of glutathione, cross-linking, and proteolysis may play some role—the binding of glutathione is an early change; oxidation of methionine seems to be a later change; deamidation and sunlight probably have no role; and the free radical and diarrhea hypotheses are entirely speculative. In my attempts to relate certain events in cataractogenesis to time, it has been most fortunate that many authors in The Netherlands, Australia, the United States, and Great Britain have classified lenses using the same system (Pirie–van Heyningen). There seems to be a clear progression in human nuclear cataracts from group II to group IV although the appearances of individual lenses differ, and this trend suggests that although there may be a variety of contributory causes, the initial changes feed at some point into a common pathway of cataractogenesis. This hypothesis is supported by animal experiments showing the synergistic action of two different factors to give cataract (218). The common pathway may include the conformational change and the decrease of glutathione.

8.8 CONCLUSION

In nuclear cataract in humans and probably in some experimental cataracts the lens proteins change conformation to expose thiol groups and peptide bonds that had been buried. The unfolding makes the proteins susceptible to oxidative reactions, first binding glutathione and forming some protein–protein disulfides to produce large aggregates of protein. Eventually other reactions occur, probably oxidations, leading to cross-linking by other covalent bonds, a deepening color, and altered fluorescence. It is important now to determine the cause or causes of the conformational changes. Several possible causes that may act singly or in concert were discussed in this chapter.

ACKNOWLEDGMENT

I am most grateful to Ruth van Heyningen and James Crabbe for their helpful criticisms.

REFERENCES

1 J. Nordmann, *Ophthal. Res.*, **3,** 323 (1972).
2 R. van Heyningen, *Exp. Eye Res.*, **13,** 155 (1972).
3 B. Philipson, *Invest. Ophthalmol.*, **8,** 271 (1969).
4 B. Philipson, *Invest. Ophthalmol.*, **8,** 281 (1969).
5 J. Papaconstantinou, *Biochim. Biophys. Acta,* **107,** 81 (1965).
6 J. Papaconstantinou, *Science,* **156,** 338 (1967).
7 E. E. Schubert, J. R. Trevithick, and M. J. Hollenberg, *Can. J. Ophthalmol.*, **5,** 353 (1970).
8 J. W. McAvoy, *J. Embryol. Exp. Morphol.*, **44,** 149 (1978).
9 J. J. Harding and K. J. Dilley, *Exp. Eye. Res.*, **22,** 1 (1976).
10 H. Bloemendal, *Science,* **197,** 127 (1977).
11 C. F. Wannemacher and A. Spector, *Exp. Eye Res.*, **7,** 623 (1968).
12 K. J. Dilley and R. van Heyningen, *Doc. Ophthal. Proc.*, **8,** 171 (1976).
13 R. Clark, S. Zigman, and S. Lerman, *Exp. Eye Res.*, **8,** 172 (1969).
14 A. T. York, C. Martinez, and W. Manski, *Ophthal. Res.*, **3,** 183 (1972).
15 G. Duncan and A. R. Bushell, *Exp. Eye Res.*, **20,** 223 (1975).
16 A. Pirie, *Invest. Ophthalmol.*, **7,** 634 (1968).
17 R. van Heyningen, *Exp. Eye Res.*, **13,** 136 (1972).
18 L. T. Chylack, *Arch. Ophthalmol.*, **96,** 888 (1978).
19 H. Mach, *Klin. Monatsbl. Augenheilkd.*, **143,** 689 (1963).
20 H. Mach, *Klin. Monatsbl. Augenheilkd.*, **145,** 827 (1964).
21 R. C. Augusteyn, *Exp. Eye Res.*, **19,** 91 (1974).
22 H. A. Kramps, H. J. Hoenders, and J. Wollensak, *Biochim. Biophys. Acta,* **434,** 32 (1976).
23 R. J. W. Truscott and R. C. Augusteyn, *Exp. Eye Res.*, **24,** 159 (1977).
24 R. van Heyningen, *Exp. Eye Res.*, **13,** 148 (1972).
25 G. Duncan and R. van Heyningen, *Exp. Eye Res.*, **25,** 183 (1977).
26 J. J. Harding, *Biochem. J.*, **117,** 957 (1970).
27 R. J. W. Truscott and R. C. Augusteyn, *Exp. Eye Res.*, **25,** 139 (1977).

28 J. J. Harding, *Exp. Eye Res.*, **17**, 377 (1973).

29 R. H. Buckingham, *Exp. Eye Res.*, **14**, 123 (1972).

30 J. A. Kramps, H. J. Hoenders, and J. Wollensak, *Exp. Eye Res.*, **27**, 731 (1978).

31 R. C. Augusteyn, *Ophthal. Res.*, **7**, 217 (1975).

32 R. J. W. Truscott and R. C. Augusteyn. *Biochim. Biophys. Acta*, **492**, 43 (1977).

33 K. M. Rogers and R. C. Augusteyn, *Exp. Eye Res.*, **27**, 719 (1978).

34 P. M. Masters, J. L. Bada, and J. S. Zigler, *Nature*, **268**, 71 (1977).

34a J. Cuthbert, R. M. Clayton, D. E. S. Truman, C. I. Phillips, and R. S. Bartholemew, *Interdiscipl. Topics Gerontol.*, **13**, 183 (1978).

35 P. J. Ringens, K. N. Liem-The, H. J. Hoenders, and J. Wollensak, *Interdiscipl. Topics Gerontol.*, **13**, 193 (1978).

36 D. C. Wood, D. Sweet, J. van Dolah, J. C. Smith, and I. Contaxis, *Ann. N.Y. Acad. Sci. U.S.*, **141**, 346 (1967).

37 R. van Heyningen and J. J. Harding, *Exp. Eye Res.*, **14**, 91 (1972).

38 V. D. Rowe, S. Zigler, A. E. Andersen, J. B. Sidbury, and G. Guroff, *Exp. Eye Res.*, **17**, 245 (1973).

39 S. K. Brahma and J. Bours, *Exp. Eye Res.*, **23**, 57 (1976).

40 K. N. Liem-The, P. J. Ringens, J. C. W. M. Holtslag, and H. J. Hoenders, *Interdiscipl. Topics Gerontol.*, **12**, 105 (1978).

41 F. J. Giblin, B. Chakrapani, and V. N. Reddy, *Exp. Eye Res.*, **26**, 507 (1978).

42 J. Bours, L. Gruber, O. Hockwin, and J. Harris, *Interdiscipl. Topics Gerontol.*, **12**, 127 (1978).

43 P. Russel, D. Carper, and J. H. Kinoshita, *Exp. Eye Res.*, **27**, 673 (1978).

44 B. J. Wagner and S.-C. J. Fu, *Exp. Eye. Res.*, **26**, 255 (1978).

45 J. Piatigorsky, H. N. Fukui, and J. H. Kinoshita, *Nature*, **274**, 558 (1978).

46 G. E. Bunce, J. L. Hess, and G. M. Fillnow, *Exp. Eye Res.*, **26**, 399 (1978).

47 J. Francois and M. Rabaey, *Acta Ophthalmol.*, **36**, 837 (1958).

48 O. Hockwin, N. Poonawalla, E. Noll, and W. Licht, *Arch. Ophthalmol.*, **188**, 175 (1973).

48a H. O. Sandberg and O. Closs, *Exp. Eye Res.*, **28**, 601 (1979).

49 J. François, M. Rabaey, and L. Stockmans, *Exp. Eye Res.*, **4**, 312 (1965).

50 E. J. Sheridan and S. Zigman, *Exp. Eye Res.*, **12**, 33 (1971).

51 L. R. Croft *The Human Lens in Relation to Cataract, Ciba Found. Symp.*, Vol. 19, Elsevier, Amsterdam, 1973, p. 207.

52 J. J. Harding, *Exp. Eye Res.*, **13**, 33 (1972).

53 H. O. Sandberg and O. Closs, *Exp. Eye Res.*, **27**, 701 (1978).

54 A. Jess, *Z. Biol.*, **61**, 92 (1913).

55 W. Manski, M. Behrens, and C. Martinez, *Exp. Eye Res.*, **7**, 164 (1968).

56 Z. Dische, E. Borenfreund, and G. Zelmenis, *A.M.A. Arch. Ophthalmol.*, **55**, 471 (1956).

57 S. Lerman, *Surv. Ophthalmol.*, **12**, 112 (1967).

58 S. Zigman, J. Schultz, and T. Yulo, *Biochem. Biophys. Res. Commun.*, **35**, 931 (1969).

59 R. W. Dapson, J. G. Otero, and W. R. Holloway, *Nature*, **218**, 573 (1968).

60 P. D. Mehta and H. Maisel, *Biochemistry of the Eye,* Karger, Basel, 1968, p. 271.

61 K. Satoh, *Exp. Eye Res.*, **14,** 53 (1972).

62 A. Spector, D. Roy, and J. Stauffer, *Exp. Eye Res.*, **21,** 9 (1975).

63 S. Zigman, J. Groff, T. Yulo, and G. Griess, *Exp. Eye Res.*, **23,** 555 (1976).

64 S. D. Coghlan and R. C. Augusteyn, *Exp. Eye Res.*, **25,** 603 (1977).

65 W. H. Garner and A. Spector, *Proc. Natl. Acad. Sci. U.S.*, **75,** 3618 (1978).

66 R. van Heyningen, *The Eye* 2nd ed., Vol. 1, Academic Press, New York-London, 1969, p. 381.

67 S. Zigman, J. Schultz, and T. Yulo, *Exp. Eye Res.*, **10,** 58 (1970).

68 G. Auricchio and M. Testa, *Ophthalmologica,* **164,** 228 (1972).

69 K. J. Dilley, *Exp. Eye Res.*, **20,** 73 (1975).

70 J. F. R. Kuck, in J. G. Bellows, Ed., *Cataract and Abnormalities of the Lens,* Grune and Stratton, New York, (1975) p. 69.

71 A. C. Krause, *The Biochemistry of the Eye,* John Hopkins Press, Baltimore, (1934) p. 186.

72 C. T. Mőrner, *Z. Physiol. Chem.*, **18,** 61 (1894).

73 Z. Dische, *Invest. Ophthalmol.*, **4,** 759 (1965).

74 A. Lasser and E. A. Balazs, *Exp. Eye Res.*, **13,** 292 (1972).

75 J. J. Harding, *Exp. Eye Res.*, **14,** 289 (1972).

76 J. Alcalá, N. Lieska, and H. Maisel, *Exp. Eye Res.*, **21,** 581 (1975).

77 R. M. Broekhuyse and E. D. Kuhlmann, *Exp. Eye Res.*, **26,** 305 (1978).

78 S. Zigman and S. Lerman, *Biochim. Biophys. Acta,* **154,** 423 (1968).

79 J. J. Harding, *Exp. Eye Res.*, **8,** 147 (1969).

80 Z. Dische and H. Zil, *Am. J. Ophthalmol.*, **34,** 104 (1951).

81 M. Testa, C. Fiore, N. Bocci, and S. Calabrò, *Exp. Eye Res.*, **7,** 276 (1968).

82 R. C. Augusteyn, *Exp. Eye Res.*, **28,** 665 (1979).

83 K. J. Dilley and A. Pirie, *Exp. Eye Res.*, **19,** 59 (1974).

84 J. A. Kramps, W. W. de Jong, J. Wollensak, and H. J. Hoenders, *Biochim. Biophys. Acta,* **533,** 487 (1978).

85 P. J. Stein, R. W. Henkens, B. S. Yamanashi, and M. Wolbarsht, *Ophthal. Res.*, **8,** 388 (1976).

86 R. van Heyningen, *Invest. Ophthalmol.*, **15,** 685 (1976).

87 J. S. Zigler, J. B. Sidbury, B. S. Yamanashi, and M. Wolbarsht, *Ophthal. Res.*, **8,** 379 (1976).

88 R. J. W. Truscott, K. Faull, and R. C. Augusteyn, *Ophthal. Res.*, **9,** 263 (1977).

89 S. Garcia-Castineiras, J. Dillon, and A. Spector, *Science,* **199,** 897 (1978).

90 J. Dillon, A. Spector, and K. Nakanishi, *Nature,* **259,** 422 (1976).

91 A. Pirie, *Biochem. J.*, **125,** 203 (1971).

92 P. M. M. van Haard, J. A. Kramps, H. J. Hoenders, and J. Wollensak, *Interdiscipl. Topics Gerontol.*, **13,** 212 (1978).

93 A. Spector, L.-K. Li, R. C. Augusteyn, A. Schneider, and T. Freund, *Biochem. J.*, **124,** 337 (1971).

94 H. J. Hoenders and G. J. van Kamp, *Acta Morphol. Neerl-Scand.*, **10,** 215 (1972).

95 S. Zigman, J. Groff, T. Yulo, and G. Griess, *Exp. Eye Res.*, **23,** 555 (1976).

96 D. Roy and A. Spector, *Exp. Eye Res.*, **22,** 273 (1976).

97 J. A. Jedziniak, J. H. Kinoshita, E. M. Yates, L. O. Hocker, and G. B. Benedek, *Exp. Eye Res.*, **15,** 185 (1973).

98 A. Spector, J. Stauffer, and J. Sigelman, in *The Human Lens in Relation to Cataract, Ciba Found. Symp.*, Vol. 19, Elsevier, Amsterdam, (1973), p. 185.

99 J. A. Jedziniak, J. H. Kinoshita, E. M. Yates, and G. B. Benedek, *Exp. Eye Res.*, **20,** 367 (1975).

100 J. J. Harding, *Ophthal. Res.*, **9,** 123 (1977).

101 D. Roy and A. Spector, *Biochemistry,* **15,** 1180 (1976).

102 D. Roy and A. Spector, *Invest. Ophthalmol.*, **15,** 394 (1976).

102a P. Russell, S. G. Smith, D. A. Carper, and J. H. Kinoshita, *Exp. Eye Res.*, **29,** 245 (1979).

103 A. R. Bushell and G. Duncan, *Exp. Eye Res.*, **26,** 223 (1978).

104 L. J. Takemoto and P. Azari, *Exp. Eye Res.*, **24,** 63 (1977).

105 L. J. Takemoto, P. Azari, and W. C. Gorthy, *Exp. Eye Res.*, **20,** 1 (1975).

106 A. Spector and D. Roy, *Proc. Natl. Acad. Sci. U.S.*, **75,** 3244 (1978).

107 S. G. Waley, *The Eye,* 2nd ed., Vol. 1, Academic Press, New York-London, 1969, p. 299.

108 G. B. Benedek, *Appl. Optics,* **10,** 459 (1971).

109 R. A. Schachar and S. A. Solin, *Invest Ophthalmol.*, **14,** 380 (1975).

110 B. Philipson, *Exp. Eye Res.*, **16,** 29 (1973).

111 A. Spector, S. Li, and J. Sigelman, *Invest. Ophthalmol.*, **13,** 795 (1974).

112 T. Tanaka and G. B. Benedek, *Invest. Ophthalmol.*, **14,** 449 (1975).

113 J. A. Jedziniak, D. F. Nicoli, H. Baram, and G. B. Benedek, *Invest. Ophthalmol.*, **17,** 51 (1978).

114 J. J. Harding, *Biochem. J.*, **129,** 97 (1972).

115 G. Duncan and R. van Heyningen, *Doc. Ophthalmol. Proc.*, **8,** 229 (1976).

116 J. H. Kinoshita and L. O. Merola, in K. Elliott and D. W. Fitzsimons, Eds., *Ciba Found. Symp.*, Vol. 19, Elsevier, Amsterdam, 1973, p. 173.

117 P. M. Masters, J. L. Bada, and J. S. Zigler, *Proc. Natl. Acad. Sci. U.S.*, **75,** 1204 (1978).

118 V. J. Stevens, C. A. Rouzer, V. M. Monnier, and A. Cerami, *Proc. Natl. Acad. Sci. U.S.*, **75,** 2918 (1978).

118a V. N. Orekhovich, K. F. Firfarova, E. M. Kedrova, and G. A. Levdikova, *Exp. Eye Res.*, **1,** 324 (1962).

118b J. J. Harding, in F. Regnault, O. Hockwin, and Y. Courtois, Eds., *Aging of the Lens,* Elsevier, Amsterdam, (1980), p. 71.

119 R. B. Kurzel, M. Wolbarsht, B. S. Yamanashi, G. W. Staton, and R. F. Borkman, *Nature,* **241,** 132 (1973).

119a P. Rácz, K. Tompa, and I. Pócsik, *Exp. Eye Res.,* **28,** 129 (1979).

120 S. K. Srivastava and E. Beutler, *Exp. Eye Res.,* **17,** 33 (1973).

121 E. I. Anderson and A. Spector, *Exp. Eye Res.,* **26,** 407 (1978).

122 V. N. Reddy and R. F. Han, *Doc. Ophthalmol. Proc.,* **8,** 153 (1976).

123 K. J. Dilley and J. J. Harding, *Biochim. Biophys. Acta,* **386,** 391 (1975).

124 H. Bloemendal, A. J. M. Berns, F. van der Ouderaa, and W. W. W. de Jong, *Exp. Eye Res.,* **14,** 80 (1972).

125 J. Delcour and J. Papaconstantinou, *J. Biol. Chem.,* **247,** 3289 (1972).

126 T. Flatmark and K. Sletten, *J. Biol. Chem.,* **243,** 1623 (1968).

127 C. F. Midelfort and A. H. Mehler, *Proc, Natl. Acad. Sci. U.S.,* **69,** 1816 (1972).

128 A. B. Robinson and C. J. Rudd, *Curr. Topics Cell. Reg.,* **8,** 247 (1974).

129 J. J. Harding, *Colloq. d'INSERM,* **60,** 215 (1976).

130 J. L. Bada and R. Protsch, *Proc. Natl. Acad. Sci. U.S.,* **70,** 1331 (1973).

131 P. M. Helfman and J. L. Bada, *Proc. Natl. Acad. Sci. U.S.,* **72,** 2891 (1975).

132 P. M. Helfman and J. L. Bada, *Nature,* **262,** 279 (1976).

133 A. Cerami and R. J. Koenig, *Trends Biochem. Sci.,* **3,** 73 (1978).

134 A. J. Bailey, S. P. Robins, and G. Balian, *Nature,* **251,** 105 (1974).

135 A. Cerami, V. J. Stevens, and V. M. Monnier, *Metabolism,* **28,** 431 (1979).

135a V. M. Monnier, V. J. Stevens, and A. Cerami, *J. Exp. Med.,* **150,** 1098 (1979).

135b A. Pande, W. H. Garner, and A. Spector, *Biochem. Biophys. Res. Commun.,* **89,** 1260 (1979).

136 S. K. Srivastava, *Exp. Eye Res.,* **22,** 577 (1976).

137 R. van Heyningen and S. G. Waley, *Exp. Eye Res.,* **1,** 336 (1962).

138 S. G. Waley and R. van Heyningen, *Biochem. J.,* **83,** 274 (1962).

139 A. M. J. Blow, R. van Heyningen, and A. J. Barrett, *Biochem. J.,* **145,** 591 (1975).

140 P. Trayhurn and R. van Heyningen, *Exp. Eye Res.,* **22,** 251 (1976).

141 R. van Heyningen and P. Trayhurn *Exp. Eye Res.,* **22,** 625 (1976).

142 R. van Heyningen, *Interdiscipl. Topics Gerontol.,* **12,** 232 (1978).

143 R. van Heyningen, J. E. Shirley, and S. van Heyningen, *Doc. Ophthal. Proc.,* **18,** 101 (1979).

144 F. S. M. van Kleef, M. J. C. M. Nijzink-Maas, and H. J. Hoenders, *Eur. J. Biochem.,* **48,** 563 (1974).

145 J. Stauffer, C. Rothschild, T. Wandel, and A. Spector, *Invest. Ophthalmol.,* **13,** 151 (1974).

146 K. N. Liem-The and H. J. Hoenders, *Exp. Eye Res.,* **19,** 549 (1974).

147 K. N. Liem-The, A. L. H. Stols, P. H. K. Jap, and H. J. Hoenders, *Exp. Eye Res.,* **20,** 317 (1975).

148 R. van Heyningen, *Invest. Ophthalmol.,* **15,** 685 (1976).

149 Sir W. S. Duke-Elder, in *Text-book of Ophthalmology,* Vol. III, *Diseases of the Inner Eye,* Henry Kimpton, London, 1940, p. 3102.

150 A. Fuchs, *Am. J. Ophthalmol.,* **49,** 1039 (1960).

151 R. Hiller, L. Giacometti, and K. Yuen, *Am. J. Epidemiol.,* **105,** 450 (1977).

151a S. Zigman, M. Datiles, and E. Torczynski, *Invest. Ophthalmol.,* **18,** 462 (1979).

152 S. Franken and K. R. Mehta, "A Survey to Determine the Ophthalmic Morbidity in the Population Living in the Dry Belt of Southern Panjab and Haryana." Christian Medical College, Ludhiana, Punjab, 1968.

153 A. Chatterjee, in *The Human Lens-in Relation to Cataract, Ciba Found. Symp.,* 19, Elsevier, Amsterdam, (1973), p. 265.

154 C. O. Peckar, *Doc. Ophthalmol.,* **31,** 141 (1972).

155 S. W. Jamieson, *Postgrad. Med. J.,* **46,** 557 (1970).

156 H. S. Halevi and J. Landau, *Brit. J. Ophthalmol.,* **46,** 285 (1962).

157 A. Pirie, *Isr. J. Med. Sci.,* **8,** 1550 (1972).

158 A. Bachem, *Am. J. Ophthalmol.,* **41,** 969 (1956).

159 D. MacKeen, S. Fine, and B. S. Fine, *Ophthal. Res.,* **5,** 317 (1973).

160 J. A. Zuclich and J. S. Connolly, *Invest. Ophthalmol.,* **9,** 760 (1976).

161 A. Pirie, *Biochem. J.,* **128,** 1365 (1972).

162 R. H. Buckingham and A. Pirie, *Exp. Eye Res.,* **14,** 297 (1972).

163 S. Zigman, *Science,* **171,** 807 (1971).

164 D. Grover and S. Zigman, *Exp. Eye Res.,* **13,** 70 (1972).

165 S. Zigman, *Isr. J. Med. Sci.,* **8,** 1590 (1972).

166 S. Zigman, G. Griess, T. Yulo, and J. Schultz, *Exp. Eye Res.,* **15,**255 (1973).

167 K. J. Dilley, *Biochem. J.,* **133,** 821 (1973).

168 R. van Heyningen, *Nature,* **230,** 393 (1971).

169 R. van Heyningen, in K. Elliott and D. W. Fitzsimons, Eds., *The Human Lens-in Relation to Cataract, Ciba Found. Symp.,* Vol. 19, Elsevier, Amsterdam, (1973), p. 151.

170 R. van Heyningen, *Exp. Eye Res.,* **17,** 137 (1973).

171 R. van Heyningen, *Exp. Eye Res.,* **15,** 121 (1973).

172 S. Zigman, J. Schultz, and T. Yulo, *Exp. Eye Res.,* **15,** 201 (1973).

173 J. J. Weiter and E. D. Finch, *Nature,* **254,** 536 (1975).

174 S. Lerman and R. Borkman, *Ophthal. Res.,* **8,** 335 (1976).

175 R. van Heyningen, *Sci. Am.,* **233,** 70 (1975).

176 K. J. Dilley, *Nature,* **257,** 71 (1975).

177 R. van Heyningen, *Invest. Ophthalmol.,* **15,** 685 (1976).

178 J. J. Weiter and E. D. Finch, *Nature,* **257,** 71 (1975).

179 S. Zigman, *Photochem, Photobiol.,* **26,** 437 (1977).

180 S. Duke-Elder, in *System of Ophthalmology,* Vol. VIII, Henry Kimpton, London, 1965, p. 573.

181 A. Freedman, *A.M.A. Arch. Ophthalmol.,* **89,** 193 (1973).

182 F. C. Rodger, *Brit. J. Ophthalmol.,* **57,** 657 (1973).

183 F. C. Rodger, J. A. Cuthill, P. J. Fydelor, and A. P. Lenham, *Acta Ophthalmol.,* **52,** 777 (1974).

184 J. D. H. Young and R. D. Finlay, *Am. J. Ophthalmol.,* **79,** 129 (1975).

185 T. M. Cloud, R. Hakim, and A. C. Griffin, *A.M.A. Arch. Ophthalmol.,* **64,** 346 (1960).

186 T. M. Cloud, R. Hakim, and A. C. Griffin, *A.M.A. Arch. Ophthalmol.,* **66,** 689 (1961).

187 J. R. Siddall, *A.M.A. Arch. Ophthalmol.,* **74,** 460 (1965).

188 T. C. Stancliffe and A. Pirie, *FEBS Lett.,* **17,** 297 (1971).

189 S. D. Varma, T. K. Ets, and R. D. Richards, *Ophthal. Res.,* **9,** 421 (1977).

190 K. C. Bhuyan and D. K. Bhuyan, *Biochim. Biophys. Acta,* **542,** 28 (1978).

191 R. Crouch, D. G. Priest, and E. J. Duke, *Exp. Eye Res.,* **27,** 503 (1978).

192 N. J. Holmberg, *Exp. Eye Res.,* **7,** 570 (1968).

193 K. C. Bhuyan and D. K. Bhuyan, *Biochim. Biophys. Acta,* **497,** 641 (1977).

194 A. Pirie and J. R. Rees, *Exp. Eye Res.,* **9,** 198 (1970).

195 R. B. Kurzel, M. L. Wolbarsht, and B. S. Yamanashi, *Exp. Eye Res.,* **17,** 65 (1973).

196 D. S. McLaren, *Malnutrition and the Eye,* Academic Press, New York, (1963).

197 D. Singh, in *Proceedings of the Seventh Asia-Pacific Academy of Ophthalmology Congress,* (1980) (in press).

198 J. W. McAvoy, L. J. Palfrey, and R. van Heyningen, *Exp. Eye Res.,* **28,** 533 (1979).

199 A. Pirie and M. Overall, *Exp. Eye Res.,* **13,** 105 (1972).

200 H. S. Mitchell, G. M. Cook, and M. D. Henderson, *A.M.A. Arch. Ophthalmol.,* **24,** 990 (1940).

201 A. M. Yudkin and H. A. Geer, *A.M.A. Arch. Ophthalmol.,* **23,** 28 (1940).

202 M. Weinstock and J. D. Scott, *Exp. Eye Res.,* **6,** 368 (1967).

203 F. T. Fraunfelder and R. P. Burns, *Exp. Eye Res.,* **10,** 19 (1970).

204 F. H. Adler, *A.M.A. Arch. Ophthalmol.,* **10,** 11 (1933).

205 E. Moore, H. G. Scheie, and F. H. Adler, *A.M.A. Arch. Ophthalmol.,* **27,** 317 (1942).

206 F. Hollwich, A. Boateng, and B. Kolck, in J. G. Bellows, Ed., *Cataract and Abnormalities of the Lens,* Grune and Stratton, New York, 1975, p. 231.

207 J. Orth, H-R. Koch, H-U. Klehr, and M. Siedek, *Interdiscipl. Topics Gerontol.,* **13,** 109 (1978).

208 H. Laqua, *Klin. Monatsbl. Augenheilk.,* **160,** 346 (1972).

209 J. Carreras, A. Chabás, and D. Diederich, in S. Grisolia, Ed., *The Urea Cycle,* Wiley, 1976, p. 501.

210 P. Hagel, J. J. T. Gerding, W. Feiggen, and H. Bloemendal, *Biochim. Biophys. Acta,* **243,** 366 (1971).

211 J. J. T. Gerding, A. Koppers, P. Hagel, and H. Bloemendal, *Biochim. Biophys. Acta,* **243,** 374 (1971).

212 P. Dirnhuber and F. Schutz, *Biochem. J.,* **42,** 628 (1948).

213 J. H. Kinoshita and L. O. Merola, *Invest. Ophthalmol.,* **12,** 544 (1973).

214 D. H. Nicholson, D. R. Harkness, W. E. Benson, and C. M. Peterson, *A.M.A. Arch. Ophthalmol.*, **94,** 927 (1976).

215 H. L. Kern, R. W. Bellhorn, and C. M. Peterson, *J. Pharmacol. Exp. Ther.*, **200,** 10 (1977).

216 J. J. Harding and K. C. Rixon, *Exp. Eye Res.*, **31,** 567 (1980).

217 B. E. Glader and M. E. Conrad, *Nature,* **237,** 336 (1972).

218 O. Hockwin and H-R. Koch, in J. G. Bellows, Ed., *Cataracts and Abnormalities of the Lens,* Grune and Stratton, 1974, p. 243.

CHAPTER

9

Enzymes in Normal, Aging and Cataractous Lenses

Otto Hockwin and Christian Ohrloff

Division Biochemistry of the Eye, Institute of Experimental Ophthalmology, University of Bonn, Bonn, Fed. Rep. Germany

9.1 INTRODUCTION

Within the scope of this chapter it is not possible to discuss every kind of investigation so far conducted on enzymes of the eye lens. Among the numerous publications, preference is given to those describing the interrelationships between enzyme properties and behavior, and the aging process. This restriction may help to emphasize the importance of lens research, since senile cataract is still one of the main causes of blindness in man.

9.2 GENERAL REMARKS ON ENZYMES

9.2.1 Enzyme Functions

Enzymes are proteins of specific catalytic functions with the following properties in common:

1 They accelerate the rate of a thermodynamically possible biochemical reaction.
2 They themselves are not changed by the reaction they catalyze (biocatalysts).
3 Even minimum amounts are active.

Further, the enzymes have the characteristic properties of proteins, and the lability of the enzyme (protein) structure is of special importance in enzymology.

9.2.2 Enzyme Heterogeneity

Improved test methods such as protein separation and kinetic investigations show that a considerable amount of enzyme variants may be present; for example, in single cells or in different tissues of individuals of the same species, enzymes of different protein properties but similar substrate specificity are found. They are the so-called multimolecular enzyme forms. This enzyme heterogeneity, to which the isoenzymes and metazymes belong, may be due to various causes.

9.2.2.1 Isoenzymes

Several genetically coded enzyme forms catalyze the same reaction in a certain species. They differ, however, in their kinetic properties such as substrate affinity and inhibitor sensitivity. Since their amino acid structure is not the same, they differ in physicochemical properties such as heat lability, isoelectric point, electrophoretic mobility, and, although less frequently, their molecular weight (1). This form of enzyme heterogeneity is represented by the isoenzymes.

Today multiple forms of about 100 enzymes are known (2) In this context, the age-related isoenzyme modifications should be mentioned. LDH5 in the aorta, for instance, increases with age, whereas LDH1 decreases (3). Wurster and Hoffmann (4) found an age-related decrease in LDH5 in the retina of rat and cattle. This biological importance of the isoenzymes is at present only partly understood.

The different kinetic behavior of the single isoenzymes indicates their role as metabolic regulators.

9.2.2.2 Metazymes

Metazymes are variants of enzymes that are not primarily genetically coded but occur as secondary enzyme modifications. They, too, show the heterogeneity of enzymes (5). They are found less frequently than isoenzymes and occur mostly in the aging tissue, particularly after completed protein synthesis.

Some LDH-isoenzymes, for instance, show after electrophoretic separation several bands, that is, further microheterogeneity insofar as LDH5 not only consists of the MMMM tetramer, but also of a MMMM' tetramer. The subunit M' may be due to a change in the protein structure of the original M subunit which changes the electrophoretic mobility. This was also found with the LDH in human erythrocytes (6) and in the rabbit retina (7). Similar changes are known for aldolase in aging muscle fibers, which is also present as a tetramer (8), and erythrocytes (9).

Enzymes coded by only one gene also show changed electrophoretic mobility in older erythrocytes. These are glucose-6-p dehydrogenase (10), triose-phosphate isomerase, and nucleoside phosphorylase (11). A similar behavior of these three enzymes was also found in aging lens tissue (12).

Separation on Sepharose-6B showed that sorbitol dehydrogenase, aldose reductase, and ketohexokinase behave differently, as far as molecular weight is concerned, in old lenses, which also supports the heterogeneity of these enzymes (13). This may mainly be due to deamidation of glutamine and asparagine as could be verified for aldolase in rabbit muscle (14). Further support for this assumption is the uniform occurrence of the increased anodic mobility of the postsynthetically found metazymes, which have a more negative charge than the original molecule.

9.2.3 Kinetics of Enzyme-catalyzed Reactions

Metabolic processes are regulated by the velocity of metabolization per time of the individual reaction steps. This rate depends on the temperature, the concentration of the initial substrate, the concentration of the catalyst (enzyme concentration, coenzyme concentration), and the affinity of the catalyst for the substrate. Among other factors, the pH, the ionic strength, and the inorganic catalysts are often important.

For characterization of the turnover rate through a certain enzyme, the Michaelis–Menten constant (mol/l), which gives the substrate concentration of the enzyme at half maximum velocity ($v = \frac{1}{2} V_{max}$), is determined. This value enables calculation of the possible *in vivo* metabolization, if the concentration of the substrate is known. In case of substrate competition of several enzymes (as happens, for instance, with glucose-6-phosphate), the metabolic flow into bypaths may be calculated from the K_m values of the enzyme involved (15).

If an enzyme is present as two isoenzyme or metazyme forms the K_m determination is rather difficult if it is not possible to separate the compound. According to Reiner (16), it is possible to determine the K_m value of each enzyme form. Bous (17) succeeded in determining the K_m value for the phosphofructokinase forms from bovine lenses; Jeschke (18) determined the different K_m values of the two hexokinase forms in calf and bovine lenses.

9.2.4 Enzyme Activity (Specific Activity)

The velocity (v) of a reaction catalyzed by a certain enzyme is called enzyme activity and is calculated in micromoles of metabolized substrate/time.

For many years a standard unit for all enzymes has been used. It is called the international unit (I.U.) and is defined as follows: one unit (*U*) corresponds to an enzyme activity catalyzing the metabolization of 1 μmol substrate per minute, under clearly defined assay conditions. Formerly, the enzyme activity was arbitrarily referred to various criteria, for example, to the dry weight of the tissue investigated (e.g., the whole lens), a procedure which made comparison difficult or even impossible. For several years now, the activities have been related to the water-soluble protein, of which the enzyme protein forms a part. This unit is the "specific activity" of an enzyme, its dimension being micromoles of substrate metabolized per minute per milligram protein (U = μmol/min/mg).

Activity measurements must be performed under the strictest assay conditions as to temperature, pH, buffer, substrate concentration, ions, and coenzymes. Enzymes are mostly measured *in vitro* under optimum conditions (e.g., substrate saturation, reaction zero order). These conditions are practically never found *in vivo*.

9.2.4.1 Histochemical Determination of Enzymes

With histochemical methods it is possible to demonstrate metabolic processes in the tissue structure. Enzyme activities may be coupled with the corresponding staining reaction, for example, formazan staining for hydrogenases, thus enabling a better insight into the topography of biochemical processes. A certain disadvantage lies in the rather difficult quantitative evaluation of the stain intensities.

Although the lens, owing to its heterogeneous consistency, is not a suitable object for the preparation of frozen sections, some interesting results on enzyme localization could be obtained. In these studies the lens capsule appeared to have no enzyme activity for all species investigated. The most intense staining was always found in the lens epithelium, and decreased toward the nucleus where, in some cases, no enzyme activity could be determined. Histochemical studies on enzymes in lens have been reported by various authors (19–35). In some papers the histochemical assays revealed the occurrence of species difference more clearly than other enzyme determinations (33–35).

9.2.5 Heat Lability

Heat lability means the loss of enzyme functions with increased temperature, whereby denaturation of the enzyme proteins is induced. The proteins reach maximum stability at a certain temperature, which for most of them is in the cold; only a few are more stable at room temperature. At temperatures above 40 to 50°C an increasing instability of proteins is observed, and the higher the temperature, the more randomized the configuration of the disordered chains. This implies denaturation with loss of functional biological properties, such as enzymatic or hormonal effects. The phenomenon of denaturation of an enzyme by heat, the so-called heat lability, and the loss of enzymatic activity can be determined for each enzyme and is an indicator for the stability of its conformation. There is an exponential relationship between the kinetics of the heat inactivation of a certain enzyme and the time of heat influence, provided the molecules are uniform.

9.3 ENZYMES AND NORMAL LENS METABOLISM

As already mentioned, the main function of lens metabolism is to ensure the growth of the lens over the whole life-span and the maintenance of transparency (36, 37). How far a "metabolism of functions" is also active in the lens has not yet been ascertained. Investigations of Kleifeld et al. (38, 39), however, indicate the existence of such metabolism of functions, since the lens is actively involved in the process of accommodation.

Besides the biosynthetic processes (protein synthesis) most of the energy is used for two transport processes, namely, the uptake of amino acids and the Na^+, K^+ pump.

The necessary metabolic energy in the form of energy-rich phosphate is supplied by the carbohydrate metabolism, particularly anaerobic glycolysis.

9.3.1 Carbohydrate Metabolism

A schematic representation is given in Figure 1.

9.3.1.1 General Remarks

The anaerobic glycolysis (2 mol ATP) is not very efficient compared with the aerobic glucose breakdown (38 mol ATP); nevertheless it plays an important role in energy metabolism. Although 3 to 5% of glucose (40) is further catabolized within the citric acid cycle, this is of importance only in the outermost periphery, particularly in the lens epithelium (41, 42). It is only in this region that the respiratory chain phosphorylation is possible, since in lens fibers mitochondria are no longer present.

About 5% of the glucose in the lens is metabolized through the pentose shunt (41). Via this pathway the pentoses necessary for DNA and RNA synthesis are supplied, and further, NADP is reduced. Last not least, glucose can be stored as glycogen, particularly in the lens nucleus (43). The enzymes of glycogen synthesis and glycogen metabolism were determined biochemically by Ohrloff et al. (44, 45) and histochemically by Kaskel et al. (33, 35). Comparison of the capacities of the three initial enzymes of these

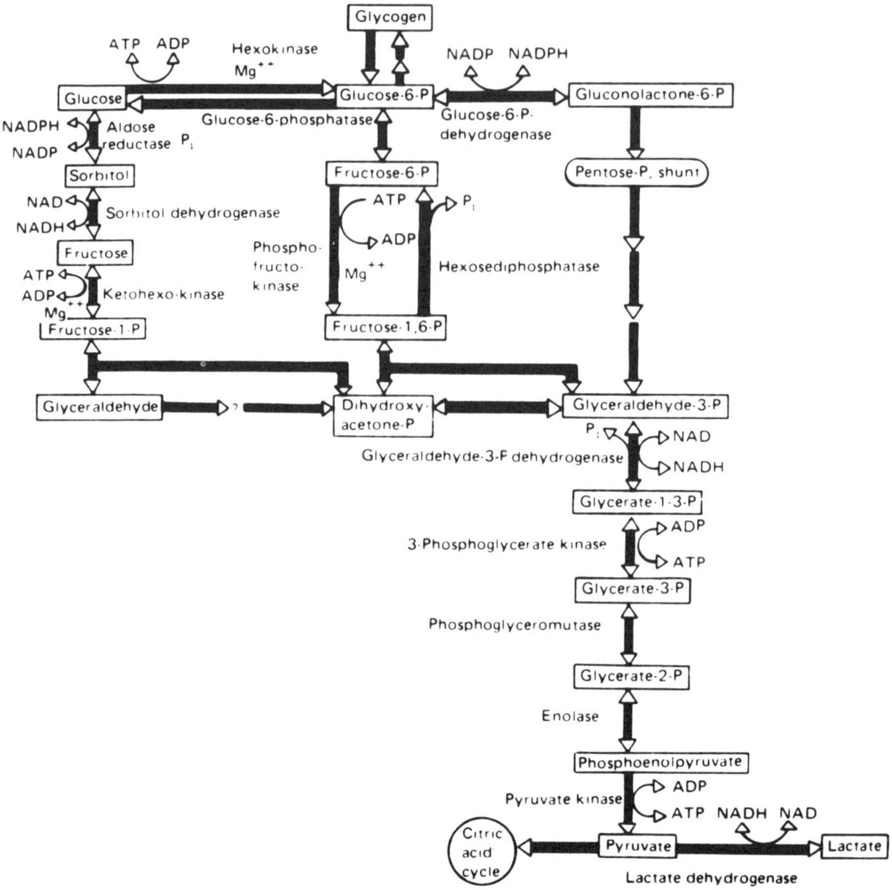

Figure 1 Diagram representing the carbohydrate metabolism.

metabolic pathways competing for glucose-6-phosphate as substrate gave a ratio of 100 (glycolysis) to 4 (glycogen metabolism) to 3 (pentose-phosphate shunt) (46).

Of special importance is the sorbitol pathway in the lens, which was first determined by Hers (47) in the seminal vesicle and is also present in placenta and neural tissue. It enables the enzymatic transformation of glucose to fructose via the polyol sorbitol. The sorbitol pathway was found on investigation of the pathogenesis of lens opacities in cases of diabetes and galactosemia or excessive supply of galactose (48, 49). At present there are three physiological interpretations. (*1*) The sorbitol pathway serves as regenerating system for NADP. This coenzyme, being formed during the first step of the pentose shunt, is oxidized with sorbitol formation and is then again available as coenzyme of G-6-PDH. An interesting phenomenon is the observation that NADPH-consuming reactions other than the reduction of GSSG (glutathione reductase system) obviously do not exist (50). (*2*) The sorbitol pathway serves as pyridine-nucleotide transhydrogenase system. NADPH is oxidized during the first step, and the second step reduces NAD to NADH. This is then available for the oxidative phosphorylation (cf. ref. 51). (*3*) The sorbitol pathway is a bypath of glycolysis. The fructose formed in the sorbitol path is partly transferred into the aqueous (52). It may, however, also be phosphorylated to fructose-1-p by ketohexokinase (53, 54), and it may also be split by aldolase into glyceraldehyde and dihydroxyacetone phosphate (55). For ATP formation, dihydroxyacetone phosphate is directly infiltrated into glycolysis. The phosphorylation of glyceraldehyde may, in some tissues, take place via different metabolic pathways (Figure 1). In the lens, phosphorylation to glyceraldehyde-3-phosphate through the enzyme glyceraldehyde kinase seems to be the most likely pathway (56).

The reversible transformation of glycerol-3-p and dihydroxyacetone-p ("α-glycerophosphate shuttle") allows a hydrogen transfer from pyridine coenzymes of the cytoplasm to the corresponding coenzymes of the mitochondria. This system is also active in the lens (57, 58). Compared with other tissues, the lens consumes little oxygen (59, 60). It seems, however, that particularly the growth of the epithelial cells (61) as well as the maintenance of the Na^+ and K^+ gradients between lens and aqueous humor depend on oxygen supply (62).

The dependence of ATP formation on oxygen consumption *in vitro* has been described by Hockwin et al. (42). With anoxia or KCN inhibition the ATP content decreased by 20% and a simultaneous increase in lactic acid took place. Neumann (63) found an interaction between oxygen and ascorbic acid in the aqueous of rabbit and human. Pirie (64) was able to further elucidate this mechanism where molecular oxygen spontaneously oxidizes ascorbic acid to dehydroascorbic acid with formation of H_2O_2. The latter permeates into the lens, and oxidizes reduced glutathione with the aid of the enzyme glutathione peroxidase. By glutathione reductase and NADPH, the GSSG is reduced, forming NADP, which is then reduced by glucose-6-p dehydrogenase.

Since the lens contains virtually no catalase and little superoxide dismutase as a barrier against free oxygen radicals, this redox system evidently exerts a protective effect for the SH groups of the proteins. The same mechanism is also active in erythrocytes as protection against oxidizing substances. If this mechanism is eliminated, as in G-6-PDH deficiency, for instance, hemolytic anemia may occur (primaquine-sensitive erythrocytes). Kinoshita (65) pointed out that certain similarities, including the above-mentioned mechanism, exist between the lens metabolism and erythrocytes.

The main substrate for lens glycolysis is glucose. However, in case of substrate deficiency, the lens is able to metabolize other substances to supply energy (66–68).

9.3.2 Enzymes of the Carbohydrate Metabolism

Reviews on earlier reports have been conducted by Nordmann (69, 70), van Heyningen (71, 72), Kuck (73), and Hockwin (36). A rather complete review on all enzyme investigations hitherto existing, based on the "Enzyme Commission List of Enzymes," was published in 1975 by Kuck (74), a work of reference that should be continually supplemented.

Comparison of today's enzyme data with those of earlier years is hardly possible since former studies did not consider the age of the lenses under investigation. Today it is an accepted fact that enzyme activities in particular show a marked age dependence. This makes the quantitative evaluation of former studies, where enzyme activities are given without mentioning the age or the lens fresh weight, almost impossible.

Generally all enzymes so far detected may also be found in the lens if the appropriate method is applied. Urease (75) and pyridine-nucleotide transhydrogenase (51), however, seem to be exceptions to this rule.

9.3.3 Protein Metabolism

Reviews on this subject have been published by Waley (76) and Harding and Dilley (77) (compare also Chapters 1 and 7).

9.3.3.1 General Remarks

The lens proteins (crystallins) have for a rather long time been considered to be organ-specific. However, proteins of identical properties have also been found in other intraocular tissues than the lens (78). Immunological investigations of mammalian lens crystallins showed a certain species specificity (79, 80). If, however, the lens crystallins are considered with a view to their physicochemical properties (for instance, isoelectric focusing), the results indicate considerable species specificity (81, 82).

9.3.3.2 Enzymes Involved in Protein Biosynthesis

Investigations on all the enzymes involved in protein biosynthesis have not been performed so far. Virmaux and Mandel (83) were able to demonstrate the presence of aminoacyl synthetases in the cortex of calf lenses, whereas in the lens nucleus no indication of this enzyme was found.

Determinations of the parameters involved allow conclusions to be drawn about the rate of synthesis. In this context, the investigations on the incorporation of labeled amino acids into the crystallins should be mentioned, which were conducted with total lenses *in vitro* by Merriam and

Kinsey (84), Mandel et al. (85), Spector and Kinoshita (86), Fukui et al. (87), Klethi and Nordmann (88), and Strous et al. (88a); and *in vivo* by Weber (89) and Young and Fulhorst (90).

9.3.3.3 Proteolytic Enzymes

Enzymes of class 3 of the Enzyme Commission List (hydrolases, esterases) have been investigated often. Here the valuable contributions from the Halle group should be mentioned (91). They have also been included in a review by Kuck (74). Native lens proteins as substrate have also been investigated in a very few cases (92–97). Hahn (98) preceded his report on age changes of proteolytic enzyme activities in bovine lenses (99) with an introductory chapter on the characteristics of all known enzymes, listing the substrate affinity, optimum temperature, pH maximum, inhibitors, and effect of metal ions.

9.3.4 Other Enzyme Systems

9.3.4.1 Biosynthesis and Catabolism of Lens Glutathione

The functional importance of glutathione with respect to the lens has not yet been fully understood, although many correlations and interrelationships have been put forward (76, 100–103). It has been accepted that this tripeptide, which is present in a remarkably high concentration, is synthesized in the lens. The *in vitro* rate of synthesis in rabbit lenses is about 1.8%/hour (104). According to Rathbun and Wicker (105), about 11% of the ATP formed in glycolysis is used. The enzymes of the γ-glutamyl cycle for the biosynthesis and the enzymes involved in catabolism could be determined and have mostly been characterized. A detailed review has been published by Rathbun (102), who also reported findings on differences in the activities of GSH peroxidase and reductase between normal and triparanol-induced cataract lenses (106).

9.3.4.2 ATPase, EC 3.6.1.3

Detailed studies on the Na-K-activated ATPase have been presented by Bonting and his group (107–112). Referring their results to the parameters of cation transport in whole lenses, they assumed that the enzyme shows a close relationship to the active transport system (72). The Mg-activated ATPase is also present in the lens, although less active than Na-K-activated ATPase. It is evident from various investigations that ATPase activity is higher in the region of lens epithelium and cortex than in the nucleus. With increasing age the activity declines in the whole lens (113). In the nucleus of old bovine lenses, Korte and Hockwin (114) found a higher activity of the enzyme than in the nucleus of the young calf lens. The Na-K-activated ATPase in the albuminoid fraction of human and bovine lenses (i.e., in plasma membranes) decreases with age (88, 115). Changes in enzyme activity are found in human senile cataracts as well as in experimental cataracts of laboratory animals (113, 116). This change is always a decrease in activity. In the case of the Nakano mouse cataract, the decrease is due to an enzyme inhibitor (117–119). Russel et al. (120) showed that the Na-K -ATPase inhibitor responsible for this form of cataract could also be determined in tissue culture of the Nakano mouse lens epithelium. In assays with normal mice the inhibitor in question was not found. The ability to synthesize the inhibitor decreased with the passage number of the cell culture.

Nordmann (121) published a detailed discussion of the dependence of Na-K-ATPase on localization, age, and cataract formation. A recent review by Bonting and de Pont (122) deals with ATPase and its importance in the cation transport.

9.4 ENZYMES IN THE AGING LENS

9.4.1 General Remarks

Gershon and Gershon (123) qualified the age-related accumulation of changed protein molecules as a common phenomenon of the aging organism, which means, with respect to the enzymes, that a great number of molecules of similar structure is found, the catalytic function of which is either completely intact, only partially intact, or completely lost. The cause of such changed and occasionally faulty proteins may be an error in protein synthesis, the presumed mechanism of which is best described by the "error theory."

Various further changes may occur once the proteins (structural proteins and enzymes) have been synthesized. They may be considered as expressions of post-translational molecular aging processes.

The molecular aging was first investigated in a model assay on collagen (124), the main protein of the connective tissue, which does not regenerate in the course of life.

With aging, insertion of strong covalent links occurs, which leads, in the case of chemical denaturation, for instance, to changes in physical behavior. Owing to this age-dependent behavior of collagen, a decrease in permeability of blood vessels and disturbances in the blood supply may occur.

The DNA of the cell nucleus also shows age-related changes with respect to the relationship between DNA and its associated proteins, such as the formation of tight bonds between DNA and histones. These bonds irreversibly block the respective genes so that certain genetic factors are lost (125). Particularly the postmitotic cells (e.g., nervous tissue), the DNA of which does not regenerate, might be affected by such processes. Further, enzyme proteins may be subjected to age-related changes of various degrees (123, 126–128). Among these changes are differences in the behavior with respect to an external electric field (electrophoresis, isoelectric focusing), which are due to a modified primary structure and consequent changes in net charge. Further, an increase in heat lability as expression of instable conformation may occur, as well as the formation of inactive enzyme molecules, which, however, still show the immunological properties of the active molecule. These inactive molecules are called "cross-reacting material" (CRM).

The above-mentioned age-related modifications particularly concern cells without autonomous protein biosynthesis (e.g., erythrocytes), and cells that seldom (liver cells) or never (nerve or muscle cells) divide. To these nondividing tissues belongs also the nematode *(Turbatrix aceti)* investigated by Gershon and Gershon (129), who found an age-related increase of CRM and heat lability of the isocitratelyase. These results, obtained with the crude extract, were verified by Reiss and Rothstein (130) with the purified enzyme. They also found an age-related decrease of the specific activity per milligram and concluded that the old enzyme must be particularly inactivated by protein changes. A decline in specific activity per milligram of purified superoxide dismutase was found by Reiss and Gershon (131) in old rat liver.

Age changes of some enzymes in erythrocytes have also been reported (11). Among these are differences in the electrophoretic behavior (anodization), and in thermolability as well as an increased CRM in the enzymes G-6-PDH and aldolase (9, 10, 132) and hexokinase (133). For LDH, microheterogeneity was reported (6), whereas Schapira et al. (134) observed an increase of the "cross-reacting material" (CRM) in the liver tissue of old rats. When, after hepatectomy, increased cell growth takes place, less CRM is found in the newly regenerated liver tissue. This indicates that, owing to the postsynthetic changes, an accumulation of modified enzyme molecules in the aging liver tissue occurs, while transcription and translation are still operative, ensuring the formation of normal enzyme molecules in the newly regenerated tissue (decrease of CRM).

In the older lens tissue, particularly in the lens nucleus, the enzymes G-6-PDH, triose-phosphate isomerase, and nucleotide phosphorylase show an increased anodic mobility. Similar findings have been reported by Banroques et al. (135) for pyruvate kinase and aldolase.

The age-related protein changes cause the formation of modified enzymes and are in part responsible for the heterogeneity of the enzymes. These modified enzymes are called "metazymes" to indicate their secondary origin and to distinguish them from the isoenzymes, which are primarily genetically coded (5). As yet, knowledge of the possible mechanisms involved in protein alterations is scarce. However, it is generally accepted that with aging nonenzymatic deamidation of asparagine and glutamine takes place. The two amino acids thus formed have a more negative charge; consequently proteins containing asparagine and/or glutamine are more negative after deamidation and, with electrophoretic separation, their migration rate towards the anode (anodization) is considerably higher. For a number of enzymes, anodization may thus be considered as an expression of postsynthetic aging. A direct proof for such a deamidation has so far been possible only in the case of aldolase in rabbit muscle (8, 14), rat cytochrome (136), and α-crystallin (137, 138). Van Kleef et al. (138) were able to show that nonenzymatic splitting of amino acids leads to the formation of postsynthetic products.

9.4.2 Biochemical Age Changes in the Lens

Reviews of investigations performed on age-related changes of lens biochemistry have been published by Hockwin (36), Nordmann (139), Barber (140), Kuck (74), Harding and Dilley (77), and Hockwin (141, 142). Studies conducted during the last decade have shown that the pathogenesis of the cataracta senilis is closely related to changes in the lens metabolism, which must be considered as expressions of physiological aging, causing lens opacities sooner or later (37). This means that investigations on the genesis of the cataracta senilis in general imply investigations of the salient problem of aging as such. Lens research is therefore of equal importance for clinical ophthalmology and basic research in gerontology. For these studies, the eye lens is an ideal model since its continuous growth over the whole life-span allows a comparison of tissues of different age which are biologically of the same origin. It is, for instance, possible to separate the lens nucleus (old tissue) from the lens cortex (young tissue) for comparative studies (143). Such investigations are possible with various animals, since for many species the age can be calculated from the lens fresh weight (Figure 2). However, it was realized rather early that comparative studies involving a group of young, transparent lenses and a group of old, opacified lenses are not very useful. One should rather try to gain insight into the metabolic processes in the aging but still transparent lens. Since opacities are to be considered as the final stage of an aging process in the lens, it is obvious that one should elucidate the biochemical changes and underlying mechanisms at the earliest possible stage.

9.4.2.1 Protein Metabolism

With their 35% content in the lens, the proteins have of course attracted the interest of various investigators. The predominance of proteins presumably in an ordered structure which ensures the transparency of the organ,

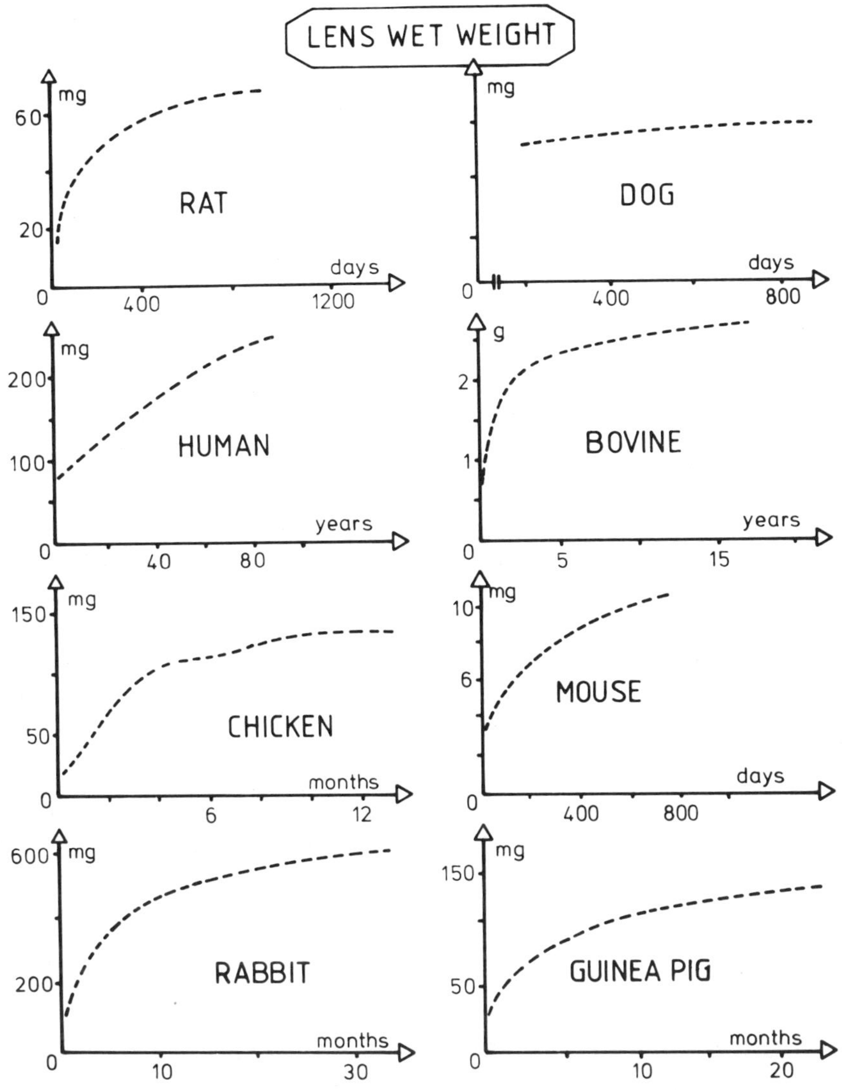

Figure 2 Lens wet weight of various species independence on age.

rendered the problem of the composition and the probable age-related changes of the protein a matter of primary importance.

In general, there are differences in the composition of the water-soluble crystallins and the water-insoluble lens protein (albuminoid), between the different regions of the lens as well as between lenses of different age. There is a remarkable increase in water-insoluble proteins, which in the postnatal stage is only 5% of the total protein (human and rat) but increases with age to 40% (human) and 50% (rat) (82, 144). Rough estimations indicate that about 1 to 2% of the water-soluble lens protein is present as enzyme protein. The problem as to whether enzyme proteins show similar age changes as the structural proteins has been investigated in detail and is discussed later.

9.4.2.2 Energy Metabolism

The transparency of the lens depends on an intact energy metabolism. Inhibition of glycolysis, for instance, by monoiodacetic acid, an inhibitor of glyceraldehyde phosphate dehydrogenase (145, 146), by NaF, an inhibitor of enolase (147), or by 2-deoxyglucose, a competitive inhibitor of hexokinase (148), leads to lens opacification, as could be proved by *in vitro* and *in vivo* experiments.

The elucidation of age-related changes in the energy metabolism of the lens belongs to the early history of lens biochemistry. H. K. Müller in Bonn and J. Nordmann in Strasbourg were among the first to realize its importance, and they ascertained that the energy metabolism is subjected to age-related changes with decreased energy supply. They assumed a direct relationship between loss in energy and cataract formation. Nordmann (121) modified his former statement after considerable ATP concentrations had been found in opacified human lenses.

In accordance with the rapid development of biochemistry and its analytical methods the knowledge on the energy metabolism of the lens increased greatly. In the meantime it could be ascertained that the intensity of the metabolism declines considerably with aging (36, 69, 70, 73, 74, 79, 121, 139, 149). In all species investigated a considerable difference in enzyme activities and energy-rich phosphate (ATP) was discerned between young and old lenses. Further, differences between cortex and nucleus with respect to enzyme activities and ATP were also found (36, 143, 150–154). Measurements mostly involved the following enzymes, the activity of which decreased in the order shown:

$$\text{LDH} > \text{MDH} > \text{GOT} > \text{ALD} > \text{G-6-PDH} > \text{GPT} > \text{HK} > \text{PK}$$

The activity of all enzymes investigated is lower in the nucleus than in the equator. In whole lenses the enzymes G-6-PDH, ALD, LDH, and MDH in particular are less active with aging. The enzymes of the sorbitol pathway are the only exception to this age-related behavior. In whole lenses they lose hardly any activity at all during the aging process, as has been reported by Hayman and Kinoshita (155) for aldose reductase, by Hockwin et al. (150) for sorbitol dehydrogenase, and by Ohrloff et al. (54) for ketohexokinase.

Absolute data on the activity of these enzymes differ widely and their interpretation is often rather difficult, since the authors refer their results to various parameters, such as the whole lens, the lens fresh weight, the lens dry weight, or the water-soluble protein.

Investigations on intermediates of glycolysis after incorporation of ^{32}P-labeled phosphate complemented the rather few results on enzyme investigations and, by evaluation of the radioactivity incorporation rate, led to certain results with respect to the dynamics of glycolysis in the lens (57, 156, 157). The following age-dependent changes have been found: (1) a decreased flow rate of glycolysis, (2) a bottleneck between fructose-6-p and fructose-1.6-p, (3) a decline of the ATP/ADP ratio, and (4) an individual though low metabolism of the lens nucleus and with some metabolic activity contrary to the general assumption.

Early investigations on age-dependent changes were mostly performed with only two or three groups of test animals of different age. One did not take into account that "aging" is not an abrupt event but a rather complex and protracted process. This knowledge spread gradually so that problems of lens gerontology are now tackled in longitudinal studies (158), which imply that analyses are conducted at as many points as possible over the whole life-span. The difficulty with this procedure, as may well be imagined, lies in procuring the necessary material covering the period of life expectancy (Figure 3). In the meantime results of such investigations on age-dependent changes have been reported in guinea pig (158), in chicken (159), in rat (160), and in bovine lenses (152–154). These longitudinal studies yield a great number of data on the age dependence of lens metabolism which should, however, be considered rather as phenomena accompanying the aging process than as the cause or mechanisms of aging. For a better interpretation, the data on chick and bovine lenses were subjected to a multivariate analysis including factor analysis (153, 154, 161–163). By such a combined assay involving biometry and analytical biochemistry, the data obtained may give some information on the causes of aging, and this might in turn result in practical consequences for lens gerontology research. Without going into detail, it seems worthwhile to pay particular attention to the qualitative changes in enzymes still operative with age, as already postulated by Müller et al. (156).

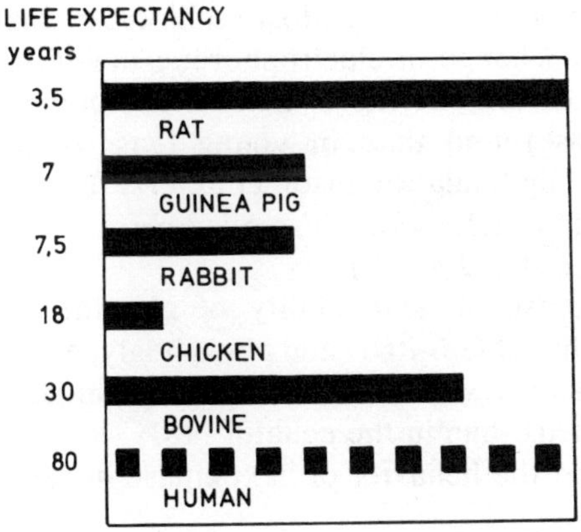

Figure 3 Longitudinal studies on age changes of lens metabolism thus far performed on rat (160), guinea pig (158), rabbit, chicken (159), and bovine (152 & & 6154). Taking the mean life expectancy as 100%, the columns represent the period investigated. The interrupted column for human lenses indicates that normal human lenses are rarely available for investigations.

In the meantime some problems have been elucidated. Kochmann et al. (164) observed in developing rabbit lenses (prenatal and postnatal) a shift from the aldolase A predominance to that of aldolase C. Banroques et al. (135, 165) investigated the isoenzymes of aldolase in rabbit lenses during aging. These authors found in the nucleus of the lenses an increase of enzymatically inactive aldolase which was, however, still immunologically identical with the active enzyme (cross-reacting material, CRM). They observed further that in the lens nucleus two forms of aldolase are present, the electrophoretic behavior of which differed from that of the young tissue. Ohrloff et al. (166) determined the percentage of distribution of the aldolase isoenzymes A, B, and C in lenses of humans, rats, and rabbits by quantitative immunological precipitation with sheep antibodies directed against human aldolase. Their antibodies proved to be hardly species-specific. In rat and rabbit lenses aldolase C prevails, whereas in human lenses aldolase A is present in considerable amounts. Small amounts of aldolase B were found in human lenses (ALD B $< 5\%$ of the total aldolase activity), and in older rabbit lenses, but not in rat lenses.

Investigations on lens cortex and lens nucleus showed that the latter contains mostly enzymatically inactive but immunologically active aldolase C. In the nucleus of rabbit lens, about twice the amount of aldolase C antibodies per unit enzyme activity is necessary for the immunoprecipitation than in the lens cortex. In rat lenses, this difference is less striking.

Isoenzymes have further been determined for LDH in chicken (167) and LDH and glucose-6-p dehydrogenase in bovine lenses of different age (168). The heterogeneous behavior of the enzymes in the bovine lens during cell differentiation has been reported in detail by Glässer (151), for LDH, MDH, G-6-PDH, SDH, and leucine aminopeptidase. The presence of isoenzymes could be verified for LDH but not for SDH. Bours et al. (7) determined the isoenzyme forms 3, 4, and 5 for lactate dehydrogenase in rabbit lens. LDH shows microheterogeneity and decreases in the nucleus, its cathodic form having nearly vanished. Isoelectric focusing of malate dehydrogenase from the cortex of rabbit lenses results in three isoenzyme bands whereas the nucleus shows only one band (169). Poenaru et al. (170) found age-related changes of glycosidases in human and bovine lenses. Most remarkable was a difference in substrate specificity, a change in electrophoretic mobility, and a difference in pH dependence. In the nucleus of human and bovine lenses a higher anodic mobility was found than in young lens tissue (periphery), as could be demonstrated by Skala-Rubinson et al. (12) for the enzymes glucose-6-p dehydrogenase (EC 1.1.1.49), triose-p isomerase (EC 5.3.1.1.) and nucleoside phosphorylase (EC 2.4.2.1.).

Harding (171, 172) found an increase in heat lability of glutathione reductase in aging human lenses, but that of G-6-PDH and phosphoglycerate kinase remained unchanged. Glycyl leucinase and leucine aminopeptidase show a higher heat lability in the nucleus than in the equator (151).

Chylack et al. (173–175) investigated the behavior of hexokinase in calf, rabbit, and rat lenses. They found two types of hexokinase, one of them showing higher heat lability, lower substrate affinity, and increased anodic mobility with aging. The LDH shows differences in substrate affinity between the equator of calf and cow lenses, respectively. Further, heterogeneous enzyme forms of LDH and MDH with different K_m values have been found in bovine lens (176). A detailed study by Lou and Kinoshita (177) showed that similarly as in other organs, the glycolysis in the lens is controlled by hexokinase and phosphofructokinase.

Cheng and Chylack (178) determined the rate of glycolysis by determining lactate in homogenates of rat lenses. According to their results the hexokinase is the pacemaker for the glycolysis in young lenses. They also ascertained findings by Müller et al. (156) that phosphofructokinase in the aging lens has a limiting influence on this metabolic pathway, although its specific activity is practically identical in young and old lenses.

9.4.3 Age Changes of Enzymes

9.4.3.1 Findings in Whole Lenses

To give a general idea about the lens enzymes and to enable comparison, the specific activities of the enzymes investigated in bovine lenses of about 5 years are listed in Table 1. The activities differ widely and range between 0.02 mU/mg protein for pyruvate kinase and 680 mU/mg protein for triosephosphate isomerase (TIM). Independent of the initial value, the enzyme activity decreases rather uniformly in young lenses during the first 5 to 6 years, then remains constant until old age, as illustrated in Figure 4 for glyceraldehyde-3-phosphate dehydrogenase, phosphoglyceromutase, 3-

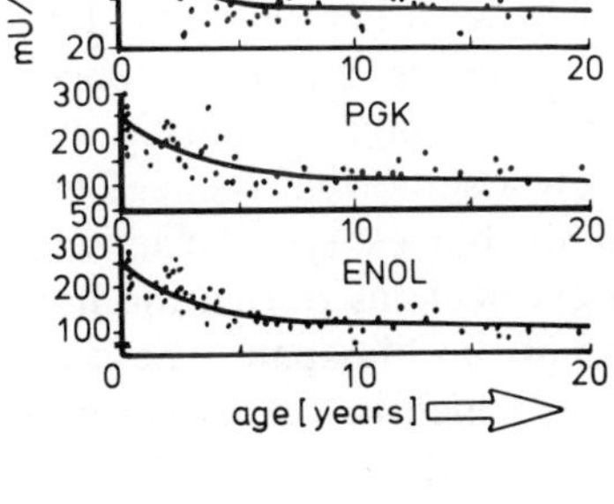

Figure 4 Age relationship of the specific activity of the enzymes glyceraldehyde-3-phosphate dehydrogenase (GADPH), phosphoglyceromutase (PGM), 3-phosphoglycerate kinase (PGK), and enolase (ENOL) in whole bovine lenses.

Table 1 Specific Activity of Some Enzymes of the Carbohydrate Metabolism in Whole Bovine Lenses (Age about 5 Years)

Enzyme (abbreviation)	E.C. designation	Activity (mU/mg protein)
Pyruvate kinase (PK)	2.7.1.40	0.02
Hexokinase (HK)	2.7.1.1	0.2
Ketohexokinase (KHK)	2.7.1.3	0.2
Phosphorylase A	2.4.1.1	0.3
Glucose-6-phosphate dehydrogenase (G-6-PDH)	1.1.1.49	0.4
Fructose-diphosphate aldolase (ALD)	4.1.2.13	0.5
Sorbitol dehydrogenase (SDH)	1.1.1.14	0.6
Glyceraldehyde dehydrogenase (GD)	1.2.1.3	1.4
Aldose reductase (AR)	1.1.1.21	2.0
Glycerol-3-phosphate dehydrogenase (GDH)	1.1.1.8	2.5
Phosphoglucomutase (PGluMu)	2.7.5.1	2.5
Phosphofructokinase (PFK)	2.7.1.11	3.0
UDPG pyrophosphorylase (UDPG-P)	2.7.7.9	4.0
Malate dehydrogenase (MDH)	1.1.1.37	7.0
Lactate dehydrogenase (LDH)	1.1.1.27	7.5
Glyceraldehyde-3-phosphate dehydrogenase (GAPDH)	1.2.1.9	25
Phosphoglyceromutase (PGM)	2.7.5.3	40
Enolase (ENOL)	4.2.1.11	140
Phosphoglycerate kinase (PGK)	2.7.2.3	180
Glucose-phosphate isomerase (PGI)	5.3.1.9	200
Triose-phosphate isomerase (TIM)	5.3.1.1	680

phosphoglycerate kinase, and enolase, as well as for the phosphofructokinase shown in Figure 5.

9.4.3.2 Findings in Single Lens Parts

On comparison of the specific activity of the single lens parts, it is evident that it is lowest in the nucleus and highest in the equator. Table 2 shows that the ratio of enzyme activity nucleus/equator differs to some extent, and ranges between 1 : 2 for PGK and 1 : 16 for PFK. It is remarkable that G-6-PDH is present only in the nucleus of young lenses, and is no longer found in the lenses older than 4 years (179). Phosphoglucomutase and glucosephosphate isomerase as well as PFK show a somewhat reduced but still measurable activity during aging. Here, too, the highest specific activity is present in the equator and, as in whole lenses, declines during the first 5 years, remaining rather constant for the rest of the life-span. The only exception is the activity of phosphoglucomutase, which increases particu-

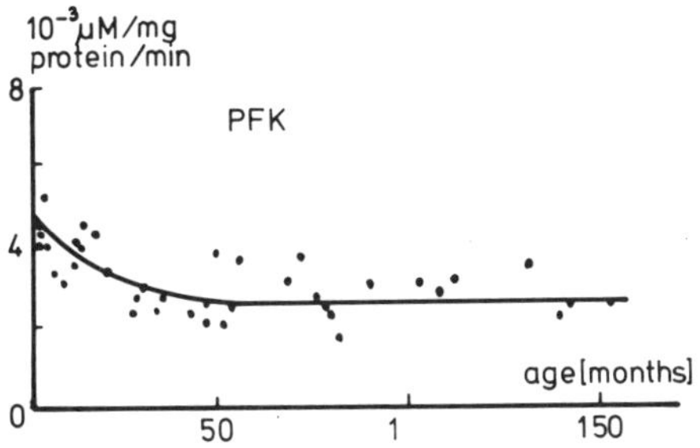

Figure 5 Specific activity of phosphofructokinase in whole bovine lenses in dependence on age.

Table 2 Enzyme Activities of Bovine Lenses after Separation into Lens Nucleus (Old Tissue) and Lens Equator (Young Tissue)

	Specific activity (mU/mg protein)			
Enzyme[a]	Equator	Nucleus	Age of Animals	Ratio eq/nucl
GK	0.005	0	4	—
HK	0.3	0.02	5	15
KHK	0.3	0.06	2	5
ALD	0.6	0.03	5	20
G-6-PDH	0.7	0	4	—
SDH	1	0.2	4	5
Phosphorylase	3	0.1	5	30
AR	3	0.9	4	3.3
GD	3	0.5	6	6
PGluMu	4	2	4	2
PFK	8	0.5	4	16
MDH	8.3	2.8	5	3
GDH	8.4	0.15	6	56
LDH	8.5	4	5	2.1
UDPG-P	11	2	5	5.5
PGM	90	20	1	4.5
GAPDH	116	14	1	8.2
ENOL	204	47	11	4.3
PGI	300	120	4	2.5
PGK	314	150	1	2.1
TIM	360	170	6	2.1
Glutathione-				
reductase	1.5	0.7	5	2
peroxidase	7.5	6.0	5	1.2

[a]GK = glyceraldehyde kinase = triose kinase (E.C. 2.7.1.28). For other enzyme abbreviations, see Table 1.

Table 3 Enzyme Activities in Equator and Nucleus of Bovine Lenses (Age 5 Years)

Sample	Extract volume (ml)	Enzyme activity (ΔE_{366}/time)				
		LDH	MDH	ALD	PFK	ENOL
Equator (1)	0.1	0.064	0.067	0.065	0.117	0.012
Nucleus (2)	0.1	0.031	0.023	0.005	0.008	0.009
Mixture of (1) + (2)[a]	0.2	0.094	0.092	0.076	0.132	0.023

[a]Equal portions of the extracts from equator and nucleus were mixed. Note that the sums of activities of equator and nucleus approximately equal the sums of the mixtures.

larly in the equator during the first 5 years of life, and after reaching a maximum, decreases slowly (179). To elucidate whether possible inhibitors may influence the age-related low activities in the nucleus, a mixed assay was conducted. The activities in nucleus and cortex of 5-year-old bovine lenses (Table 3) were measured in a mixture of nucleus and cortex extracts. The results of these measurements correspond to the sum of the single measurements of nucleus and equator extracts. This means that a possible influence of inhibitors of the enzyme activity may be ruled out.

9.4.3.3 Substrate Affinity

Bous et al. (180) and Ohrloff (179) showed with PFK that by measuring the substrate affinity the age-dependent qualitative changes of a certain enzyme may be determined. The dependence of the activity of PFK on substrate concentration was determined in lenses of 3.8 to 75 months. Figure 6 shows the results obtained for whole calf and cow lenses as well as for the nucleus of both age groups. The values from the whole calf lenses showed a typical Michaelis–Menten curve, the Lineweaver–Burk graph presenting a straight line with a K_m of $7.0 \times 10^{-5}\ M$ (a rather high substrate affinity).

In the whole bovine lens the course of the Michaelis–Menten curve is no longer hyperbolic and does not reach a maximum velocity, a phenomenon that is even more pronounced in the lens nucleus of calf and cow. The Lineweaver–Burk graph takes a convex course, thereby indicating a certain enzyme heterogeneity. Neither for the whole cow lens nor for calf and cow lens nucleus could a single K_m value be determined (Figure 6). In such a case, two K_m values can be obtained in the nucleus of the cow lens (16); that is, $K_{m1} = 5 \times 10^{-5}\ M$, $K_{m2} = 5 \times 10^{-4}\ M$ (Table 4). If these values are compared with that found in calf lenses ($7.0 \times 10^{-5}\ M$), it is evident that in the old tissue two enzyme forms must be present, namely, the already existent enzyme form E_1 with high substrate affinity and a second form, E_2, with low substrate affinity (Figure 7).

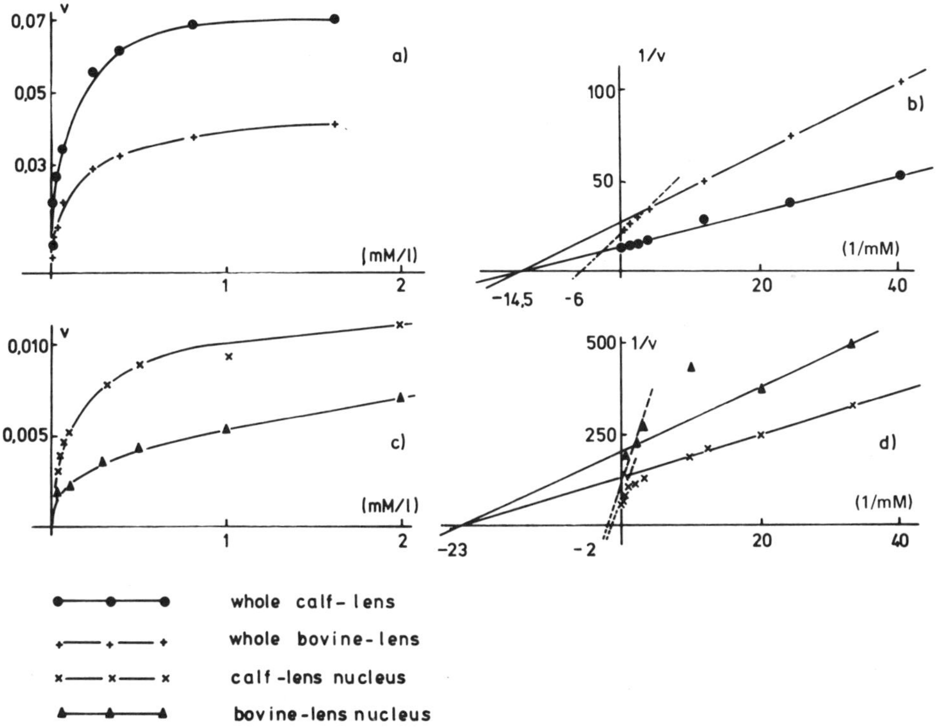

Figure 6 Kinetic investigations of phosphofructokinase in dependence on substrate concentration of fructose-6-p in whole calf lenses of 3.8 months (● 8 8●), in whole bovine lenses of 75.8 months (+ 8 8+) (*a, b*), in the nucleus of calf lenses of 3.2 months (× 8 8×), and in the nucleus of bovine lenses of 22 months (△ 8 8△). Results are plotted after Michaelis-Menten (*a, c*) and Lineweaver-Burk (*b, d*).

A similar heterogeneous behavior, which is particularly evident in the lens nucleus, could be demonstrated for hexokinase in 5-year-old bovine lenses (18, 163), where two enzyme forms with different substrate affinity also were found. A uniform enzyme form (E_2 with a K_m of 6.5×10^{-5} *M*) was present only in the anterior cortex. In all other lens parts a second enzyme form with low substrate affinity ($K_m = 3.6 \times 10^{-4}$) exists, the

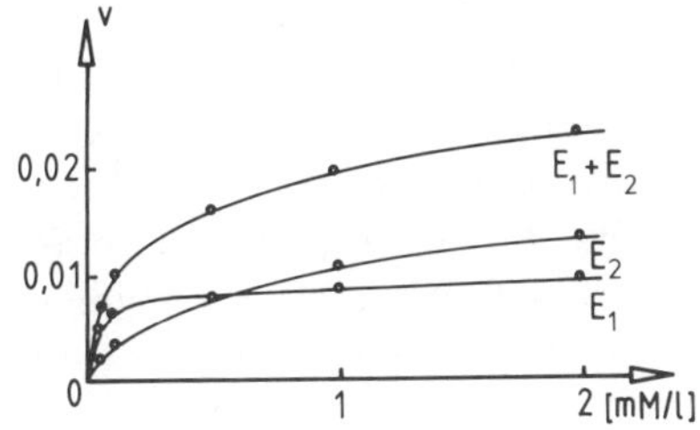

Figure 7 Distribution of the two enzyme forms of PFK in the nucleus of bovine lens (age 18 months). Enzyme E_1 with high substrate affinity is responsible for 40%of the total activity; enzyme E_2 with lower affinity is responsible for 60%.

Table 4 Substrate Affinities (K_m Values) of Enzymes in Bovine Lenses

Enzyme	(Substrate)	K_m for young lenses (M) Equator	Nucleus	K_m for old lenses (M) Equator	Nucleus
GAPDH	(Glyc-3-P)				
K_{m1}		1.6×10^{-3}	1.6×10^{-3}	$2.0 \cdot 10^{-3}$	$1.43 \cdot 10^{-3}$
K_{m2}		2.66×10^{-5}	2.7×10^{-5}	$3.0 \cdot 10^{-5}$	$2.82 \cdot 10^{-5}$
PGK	(Glyc-3-P)	1.18×10^{-4}	1.09×10^{-4}	$1.05 \cdot 10^{-4}$	$1.43 \cdot 10^{-4}$
PGM	(Glyc-3-P)				
K_{m1}		5.0×10^{-4}	5.0×10^{-4}	$6.6 \quad 10^{-4}$	$5.0 \cdot 10^{-4}$
K_{m2}		0	0	0	$3.0 \cdot 10^{-5}$
ENOL	(Glyc-2-P)				
K_{m1}		3.70×10^{-5}	3.70×10^{-5}	$3.77 \cdot 10^{-5}$	$4.26 \cdot 10^{-5}$
K_{m2}		0	0	0	$1.60 \cdot 10^{-4}$
HK	(Glucose)				
K_{m1}		6.4×10^{-5} Anterior cortex			6.5×10^{-5}
K_{m2}		0			6.3×10^{-4}
PFK	(F-6-P)				
K_{m1}		7.0×10^{-5} Whole lens			5.0×10^{-5}
K_{m2}		0			5.0×10^{-4}
PGI	(F-6-P)	1.2×10^{-4} Whole lens		1.2×10^{-4} Whole lens	
PGluMu	(G-1-P)	3.5×10^{-5} Whole lens		3.5×10^{-5} Whole lens	
G-6-PDH	(G-6-P)	1.5×10^{-5} Whole lens		1.5×10^{-5} Whole lens	

portion of which increases in the older lens tissue, and, in the nucleus, finally predominates, with 75% of the total activity (179).

Enolase in the old lens is also present in two enzyme forms, one of which with a lower substrate affinity (179, 181). In young lenses (1.5 years) a uniform K_m value of 3.7×10^{-5} M (high substrate affinity) is found. This value could still be determined in the equator of an old bovine lens of 25.4 years. In the nucleus of the old lens, however, a second K_m value with low affinity ($K_{m2} = 1.6 \times 10^{-4}$ M) is found besides the K_m value of 4.2×10^{-5} M (Table 4). The phosphoglyceromutase in the aging lens tissue also shows a heterogeneous behavior. In contrast to the above-mentioned three enzymes, however, the substrate affinity does not decline but rather increases (181).

Nucleus and equator of young lenses (1 year) have a uniform K_m value of 5×10^{-4} M (Figure 8). The value for the equator of the old lens (11.3 years) is rather similar to that of the young lens, whereas from the Lineweaver–Burk graph of the nucleus a second K_m value of 3×10^{-5} M can be calculated (Figure 8 and Table 4). With regard to the other enzymes investigated, there was no difference between old and young tissue with

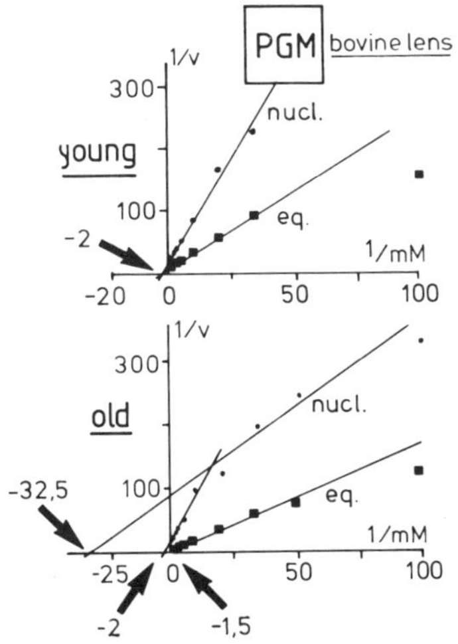

Figure 8 Kinetic investigation on phosphoglyceromutase in dependence on substrate concentration (glycerate-3-p) plotted after Lineweaver-Burk, in the equator and nucleus of bovine lenses of 1 year (top) and of 13.3 years (bottom).

respect to substrate affinity (Table 4). It is remarkable that GADPH in young lenses showed a substrate affinity which yielded two K_m values, that is, K_{m1} about $1.6 \times 10^{-3}\ M$ and K_{m2} about $2.6 \times 10^{-5}\ M$ (Figure 9). These values were also determined in equator and nucleus of an old lens (13.3 years). No change in substrate affinity, either in young or old lenses, was shown for the following enzymes: glucose-phosphate isomerase ($K_m = 1.2 \times 10^{-4}\ M$), phosphoglucomutase ($K_m = 1.5 \times 10^{-5}\ M$), and glucose-6-p dehydrogenase ($K_m = 1.5 \times 10^{-5}\ M$) (Figure 10), as well as 3-phosphoglycerate kinase ($K_m = 118 \times 10^{-4}\ M$) (Figure 11). Substrate affinities (K_m values) of the enzymes investigated in bovine lenses are listed in Table 4.

An age-related decrease in enzyme activities occurs in various organs (182); however, disturbances in metabolism do not always result, since most of the enzymes are present in sufficient amounts, so that even a certain loss in activity leaves enough catalytic capacity. However, the metabolic condition gets serious if, in addition to the decrease in activity, qualitative enzyme changes occur, involving a decreased substrate affinity. This is the case with the enzymes PFK, HK, and enolase in the aging lens.

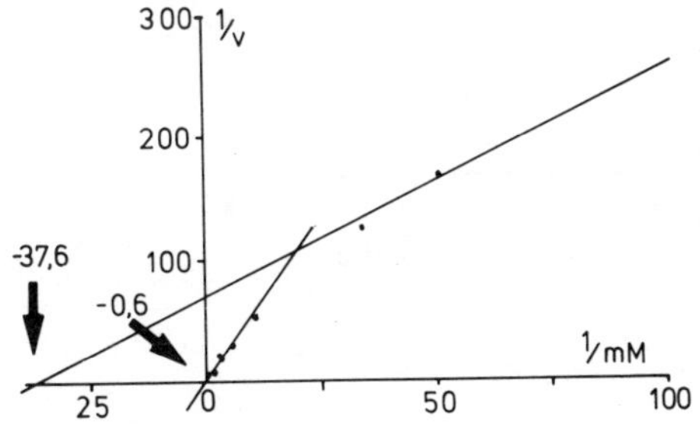

Figure 9 Kinetic investigatin on glyceraldehyde-3-phosphate dehydrogenase in dependence on substrate concentration (glycerate-3-p or glycerate-1,3-diphosphate, respectively), plotted after Lineweaver-Burk, in the equator and nucleus of bovine lenses of 1.1 and 13.3 years.

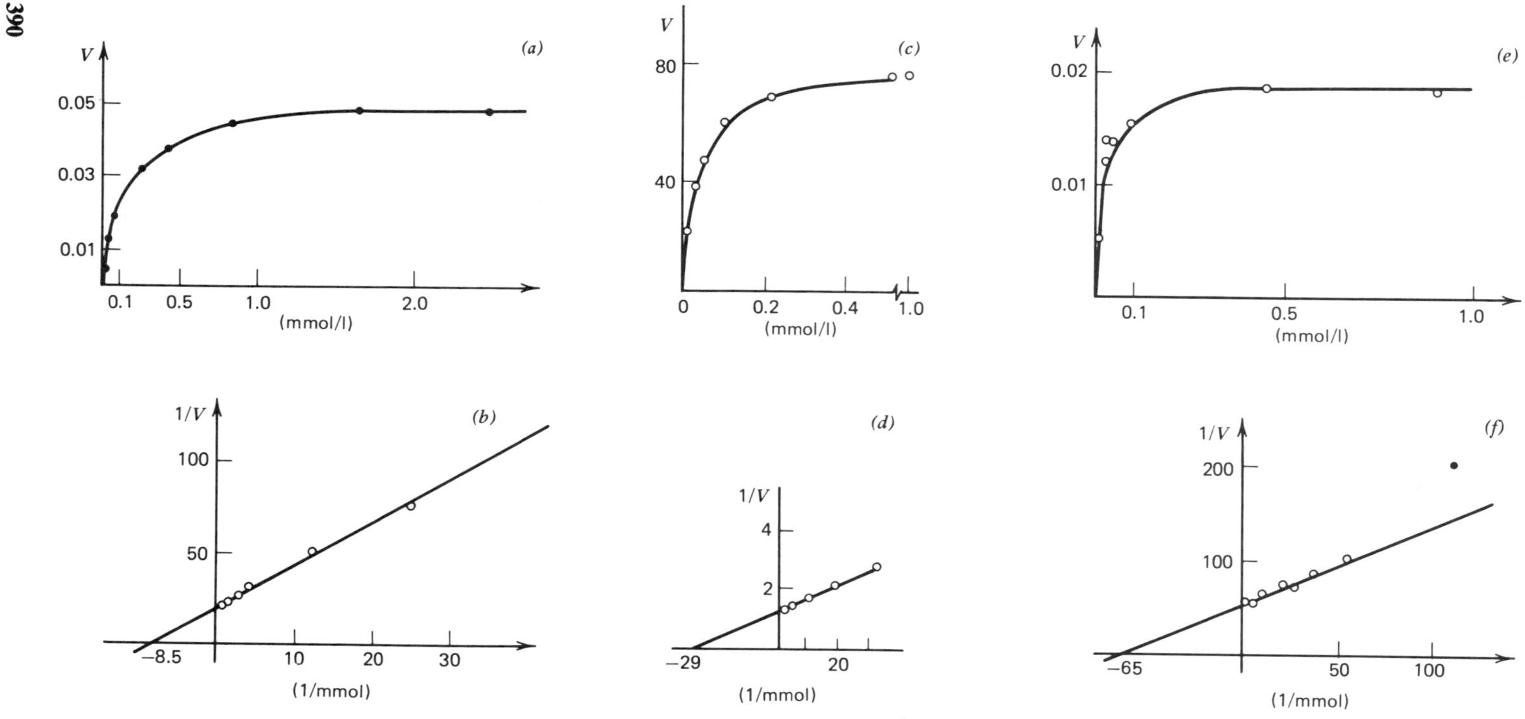

Figure 10 Enzyme activities in dependence on substrate concentration in whole bovine lenses of 3 months and about 3.5 years, plotted after Michaelis-Menten (*a*, *c*, *e*) and Lineweaver-Burk (*b*, *d*, *f*). (*a*, *b*) Glucosephosphate isomerase, $K_m = 1.2 \times 10^{-4}\ M$ (F-6-p); (*c*, *d*) phosphoglucomutase, $K_m = 3.5 \times 10^{-5}\ M$ (G-1-p); (*e*, *f*) glucose-6-phosphate dehydrogenase, $K_m = 1.5 \times 10^{-5}\ M$ (G-6-p).

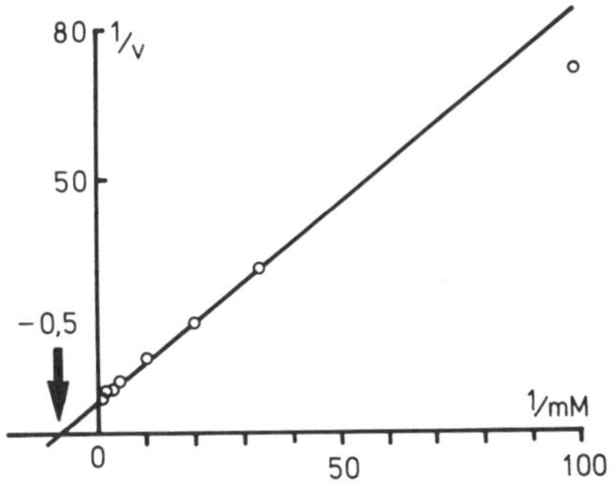

Figure 11 Kinetic investigations on 3-phosphoglycerate kinase in dependence on substrate concentration (glycerate-3-p), plotted after Lineweaver-Burk, in the equator and nucleus of bovine lenses of 1.5 and 13 years.

The PFK in the mammalian organism in particular represents a classical, allosteric enzyme. Under certain experimental conditions, however, it is possible that the sigmoidal curve for substrate saturation graduates into a hyperbolic Michaelis–Menten curve (183). With a pH of 6.8 to 7.2 and in the absence of 5′-AMP a sigmoidal substrate saturation curve may be expected, whereas at pH 8.0, and in the presence of 2 m*M* 5′-AMP, a typical Michaelis–Menten curve is found. Under the latter conditions, the PFK activity is only slightly affected by changes in pH, temperature, or concentration of activators and inhibitors. This is of special advantage when comparing PFK activities in different lens parts.

In young lens tissue the three enzymes PFK, HK, and enolase show a homogeneous behavior of their substrate affinity. With increasing age, enzyme heterogeneity occurs more frequently. The Lineweaver–Burk graph for substrate saturation takes a more and more convex course, resulting in a second K_m value, which indicates a decrease in substrate affinity.

The importance of such a decrease in affinity on the flow equilibrium of glycolysis may be evaluated by comparing particularly those *in situ* substrate concentrations that play a role with respect to the enzymes investigated and with the affinity measured. Fructose-6-p, for instance, the substrate of PFK, is present in a concentration of about 5×10^{-5} *M* (36). With such a concentration, not even in young lenses does the enzyme ($K_m = 7 \times 10^{-5}$ *M*) reach maximum velocity, which is to be expected at 10^{-3} *M*. It is not even above ½ V_{max}.

Owing to the heterogeneity in old lens tissue, part of the enzyme shows a K_m of only 50×10^{-5} *M* ($^1/_7$ of K_{m1}) and is therefore seven times slower. With respect to the activity of the enzyme in the carbohydrate metabolism, this means that with a specific activity of 3 mU/mg protein measured *in vitro* (in whole lenses of 5 years) there is an *in vivo* activity of about 1.4 mU/mg protein. This is reduced by an age-dependent decrease in affinity to about 0.2 mU/mg protein. These considerations suggest that the age-related decrease from 5 to 3 mU/mg protein during the first 5 years of life (Figure 5) is less responsible for the limited metabolic conditions than the loss in substrate affinity occurring at this period.

This also holds true for the enzyme HK (K_m young lenses $= 6.4 \times 10^{-5}$ M; K_m old lenses: $K_{m1} = 6.5 \times 10^{-5}\,M$, $K_{m2} = 6.3 \times 10^{-4}\,M$). The substrate glucose is present in a concentration of about $3 \times 10^{-4}\,M$ (36). Enolase, with its substrate glycerate-2-phosphate in a concentration of $4 \times 10^{-6}\,M$ (184), has a K_m of $1.6 \times 10^{-4}\,M$ and a second K_m of $4.2 \times 10^{-5}\,M$ in the nucleus of an old lens, whereas a uniform K_m of $3.7 \times 10^{-5}\,M$ is present in the young lens. This means that the enzyme in question never reaches maximum velocity.

It seems that age-related enzyme heterogeneity does not always lead to a decrease in substrate affinity, as can be demonstrated with phosphoglycerate mutase, the affinity of which increases in the nucleus of the old lens. It is remarkable that other enzymes such as PGI, PGluMu, G-6-PDH, GADPH, and PGK show a virtually identical affinity in both young and old tissues.

9.4.4 Causes of Age-related Enzyme Changes

Investigations thus far show that the enzymes of the energy metabolism of the lens are subjected to changes in their kinetic properties during aging. The specific activity in general as well as the substrate affinity of certain enzymes decreases, which results in a limited energy metabolism. Most remarkable in this respect is the age-dependent increase in the heterogeneity of some enzymes. This phenomenon gave rise to the question if modifications of the enzyme occur with aging, affecting the catalytic (kinetic) properties, and which of the aging mechanisms discussed might be responsible for such modifications.

9.4.4.1 Changes in the Conformation of Enzyme Molecules

At present little is known about the conformation and protein structure of lens enzymes and their possible age-related changes. Investigations on the heat lability of lens enzymes seemed a suitable method to elucidate changes in conformation. Heat lability was determined for the enzymes lactate dehydrogenase, malate dehydrogenase, fructose-diphosphate aldolase, phosphofructokinase, and enolase (185). Measurements involved the crude extract of nucleus and equator of bovine lenses and were conducted with three age groups (1.3, 4–6, and 15 years). Denaturation by heat (heat lability) with loss of enzymatic activity is a determinable value, expressing the stability of the conformation of a certain enzyme. Since conformation and structure of proteins are closely related, substitution, loss, or changes in single amino acids of an enzyme may affect its heat lability.

If the enzyme under investigation shows a homogeneous behavior, the

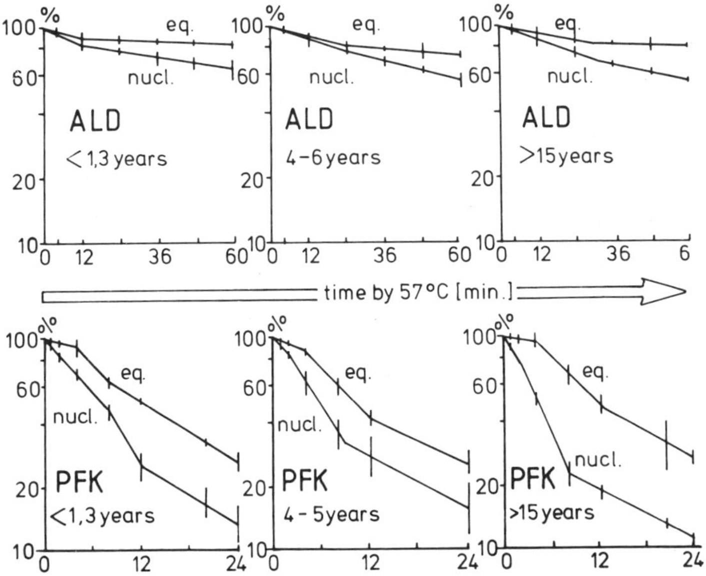

Figure 12 Inactivation at 57°C of aldolase and phosphofructokinase in the nucleus and equator of bovine lenses of different age. Ordinate, percentge of initial activity; abscissa; duration of heat treatment, $n = 5$.

graph for the dependence of its inactivation on time of heating is represented by a straight line. In the case of heterogeneous behavior there are two possibilities to be considered: (*1*) the curve shows a sharp decline in the first section and then flattens (Figure 12: ALD, PFK, nucleus). This happens if two different enzyme forms (isoenzymes or metazymes) with different heat lability are present. In this case the less stable part is inactivated first, its activity decreasing rather suddenly whereas the inactivation of the more stable part occurs much more slowly. Both curve sections allow, by means of extrapolation, the approximation of the half-life of inactivation for the single enzyme forms as well as their respective ratio. (*2*) The curve first takes a flatter course and declines only when representing a longer heat exposure (Figure 12: PFK eq.). This course may be explained by the multi-hit theory known from radiobiology (186): a short-time heat exposure causes only sporadic molecular changes effecting denaturation, whereas under long-term heat incubation the ''hits'' accumulate, leading to increased denaturation and inactivation. A third possibility is a mixed form of both inactivation phenomena mentioned above.

An increase in heat lability has been reported for enzymes in various aging tissues (liver, muscle, fibroblasts, nematodes, erythrocytes). Studies have mostly been performed with the crude extract (123, 187, 188) and incidentally with a purified enzyme (130).

In the aging human lens, an increased heat lability of glutathione reductase was found, but this phenomenon could not be determined for G-6-PDH and phosphoglycerate kinase (171). Glycyl leucinase and leucine aminopeptidase in the nucleus of bovine lenses are less stable against heat influences (151). Chylack et al. (173) reported that hexokinase is present in two forms, one of which shows increased heat lability with aging.

The results with bovine lenses may be summarized as follows:

(1) Heat lability is significantly higher in old than in young tissue.
(2) With aging, the heat lability of the enzymes MDH, PFK, and enolase increases in the nucleus but remains unchanged in the equator. This holds true for all enzymes in the age groups investigated.
(3) The curves plotted for the dependence of heat inactivation on time indicate that there are at least two enzyme forms for ALD, MDH, and PFK, and also that the less stable forms of MDH and PFK increased with age.

To eliminate the possibility that differences in the initial activities in the crude extract of nucleus and equator of the enzymes investigated might affect the results, control tests were performed by measuring activities in the extract of nucleus and equator separately. Dilution with double-distilled water was performed to obtain identical initial activities (μmol/ml/min) for the subsequent heat lability tests. For PFK measurements, dilution with inactivated (enzymatically inactive) lens proteins was preferred, since dilution with water led to a considerable loss in activity even in the nonincubated extract. The subsequent incubation was performed as described above.

No difference in heat lability of the enzymes investigated with this procedure as compared to the aforementioned results was found. This means that age-related modifications of the enzyme proteins, affecting the stability of the conformation, that is, its denaturability, occur in the lens.

The question arises, what (age-related) mechanisms might be responsible for such enzyme modifications?

Are errors in protein synthesis caused as proposed by the error catastrophe theory (189) or are they rather due to postsynthetic changes of the enzyme molecule? Since—independent of age—the heat lability of all enzymes remains equal in the equator as the site of protein synthesis, it may be assumed that protein synthesis in the aging lens does not involve the production of altered enzyme molecules. The heat lability of all enzymes investigated was significantly higher in the nucleus than in the equator and even an age-related increase with respect to the enzymes MDH, PFK, and enolase could be observed (Figure 13). Since no protein synthesis takes place in the lens nucleus, neither errors in synthesis nor synthesis-dependent

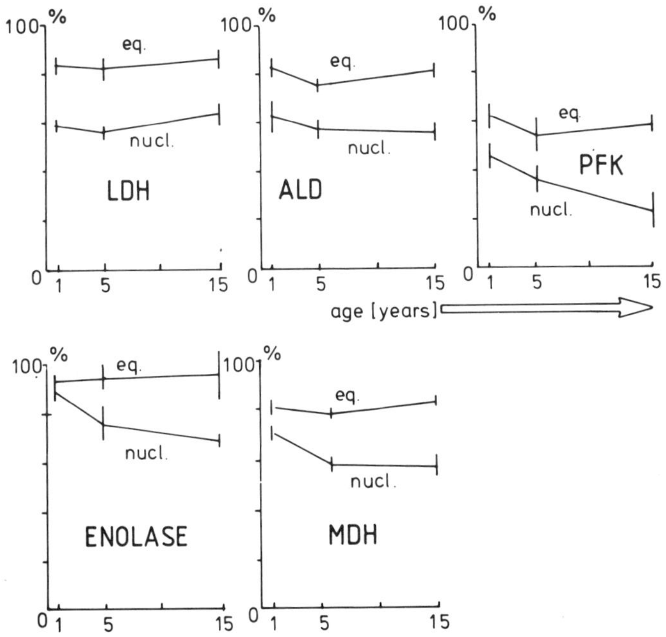

Figure 13 Heat treatment (57°C) of the extract of bovine lenses of <1, 4 & & 66, and > 15 years. The duration of the treatment is given. Values are plotted in percentage of initial activity.

isoenzyme shifts are responsible for the changes in heat lability. The more probable solution is that the enzyme proteins investigated are also subject to postsynthetic changes, for instance, nonenzymatic deamidation and/or amino acid splitting, as already has been found for α-crystallin (137, 138). In this case enzyme variants occur that, as do isoenzymes, increase the heterogeneity of the enzymes. To discriminate between these enzyme variants and the primary genetically coded isoenzymes, they are called metazymes. They catalyze the same reaction as the enzyme from which they have been derived by postsynthetic modification; they differ, however, in their physicochemical properties such as heat lability or electrophoretic mobility, as well as in their kinetic properties, particularly in their substrate affinity.

Obviously the enzyme proteins of the lens are subjected to these postsynthetic mechanisms of aging to a different degree. Slight changes in proteins, for instance, affecting the stability of conformation, may be detected by the increase in heat lability. According to the extent and site of the possible changes in conformation, either changes in substrate affinity or total loss of catalytic ability may occur.

9.4.4.2 Age Changes of Enzymes Involving Size and Molecular Weight

Separation of the water-soluble protein into the characteristic lens crystallins α_{HM}, α_L, β_H, β_L, and γ allows the determination of certain enzymes of the carbohydrate metabolism in one or several crystallin fractions (143, 190–192). Reports on some enzymes in the water-soluble crystallin fractions of young and old lens tissue have already been presented (45, 116). They suggest that enzymes associated with the α-crystallin fraction are subjected to a considerably higher loss of activity during aging than the enzymes associated with β-crystallin or with both fractions.

Since the water-soluble proteins are in general subjected to age changes that may lead to an increase of the high molecular weight fraction and also of the water-insoluble part, for instance, by aggregate formation, it is possible that the enzyme proteins are also affected.

With respect to further investigations the question arose whether physicochemical changes of enzyme molecules occur during the aging process which might affect size and weight of the molecules. By means of gel filtration, separations were achieved to render studies on enzyme distribution patterns in the water-soluble crystallin fractions possible. This enables elucidation of the age-related heterogeneity of single enzymes. Such investigations involved:

1. Aldose reductase, sorbitol dehydrogenase, and ketohexokinase in the water-soluble protein of bovine lenses (0.5–11.5 years).
2. Phosphofructokinase in the equator and nucleus of bovine lenses (3 years).
3. Phosphoglycerate kinase, phosphoglyceromutase, and enolase in the equator and nucleus of bovine lenses (5–6) years.

9.4.4.3 Determination of the Molecular Weight of Protein Fractions

The molecular sieving procedure by means of gel filtration allows determination of the molecular weight of proteins between 1×10^4 and 1×10^6 dalton; there exists a linear correlation between the logarithm of molecular weight and the volume eluted V_e (Lehmann, 193).

For enolase, 3-phosphoglycerate kinase, and phosphoglyceromutase, determination of the molecular weight was performed using Sepharose-6B gel according to Bours and Brahma (194). In this assay 150 to 200 mg protein was applied; elution was performed with 0.1 *M* Tris–HCl buffer, pH 8, at 4°C. Protein determination was conducted according to Kalckar (195). Aldose reductase (AR) and ketohexokinase (KHK) showed activities that indicate the presence of two different enzyme forms. Although the range of activity of AR is less influenced by aging, there was a considerable difference in KHK activities. Evidently phosphofructokinase (PFK) is also affected. Kinetic measurements of substrate affinity resulted in two K_m values, suggesting two enzyme components, one showing lower substrate affinity. This enzyme is present only in the aging tissue. For the enzyme fraction associated with α-crystallin a K_m value representing higher substrate affinity—a characteristic of young tissue—could be determined. Although the activity in the β-crystallin fraction was too low to allow determination of any K_m value, it may be assumed that the appearance of a metazyme with a low substrate affinity indicates certain modifications within β-crystallin. Cheng and Chylack (196) described two forms of PFK in young rat lenses which were interconvertible upon pH changes. The only difference between these forms was supposed to be their conformation. According to the results of molecular separation, however, it seems more probable that age-related postsynthetic changes of the primary synthesized phosphofructokinase occur in the α-crystallin fraction, effecting a second component of activity, which can then be found in β_L-crystallin and which might be responsible for the decrease in substrate affinity.

Phosphoglyceromutase and enolase are also subject to changes of their activity profile in the water-soluble protein of the older tissue (Table 5).

Table 5 Elution Volume of Maxima of Activity of PGK, PGM, and ENOL after Separation of the Water-soluble Protein of Nucleus and Equator (bovine lenses of 5.7 years) on Sepharose 6B

	Elution-volume (ml) with max. activity	Protein content of fraction (mg/ml)	Enzyme activity (mU/mg prot.)	Mol. wt.
Equator				
PGK	370	2.9	174	59,500
PGM	350	2.6	186	87,000
ENOL				
1st peak	340	1.65	364	105,000
2nd peak	360	3.25	162	72,000
Nucleus				
PGK	370	1.76	91	59,500
PGM				
1st peak	340	1.7	7	105,000
2nd peak	380	1.6	3	49,000
ENOL				
1st peak	340	1.7	129	105,000
2nd peak	370	1.76	43	59,500

Phosphoglycerate kinase, on the other hand, showed a uniform behavior in the same range of elution with a molecular weight of 59,000. In rabbit muscle, a comparable molecular weight (47,000 dalton) was determined for this enzyme (197).

The data reported for phosphoglyceromutase (PGM) in mammalian organism ranges from 54,000 to 65,000 (198). The rather high value of 87,000 found in young tissue suggests that the enzyme is attached to structural proteins. In the old tissue, association processes lead to one form of activity in the high molecular protein fraction at 105,000, whereas dissociation processes lead to a second form of activity in the low molecular protein fraction (49,000 dalton), which probably corresponds to the nonattached enzyme.

As is the case with PFK, such processes might be the cause of age-related changes in the kinetic behavior of the enzyme, particularly with respect to substrate affinity.

According to studies in various mammalian tissues, enolase is a dimer and consists of two identical polypeptide chains. Each of the monomers has an intact, active center; however, the dimer seems to be the only natural form of enolase with a molecular weight of 82,000 to 100,000 dalton (199, 200). The molecular weight of 105,000, determined in lens and which is found in both young and old tissue, is in good agreement with these values. In the equator, however, the distribution of the enzyme in the water-soluble protein fractions is no longer uniform and a second maximum of activity with a molecular weight of 72,000 is found. In the lens nucleus there is a further shift of this activity into the low molecular region down to a molecular weight of 59,000.

Although enzymatically active monomers have so far not been found, such a possibility can not be excluded. However, a more probable solution is that little by little parts of the molecule are split off in a gradual process, beginning in young tissue and leading to modified enzyme molecules. This phenomenon would explain why this enzyme also shows changes in kinetic properties, such as decrease of specific activity and substrate affinity, with increasing age.

Results obtained with gel filtration of enzymes show that postsynthetic mechanisms are operative in the lens which—upon aging—may also modify the macromolecular structure of the enzyme molecules.

There may be several reasons for this phenomenon:

1 The presence of subunits of an oligomerous enzyme (201).
2 The loss of larger protein chains (dissociation to smaller enzyme molecules) or association to larger enzyme complexes, respectively.
3 Association of the enzymes with structural proteins and subsequent effects of changes (aggregate formation) in these proteins.

9.4.5 Molecular and Functional Enzyme Properties of the Aging Lens in View of Present Theories on Aging

Aging of each organism depends on the aging process of its individual cells. Since many different cell types exist in an organism, it is rather difficult to postulate a general aging theory. It is necessary for instance, to know whether the cells in question are of the short-lived type and subjected to further cell division (e.g., fibroblasts) or of the long-lived type that does not divide further (e.g., nerve cells). It should also be considered whether nondividing cells are nucleated, such as nerve cells.

Lens fibers belong also to the nondividing, denucleated cells. The life-span of erythrocytes, however, amounts to only about 100 days, whereas the life of the lens fiber—at least in the nucleus—corresponds to the age of the respective individual.

At present, the following main theories on aging, based on the latest genetic and biochemical findings, are under discussion:

1 The genetic program.

2 The error catastrophe theory.

3 The theory of postsynthetic changes.

With respect to the long life expectancy of the lens fiber, however, the toxication theory, which formerly was a focal point of investigations, should also be considered.

The diagram in Figure 14 illustrates the mode of action on which each of the three aging theories is based (50).

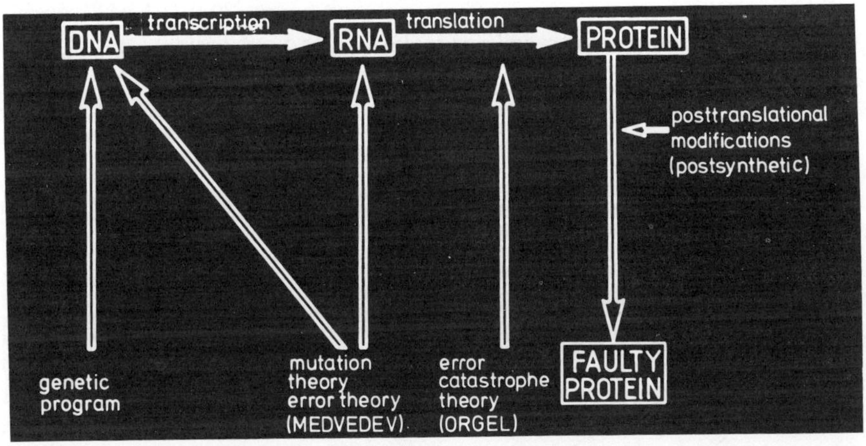

Figure 14 Diagram of the genetic and molecular relationsphis involved in the aging process from the viewpoint of the present theories on aging.

1 According to the genetic theory, aging (*a*) is a programmed process (program theory) wherein all phases of development including senescence are subject to a specific genetic control, or (*b*) is due to mutation (mutation theory) leading to a certain regression phase. It is assumed that particularly errors on the DNA–RNA level (transcription) occur which may result in mutations.

2 The error theory suggests two possibilities. (*a*) The reliability of the system of protein biosynthesis is affected by aging, which leads to the formation of faulty proteins. These errors cumulate and as soon as vital metabolic processes are affected, cell death is inevitable (error catastrophe theory) (189). (*b*) It is supposed that in accordance with the mutation theory errors in DNA and/or RNA occur, so that false information with disastrous consequences for the cells is transmitted (202, 203).

3 After normal, undisturbed protein synthesis there occur various post-translational modifications of the proteins leading to "faulty" proteins. With respect to the enzymes, those processes increase their heterogeneity and favor the formation of metazymes.

The question arises to which degree these theories apply to the age-related functional and structural enzyme modifications such as the following:

- The decrease in specific activity.
- The decrease in substrate affinity.
- The increase in heat lability.
- The changes in size and weight of molecules, which finally lead to the formation of senile cataracts.

Taking into account the toxication theory of aging (204) [which suggests that an accumulation of intracellular metabolic end products or transformed products that cause metabolic disturbances takes place], assays with a mixture of young and old lens tissue were conducted to avoid enzyme inhibition by accumulation of inhibitors in the old tissue that might influence the results (Table 3).

In those tissues, where cells are nucleated and synthesize protein even in old age it is rather difficult to discern whether the age-related accumulation of changed protein molecules is due to a faulty synthesis or to postsynthetic changes. The particularity of the lens metabolism that protein synthesis ceases at a rather early stage, and the lack in proteolysis in the normal lens, which implies the longevity of the proteins, lead to the conclusion that post-translational changes are responsible for the enzyme modifications observed in the aging lens. In contrast to the postsynthetic changes of the structural proteins, the alterations of the enzyme proteins imply an interference in their catalytic functions in a complex, interrelated reaction system. In consequence, disturbances in the equilibrium occur, and the energetic requirements necessary for the maintenance of physiological conditions are no longer guaranteed. This age-dependent process, starting relatively early and probably caused by changes of hexokinase and phosphofructokinase, might be an essential part of the mechanisms leading to lens opacities.

Investigations on identical twins indicated that an hereditary disposition may also be involved in the formation of senile cataracts (205). The phenomenon that the incidence of senile lens opacities in climatically and sociologically similar countries differs considerably may also be due to constitutional factors. In Pakistan, Sri Lanka, and India, for instance, senile cataract occurs frequently, whereas it is less often found in China and Thailand (206). The genetically defined premature aging and occurrence of senile cataract in *Progeria adultorum* support the assumption that the aging of the lens and cataractogenesis are closely related to genetic influences.

Aging and senile cataract, however, are not only genetically predetermined, but also depend on environmental influences and diseases. This is quite evident from the early onset of senile cataract in the presence of certain predisposing factors (207). These factors, including particularly diabetes, sprue, pulseless disease, and favism (inherent G-6-PDH deficiency), may affect the postsynthetic processes. It could be demonstrated, for instance, that in the case of diabetes an increased intralental concentration of glucose accelerates the formation of aggregates to the higher molecular weight protein (208). The rare lack of the enzyme G-6-PDH causes a decrease in NADPH with the relevant increase in oxidized glutathione (GSSG), and supports the formation of the -SS- bonds of protein. Nutritional disorders like sprue or pulseless disease probably affect the energy metabolism, inducing protein modifications.

The pathogenesis of senile cataract is certainly a multifactorial process, and biological age changes are, although of great importance, only one factor among others (207, 209).

9.5 ENZYMES IN CATARACTOUS LENSES

9.5.1 Human Senile Cataract

Investigations on the causes of senile lens opacities in man are rather difficult to conduct. Although some results on biochemical and physical parameters of the human cataractous lens have been reported (69, 70, 121, 139, 140, 210), their interpretation is not very satisfactory since opacified lenses extracted for medical indications were used in these studies, and normal lenses of similar age were not available as controls. Of course, there is also no possibility of comparing normal human lenses of different age. The use of eye bank material is rather doubtful, since postmortal changes commence rather early. This also holds true for eyes from legal sections (211).

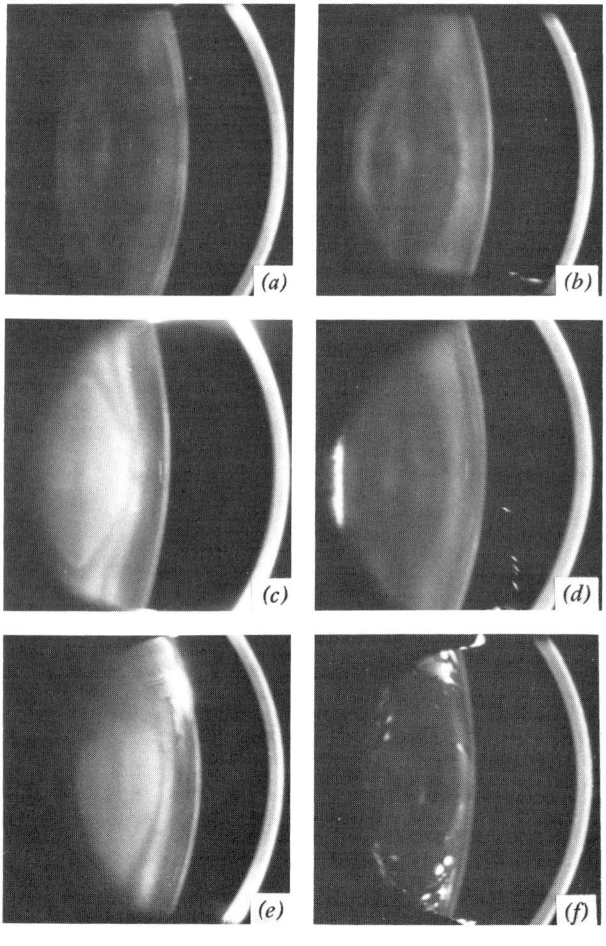

Figure 15 Photographs of the anterior segment of human eyes, obtained with the TOPCON densitograph. (*a*) normal lens, 19 years; (*b*) normal lens, 55 years; (*c*) nuclear opacity; (*d*) posterior subcapsular cortical cataract; (*e*) nuclear opacity, anterior cortical spokes and water clefts; (*f*) cataracta coronaria and circumscribed opacity in the anterior fetal nucleus.

9.5.1.1 Lens Classification

The evaluation of such lenses is made more difficult because in most cases classification by means of the slit lamp has not been performed before extraction, thus neglecting the possibility of correlating biochemical and morphological findings (139, 209, 212). The introduction of a special camera for photographic follow-up documentation of lens opacities based on the

Scheimpflug principle (213–215) allows the taking of sharply focused photographs of the optical section of the whole lens from every desired angle (216–218), a prerequisite for the objectivation of cataract findings.

The clinically irrelevant evaluations based on lens coloration (219, 220) or the rather subjective evaluation of photographs of extracted lenses (221) can now be substituted by a procedure of exact measurements. Photographs of the anterior eye segment (Figure 15) obtained with the Topcon Densitograph (216–218, 222) allow not only a densitometric analysis of lens transparency but also an exact topographic determination of areas which, after isolation with the aid of a microtome, for instance, can be subjected to a pinpoint analysis (e.g., electron microscopy, biochemical microanalysis).

9.5.1.2 Enzyme Activities and Cataract Forms

If the individually different forms of lens opacities, indicating the multifactorial mode of formation, are taken into account, the concept that senile cataracts are to be considered as "normal" age symptoms seems doubtful (37, 212). Hockwin and Berghoff (223) investigated some enzyme activities in normal and senile cataractous lenses extracted for medical reasons, where an exact classification of opacities had been performed previous to extraction. In 15 cases both lenses of a patient were available for investigations. In nine pairs, opacities were of equal degree; the other six pairs showed side differences with respect to form and localization. These studies were evidence that, with the exception of the kinases investigated, enzyme activities depend on the localization of opacities.

Studies conducted by Maraini and Friedburg and co-workers also emphasize the imperative necessity of a slit lamp-microscopical *in situ* examination of the lenses with a view to the biochemical parameters (224–235). Friedburg summarizes his results with the statement: "Enzyme patterns indicate differences between different types of cataract. These data also suggest that 'human senile cataract' is the result of several different processes leading to a loss of transparency of the lens" (232).

9.5.2 Animal Models for Senile Cataract Research

The lack of normal human lenses as well as of lenses that had been subjected to an objective classification necessitated the use of animal lenses for the study on cataractogenesis. The availability of really old laboratory animals, however, is rather problematical. Neither cows, rats, nor rabbits are in general kept over their whole life-span. This is the case only with dogs, where accordingly senile cataracts could be found (236).

For the first time a longitudinal study, supported by the Zentralinstitut für Versuchstierzucht in Hannover and in cooperation with several other institutes was started at the Institut für experimentelle Ophthalmologie der Universität Bonn. For these studies a considerable number of Wistar rats were kept over the whole life-span and made available for extensive studies on senile cataracts (160). Characteristic age changes were found after the second year. Water clefts, spokes, and cuneiform opacities in the anterior cortex as well as total opacities of the posterior cortex were often found (237). Gorthy (238) reported the finding of mostly posterior subcapsular and supranuclear opacities in the lenses of old Wistar rats.

A suitable animal model for the elucidation of presenile cataract formation might be a certain strain of Wistar rats which developed a spontaneous hypertension. They showed opacities of the anterior and posterior cortex as well as an inherent opacity in the anterior embryonic nucleus after about 21 to 50 weeks (239). The unavailability of either normal human lenses or animals of an age where senile cataracts might be expected inevitably led to investigations involving experimental cataract models. The water cleft–spoke opacity in man, for instance, is comparable to the cataract found in rats and rabbits after peroral application of naphthalene (240–241).

The symptoms of a posterior subcapsular cortical opacity are observed in rat lenses after a tryptophan-free diet (243–245) and are also found after X-ray irradiation (246–250). The analogous behavior of senile and X-ray cataract is by no means restricted to the localization of opacities. Strong similarities in the metabolic processes were also observed which led to the expression "artificial aging by X-ray irradiation." This term, however, is still a much disputed matter among radiobiologists (248, 251–253). The formation of the posterior subcapsular cortical opacity due to X-ray irradiation may be accelerated by an additionally performed ligature of the A. carotis communis (254). This is the experimental equivalent to the premature incidence of senile opacities in the case of pulseless disease. In this context, it should be mentioned that the cataracts due to diabetes have been clinically and experimentally examined to some extent. Their pathogenesis has been elucidated (48, 255).

The discovery of the risk factor involved has considerably inspired the search for a suitable drug treatment. First results were reported by Kinoshita (119), Kinoshita et al. (256), Varma et al. (257, 258), Norma et al. (259), and Korte et al. (260), who tried to inhibit the reaction leading to the accumulation of polyol.

With the aforementioned animal models, studies on lens enzymes have been undertaken repeatedly. Within the scope of the longitudinal study on rat lenses (160), for instance, 10 enzymes of carbohydrate metabolism were investigated as to their specific activity. Evaluation by statistical multivariate procedures, taking into account the biochemical data of slit lamp microscopical findings (237), is still under way. The few examples presented from experimental cataract research show that models are already available simulating various mechanisms that might be involved in the formation of senile cataracts. Although investigations on normal or opacified animal lenses can never substitute for those performed with human lenses of any kind, they nevertheless facilitate the search for basic information on lens metabolism and its age-related changes, so that the valuable human lens material can be used for more specialized studies.

REFERENCES

1 J. H. Wilkinson, *Isoenzymes,* 2nd ed., Chapman and Hall, London, 1970.

2 E. S. Vesell, *Ann. N.Y. Acad. Sci.,* **151,** 5 (1968).

3 U. Gerlach und W. Fegeler, in D. Platt, Ed., *Experimental Gerontology,* Fischer, Stuttgart, 1974, p. 49.

4 U. Wurster and K. Hoffmann, *Interdiscipl. Topics Gerontol.,* **13,** 26 (1978).

5 H. Aebi, S. R. Loyss, und B. Scherz, *Int. Symp. on Structure and Function of Erythrocytes VIII, Berlin, 1976,* p. 204 (abstr.).

6 A. P. Kraus and C. L. Neely, *Science,* **145,** 595 (1964).

7 J. Bours, H. Neuhaus, and O. Hockwin, *Arch. Ophthalmol.,* **203,** 9 (1977).

8 M. Koida, C. Y. Lai, and B. L. Horecker, *Arch. Biochem.,* **134,** 623 (1969).

9 F. Mennecier and J. C. Dreyfus, *Biochim. Biophys. Acta,* **364,** 320 (1974).

10 A. Kahn, P. Boivin, M. Vibert, D. Cottreau, and J. C. Dreyfus, *Biochimie,* **56,** 1395 (1974).

11 B. M. Turner, R. A. Fischer, and H. Harris, in Markert, Ed., *Isoenzymes,* Vol. I, *Molecular Structure,* Academic Press, New York, 1975, p. 781.

12 H. Skala-Rubinson, M. Vibert, J. C. Dreyfus, *Clin. Chim. Acta,* **70,** 385 (1976).

13 C. Ohrloff, O. Hockwin, C. Jung, and J. Bours, *Ophthal. Res.* **9,** 181 (1977).

14 C. F. Midelfort and A. M. Mehler, *Proc. Natl. Acad. Sci. U.S.,* **69,** 1816 (1972).

15 C. Ohrloff, F. Bous, and O. Hockwin, *Exp. Eye Res.,* **22,** 292 (1976).

16 J. M. Reiner, *Behaviour of Enzyme Systems,* 2nd ed., Van Nostrand Reinhold, New York-Toronto-Melbourne, 1969, pp. 127ff.

17 F. Bous, Thesis, University of Bonn, 1977.

18 G. Jeschke, Thesis, University of Bonn, 1977.

19 J. Forgacs, *Experientia,* **18,** 339 (1962).

20 J. Forgacs, *Ophthalmologica,* **145,** 272 (1963).

21 E. Cotlier, *J. Histochem. Cytochem.,* **12,** 419 (1964).

22 S. Lessel and T. Kuwabara, *Arch. Ophthalmol.,* **71,** 851 (1964).

23 E. Korhonen and L. K. Korhonen, *Acta Ophthalmol.,* **44,** 577 (1966).

24 T. R. Shanthaveerappa, M. B. Waitzman, and G. H. Bourne, *Histochemie,* **7,** 80 (1966).

25 A. A. Swanson, J. Jeter, and K. W. Raley, *Exp. Eye Res.,* **6,** 351 (1967).

26 W. Emmes, Thesis, University of Bonn, 1968.

27 O. Kleifeld, W. Emmes, und O. Hockwin, *Ber. Dtsch. Ophthalmol. Ges.,* **69,** 462 (1969).

28 H. Ozawa, *Acta Soc. Ophthalmol. Jap.,* **73,** 1155 (1969).

29 Z. Albert, Z. Rzucidlo, and H. Starzyk, *Acta Histochem.,* **37,** 34 (1979).

30 K. D. Reinel, Thesis, University of Bonn, 1972.

31 M. Palva and A. Palkama, *Exp. Eye Res.,* **19,** 117 (1977); **22,** 229 (1976).

32 J. G. Collins and C. N. Corder, *Invest. Ophthalmol. Visual Sci.,* **16,** 242 (1977).

33 D. Kaskel, C. M. Schedtler, O. Hockwin, and S. Cioli, *Arch. Ophthalmol.,* **191,** 247 (1974).

34 D. Kaskel, H. Muno, O. Hockwin, and C. M. Schedtler, *Interdiscipl. Topics Gerontol.,* **12,** 50 (1978).

35 D. Kaskel, U. Herrmann, C. M. Schedtler, and O. Hockwin, *Interdiscipl. Topics Gerontol.,* **12,** 63 (1978).

36 O. Hockwin, "Age Changes of Lens Metabolism," in H. Bredt and J. W. Rohen, Eds., *Ageing and Development,* Vol. I, F. K. Schattauer, Stuttgart-New York, 1971, p. 95.

37 O. Hockwin, H.-R. Koch, C. Ohrloff, und J. Bours, *Klin. Monatsbl. Augenheilk.,* **169,** 165 (1976).

38 O. Kleifeld, O. Hockwin, and P. Arens, *Arch. Ophthalmol.,* **156,** 467 (1955).

39 O. Kleifeld, *Doc. Ophthalmol.,* **10,** 132 (1956).

40 O. Hockwin, G. Blum, I. Korte, T. Murata, W. Radetzky, and F. Rast, *Ophthal. Res.,* **2,** 143 (1971).

41 O. Hockwin and I. Korte, in J. Nordmann and M. U. Dardenne, Eds., *Symp. Biochem. of the Eye,* S. Karger, Basel-New York, 1968, p. 216.

42 O. Hockwin, I. Korte, G. Blum, T. Murata, W. Radetzky, and F. Rast, *Ber. Dtsch. Ophthalmol. Ges.,* **70,** 350 (1969).

43 O. Hockwin, *Exp. Eye Res.,* **15,** 235 (1973).

44 C. Ohrloff, O. Hockwin, and D. Kaskel, *Ophthal. Res.,* **5,** 222 (1973).

45 C. Ohrloff, A. Holstege, and O. Hockwin, *Ophthal. Res.,* **8,** 227 (1976).

46 C. Ohrloff, F. Bous, and O. Hockwin, *Doc. Ophthalmol. Proc. Ser.,* **8,** 177 (1976).

47 H. G. Hers, *Biochim. Biophys. Acta,* **22,** 202 (1956).

48 R. van Heyningen, *Nature,* **184,** 194 (1959).

49 J. H. Kinoshita, *Invest. Ophthalmol.,* **4,** 786 (1965).

50 C. Ohrloff, *Altern der Augenlinse, Habilitationsschrift Med. Fak. der Univ. Bonn,* 1979.

51 S. Lerman, *Arch. Ophthalmol., (Chicago),* **65,** 334 (1961).

52 J. F. R. Kuck, Jr. and H. H. Crosswell, Jr., *Ophthal. Res.,* **6,** 189 (1974).

53 C. Ohrloff and O. Hockwin, *Ophthal. Res.,* **5,** 121 (1973).

54 C. Ohrloff, O. Hockwin, and W. Licht, *Ophthal. Res.,* **6,** 223 (1974).

55 W. Schmack, in J. Nordmann and M. U. Dardenne, Eds., *Symp. Biochemistry of the Eye,* S. Karger, Basel-New York, 1968, p. 243.

56 C. Ohrloff, S. Zierz, and O. Hockwin, in preparation.

57 O. Hockwin, *Arch. Ophthalmol.*, **162,** 339 (1960).

58 A. Pirie, *Exp. Eye Res.*, **1,** 427 (1962).

59 W. Hans, O. Hockwin, und O. Kleifeld, *Arch. Ophthalmol.,* **157,** 72 (1955).

60 O. Hockwin, Thesis, University of Bonn, 1956.

61 B. Schwartz, B. Danes, and P. J. Leinfelder, *Am. J. Ophthalmol.,* **38**(II), 182 (1954).

62 J. E. Harris, J. D. Hauschildt, and L. T. Nordquist, *Am. J. Ophthalmol.,* **38**(II), 148 (1954).

63 H. G. Neumann, Thesis, University of Bonn, 1958.

64 A. Pirie, *Biochem. J.*, **96,** 244 (1965).

65 J. H. Kinoshita, *Arch. Ophthalmol. (Chicago),* **72,** 554 (1964).

66 O. Hockwin, I. Korte, R. Breuer, G. Schmidt, und F. Rast-Czyborra, *Arch. Ophthalmol.,* **207,** 169 (1978).

67 I. Korte, O. Hockwin, and D. Kaskel, *Doc. Ophthalmol. Proc. Ser.,* **18,** 163 (1979).

68 O. Hockwin, M. Günter, and E. Noll, *Doc. Ophthalmol. Proc. Serv.,* **18,** 107 (1979).

69 J. Nordmann, *Biologie du cristallin,* Masson, Paris, 1954.

70 J. Nordmann, "Acquisitions recentes dans le domaine de la biologie du cristallin," *Fortschr. Augenheilk.,* **12,** 1 (1962).

71 R. van Heyningen, "The Lens," Chapter 5 in H. Davson, *The Eye,* Vol. 1, Academic Press, New York-London, 1962.

72 R. van Heyningen, "The Lens: Metabolism and Cataract," Chapter 6 in H. Davson, Ed., *The Eye, Vegetative Physiology and Biochemistry,* Vol. I, Academic Press, New York-London, 1969.

73 J. F. R. Kuck, Jr., in C. N. Graymore, Ed., *Biochemistry of the Eye,* Academic Press, London-New York, 1970, Chapters 3–5.

74 J. F. R. Kuck, Jr., "Composition of the Lens," in J. G. Bellows, Ed., *Cataract and Abnormalities of the Lens,* Grune and Stratton, New York-San Francisco-London, 1975.

75 M. U. Dardenne and G. Kirsten, *Exp. Eye Res.,* **1,** 415 (1962).

76 S. G. Waley, "The Lens, Function and Macromolecular Composition," in H. Davson, Ed., *The Eye,* Vol. 1, Academic Press, New York-London, 1969, p. 299.

77 J. J. Harding and K. J. Dilley, *Exp. Eye Res.,* **22,** 1 (1976).

78 J. Bours, *Exp. Eye Res.,* **15,** 299 (1973).

79 P. T. Uhlenhuth, *R. Koch Festschrift,* Fischer, Jena, 1903, p. 49.

80 H. Maisel, *Am. J. Ophthalmol.,* **55,** 1208 (1963).

81 J. Bours und O. Hockwin, *Berl. Muench. Tieraerztl. Wochenschr.,* **89,** 147 (1976).

82 J. Bours, (personal communication).

83 N. Virmaux and P. Mandel, *Nature,* **201,** 297 (1964).

84 F. C. Merriam and V. E. Kinsey, *Arch. Ophthalmol. (Chicago)*, **44,** 651 (1950).

85 P. Mandel, U. Dardenne, and A. Lessinger, *C. R. Acad. Sci.*, **245,** 985 (1957).

86 A. Spector and J. H. Kinoshita, *Invest. Ophthalmol.*, **3,** 517 (1964).

87 H. N. Fukui, H. Obazawa, and J. H. Kinoshita, *Invest. Ophthalmol.*, **15,** 422 (1976).

88 J. Klethi and J. Nordmann, *Interdiscipl. Topics Gerontol.*, **13,** 143 (1978).

88a G. J. A. M. Strous, H. van Westreenen, J. van der Logt, and H. Bloemendal, *Biochim. Biophys. Acta*, **353,** 89 (1974).

90 R. W. Young and H. W. Fulhorst, *Invest. Ophthalmol.*, **5,** 288 (1966).

91 D. Glässer, in *In Memoriam Emil Abderhalden, Beiträge zur Universitätsgeschichte*, Halle/Saale, G.D.R., 1977.

92 S. G. Waley and R. van Heyningen, *Biochem. J.*, **83,** 274 (1962).

93 A. A. Swanson and J. T. Nichols, *Biochem. J.*, **125,** 575 (1971).

94 A. M. J. Blow, R. van Heyningen, and A. J. Barrett, *Biochem. J.*, **145,** 591 (1975).

95 P. Trayhurn and R. van Heyningen, *Exp. Eye Res.*, **22,** 251 (1976).

96 R. van Heyningen, *Interdiscipl. Topics Gerontol.*, **12,** 232 (1978).

97 A. A. Swanson and B. Albers, *Interdiscipl. Topics Gerontol.*, **12,** 241 (1978).

98 U. Hahn, Thesis, University of Bonn, 1977.

99 U. Hahn, A. A. Swanson and O. Hockwin, *Arch. Ophthalmol.*, **199,** 197 (1976).

100 D. V. N. Reddy, *Exp. Eye Res.*, **11,** 310 (1971).

101 D. V. N. Reddy, S. D. Varma, and B. Chakrapani, *Exp. Eye Res.*, **16,** 105 (1973).

102 W. B. Rathbun, *Doc. Ophthalmol. Proc. Ser.*, **8,** 145 (1976).

103 S. D. Varma, T. K. Ets, and R. D. Richards, *Ophthal. Res.*, **9,** 421 (1977).

104 D. V. N. Reddy, J. Klethi, and V. E. Kinsey, *Invest. Ophthalmol.*, **5,** 594 (1966).

105 W. B. Rathbun and K. Wicker, *Exp. Eye Res.*, **15,** 161 (1973).

106 W. B. Rathbun, M. Hough, L. Gruber, and J. E. Harris, *Interdiscipl. Topics Gerontol.*, **12,** 132 (1978).

107 S. L. Bonting, K. A. Simon, and N. M. Hawkins, *Arch. Biochem. Biophys.*, **95,** 416 (1961).

108 S. L. Bonting, L. L. Caravaggio, and N. M. Hawkins, *Arch. Biochem. Biophys.*, **101,** 37 (1963).

109 S. L. Bonting, L. L. Caravaggio, and N. M. Hawkins, *Arch. Biochem. Biophys.*, **101,** 47 (1963).

110 S. L. Bonting, *Invest. Ophthalmol.*, **4,** 723 (1965).

111 S. L. Bonting, *Arch. Ophthalmol., (Chicago)*, **74,** 561 (1965).

112 S. L. Bonting, *Arch. Ophthalmol., (Chicago)*, **76,** 607 (1966).

113 J. Gupta and J. Harley, *Exp. Eye Res.*, **20,** 207 (1975).

114 I. Korte and O. Hockwin, in *5th Eur. Symp. Basic Res. Gerontol.* Perimed Verlag, Erlangen, (1976), p. 623.

115 J. Nordmann and J. Klethi, *Arch. Ophtalmol. (Paris),* **36,** 523 (1976).

116 D. Fournier and J. Patterson, *Proc. Soc. Exp. Biol. Med.,* **137,** 826 (1971).

117 S. Iwata and J. H. Kinoshita, *Invest. Ophthalmol.,* **10,** 504 (1971).

118 S. Iwata, *Ophthal. Res.,* **6,** 138 (1974).

119 J. H. Kinoshita, *Invest. Ophthalmol.,* **13,** 713 (1974).

120 P. Russel, H. N. Fukui, J. Tsunematsu, F. L. Huang, and J. H. Kinoshita, *Invest. Ophthalmol.,* **16,** 243 (1977).

121 J. Nordmann, *Adv. Ophthalmol.,* **34,** 1 (1977).

122 S. L. Bonting and J. J. H. H. M. de Pont, *Doc. Ophthmol. Proc. Ser.,* **18,** 41 (1979).

123 D. Gershon and H. Gershon, *Gerontology,* **22,** 212 (1976).

124 F. Verzar, *Int. Rev. Connect. Tissue Res.,* **2,** 243 (1964).

125 H. P. von Hahn and E. Fritz, *Gerontologica (Basel),* **12,** 237 (1966).

126 S. Goldstein and E. J. Moerman, *Interdiscipl. Topics Gerontol.,* **10,** 24 (1976).

127 W. R. Pendergrass, G. M. Martin, and P. Bornstein, *J. Cell. Comp. Physiol.,* **87,** 3 (1976).

128 J. C. Dreyfus, H. Rubinson, F. Schapira, A. Weber, J. Marie, and A. Kahn, *Gerontology,* **23,** 211 (1977).

129 H. Gershon and D. Gershon, *Nature,* **227,** 1214 (1970).

130 U. Reiss and M. Rothstein, *J. Biol. Chem.,* **250,** 826 (1975).

131 U. Reiss and D. Gershon, *J. Biochem.,* **63,** 617 (1976).

132 G. Fornaini, G. Leoncini, P. Segni, G. A. Calabria, and M. Dache, *Eur. J. Biochem.,* **7,** 24 (1969).

133 G. Rijksen and G. Staal, *INSERM Colloquium Molecular and Cell Models of Ageing,* Paris, 1978.

134 F. Schapira, A. Weber, and C. Gregori, *C.R. Acad. Sci.,* **280,** 1161 (1975).

135 J. Banroques, H. Skala, F. Schapira, and J. C. Dreyfus, *Interdiscipl. Topics Gerontol.,* **12,** 180 (1978).

136 T. Flatmark and K. Sletten, *J. Biol. Chem.,* **243,** 1623 (1968).

137 L. K. Li, *Exp. Eye Res.,* **18,** 383 (1974).

138 F. S. M. van Kleef, W. Willems-Thijssen, and H. J. Hoenders, *Eur. J. Biochem.,* **66,** 477 (1976).

139 J. Nordmann, *Ophthal. Res.,* **3,** 323 (1972).

140 G. W. Barber, *Exp. Eye Res.,* **16,** 85 (1973).

141 O. Hockwin, Ed., *Lens Ageing and Development of Senile Cataracts, Interdiscipl. Topics Gerontol.,* **12,** Karger, Basel, 1978.

142 O. Hockwin, Ed., *Gerontological Aspects of Eye Research, Interdiscipl. Topics Gerontol.,* **13,** Karger, Basel, 1978.

143 O. Hockwin and O. Kleifeld, in J. W. Rohen, Ed., *Die Struktur des Auges,* F. K. Schattauer, Stuttgart, 1965, p. 395.

144 R. Clark, S. Zigman, and S. Lerman, *Exp. Eye Res.,* **8,** 172 (1969).

145 J. Nordmann, P. Mandel, and M. Achard, *Brit. J. Ophthalmol.,* **38,** 673 (1954).

146 O. Hockwin, O. Kleifeld, and P. Arens, *Arch. Ophthalmol.,* **160,** 1 (1958).

147 O. Kleifeld, O. Hockwin, und N. Ayberk, *Arch. Ophthalmol.,* **158,** 39 (1956).

148 L. T. Chylack, Jr., *Invest. Ophthalmol.,* **14,** 746 (1975).

149 H. Pau und W. Graeber, "Physiologische Chemie," in K. Velhagen, Ed., *Der Augenarzt* Vol. 1, 2nd ed., Thieme, Stuttgart, 1969.

150 O. Hockwin, L. Weimar, E. Noll, und W. Licht, *Arch. Ophthalmol.,* **170,** 99 (1966).

151 D. Glässer, *Nova Acta Leopoldina,* N. F. Nr. 197, Joh. Ambr. Barth, Leipzig, 1971.

152 O. Hockwin, W. Licht, E. Noll, und M. Glasmacher, *Ber. Dtsch. Ophthalmol. Ges.,* **70,** 368 (1969).

153 O. Hockwin, H. Fink, and D. Beyer, *Mech. Ageing Dev.,* **2,** 409 (1973/74).

154 O. Hockwin, H. Fink, and M. Glasmacher, *Z. Alternsforsch.,* **31,** 521 (1976).

155 S. Hayman and J. H. Kinoshita, *J. Biol. Chem.,* **240,** 877 (1965).

156 H. K. Müller, U. Dardenne, O. Hockwin, O. Kleifeld, and G. Schafhausen, *XVIII Concil. Ophthal. Belg. Acta,* **1,** 750 (1958).

157 S. Mann, Thesis, University of Bonn, 1959.

158 O. Hockwin, W. Licht, und E. Noll, *Arch. Ophthalmol.,* **167,** 158 (1964).

159 O. Hockwin, H. Fink, E. Noll, and W. Licht, *Doc. Ophthmol.,* **20,** 73 (1966).

160 O. Hockwin, *Interdiscipl. Topics Gerontol.,* **13,** 88 (1978).

161 H. Fink, O. Hockwin, and E. Weigelin, *Arch. Ophthalmol.,* **177,** 323 (1969).

162 H. Fink und O. Hockwin, *Arch. Ophthalmol.,* **191,** 177 (1974).

163 O. Hockwin, H. Fink, and C. Ohrloff, *5th Eur. Symp. Basic Res. Gerontol.,* Perimed Verlag, Erlangen, 1976, p. 646.

164 U. Kochmann, V. Krzywda, O. Kwiatkowski, and T. Baranowski, *Int. J. Biochem.,* **2,** 221 (1971).

165 J. Banroques, C. Gregori, and F. Schapira, *FEBS Lett.,* **65,** 204 (1976).

166 C. Ohrloff, A. Güntert, W. Braun, and O. Hockwin, *Z. Gerontol.,* **12,** 40 (1979).

167 H. Maisel, M. Kerrigan, and F. Syner, *Invest. Ophthalmol.,* **4,** 362 (1965).

168 I. Korte, O. Hockwin, und C. L. Bannermann, *Ber. Dtsch. Ophthal. Ges.,* **69,** 458 (1968).

169 M. Garbers and J. Bours, *Exp. Eye Res.,* **25,** 537 (1977) (abstr.).

170 L. Poenaru, H. Skala, Y. Courtois, and J. Dreyfus, *Exp. Eye Res.,* **25,** 39 (1977).

171 J. J. Harding, *Biochem. J.,* **134,** 995 (1973).

172 J. J. Harding, *INSERM,* **60,** 215 (1976).

173 L. T. Chylack, Jr., J. H. Kinoshita, and R. Kasabian, *Exp. Eye Res.,* **10,** 250 (1970).

174 L. T. Chylack, Jr., *Exp. Eye Res.,* **11,** 280 (1971).

175 L. T. Chylack, Jr., *Ophthal. Res.,* **6,** 93 (1974).

176 O. Hockwin and C. Ohrloff, *Arch. Ophthalmol.,* **187,** 235 (1973).

177 M. F. Lou and J. H. Kinoshita, *Biochim. Biophys. Acta,* **141,** 547 (1967).

178 H. M. Cheng and L. T. Chylack, Jr., *Ophthal. Res.,* **9,** 381 (1977).

179 C. Ohrloff, *Interdiscipl. Topics Gerontol.*, **12,** 158 (1978).
180 F. Bous, O. Hockwin, C. Ohrloff, and J. Bours, *Exp. Eye Res.*, **24,** 383 (1977).
181 H. Berdjis, Thesis, University of Bonn, 1979.
182 P. D. Wilson, *Gerontologia,* **19,** 79 (1973).
183 H. W. Hofer und D. Pette, *Z. Physiol. Chem.*, **349,** 1378 (1968).
184 R. van Heyningen, *Exp. Eye Res.*, **4,** 298 (1965).
185 C. Ohrloff, U. Teimann, and O. Hockwin, *Doc. Ophthalmol. Proc. Ser.*, **18,** 205 (1979).
186 H. Dertinger und H. Jung, *Molekulare Strahlenbiologie,* Vol. 57/58, Springer, Berlin-Heidelberg, 1969.
187 R. Holliday and G. M. Tarrant, *Nature,* **238,** 26 (1972).
188 E. Steinhagen-Thiessen and H. Hilz, *Mech. Ageing Dev.*, **5,** 447 (1976).
189 L. E. Orgel, *Proc. Natl. Acad. Sci. U.S.*, **49,** 517, (1963).
190 W. Jansen, Thesis, University of Bonn, 1966.
191 H. Feldmann, Thesis, University of Bonn, 1967.
192 O. Hockwin and R. Gassner, *Exp. Eye Res.*, **7,** 269 (1968).
193 F. G. Lehmann, *Clin. Chim. Acta,* **28,** 335 (1970).
194 J. Bours and S. K. Brahma, *Exp. Eye Res.*, **16,** 131 (1973).
195 H. M. Kalckar, *J. Biol. Chem.*, **167,** 461 (1947).
196 H. M. Cheng and L. T. Chylack, Jr., *Invest. Ophthalmol.*, **15,** 279 (1976).
197 W. K. G. Krietsch and T. Bücher, *Eur. J. Biochem.*, **17,** 568 (1970).
198 W. J. Ray, Jr., and E. J. Peck, Jr., in P. D. Boyer, *The Enzymes,* Vol. 6, 3rd ed., Academic Press, New York, 1976, p. 459.
199 R. Czok, in Hoppe-Seyler und Thierfelder, Eds., *Handbuch der physiologischen und pathologischen chemischen Analyse,* Vol. VIc, Springer, Berlin-Heidelberg, 1966, p. 657.
200 F. Wold, in P. D. Boyer, Ed., *The Enzymes,* Vol. V, 3rd ed., Academic Press, New York, 1976, p. 499.
201 H. Gutfreund, *An Introduction to the Study of Enzymes,* Blackwell, Oxford, 1965, p. 182.
202 Z. A. Medvedev, *Protein Biosynthesis and Problems of Heredity and Aging,* Oliver and Boyd, Edinburgh-London, 1966.
203 Z. A. Medvedev, "Error Theories of Aging," in D. Platt, Ed., *Alternstheorien,* F. K. Schattauer, Stuttgart-New York, 1976, p. 37.
204 F. Metchnikoff, *The Prolongation of Life, Optimistic Studies,* Putnam, London, 1908.
205 A. Vogt, *Klin. Monatsbl. Augenheilk.*, **100,** 497 (1938).
206 I. Mann, *Culture, Race, Climate and Eye Disease,* Charles C. Thomas, Springfield, Ill., 1966.
207 O. Hockwin and H.-R. Koch, "Combined Noxious Influences," in G. Bellows, Ed., *Cataract and Abnormalities of the Lens,* Grune and Stratton, New York, 1975.
208 A. Spector, *Exp. Eye Res.*, **11,** 376 (1971).

209 O. Hockwin and H.-R. Koch, "Fortschritte auf dem Gebiet der experimentellen Kataraktforschung," in H. J. Küchle, Ed., *Aktuelle Ophthalmologie, Fachalmanach für Augenheilkunde,* Lehmann, Munich, 1976.

210 A. Pirie, *The Human Lens—in Relation to Cataract, CIBA Found. Symp.* **19,** (new series), Elsevier-Excerpta Medica—North Holland Associated Scientific Publ., Amsterdam-London-New York, 1973.

211 J. Kmonicek and J. Tovarek, *Csl. Oftal.,* **17,** 102 (1961).

212 O. Hockwin, *INSERM Symp. Lens Dev. Aging,* **60,** 263 (1976).

213 C. Drews, *Ophthalmologica,* **148,** 143 (1964).

214 P. Niesel, *Ophthalmologica,* **151,** 489 (1966).

215 N. Brown, *Trans. Ophthalmol. Soc. U.K.,* **89,** 397 (1969).

216 V. Dragomirescu, O. Hockwin, H.-R. Koch, and K. Sasaki, *Interdiscipl. Topics Gerontol.,* **13,** 118 (1978).

217 O. Hockwin, V. Dragomirescu, H.-R. Koch, and K. Sasaki, *XXIII. Int. Congr. Ophthalmol., Kyoto, 1978,* Excerpta Medica No. 442, ICS., p. 182.

218 O. Hockwin, V. Dragomirescu, H.-R. Koch, and K. Sasaki, *Ber. Dtsch. Ophthal. Ges.,* **76,** 827 (1979).

219 A. Pirie, *Invest. Ophthalmol.,* **7,** 634 (1968).

220 R. van Heyningen, *Exp. Eye Res.,* **13,** 136 (1972).

221 L. T. Chylack, Jr., *5th Conf. Biochem. of the Eye, 1977, Oakland Univ., Rochester, Mich.; Arch. Ophthalmol. (Chicago),* **96,** 888 (1978).

222 V. Dragomirescu, O. Hockwin, and H.-R. Koch, *Ophthal. Res.,* **10,** 333 (1978).

223 O. Hockwin and P. Berghoff, *Klin. Monatsbl. Augenheilk.,* **149,** 87 (1966).

224 G. Maraini and A. Pescatori, *Ophthal. Res.,* **3,** 108 (1972).

225 G. Maraini and R. Mangili, *Exp. Eye Res.,* **16,** 123 (1973).

226 G. Maraini and R. Mangili, in ref. (210), p. 79.

227 G. Maraini, F. Carta, A. Pescatori, and L. Prosperi, *Exp. Eye Res.,* **11,** 83 (1971).

228 D. Friedburg, *Arch. Ophthalmol.,* **170,** 365 (1966).

229 D. Friedburg, *Arch. Ophthalmol.,* **173,** 309 (1967).

230 D. Friedburg, *Ber. Dtsch. Ophthal. Ges.,* **69,** 446 (1968).

231 D. Friedburg, *Ophthal. Res.,* **3,** 224 (1972).

232 D. Friedburg, in ref. (210), p. 117.

233 D. Friedburg and W. Hammerstein, *Arch. Ophthalmol.,* **181,** 154 (1971).

234 D. Friedburg and K. F. Manthey, *Exp. Eye Res.,* **15,** 173 (1973).

235 D. Friedburg and U. Meyer, *Arch. Ophthalmol.,* **174,** 367 (1968).

236 R. Kistler, *Klin. Monatsbl. Augenheilk.,* **80,** 181 (1928).

237 H.-R. Koch, O. Hockwin, and F. Kremer, *Interdiscipl. Topics Gerontol.,* **13,** 95 (1978).

238 W. C. Gorthy, *Ophthal. Res.,* **9,** 329 (1977).

239 H.-R. Koch, A. Fischer, and H. Kaufmann, *Ophthal. Res.,* **9,** 189 (1977); *Interdiscipl. Topics Gerontol.,* **12,** 90 (1978).

240 O. Salffner, *Arch. Ophthalmol.,* **59,** 520 (1904).

241 A. Pirie and R. van Heyningen, in J. Nordmann and M. U. Dardenne, Eds., *Symposium on Biochemistry of the Eye,* S. Karger, Basel-New York, 1968, p. 410.

242 H.-R. Koch, K. Doldi, and O. Hockwin, *Doc. Ophthalmol. Proc. Ser.,* **8,** 293 (1976).

243 W. Buschke, *Arch. Ophthalmol. (Chicago),* **30,** 735 (1943).

244 C. Ohrloff, C. Stoffel, H.-R. Koch, U. Wefers, J. Bours, and O. Hockwin, *Exp. Eye Res.,* **20,** 179 (1975).

245 C. Ohrloff, C. Stoffel, H.-R. Koch, U. Wefers, J. Bours, and O. Hockwin, *Arch. Ophthalmol.,* **205,** 73 (1978).

246 H. Goldman und A. Liechti, *Arch. Ophthalmol.,* **138,** 722 (1938).

247 L. von Sallmann, *Arch. Ophthalmol. (Chicago),* **45,** 149 (1951).

248 O. Hockwin, *Exp. Eye Res.,* **1,** 422 (1962).

249 H.-R. Koch, *Klinische und experimentelle Untersuchungen über den Einfluss von Corticosteroiden auf die Augenlinse,* Habilitationsschrift Med. Fak. der Univ. Bonn, 1976.

250 H.-R. Koch and O. Hockwin, "Effects of Ionizing Radiation on the Lens," in S. Lerman, Ed., *Radiant Energy and the Eye,* Macmillan, New York, 1980.

251 O. Hockwin, *Fortschr. Med.,* **80,** 659 (1962).

252 O. Hockwin, H. D. Bergeder, und L. Kaiser, *Ber. Dtsch. Ophthalmol. Ges.,* **68,** 135 (1967).

253 O. Hockwin, H. D. Bergeder, U. Ninnemann, und H. Fink, *Arch. Ophthalmol.,* **189,** 171 (1974).

254 H.-R. Koch, F. Kremer, E. Linner, O. Hockwin, H. Kaufmann, W. Breull, and H. W. Dahmers, *Ophthal. Res.,* **6,** 175 (1974).

255 J. H. Kinoshita, L. O. Merola, and E. Dikmack, *Exp. Eye Res.,* **1,** 405 (1962).

256 J. H. Kinoshita, D. Dvornik, M. Krahml, and K. H. Gabbag, *Biochim. Biophys. Acta,* **158,** 472 (1968).

257 S. D. Varma, I. Mikuni, and J. H. Kinoshita, *Science,* **188,** 1215 (1975).

258 S. D. Varma and J. H. Kinoshita, *Doc. Ophthalmol. Proc. Ser.,* **8,** 305 (1976).

259 T. Noma, Y. Okubo, Y. Yamamoto, F. Ikemoto, and S. Iwata, *Senju Res. Lab. Rep.,* **328,** (1974).

260 I. Korte, O. Hockwin, and C. Ohrloff, *Ophthal. Res.,* **11,** 123 (1979).

CHAPTER 10

The Developmental Mechanics of the Lens

W. J. van Doorenmaalen

Department of Medical Anatomy and Embryology, The State University of Utrecht, Utrecht, the Netherlands

10.1 INTRODUCTION

The development of the lens has been a subject of research for many investigators because of the attractiveness of its problems, the very suitable experimental circumstances, and its clinical implications. An overwhelming number of articles have appeared on lens. Many of these have been referred to in the foregoing chapters. Therefore a selection has been made in this review.

Lens development and differentiation are the result of many developmental steps and/or tissue interactions in both the embryo and in the eye *per se*. These processes and the relevant experiments can be classified as follows:

1 Chains of sequential and synergic tissue interactions in the egg and embryo which lead to the conditioning of both ectoderm and optic vesicle on behalf of the induction and differentiation of the lens;
2 The induction process itself where the presumptive ectoderm is conditioned by the optic vesicle to form the placode, lens vesicle, and primary lens fibers;
3 The influence of the retina on further differentiation of the lens vesicle in which lens fiber production is assured and maintenance of the condition of the lens is established;
4 The interactions of the tissues of the eye itself, including the lens, in order to develop into an adult, functioning eye as an organ.

Furthermore, the results of experimental work *in vitro* have to be mentioned because they are performed under conditions quite different from those of the developing eye in the embryo. The age of the tissues, the composition of the culture medium or intercellular matrix, and the kind of inductor or responding tissue, for example, differ markedly from those in the embryo at that relevant stage.

The results of these tissue interaction experiments will contribute more to the knowledge about the conditions by which information is initiated, transferred, and received than they will provide us with knowledge about the kind of information and its working mechanisms in genuine lens development.

Another complicating factor is that research has been done on a great variety of animals in three classes: amphibians, birds (chick), and mammals (mouse, occasionally rabbit). Within these classes various species from different orders and suborders have been used. From the results it can be concluded that many differences exist in developmental mechanics between those test animals. Especially amphibians have quite different early developmental channels; even among them variations in eye development and lens regeneration exist. These developmental and metabolic differences render it difficult to put forward a general theory of eye development and differentiation for all the classes. In order to understand the views of lens development two kinds of embryological processes are reviewed, those establishing the first tissues in the embryo and those of the development of the eye itself.

10.2 PRIMARY TISSUE INTERACTIONS

During the developmental "gastrulation" stage the shape of the amphibian embryo can be compared with a sphere. At the lower part of this sphere, at one side, the tissues forming the surface layer start moving to one point and curve themselves inward along a circular groove, the "blastopores." An inner layer is therefore established under the surface layer. The tissues above the blastopores move faster and in greater quantity inward through the dorsal part (dorsal "lip") of the blastoporal groove. This results in a mesoderm layer underneath the surface layer of the egg, radiating out from the blastopores to the future cranial part and lateral parts of the embryo. The tissue in the median line is named the chorda mesoderm. In the surface layer, called the gastrula ectoderm, the neural plate is induced by the transfer of a chemical compound (cf. refs. 1, 2) originating from the chorda mesoderm. From the base of the sphere, where a mass of cells lies near the ventral lip of the blastopores, another, third layer is formed, proliferating from the blastopores sideward, upward, and in cranial direction. However, it does not stretch itself out as a complete layer under the mesoderm but leaves an oval space open, situated medially under the chorda mesoderm layer and not quite reaching the cranial end of the egg.

In the chick the shape of the embryo in that comparable stage can be seen as a disk, formed by a layer of primitive ectoderm cells.

The primitive entoderm is not migrating via a "gastrulating" opening but is "delaminating" from the primitive ectoderm layer, called the epiblast. Cells from the epiblast loosen themselves from this layer to form a second layer. As soon as the two layers are formed, the second is called hypoblast and can be considered as primitive entoderm. The formation of one layer starts at the caudal end and progresses cranially. During development of the "primitive streak"—a linear fusion of both layers in the midline of the blastoderm—the mesoderm is proliferating and moving by invagination through the streak laterally and cranially forming a layer between hypoblast and epiblast. Notochord primitive streak (containing also hypoblast cells) and the hypoblast layer are responsible for inducing the neural plate in the chick (3).

The influences of the early embryological tissues on presumptive lens ectoderm, prior to the genuine lens induction by the optic vesicle, has been investigated extensively in salamanders (urodeles) by Jacobson (4). In a great number of experiments he combined ectoderm, heart mesoderm, and retina, a derivative of the optic vesicle, with lens epidermis, and could express the inductive power in percents of lenses formed. The results are depicted in Figure 1. In the diagram the extent of participation of the three "inductors" is seen, whereas in the dashed lines the decrease in the number of lenses to 92 and 75% is given after excluding entoderm from induction at stages 14 and 22, respectively. The participation of entoderm is apparently

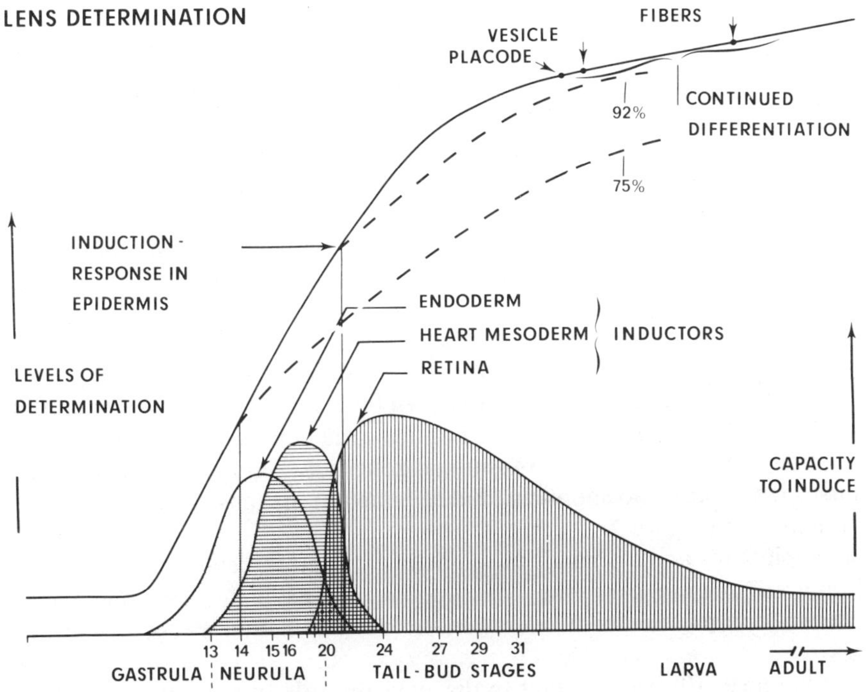

Figure 1 Diagram modified after Jacobson (4) presenting the inductive power of the three inductors in lens development and differentiation.

a limited process and depends on age, whereas the retina (inner layer of optic vesicle) is necessary for further development and differentiation, which latter phenomenon has already been found by King (5), LeCron (6), and Lewis (7), respectively.

Jacobson's extensive work suggests strongly the participation of entoderm and mesoderm in the development of the lens but does not inform us as to the kind of influence under these experimental conditions.

10.3 INTERACTION BETWEEN OPTIC VESICLE AND LENS

Notwithstanding the fact that several authors presume that in amphibians earlier interactions between other embryological tissues and presumptive lens ectoderm are of significance in directing the influence of lens development, it is quite obvious that the genuine developmental processes in the embryological eye are important links in the chain of tissue interactions that lead to lens differentiation.

The tissue interaction processes in the eye itself start with the induction of the lens placode.

From the cranial end of the neural tube two evaginations bulge laterally; these are named optic vesicles. As a consequence of growth they approach and apply themselves to the ectoderm at stage 9 of the chick. O'Rahilly and Meyer (8), in a detailed description of the development of the chick eye, assume that at stage 10 the first reaction of the ectoderm consists of a slight thickening of the ectoderm (see Figure 2). Later on, the lens placode, a disk of columnar ectoderm cells, is formed. About this time the ectoderm and vesicle are firmly attached to each other for a number of hours (stages 10–20). Then vesicle wall and ectodermal placode invaginate and separate completely, and the optic vesicle further invaginates to form the optic cup and the placode curves the lens vesicle, which becomes free from the ectoderm (Figure 2).

The cells of the inner wall of the lens vesicle start to elongate and form the first lens fibers; those of the outer wall keep their cubical shape, become lens epithelium, and remain as such during the whole lifetime. Owing to the continued influence of the inner wall of the optic cup (the retina), lens fiber formation proceeds from the transitional zone between lens epithelium and primary lens fibers; the annular pad. The lumen of the lens vesicle decreases to a minimum (Figure 2).

In amphibians and mammals the same events can be seen with only slight differences in time or tissue constituents, for example. In amphibians, however, one meets a controversy in the developmental mechanics of the lens.

Spemann (9, 10) was the first to discover the role of the optic vesicle as an inductor of the lens and demonstrated that in most amphibians the head ectoderm was not able to form a lens in absence of the optic vesicle. Lewis

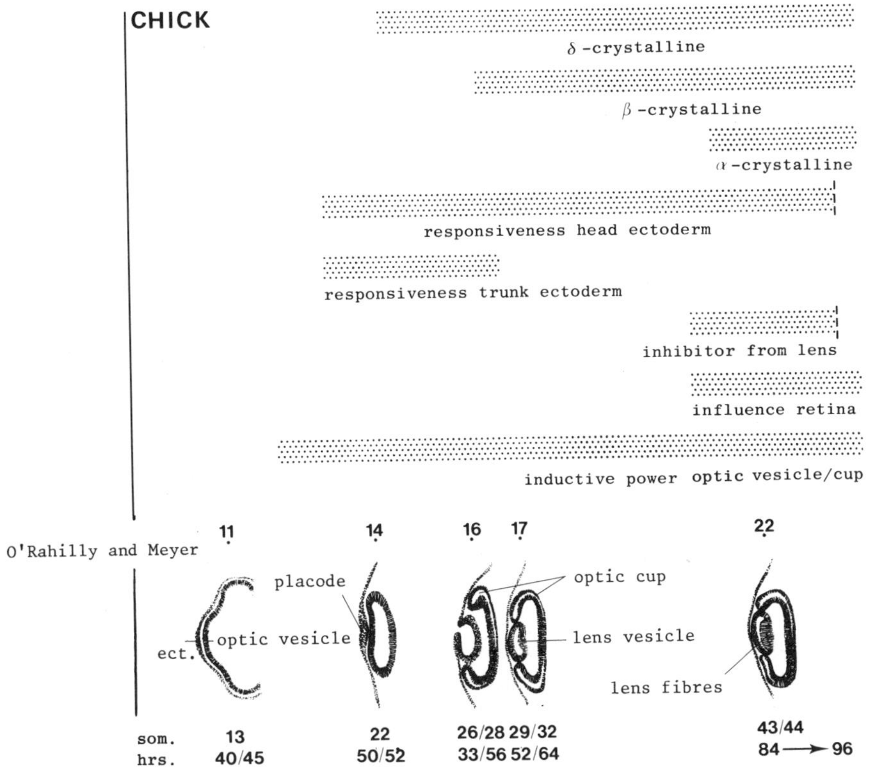

Figure 2 Diagram of the morphology of the developmental stages of the eye, correlated to tissue interactions and crystallin synthesis. For comparison, stages are given according to O'Rahilly and Meyer (8), in number of somites and in hours of incubation.

(7) and LeCron (6) also shared his opinion. On the contrary, King (5) postulated that the ectoderm in *Rana palustris* could differentiate into a lens without the inductive stimulus from the optic vesicle. But since she does not mention clearly the stages of the experiments and applied thermal cauterizing to destroy the optic cup, either the ectoderm was already conditioned to form a placode or the operation itself could be considered as a "disturbed" stimulus for lens formation. Brahma (11), however, found in a set of varied experiments on *Xenopus laevis* that a causal relationship does exist between optic vesicle and lens development and that the presumptive ectoderm is incapable of developing a lens after long term culturing *in vitro* without an optic cup. On the other hand, investigators such as Spemann (9, 10), Balinsky (12), Tahara (13), Reyer (14–16), and Jacobson (4) demonstrated that of certain amphibian species ectoderm could form a lens in the absence of optic vesicle and under the influence of

other head tissues. In these experiments either ectoderm *after* contact with tissues such as entoderm and mesoderm and before contact with optic vesicles or ectoderm *before* contact with entoderm was used. Only in the presence of or after contacting entoderm did lenses form.

For birds, Alexander (17), Waddington and Cohen (18), van Deth (19), McKeehan (20, 21) and van der Starre (22, 23) demonstrated the strong dependence of lens development on the optic vesicle. Muthukkaruppan (24) confirmed the same relationship in the mouse.

The results of Mizuno's work (25–27) and that of Karkinen–Jääskeläinen (28), which is discussed later, suggest that chick lens developmental mechanics is applicable to some amphibians. This has not yet been confirmed by others, however.

McKeehan (21), using cellophane membranes between chick ectoderm and optic vesicle, demonstrated the necessity of the optic cup at that stage of development because no lens was formed. However, at the edges of the membrane, where both tissues could communicate, a reaction of the ectoderm could be observed. Using agar strips instead of cellophane membranes, McKeehan (21) did achieve responsiveness of the ectoderm to the optic vesicle. Also van der Starre (22, 23) could demonstrate the transfer of a signal from optic cup, through an agar layer, toward the ectoderm.

The suggestion of a compound being transferred from optic vesicle or cup to the ectoderm has been confirmed by Sirlin and Brahma (29), who could estimate a transfer of [^{14}C]-phenylalanine from eye cups into the developing lens *in vitro*.

Hunt (30) demonstrated a significant transfer of RNA containing particles from optic cup to ectoderm of the chick. A transfer of antigenic components from bone marrow could be demonstrated by immunofluorescence. Thus there are many indications of a transfer of biochemical compounds transported either through a substantial matrix or via direct or close cell contact.

Obviously in order for placode formation to be initiated in the ectoderm a substance is transferred. In other tissue interaction systems the transfer of a substance by which a tissue starts to differentiate has been demonstrated as well (1, 2, 31–33). Niu and Twitty (34) achieved differentiation of gastrula-ectoderm while culturing in a medium that was "conditioned" by dorsal lip tissue. The lens for this was thoroughly investigated by van der Starre (23, 35) *in vitro* under conditions as close as possible to *in vivo* circumstances. Ectoderm and optic vesicles, with small intervals in age, were brought together in culture. With the first series of experiments van der Starre established an experimental model by which the onset of inductor production, the stage of maximum inductive power, and the end of the induction period are known. (see Figure 2, 3*a*, *b*). Also the responsiveness of trunk ectoderm and head ectoderm as a function of age was estimated. The competence of trunk ectoderm to react on eye cup decreases more rapidly with age than that of head ectoderm (Figure 2). These properties of both ectoderm and eye cup have been confirmed by Liedke (36) in *Ambystoma punctatum* and by Takeuchi (37) in *Triturus*. Contrary to other investigators, van der Starre thoroughly demonstrated a gradient in responsiveness of head and trunk ectoderm to the optic vesicle/cup and never found in his control experiments any development of lenses in ectoderm deprived of the optic cup.

In another experiment only the slice of agar was placed on an optic cup at the time of strongest inductive power. After some time this slice of agar was removed to another culture dish, and a piece of ectoderm was laid on it. After 4 hr of culturing a lens starts to differentiate in the ectoderm (Figure 3*c*). Then this ectoderm was replaced by another piece and induction could be repeated (Figure 3*d*). These results support the idea that the agar slices have received a substance from the optic vesicle in a sufficient amount to activate ectoderm to form a lens. This can be repeated several times (until the substance exhausted, apparently), which really suggests that the inductor consists of a chemical, diffusing substance.

Van der Starre (35) was able to collect a sufficient amount of "induced" or "conditioned" agar slices to produce an antiserum against the contents of the agar slice. With this antiserum induction could be inhibited in *all* tissue culture experiments. This has never been achieved with other antisera against lens crystallins, by which lens induction could be prevented by only a small percentage (30–41). Moreover, other developmental processes were disturbed, which suggests a more general toxic activity of the serum. Flickinger et al. (42), however, could not find any pathological changes after culturing ectoderm/cup combinations in media containing lens antisera.

In the control experiments, in which lens induction was investigated in culture media containing antibodies against substances excreted by embryological organs other than optic cups, no inhibition of induction was observed at all (35).

The absolute inhibition of the induction by antisera against the substances passing through the agar strongly suggests that there is a specific substance which is responsible for lens induction in ectoderm. In my laboratory attempts have been made (Janssen, unpublished) to isolate this substance in order to analyze it for future developmental work, but owing to the extreme low concentration of this substance and its low molecular weight the various microanalytical chemical methods were not adequate. At this moment it is known by micro-membrane dialysis that the molecular weight lies between 10,000 and 3,000. Also van der Starre (35) and Karkinen-Jääskeläinen (43) got a molecular weight lower than 10,000 by inserting Millipore filters between ectoderm and eye cup.

Karkinen-Jääskeläinen (28), however, performed culture experiments with optic cup/ectoderm of the chick and the mouse and demonstrated that lens development (differentiation is not mentioned) occurs in the ectoderm of both species without the presence of the optic cup. In 80% of the cultures

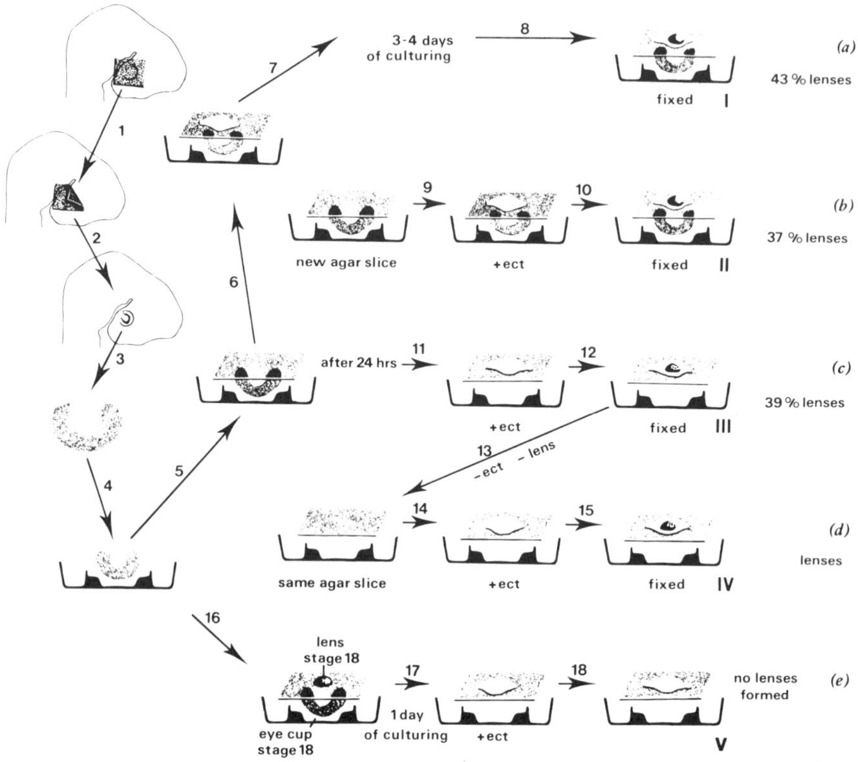

Figure 3 Scheme of culture experiments after van der Starre (35). Four experimental modifications are given. On the left side (arrows 1 & & 64) the dissection of a square piece of ectoderm is demonstrated and the extirpation of the eye cup is shown, after which it is brought into the culture dish. (*a*) An agar slice is laid on top of the eye cup (arrow 5) and finally the transplantation of ectoderm on the agar slice is shown (arrows 6–8). After culturing the development of the lens is seen (top right). (*b*) After experiment (*a*) the agar slice with tissues on top of it is taken away; then the eye cup is used again by laying a fresh agar slice and new ectoderm on top of it (arrows 9–10). After some time a second lens develops. This can be repeated several times tuntil the eye cup stops producing inductor, to an advanced age. (*c*) The eye cup is cultured 24 hours with only an agar slice lying on it. Then the agar slice is put into a culture dish with ectoderm transplanted on it and a lens develops owing to the presence of inductor (arrows 11–12). (*d*) After experiment (*c*) the tissues can be taken away from the agar (arrow 13) and the slice cultured for the second time with fresh ectoderm in a culture dish; a second lens is formed owing to the presence of enough inductor left from experiment (*c*) (arrows 14–15). This can be repeated several times until the slice has lost its inductor activity. (*e*) The eye cup is cultured with the agar layer with a lens of stage 18 laid onto it. After 24 hours the lens is taken away. The agar slice, without optic cup, is brought into culture with fresh ectoderm lying on it. In this experiment no lens is formed because the optic cup is inhibited by the lens from producing the inductor (arrows 16–18).

a lens was formed, but in my opinion the ectoderm was taken at stages where contact with the eye vesicle had already occurred and the synthesis of inductor had begun some time before (23, 24, 35). The ectoderm might have received the signal before the culture experiments took place, for production of substances as "inductor" starts quite early. When the ectoderm was removed early enough to prevent contact no lens was formed, *except* that in the *mouse* experiments, a small percentage of lens formation was achieved. However, in the mouse detailed investigation about the gradient of inductive power of the optic vesicle and responsiveness of ectoderm has not been completed, and isolating this tissue is complicated because of its tenderness compared with that of chick. It can, therefore, not be excluded that sometimes "induced" ectoderm has been taken.

Karkinen-Jääskeläinen (43) also brought together young "precontact" mouse ectoderm and mouse mesenchyme from head, metanephros, salivary gland, long tissue, and chick optic cup or human retinal tissue, resulting in lens formation. Van der Starre and van Doorenmaalen (unpublished), however, cultured more than 100 combinations of chick and mouse optic cup/ectoderm, respectively, and vice versa, with several variations in age (within the period of time of inductive activity and responsiveness) and in culture media of several compositions, but they never observed the establishment of a placode or lens in these culture combinations. Karkinen-Jääskeläinen in a similar set of experiments as mentioned above, but using *quail* trunk ectoderm, obtained quail lenses that were induced by *chick* optic cup. With the other "inductor" tissues that were used for the experiments with chick and mouse, a lens could not be induced in the *quail* trunk ectoderm.

10.4 HISTOLOGICAL AND CYTOLOGICAL ALTERATIONS DURING INDUCTION

It is interesting to know whether morphological changes provide information about the chemical substrate of these tissue interactions. The cells of the optic vesicle do not differ very much from those of the neural tube. Besides the normal cytological structures such as numerous ribosomes and mitochondria the presumptive retina cells show many mitoses from stage 12 on and the retina becomes thick in the region adjacent to the ectoderm to form a kind of retinal placode.

According to Porte et al. (44) these cells also carry zonulae adhaerentes in the apical parts of the cells with fine filaments crossing the apex of the cell in the same plane. They might contribute to invagination of the optic vesicle, as suggested for the mouse lens vesicle (45).

The lens ectoderm during the induction process (stages 12–18) in the chick has been studied by McKeehan (46). Cells become tall and wedge-shaped, with the nuclei smooth and oval, and they orientate themselves at the basal side of the cell with the length axis perpendicular to the vesicle/ectoderm plane of junction. This is called the "palisade" effect. Zwaan et al. (47, 48) demonstrated that these changes are due to a more complex movement of the nucleus during DNA synthesis in which it moves from the apex of the cell to the base and back again to the free surface. Byers and Porter (49) and Zwaan et al. (48) described microtubules in the presumptive lens cells during the early differentiation process. Also Beebe et al. (50) described these tubules and experimentally proved that cell volume changes rather than these structures are responsible for lens cell elongation. Interesting is the demonstration of a layer of fine filaments in the apexes of the cells in the lens placode of the chick as well as in that of the mouse. These filaments run parallel to the surface and are connected with each other via cell membrane junctions, the zonulae adhaerentes. Porte et al. (44) and Wrenn et al. (45) attribute contractility to these filaments, which causes the invagination of the placode into a vesicle by reducing the surface diameter of the cells. This suggestion corresponds with the results of a scanning electron microscopy carried out by Wakely (51) and Bancroft and Bellairs (52). They described irregular bubbling of cell surfaces during vesicle formation at the lumen of the cells which is caused by the escaping cytoplasm. Blebs are probably visible in histological sections as "amorphous, granular material with disruptions," also mentioned by Hendrix and Zwaan (53).

Contacts between inducing and responding tissues for transmitting inducting substances are quite different; in some instances direct contacts between cells are consistently found (1, 31, 33, 54, 55). In other induction systems direct contact does not seem obligatory. McKeehan (20, 21), van der Starre (23, 35), Muthukkaruppan (24), and Karkinen-Jääskeläinen (43) proved that induction of the lens takes place through agar or Millipore filter, which also excludes the necessity of direct contact for lens induction. In relation to the transfer of the inducting substances the structure of the space between presumptive lens ectoderm and optic vesicle is important. Several investigations have been focused on this epithelial "interlayer." O'Rahilly and Meyer (8) provided information about two clearly visible PAS-positive basement membranes, confirmed by Cohen in mouse (56). Hunt (30) and Weiss and Fitton-Jackson (57) demonstrated these interlayers by electron microscopy.

Hendrix and Zwaan (53, 58) developed a very interesting hypothesis based on the results of histochemical investigations and tissue culture experiments with labeled precursors such as [^{3}H]-glucosamine and [^{3}H]-proline. These authors showed that the interepithelial matrix is formed by epithelial cell layer and consists of glycoprotein and glycosaminoglycans oriented in a structural way, whereas three layers of collagen fibers exist. Their appearance correlates quite well with the cytological changes during the important inductional steps. This complicated structure of the matrix may control the transfer of activating substances. Moreover, these authors suggest that it connects and influences both epithelial sheets to establish the localization and extent of the lens and other eye tissues and even the position of the optical axis for all the parts of the eye, a very interesting supposition.

Another aspect, accompanying the differentiation processes of the lens, is the synthesis of lens-specific proteins. Originally the lens proteins were assumed to be responsible for the induction process of the lens; later on they were investigated mainly for their importance in fiber differentiation.

Since crystallins can be demonstrated before the specific morphological features can be observed their appearance is an important sign of the result of the induction process. Ten Cate and van Doorenmaalen (59) demonstrated lens proteins in extracts of ectoderm of chick and frog at the stage of invaginating lens placode with the use of an antiserum against extracts of the whole lens. Van Doorenmaalen (60, 61), using fluorescent antisera against the whole chick lens, could demonstrate lens proteins only from the fifth day on, owing to heavy background staining. Since antisera against whole lens extracts are expected to contain many different antibodies, there have been numerous investigations to separate lens proteins and prepare isolated fractions (see Chapter I). Also highly specific antisera were made against those fractions, for example, by van Doorenmaalen et al. (61–63). Zwaan and Ikeda (64) demonstrated by immunofluorescent reactions in the chick that α-crystallin appears at stage 19 HH in the primary lens fibers, β-crystallin in the open lens vesicle at stage 16 HH, and δ-crystallin or FISC (chick) at stage 14 (60 hours) in the invaginating lens placode. Brahma and van Doorenmaalen (63) arrived almost at the same conclusions for chick lens, but the distribution of those proteins in the lens was somewhat different (see Figure 4*b*, *c*).

In the mouse van de Kamp and Zwaan (65) demonstrated lens crystallins by immunofluorescent staining of the cells of the invaginating placode, first in those cells of which the nuclei had migrated to the base of the cells. McAvoy (66) was able to demonstrate the appearance and distribution of α, β-, and γ-crystallins in the lens of the rat.

It is clear that the synthesis of most crystallins starts after induction from optic vesicle has taken place and before specific morphological structures are established. Also some crystallins appear quite late in comparison to others. Species differences in composition of proteins exist such as, in birds, FISC (first important soluble crystallin or δ-crystallin) and, in amphibians, the protein dominating in quantity, γ-crystallin.

McDevitt et al. (67–70), carried out investigations on the appearance and distribution of lens crystallins in anurans, and urodeles with some slight topographical differences compared to the chick. In general the proteins appear at slightly later stages (Figure 4*a*).

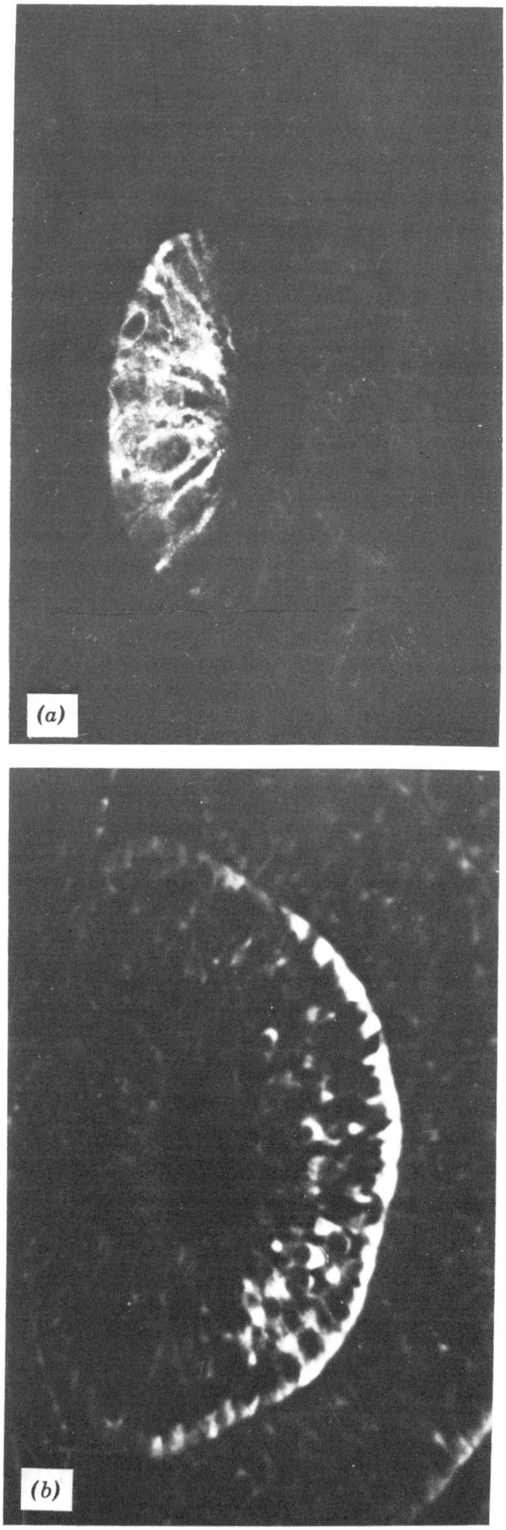
(a)
(b)

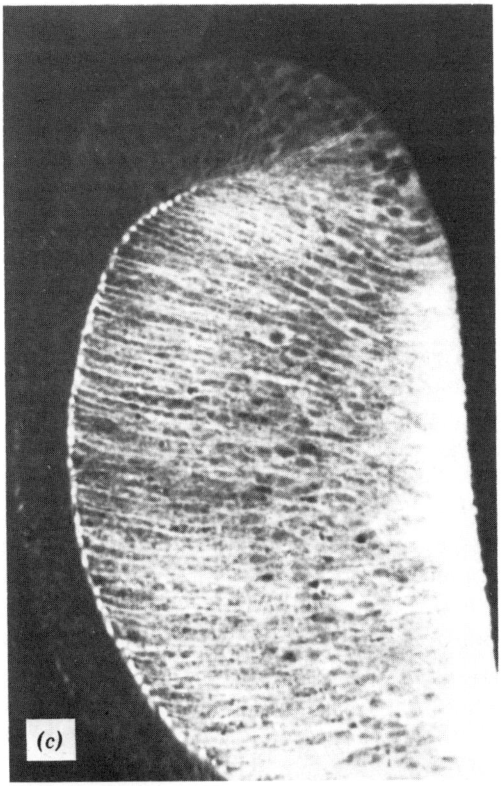

Figure 4 (*a*) Section through the amphibian eye at the stage of the early lens vesicle staining for γ-crystallins. (*b*) Section through the chick eye at stage 14 (60 hours) staining for δ-crystallins. The first staining starts at the retinal side of the cells of the vesicle. (*c*) Section through the chick eye at 3½ days staining for α-crystallins.

10.5 RETINAL INFLUENCES ON THE LENS

The retinal role in development does not end with the induction of the lens. In the early 1920s (1, 7, 71) and later (72, 73), investigators postulated that without the retina the lens does not grow or develop normally. Lenses are smaller in size when grown without a retina and even a correlation between the size of the optic cup and that of the lens was found (74). Also in early experiments it was seen that fibers always start to differentiate next to the retina but Coulombre (75), Coulombre and Coulombre (76), Génis-Gálvez (77), Génis-Gálvez et al. (74), and Clavert (78) performed experiments in which the chick lens was surgically reversed so that its fibers were facing the presumptive cornea. After only 2 days it was seen that the epithelium cells, now situated against the retina, were elongating and differentiating into lens fiber cells, whereas fiber formation at the other side was stopped and a new epithelial layer was formed. Also the equatorial zone of mitosis was reversed so that after some time the polarity of the lens was restored, which was ascribed to the retinal influence. Reyer (15, 79) got the same results in experiments with amphibian lenses.

Takeichi (80), transplanting [^{3}H]-thymidine-labeled embryonic lenses on the chorioallantoic membrane, could demonstrate a metabolism and fiber-promoting influence of the retina but not a maintenance of the condition of the lens. Moreover, Yamamoto (81), transplanting reversed lenses into the eye of the mouse, observed distinctly the same influence of the retina on fiber formation and also maintenance of the lens as an organ, as previously suggested by Muthukkaruppan (24).

This phenomenon has been confirmed by Reyer (16, 79) in *Amblystoma* and *Notophtalmus* as amphibian species. His results were strongly in favor of the importance of the retina for lens development, growth, and maintenance. From all these experiments a correlation also could be found between the size of the lens and the quantity of retinal influence. Not only is the retina of trophic influence during the growth and development of the eye, but also under experimental circumstances retina promotes transdifferentiation into lens cells (82, 83).

Last not least tissue actions between several eye tissues are reported in embryological life as well as during the adult stage. Van der Starre (25, 35) reported that in early development the lens, once formed, influences the retina by excreting a substance that signals the retina to stop the production of the inductor. Its working mechanism is located in the cells of the retina (Figure 3).

In later stages the lens is able to exert a trophic influence on the cornea as found by Zinn (84). Without the presence of a lens the cornea develops into quite another structure lacking some important elements of the normal corneal tissue and more resembling the sclera. A relationship also exists between iris and lens. Clavert (78) observed that in the rabbit the iris differentiation depends on the presence of the lens. Reyer (16, 79) reported this phenomenon also for amphibians. Also Coulombre and Coulombre, mentioned reciprocal influences of eye tissues in the chick in their many experiments.

Besides the lens formation from ectoderm in the presence or absence of the optic cup, several experiments were performed in which lens tissue was formed from tissues other than ectoderm or "induced" with other organs than optic vesicle. Lens regeneration from the iris in urodeles and amphibians is a well-known phenomenon. Reyer's (16) excellent review gives all information and references for the newt. Regeneration takes place from the dorsal iris exclusively, after the lens has been removed. But again the presence of the retina is of crucial importance. Also new lenses can be formed from retina itself after lensectomy (85). Lenses have been found to develop from the cornea also, as shown by Waggoner (86) in *Xenopus*. When cornea was transplanted into a lensectomized eye, a new lens was found originating from the cornea. Moreover, transplantation of cornea into a regenerate of an amputated hind limb also produced lenses, whereas implantation of cornea into a dorsal fin did not. Thus the environment of a regenerating limb blastema promotes the regeneration of a lens from the cornea. This was also found by Reyer et al. (87, 88) for the dorsal iris, which is also stimulated in lens production when brought into a regenerating blastema. Therefore it is conceivable that experiments have been successfully done by culturing iris epithelium of the newt (83). Indeed, iris cells transform into lens fiber cells containing the specific lens proteins, as was shown by immunofluorescence. Dorsal as well as ventral iris cells could establish this transdifferentiation, whereas the presence of retinal epithelium (of frog larvae) was necessary and the participation of stroma tissue was not. Okada et al. (89) cultured 8-day neural retina of the chick for a long time and got lens cells and fibers containing α- and δ-crystallins. The longer the cells were cultured, the larger the percentage of lens cells was found. Lens epithelial cells did not transform into lens protein synthesizing cells during culture; however, Hoperskaya (90, 91) observed lens induction power of these cells.

De Pomerai and Clayton (92) got interesting results by culturing the retinal layer of the optic cup starting at different ages of the retina. Taken into culture before 15 days of development, the cells of the nervous layer could transdifferentiate into pigment cells as well as into lens cells (lentoids). The lens cells appeared to have synthesized α-, β-, and δ-crystallins. Furthermore, another interesting phenomenon was observed. The younger the age at which the retinal cells were taken into culture, the more δ-crystallin was formed; the older, the more α- and β-crystallins were synthesized. This is more or less analogous to the sequence of production of these proteins during ontogenesis. After 15 days of culture the retina seems to have lost these properties.

Hoperskaya (91) cultured the anterior part of the neural plate of *Rana* taken out without endomesoderm and neural folds. The segregate consisted of a forebrain cell mass and retina tissue. An interlayer between the two was composed of iris tissue and typical lentoids were found, which originated from the retina tissue only.

Karkinen-Jääskeläinen reported atopic "inductors," such as lung tissue and gland, which "induced" lenses in mouse ectoderm. Another atopic, but more acceptable, inductor is formed by lens epithelium. Hoperskaya (90) brought lens epithelium of the adult frog together with gastrula and neurula ectoderm. Numerous lentoids were formed in her experiments and by immunofluorescence she could also demonstrate lens proteins. Neurula ectoderm seems to be more responsive than gastrula ectoderm. With the lyophilized extracts of lens epithelium it was also possible to get induction of lentoids. Treatment with 96% ethanol destroyed this inductive capacity and no other inductive properties developed. The inductive action of the epithelium did not pass through the lens basement membrane, whereas lens fibers did not possess this inductive power at all.

Mizuno (25, 26) got very striking results from some remarkable experiments. He combined in culture the epiblast (early ectoderm) and hypoblast (presumptive endoderm) of the cephalic region of the chick embryo at primitive streak stage with skin dermis of 6½ or 13½ days (even after alcohol denaturation) and got fiber-containing lenses or lentoids, formed in the epiblast. Even the dermis could be replaced by other tissues such as gizzard mesenchyme, liver, neural retina, mesonephros, and sclerotome, whereas the presence of the hypoblast was absolutely essential, as the "inductors" were. With fluorescent antisera directed against lens crystallins he could demonstrate the presence of these proteins in the epiblast lenses (27). Another phenomenon that resembles the development in lens was the formation and the orientation of the fibers at the side of and perpendicular to the dermis, analogous to the action of the retina.

10.6 CONCLUDING REMARKS

It may be assumed that lens development in amphibians, birds, and mammals has important tissue interactions in common, but many differences in developmental mechanics do exist, the number of which might increase when all steps, especially those of mammals, are thoroughly investigated and analyzed. Mammalian embryos, however, are difficult to work with because their developmental stage has not been extensively determined and the early embryonic tissues are very fragile for tissue culture techniques.

It seems that in amphibians differentiation of the eye tissues has not advanced as far as in birds. In amphibians regeneration of lens can occur from tissues such as iris, retina, and cornea, which is not the case in birds and mammals. Moreover, lens development in some amphibians not only is initiated by the optic vesicle but needs the participation of or even can be evoked without the optic vesicle by other tissues such as mesoderm and entoderm. This has been proved for amphibians only. What this action may be is not known yet. It may be considered as a conditioning or preparation of the ectoderm in order to make it responsive to the signal from the optic vesicle. But still it has to be explained why under experimental circumstances these tissues, on their own, are able to modify the metabolic channels of young ectoderm cells and let them transdifferentiate into lens cells.

Also in ontogenesis a progression of differentional level of tissues is seen. The more advanced in age tissues are, the more they are fixed in the metabolic channels and the less they are able to transdifferentiate or to return to young embryological functions. Trunk ectoderm of certain young stages can respond to eye cup (or other nonoptic tissues under experimental circumstances) and differentiate into a lens, but in later stages it no longer can. Retina cells in culture can transdifferentiate into lens cells up to a certain age; thereafter they seem to have lost this property. More examples can be found which illustrate that embryonic cells are submitted to a process of maturation, by which they lose polyvalency but gain other properties in which stage of differentiation they seem to be fixed. Under experimental conditions, however, they can return to former developmental stages. However, the ectoderm always develops into the species-specific lens whatever the kind of the inductor might be, pointing to a strong differential power of this tissue. Since all cells have identical genomes and eye tissues have a common origin, metabolic processes of lens cells might be still in an operational state. The demonstration of lens crystallins in other eye tissues by Bours and van Doorenmaalen (93), Brahma et al (94), Maisel and Harmison (95), and Dandrieu (41) indicates the possibility that metabolic channels of lens cells are still functioning on a low scale. The ability of other eye tissues to transdifferentiate into lens cells may be due to the activation and intensification of these metabolic channels. This may be caused by chemical substances entering the cells under experimental circumstances or even by the absence of inhibitors in those experimental environments. The younger cells are, the more they are able to transdifferentiate both morphologically and biochemically.

In tissue interactions the chemical (metabolic) information, going from one cell to another cell, involves directive and permissive actions of structures such as nucleus, cytoplasmic centers, and the cell membrane of the "inductive" cell on one side. On the other hand, the receiving or "responding" cell must be prepared to "read" the "inductive" message and such "readiness" must be directed certainly by the cell membrane, but also by the metabolic centers of the cytoplasm, and last but not least by the genetic system. More than one of these structures may have to be prepared to read this message. But in embryological cells also these structures frequently change, as do their functions, more or less in relation to age.

Between the cells a matrix is present through which the message has to be transferred. This matrix is in a strategic position and is able to control the stream of chemical substances; it can even direct them through spatial channels. It has a complicated structure and is very sensitive and changeable in the embryo. Under experimental circumstances this matrix is obviously absent. The cultured cells may be deprived of certain substances or may be confronted with others, and thus a cell can be forced to follow another metabolic line, as results of culture experiments have proved. In our laboratory we have found that in tissue culture cells start metabolizing differently.

The culture medium composition has been proved to be important for the expression and facilitation (83) or suppression of (trans)differentiation. For lens differentiation and regeneration the influence of a neural trophic factor in the medium has been demonstrated by Reyer et al. (16) and Waggoner (86). Karkinen-Jääskeläinen (28) found that a higher protein content of the medium increased the percentage of lenses formed quite substantially.

The surrounding tissues of certain developmental systems also might produce substances (metabolites) that contribute to the matrix structure and to developmental processes. They might supply key substances for important metabolic functions or even produce inhibitors for certain developmental pathways. In experimental circumstances as in tissue culture the test tissues are deprived of such substances and may take another developmental course.

Mesenchyme has been found to assist development and differentiation of the eye tissues too, but this must be considered to be indicated but not yet proved. The regenerating limb blastema does indeed initiate and promote lens regeneration as has experimentally been shown. But this action might be considered as general metabolic effect.

The contribution to or even initiation of lens differentiation by mesoderm and entoderm in amphibians does not mean that this is an effect identical to lens induction and differentiation in its proper time and place. Moreover, this phenomenon cannot be expected to be matched identically in birds because developmental time, place, quantity, and quality of these tissues differ enormously in these two classes. The nature of this influence might be a more metabolic or trophic one rather than a specific inductive activity.

Thus many factors may be involved in tissue interactions in general and in optic cup ectoderm in particular. Experimental evidence has to be handled with care for many factors dealing with development and growth can change, thus obscuring the genuine stimulus of differentiation of the lens.

REFERENCES

1 S. Toivonen, D. Tarin, and L. Saxén, *Diff.* **5,** 49 (1976).

2 S. Toivonen, D. Tarin, L. Saxén, P. J. Tarin, and J. Wartiovaara, *Diff.* **4,** 1 (1975).

3 H. Eyal-Giladi and M. Wolk, *W. Roux' Arch.,* **165,** 226 (1970).

4 A. G. Jacobson, *Science.,* **152,** 25 (1966).

5 H.D. King, *W. Roux' Arch.,* **19,** 85 (1905).

6 W. L. LeCron, *Am. J. Anat.,* **6,** 245 (1906).

7 W. H. Lewis, *Am. J. Anat.* **6,** 473 (1904).

8 R. O'Rahilly and D. B. Meyer, *Acta Anat.* **36,** 20 (1959).

9 H. Spemann, *Verh. Anat. Ges. Jena,* **15,** 61 (1901).

10 H. Spemann, *Embryonic development and Induction;* Yale University Press, New Haven, Conn., 1938.

11 S. K. Brahma, *Roux' Arch. Entw.* **151,** 181 (1959).

12 B. I. Balinsky, *Experientia*, **7**, 180, (1951).

13 Y. Tahara, *Embryologia*, **7**, 127 (1962).

14 R. W. Reyer, *J. Exp. Zool.*, **139**, 137 (1958).

15 R. W. Reyer, *J. Exp. Zool.*, **162**, 99 (1966).

16 R. W. Reyer, *Handbook of Sensory Physiology; The Visual System in Vertebrates*, Vol. VII, No. 5, Springer Verlag, Berlin, 1977, p. 309.

17 L. E. Alexander, *J. Exp. Zool.*, **75**, 41 (1937).

18 C. H. Waddington and A. Cohen, *J. Exp. Biol.*, **13**, 219 (1936).

19 J. H. M. G. van Deth, *Acta Morphol. Neerl. Scand.*, **3**, 151 (1940).

20 M. S. McKeehan, *J. Exp. Zool.*, **126**, 157 (1951).

21 M. S. McKeehan, *Anat. Rec.*, **132**, 297 (1958).

22 H. van der Starre, *Ophthal. Res.*, **3**, 20 (1972).

23 H. van der Starre, *Acta. Morph. Neerl.-Scand.*, **15**, 275 (1977).

24 V. Muthukkaruppan, *J. Exp. Zool.*, **159**, 269 (1965).

25 T. Mizuno, *Acad. Sci.*, **271**, 2190 (1970).

26 T. Mizuno, *J. Embryol. Exp. Morphol.*, **28**, 117 (1972).

27 T. Mizuno, *Proc. Jap. Acad.*, **48**(7), 522, (1972).

28 M. Karkinen-Jääskeläinen, *J. Embryol. Exp. Morphol.*, **44**, 167 (1978).

29 J. L. Sirlin and S. K. Brahma, *Dev. Biol.*, **1**, 234, (1959).

30 H. H. Hunt, *Dev. Biol.*, **3**, 175 (1961).

31 S. Toivonen and J. Wartiovaara, *Diff.* **5**, 61 (1976).

32 H. Tiedemann, *J. Act. Phys.*, **72**, Suppl. 1, 129 (1968).

33 J. Wartiovaara, S. Nordling, E. Lehtonen, and L. Saxén, *J. Embryol. Exp. Morphol.*, **31**, 667 (1974).

34 M. C. Niu and V. C. Twitty, *Proc. Natl. Acad. Sci., U.S.*, **39**, 985 (1953).

35 H. van der Starre, *Acta Morphol. Neerl.-Scand.*, **16**, 109 (1977).

36 K. B. Liedke, *J. Exp. Zool.*, **130**, 2, 353 (1955).

37 S. Takeuchi, *Embryologia*, **8**(1), 21 (1963).

38 G. Ten Cate, M. Dandrieu, and E. Gortzak, *Pallanza*, 623 (1960).

39 J. Langman, H. Maisel, and J. Squires, *J. Embryol. Exp. Morphol.*, **10**, 178 (1962).

40 Fowler and Clark, *Lens Diff., Symp. Oak Ridge Natl. Lab., 20 march 1964*, abstr.

41 M. R. Dandrieu, *Acta Morphol. Neerl.-Scand.*, **9**, 375 (1972).

42 R. A. Flickinger, E. Leve, and N. E. Sruith, *Phys. Zool.*, **28**, 79 (1955).

43 M. Karkinen-Jääskeläinen, *Diff.*, **12**, 31 (1978).

44 A. Porte, M. E. Stoeckel, and A. Brini, *Arch. Ophthal. (Paris)*, **28**, 681 (1968).

45 J. T. Wrenn and N. K. Wessels, *J. Exp. Zool.*, **171**(3), 359 (1969).

46 M. S. McKeehan, *J. Exp. Zool.*, **117**, 31 (1951).

47 J. Zwaan, P. R. Bryan, and T. L. Pearse, *J. Embryol. Exp. Morphol.*, **21**, 71 (1969).

48 J. Zwaan and R. W. Hendrix, *Amr. Zool.*, **13**, 1039 (1973).

49 B. Byers and K. R. Porter, *Proc. Natl. Acad. Sci. U.S.*, **52,** 1091 (1964).
50 D. C. Beebe, D. E. Feagans, E. J. Blanchette-Mackie, and M. E. Nau, *Am. Assoc. Anat. 92nd Session, Anat. Rec.*, **193,** 3 (1979).
51 J. Wakely, *Exp. Eye Res.*, **22,** 647 (1976).
52 M. Bancroft and R. Bellairs, *Anat. Embryol.*, **151,** 97 (1977).
53 R. W. Hendrix and J. Zwaan, *J. Embryol. Exp. Morphol.*, **33,** 1023 (1975).
54 E. Lehtonen, J. Wartiovaara, S. Nordling, and L. Saxén, *J. Embryol. Exp. Morphol.*, **33,** 187 (1975).
55 E. D. Hay and J. W. Dodson, *J. Cell Biol.*, **57,** 190 (1973).
56 A. I. Cohen, *Dev. Biol.*, **3,** 297 (1961).
57 P. Weiss, and S. Fitton-Jackson, *Dev. Biol.*, **3,** 532 (1961).
58 R. W. Hendrix, and J. Zwaan, *Diff.* **2,** 357 (1974).
59 G. Ten Cate and W. J. van Doorenmaalen, *Proc. Kon. Ned. Acad. Wet.*, **53,** 894 (1950).
60 W. J. van Doorenmaalen, *Exp. Eye Res.*, **5,** 151 (1966).
61 W. J. van Doorenmaalen, in *New Methods in Embryology,* Hermann, Paris, 1965.
62 W. J. van Doorenmaalen, S. K. Brahma, and H. J. Hoenders, *Experientia,* **24,** 30 (1968).
63 S. K. Brahma and W. J. van Doorenmaalen, *Opthal. Res.*, **2,** 344 (1971).
64 J. Zwaan and A. Ikeda, *Exp. Eye Res.*, **7,** 301 (1968).
65 M. van de Kamp and J. Zwaan, *J. Exp. Zool.*, **186**(1), 23 (1973).
66 W. J. McAvoy, *J. Embryol. Exp. Morphol.*, **44,** 149 (1978).
67 D. S. McDevitt, I. Meza, and T. Yamada, *Dev. Biol.*, **19,** 581 (1969).
68 D. S. McDevitt and S. K. Brahma, *J. Exp. Zool.*, **186,** 127 (1973).
69 S. K. Brahma and D. S. McDevitt, *Exp. Eye Res.*, **19,** 397 (1974).
70 D. S. McDevitt, *Experientia,* **28,** 1215 (1972).
71 W. H. Lewis, *Am. J. Anat.*, **6,** 473 (1907).
72 M. W. Woerdeman, *Nederl. Tijdschrift voor Geneeskunde,* **100** (II), 1756 (1956).
73 M. W. Woerdeman, *Proc. Kon. Ned. Acad. Wet., Ser. C,* **65**(2) (1962).
74 J. M. Génis-Gálvez, L. Santos, and A. Rios, *Exp. Eye Res.*, **6,** 48 (1967).
75 A. J. Coulombre, *Organogenesis,* 219 (1965).
76 A. J. Coulombre and J. L. Coulombre, *Science,* **142,** 1489 (1963).
77 J. M. Génis-Gálvez, *Bull. Assoc. Anat.*, **49,** 642 (1964).
78 A. Clavert, *Arch. Anat. Hist. Norm. Exp.*, **55,** 75 (1972).
79 R. W. Reyer, *Exp. Eye. Res.*, **24,** 501 (1977).
80 M. Takeichi, *Dev. Growth Diff.*, **12,** 21 (1970).
81 Y. Yamamoto, *Dev. Growth Diff.*, **18,** 273 (1976).
82 T. Yamada and D. S. McDevitt, *Dev. Biol.*, **38,** 104 (1974).
83 T. Yamada, D. H. Reese, and D. S. McDevitt, *Diff.*, **1,** 65 (1973).
84 K. M. Zinn, *Invest. Ophthalmol.*, **9,** 165 (1970).

85 G. Freeman, *J. Exp. Zool.*, **154,** 39 (1963).

86 P. R. Waggoner, *J. Exp. Zool.*, **186**(1), 97 (1973).

87 R. W. Reyer and R. A. Woolfit, *Anat. Rec.*, **163,** 249 (1969).

88 R. W. Reyer and L. T. Withersty, *Ophthal. Res.*, **3,** 25 (1972).

89 T. S. Okada, Y. Itoh, K. Watanaba, and G. Eguchi, *Dev. Biol.*, **45,** 318 (1975).

90 O. A. Hoperskaya, *W. Roux' Arch.*, **171,** 1 (1972).

91 O. A. Hoperskaya, *W. Roux' Arch.*, **180,** 213 (1976).

92 D. I. de Pomerai and R. M. Clayton, *J. Embryol. Exp. Morphol.*, **47,** 179 (1968).

93 J. Bours and W. J. van Doorenmaalen, *Exp. Eye. Res.*, **13,** 236 (1972).

94 S. K. Brahma, J. Bours, and W. J. van Doorenmaalen, *Exp. Eye. Res.*, **12,** 194 (1971).

95 H. Maisel and A. Harmison, *J. Embryol. Exp. Morphol.*, **11,** 483 (1963).

Author Index

Numbers in *italics* indicate pages on which full references appear. Numbers in parentheses indicate reference numbers.

Subject Index